DATE DUE

NOV 13 1985		
APR 14 1986		MAY 08 1987
		JUN 04 1987
DEC 30 1987		

DEMCO 38-297

CRC Handbook of Nutrition in the Aged

Editor

Ronald R. Watson, Ph.D.

Associate Professor for Research in
Nutrition and Immunology
Family and Community Medicine
University of Arizona
Arizona Health Services Center
Tucson, Arizona

CRC Press, Inc.
Boca Raton, Florida

Library of Congress Cataloging in Publication Data
Main entry under title:

CRC handbook of nutrition in the aged.

Bibliography: p.

Includes index.
1. Aging--Nutritional aspects. 2. Aged--Nutrition.
3. Aged--Diseases. 4. Nutrition disorders--Age factors.
I. Watson, Ronald R. (Ronald Ross) II. Title: C.R.C.
handbook of nutrition in the aged. III. Title: Handbook of nutrition in the aged. [DNLM: 1. Nutrition--In old age. 2. Geriatrics. 3. Aged. WT 100 C911]
QP86.C7 1984   612'.3'0880565   83-27291
ISBN 0-8493-2933-7

Direct all inquiries to CRC Press, Inc., 2000 Corporate Blvd., N.W., Boca Raton, Florida, 33431.

© 1985 by CRC Press, Inc.
International Standard Book Number 0-8493-2933-7

Library of Congress Card Number 83-27291

Printed in the United States

PREFACE

NUTRITIONAL PROBLEMS AND NEEDS OF THE AGED

Mature or aging bodies and physiological systems function with nutrient requirements quite distinct from the young. Recognition and understanding of the special nutrition problems of the aged may be changing with increases in the elderly in the general population. Some of these changes include decreased nutritional needs, while others reflect changes in economic conditions with age, or self-treatment of health conditions with high nutrient intake. At the other extreme, in many developing countries, adults and aging adults, particularly men, get preferential access to food and may have improved opportunities for adequate nutrient intake with increasing age. By contrast, in many developed countries, it is clear that economic restrictions and physical changes during aging can significantly reduce food intakes from those enjoyed as young adults. These are some of the changes with age which contribute to various nutritional stresses as outlined in this book.

Many disease entities including cancer are found with higher frequency in the aged and increase nutritional stresses in adults. Cancer, trauma, or infectious disease can alter intakes and/or requirements for various nutrients. For example, hyperalimentation or renutrition is being accepted as helpful in nutritional treatment to maintain optimum immune functions of hospitalized adults, including cancer patients, and deserves special attention.

A major problem in adults involves altering nutritional requirements by drugs used to treat conditions common in the aged. As drug use in the aged has been rapidly increasing, so has the recognition of modified nutritional needs by drugs. Detailed understanding of drug-nutrient interactions is critical in health care of the aged and the proper application of basic principles of nutrition.

Some deficiencies of various nutrients associated with aging are often distinct from nutritional problems of childhood. On the other hand, many adults and elderly are using levels of various nutrients well above the recommended daily allowance. For example, high intakes of supplementary vitamin E is common for many adults. On the other hand, uses of large amounts of foods such as alcohol or sugar are associated with causation or exacerbation of conditions such as alcoholism or diabetes. To some extent, treatment of these conditions is unique in the aged and represents major health problems, often associated with prolonged use or age.

The major objective of this handbook is to review in detail some of the major nutritional problems occurring with significant frequency in adults or the elderly. The effects of the aging processes, changes in social status, and financial conditions significantly affect the approaches to treatment and study of nutritional and health problems in the aging adult and the elderly. The increasing numbers and percentage of older adults and the elderly in the population requires detailed study and directed research to understand and solve their nutritionally related health problems.

Ronald Ross Watson

EDITOR

Ronald Ross Watson, Ph.D., is an Associate Professor in the Department of Family and Community Medicine (Nutrition Section) at the Arizona Health Science Center, University of Arizona.

Dr. Watson was graduated from Brigham Young University, with a B.S. (Chemistry) degree in 1966. He received the Ph.D. degree (Biochemistry) in 1971 from Michigan State University.

Dr. Watson did post-doctoral research in immunology at the Harvard School of Public Health from 1971 to 1973. He taught and did research on nutrition and immunology at the Indiana University Medical School from 1974 to 1978, then continued his research at Purdue University (1978 to 1982). He has carried out numerous human and animal studies on nutrition in young children and the elderly, including in Colombia and Egypt. He has several funded research projects on aging and the use of nutrients to prevent cancer and modify immune functions. These include nutrition research in Columbia and Egypt in adults.

Dr. Watson has won awards in nutrition research and is listed in *Who's Who in the Midwest* and in *American Men and Women of Science*. He has published more than 70 research papers and reviews.

CONTRIBUTORS

Richard A. Anderson
USDA Research Chemist
Beltsville Human Nutrition Research
  Center
Beltsville, Maryland

Joseph J. Barboriak
Professor of Pharmacology and
  Preventive Medicine
Medical College of Wisconsin &
  Research Service
VA Medical Center
Wood, Wisconsin

Hemmige N. Bhagavan
Nutrition Coordinator
Vitamins & Clinical Nutrition
Hoffman-La Roche, Inc.
Nutley, New Jersey

Dorothy L. Brooks
Educational Director, Dietetic Internship
  Program
University of Medicine & Dentistry of
  New Jersey
School of Health-Related Professions
Newark, New Jersey

Lorraine Cheng
Director of Scientific Literature
  Department
Hoffman-La Roche, Inc.
Nutley, New Jersey

Nathaniel Ching
Department of Surgery
St. Francis Medical Building
Honolulu, Hawaii

Marvin Cohen
Research Fellow
Scientific Literature Department
Hoffman-La Roche, Inc.
Nutley, New Jersey

Edmund H. Duthie, Jr.
Acting Chief
Section of Geriatrics & Gerontology
Medical College of Wisconsin
Chief, Nursing Home Care Unit
Wood VA Medical Center
Wood, Wisconsin

Gary J. Fosmire
Associate Professor of Nutrition
Pennsylvania State University
University Park, Pennsylvania

Steven R. Gambert
Professor of Medicine
Director, Center for Aging and Adult
  Development
New York Medical College
Valhalla, New York

Seetha N. Ganapathy
Professor of Nutrition
Nutrition Research Program
N.C. Agricultural & Technical State
  University
Greensboro, North Carolina

Carlo E. Grossi
Chief of Surgical Oncology
St. Vincent Hospital and Medical Center
New York, New York

Malekeh K. Hakami
Research Fellow
Department of Psychiatry
Memory & Learning Clinic
VA Outpatient Clinic
Boston, Massachusetts

A. H. Ismail
Professor of Physical Education
Purdue University
West Lafayette, Indiana

JC. Jackson
Physician in Training
College of Osteopathic Medicine of the
  Pacific
Pomona, California

Robert A. Jacob
Research Leader
Bioanalytical Methods
USDA Western Human Nutrition
  Research Center
Presidio of San Francisco, California

Peter P. Lamy
Professor & Director
The Center for the Study of Pharmacy
  and Therapeutics for the Elderly
Chairman, Department of Pharmacy
  Practice & Administrative Science
University of Maryland School of
  Pharmacy
Baltimore, Maryland

Miroslav Ledvina
Department Head & Professor
Department of Biochemistry
Charles University
Faculty of Medicine
Hradec Kralove
Czechoslovakia

Bernard S. Linn
Associate Chief of Staff for Education
Professor of Surgery
VA Medical Center
Miami, Florida

Margaret W. Linn
Director of Social Science Research
Professor of Psychiatry
VA Medical Center
Miami, Florida

Maury Massler
Professor Emeritus
Tufts University
School of Dental Medicine
Boston, Massachusetts

Mary E. Mohs
Registered Dietician, A.D.A
Department of Family & Community
  Medicine
University of Arizona
Tucson, Arizona

Emma A. Montgomery
Registered Dietician, A.D.A.
Home Hospital
Lafayette, Indiana

David M. Nathan
Director of Diabetes Clinic
Massachusetts General Hospital
Boston, Massachusetts

Lawrence C. Perlmuter
Director of Memory & Learning Clinic
VA Outpatient Clinic
Boston, Massachusetts

Arlan Richardson
Professor of Chemistry
Illinois State University
Normal, Illinois

Carol B. Rooney
Assistant Chief
Dietetic Service
VA Medical Center
Wood, Wisconsin

Robert M. Russell
Director of Human Studies at USDA
  Human Nutrition Research Center on
  Aging
Tufts University
Boston, Massachusetts

Harold H. Sandstead
Director of USDA Human Nutrition
  Research Center on Aging at Tufts
  University
Boston, Massachusetts

Helen Smiciklas-Wright
Associate Professor of Nutrition
Pennsylvania State University
University Park, Pennsylvania

Bertil Steen
Professor of Geriatric Medicine
Lund University
Department of Community Health
  Sciences
Section of Geriatric Medicine
Malmoe, Sweden

.

Clifford H. Swensen
Professor of Psychological Science
Purdue University
West Lafayette, Indiana

Shailaja D. Telang
Associate Professor of Biochemistry
M.S. University of Baroda
Baroda, India

Saroja Thimaya
Graduate Research Assistant
Department of Nutrition
Oklahoma State University
Stillwater, Oklahoma

Shobha A. Udipi
Graduate Research Assistant
Department of Foods & Nutrition
Purdue University
West Lafayette, Indiana

Ronald R. Watson
Associate Professor for Research in
  Nutrition & Immunology
Department of Family and Community
  Medicine
University of Arizona
Arizona Health Science Center
Tucson, Arizona

J. P. Wattis
Consultant in the Psychiatry of Old Age
Honorary Psychiatric Lecturer
Leed University Medical School
Leeds, United Kingdom

Ira Wolinsky
Associate Professor of Nutrition
Department of Human Development &
  Consumer Sciences
University of Houston
Houston, Texas

TABLE OF CONTENTS

Nutrition, Disease, and Body Function in Aging

PROTEIN-CALORIE MALNUTRITION IN THE ELDERLY

Steven R. Gambert and Edmund H. Duthie, Jr.

INTRODUCTION

For years, gerontologists have been intrigued over a possible relationship between aging and dietary restriction. In the 1930s, McCay, a Cornell University biologist, reported increased longevity for rats fed only 60% of controls allowed to eat *ad libitum*. Since this time, numerous studies have supported this concept in certain lower species;[1,2] however, no studies have implicated a comparable survival advantage for humans. More certainly, human malnutrition is associated with numerous alterations in body physiology. Although isolated deficiencies may be found in calories, vitamins, protein, or trace minerals, usually a combined deficit exists.

Protein-calorie malnutrition (PCM) is the most undiagnosed nutritional disorder in the world today.[3-5] Although most prevalent in young children, PCM can develop at any age whenever prolonged consumption of a diet high in carbohydrate and low in protein is consumed. Due to a variety of reasons, the elderly are particularly prone to this problem.

As one grows older, normal aging processes may result in less than optimal intake of certain nutrients. Changes in one's sense of smell, taste, and vision may result in anorexia or cause difficulty in food shopping. A decline in muscle strength may make food shopping or chewing difficult. Gastrointestinal (GI) enzyme changes may lead to poor digestion of certain food sources, and a decreased renal function may make for difficulty in eliminating food by-products. Although the causes of nutritional deficiencies are complex, socioeconomic factors are of prime importance. Poverty, isolation, and dietary ignorance are key factors. Since eating is a social behavior, it becomes apparent that the elderly who may be recently bereaved, separated from loved ones, or relocated, may find the impetus to eat reduced. The crude reality of having to live on a pension may seriously hamper one's choice of food. The loss of teeth, poor-fitting dentures, or other perioral pathology may also interfere with a proper dietary intake. It has been estimated that PCM may affect from 25 to 50% of medical and surgical patients for whom the duration of hospital stay is 2 weeks or longer. Since the elderly have a greater prevalence of chronic illness and multiple hospitalizations, negative protein balance can become a major problem.

Persons suffering from PCM show a higher incidence of concurrent illness.[6] In addition, PCM impairs mechanical, cellular, and humoral immunologic defenses.

PROTEIN METABOLISM IN THE ELDERLY

Using tracer infusion techniques, Uauy et al.[7] studied rates of whole-body protein synthesis in young and old subjects. When expressed on a weight basis, rates of protein synthesis and degradation were significantly lower in both old men and women. When expressed per unit of body cell mass however, total protein synthesis and breakdown were slightly greater in elderly subjects as compared to younger ones. This difference became even greater when expressed as grams of protein synthesized or broken down per gram of creatinine excreted. This most likely results from an age-related decrease in muscle mass, the precursor of urinary creatinine.

Using another parameter of protein turnover, Munro and Young[8] measured the excretion of 3-methylhistidine in young and old men. Since histidine is protein-derived, methylated in the three position by the liver, and not reincorporated into protein, it is an accurate indication of protein breakdown. Elderly men have less ouptut of the degradative product

Table 1
PHYSICAL SIGNS SUGGESTING NUTRITIONAL INADEQUACY

General appearance	Thin? obese?
Skin	Pallor, hyperkeratosis, bruises, dermatitis, fistulas, edema, abnormal pigmentation, subcutaneous fat, skinfold thickness
Eyes	Ophthalmoplegia, cataracts, xerosis, night blindness, Bitot's spots, retinal hemorrhage, papilledema
Mouth	Glossitis, gingivitis, cheilosis, dysgensia, agensia, peridontal disease, caries
Nose	Nasolabial seborrhea, anosmia, dysosmia
Hair	Sparseness, depigmentation, easily plucked
Nails	Friability, bands, lines
Neck	Goiter
Cardiovascular	High-output failure, cardiac enlargement, resting tachycardia
Lungs	Breathing with accessory muscles
Abdomen	Enlarged liver, ascites, varices
Skeletal	Tenderness, epiphyseal thickening, bowing, rachitic rosary, osteoporosis, frogleg position
Muscle	Atrophy, pain
Joints	Arthralgia, effusions
Neurologic	Hyporeflexia, irritability, convulsions, foot drop, decreased position and vibratory sense, confabulation, hyperreflexia

when expressed either per kilogram of weight or body cell mass. When expressed per unit of creatinine, however, no difference could be detected with age. It is apparent, therefore, that the decline in 3-methylhistidine excretion results not from a change in protein turnover but rather from a change in muscle mass.

Gersowitz et al.[9] studied the effect of age on visceral protein synthesis. The ability of the liver to synthesize and degrade albumin was studied using orally ingested $[_N^{15}]$ glycine. Although healthy aged subjects were found to have a lower plasma concentration of albumin, the fractional synthesis of the albumin pool was not affected by either increasing age or level of dietary protein ingestion. This was in direct contrast to younger subjects who decreased their rate of albumin synthesis when fed a low protein containing diet.

Although protein intake required for metabolic balance does not vary with age on a per weight basis (0.8 g/kg body weight) due to decreased intake, the required amount represents a higher percentage of the elderly person's daily diet.

Clinical Manifestations of Protein-Calorie Malnutrition

Protein-calorie malnutrition results in numerous changes to our physiology. Hepatomegaly is commonly seen and fatty infiltration of the liver may be significant. Due to other commonly associated nutritional deficiencies, the patient may also have other multiple physical signs (Table 1).

Although children with PCM have a decreased I.Q. and head circumference, little is known about the effect of PCM on adults. It is known, however, that poor nutritional intake may result in changes in brain neurotransmission and function.[10]

Pancreatic size may decrease with associated local edema and loss of normal cellular architecture. Pancreatic exocrine cells may have a decrease in cytoplasm, fewer zymogen granules, and an altered arrangement of acinar cells. Studies using animal models have shown a decrease in pancreatic islet cells.

There are numerous biochemical changes that may accompany PCM. Concentrations of blood glucose and essential amino acids may decline; concentrations of fatty acids and growth hormone may increase. Insulin levels are usually normal in PCM unless the patient also has been fasting for a prolonged period of time.[11] An impairment in glucose tolerance has been noted in patients with PCM.[12,13] In addition, glucose clearance may be delayed

Table 2
LABORATORY SCREENING TESTS FOR
NUTRITIONAL DEFICIENCY

Hemoglobin	Urinary nitrogen
Hematocrit	Serum and urinary 3-methylhistidine (?)
Serum iron	Serum and urinary ribonuclease (?)
Serum protein	Serum carotene
Serum albumin	Prothrombin time
Blood urea nitrogen	Serum vitamins
Glucose tolerance test	Growth hormone, serum-free fatty acids

despite concomitant administration of exogenous insulin,[14,15] suggesting insulin resistance. Clinically, patients with PCM may appear to be diabetic with an abnormal glucose tolerance for as long as 10 years after nutritional repletion.[12] Although pure protein deficiency has been associated with an increase in serum T_3 in animal models,[16] this is rare in actual clinical practice, and T_3 may be low, normal, or high, depending on a variety of other factors.

In screening the elderly person for PCM, it is important that the clinician take age into account when reviewing laboratory parameters. In many cases, normal values must be adjusted for age.[17] Tests that may be helpful are listed in Table 2.

Experimentally, measurements of serum and urinary 3-methylhistidine may provide an accurate assessment of protein status due to the fact that after methylation at the three position by the liver, this is not reincorporated into protein. Measuring urinary nitrogen gives only an indirect estimate of protein status due to the fact that this represents the net release following not only breakdown but also reincorporation into protein.

Nutritional Assessment

In evaluating the elderly person for nutritional deficiency, it is important to remember that a pure, isolated deficiency is rare. As with all other aspects of medical care, a thorough history and physical examination are keys not only to the detection of existing deficiencies, but also in determining potential problems.

Key historical questions include number of meals eaten per day, dietary content, special meal likes and dislikes, food allergies, cultural taboos, weight history, who prepares the meals, and where the food is obtained.

The physical examination must include an accurate height and weight. Measurements obtained should be compared to the "ideal" body weight and height as normalized for age.[18]

An accurate weight is essential with appropriate comparison to the recent and distant past. Rapid weight loss implies a change in protein status with the use of endogenous protein as a fuel source. Adipose tissue is lost more slowly due to a higher caloric content. It is important for the clinician to remember that an obese patient may still be overweight despite severe PCM. In addition, protein deficiency can still be present in a person with little change in weight; the decline in muscle mass may be compensated with increased fat stores.

Determinations of skin fold thickness at several sites may be used as a rough indicator of nutritional status. Due to variations with age, activity, etc. it is probably best not to compare results from one person to another, but for comparison over time for a given subject. This also appears to be the best way to use the measurement of mid-arm circumference reflecting muscle mass.[19]

REFERENCES

1. **Berg, B. N. and Simms, H. S.,** Nutrition and longevity in the rat. II. Longevity and onset of disease with different levels of food intake, *J. Nutr.,* 71, 255, 1960.
2. **Byung, P. Y., Masoro, E. J., Murata, I., Bertrand, A. B., and Lynd, F. T.,** Lifespan study of SPF Fischer 344 male rats fed ad libitum on restricted diets: longevity, growth, lean body mass and disease, *J. Gerontol.,* 37, 130, 1982.
3. **Bralove, M.,** Eating ourselves to death, *Nation's Health,* 43, 147, 1971.
4. **Bengoa, J. M.,** The problem of malnutrition, *WHO Chron.,* 28, 1974.
5. **McLaren, D. S.,** *Undernutrition, Nutrition and Its Disorders,* Williams & Wilkins, Baltimore, 1972, 99.
6. **Wade, N.,** Bottle feeding: adverse effects of a Western technology, *Science,* 184, 45, 1974.
7. **Uauy, R., Scrimshaw, W. S., and Young, V. R.,** Human protein requirements: nitrogen balance response to graded levels of egg protein in elderly men and women, *Am. J. Clin. Nutr.,* 31, 79, 1978.
8. **Munro, H. N. and Young, V. R.,** Urinary excretion of N gamma-methylhistidine (3-methylhistidine): a tool to study metabolic responses in relation to nutrient and hormonal status in health and disease of men, *Am. J. Clin. Nutr.,* 31, 1608, 1978.
9. **Gersowitz, M., Bier, D., Mathews, D., Udall, J., Munro, H. N., and Young, V. R.,** Dynamic aspects of whole body glycine metabolism: influence of protein intake in young adult and elderly males, *Metabolism,* 29, 1087, 1980.
10. **Wurtman, R. J., Cohen, E. L., and Fernstrom, J. D.,** Control of brain neurotransmitter synthesis by precursor availability and food consumption, in *Neuroregulatory and Psychiatric Disorders,* Usdin, E., Hambur, D. A., and Barchas, J. D., Eds., Oxford Press, N.Y., 1977, 103.
11. **Sims, E. A. H. and Horton, E. S.,** Endocrine and metabolic adaptations to obesity and starvation, *Am. J. Clin. Nutr.,* 21, 1455, 1968.
12. **Becher, D. J., Primstone, B. L., Hansen, J. D. L., and Hendricks, S.,** Insulin secretion in protein-calorie malnutrition, *Diabetes,* 20, 542, 1971.
13. **Mulner, R. D. G.,** Insulin secretion in human protein-calorie deficiency, *Proc. Nutr. Soc.,* 31, 219, 1972.
14. **Alleyne, G. A., Trust, P. M., Flores, H., and Robinson, H.,** Glucose intolerance and insulin sensitivity in malnourished children, *Br. J. Nutr.,* 27, 585, 1972.
15. **Bowie, M. D.,** Intravenous glucose tolerance in kwashiorkor and marasmus, *S. Afr. Med. J.,* 38, 328, 1964.
16. **Tulp, O. L., Horton, S., Tyzbir, E. D., Danforth, E., Jr., and Bollinger, J.,** Increased plasma triiodothyronine and caloric inefficiency in experimental protein malnutrition, *Am. J. Clin. Nutr.,* 29, 471, 1976.
17. **Gambert, S. R., Csuka, M. E., Duthie, E. H., and Tiegs, R.,** A clinician's guide to the interpretation of laboratory tests in the elderly, *Postgrad. Med.,* 72, 147, 1982.
18. **Master, A. M., Lasser, R. P., and Beckman, G.,** Tables of average weight and height of Americans aged 65 to 94 years, *JAMA,* 172, 659, 1960.
19. **Butterworth, C. E. and Blackburn, G. L.,** Hospital malnutrition and how to assess the nutritional status of a patient, *Nutr. Today,* 10, 8, 1975.

NUTRITIONAL ASSESSMENT FOR THE ELDERLY

Mary E. Mohs and Ronald R. Watson

INTRODUCTION

The number of older persons in the U.S. has increased dramatically in the last 50 years and demographers predict the continuation of this trend. In 1970, 20 million persons over the age of 65 lived in this country, representing 10% of the population. By the year 2000, in excess of 15% of the population will be over the age of 65. This figure indicates a greater than 50% increase in less than 30 years.[1,2] The definition of "elderly" in this paper will refer to persons aged 65 years and older.

It is well recognized that with advancing age there is a higher incidence of chronic diseases such as atherosclerosis, heart disease, and diabetes mellitus, and increasing evidence points to the importance of nutrition in the occurrence of and susceptibility to these diseases. Decreasing renal and respiratory function are highly associated with the aging process, ultimately leading to a high incidence of renal failure, chronic obstructive pulmonary disease, and acute respiratory failure among the aged. Medical personnel have come to recognize the need for nutritional assessment and support for patients with such problems. Chronic and acute respiratory disease produce significant deficits in serum transferrin and retinol binding proteins, creatinine height index, and total lymphocyte counts. In addition, patients with respiratory failure have been found to have a mean 19% deficit in percentage of ideal body weight, a mean 4.4-mm decrease in triceps skinfold thickness, and a deficit of 3.3 cm in arm muscle circumference when compared with control patients. The development of protein-calorie malnutrition (PCM) in these respiratory patients prior to hospitalization indicates the need for nutritional support from the onset of such illnesses.[3]

Patients with chronic renal failure present unique problems in nutritional assessment. Methods for assessing body composition such as densitometry and total body potassium may not be useful for renal failure patients. Serum amino acid composition in uremic patients resembles that in malnourished patients. Serum protein, albumin, transferrin, and some serum complement measurements are often abnormal in uremic patients.[4] A recent study of healthy dialysis patients showed them to have no reduction in relative body weight and mid-upper arm circumference. However, these dialysis patients, who were considered to be robust and healthy, showed evidence of wasting not only with a slightly decreased subcutaneous and total body fat, but also with reduced serum albumin and transferrin levels. Thus, wasting has been shown to be not necessarily indicative of malnutrition in chronically uremic patients. Therefore, nutritional status standards for renal failure patients are needed which will be realistic measures upon which to base nutritional support programs.[4]

Emergency hospital admission of the elderly occurs frequently, and measurement of nutritional status of these acute admissions is advisable. A recent Swedish study, which used weight-for-height, triceps skinfold thickness, arm muscle circumference, plasma albumin, and serum transferrin as nutritional indicators, found undernutrition in 22% and obesity in 9% of newly admitted patients. Risk factors for undernutrition included age over 75 years, lack of own teeth, and a reason for admission other than circulatory disorders or diabetes. The malnourished population of the hospital was composed partly of elderly who were already malnourished prior to admission.[5]

PCM is being recognized more and more frequently among all hospitalized patients.[6-11] Numerous surveys have shown that by standard criteria more than 15% of the patients on medical and surgical wards of various hospitals exhibited clear clinical signs of malnutrition while over 50% of the patients surveyed were found to be at various states of

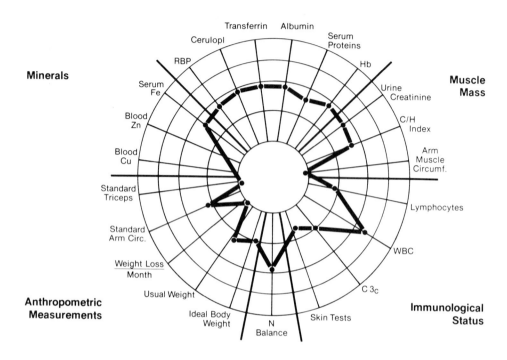

Circular Multivariate Nutrition Assessment Chart

FIGURE 1. Circular multivariate nutritional assessment chart. Sequential evaluation of nutritional status in a patient with gastrectomy and colectomy for invasive gastric cancer and treated with TPN. NA has been performed after 16 (*—-*) and 37 (.- .) of TPN. Improvements recorded in most of the parameters of visceral protein compartment and muscle mass. (From Nazari, S., Dionigi, R., Dionigi, P., and Bonoldi, A., *J. Parent. Ent. Nutr.*, 4(5), 499(15), 1980. With permission.)

starvation. Because of these findings, physicians and nutritionists have become increasingly concerned with the nutritional care of hospitalized patients, especially when a patient has a hospital stay of 2 weeks or longer or has surgery. It has been shown that a hospital stay of 2 weeks or longer induces an even further decline in malnourished or "high risk" patients.[7,8-11] Out of this concern, many hospitals have begun implementation of a standardized approach to nutritional assessment for their patients. Ongoing nutritional assessment is crucial to detect and counteract developing protein-energy malnutrition in hospitalized patients, because of the catabolic effect of injury and disease on the body's metabolism and the immune system's function.[10] Various routines for nutritional assessment have been suggested for use in clinical medicine[12,13] and among the general hospitalized population.[14,15] Basically, they rely on anthropometric measurements and selected clinical tests, including those of anergic and metabolic function. An interesting, circular multivariate chart is currently being used in an institution in Pavia, Italy, and has been suggested for general hospital use (Figure 1). This chart combines measures for muscle mass, immune status, anthropometric indexes, blood minerals (iron, copper, and zinc), and visceral protein compartments.[15] One advantage such a chart would have would be the ability of the physician to see the various types of nutritional status indicators' results simultaneously. Another would be the ability to see sequential measurements of all these indicators at once.

Nutritional assessment can be defined as the interpretation of clinical, biochemical, anthropometric, and dietary data to determine an individual's nutritional status. Thus, nutritional assessment is the first step toward developing a goal of providing optimal nutritional care for an individual. With the routine use of nutritional assessment, many problems have arisen as to the acceptability of existing standards[10,16,17] and to the lack of standards for the elderly population.[18,19]

The proportion of malnourished patients who are elderly has not been documented. However, Mitchell et al.[18] contend that many of these patients are elderly and find the proportion to be relatively high. The findings of the Swedish study support this contention.[5] The standards currently in use often do not meet the needs of these elderly, hospitalized patients. A large percentage of elderly people have been shown to have a low nutrient intake correlating with a low total calorie intake. With a low level of total food intake, essential nutrients are more likely to be deficient in the diet also, including selected vitamins and minerals.[20] Chen and Fan-Chiang[21] recently compared institutionalized and noninstitutionalized elderly in Kentucky for riboflavin and vitamin B_6 status.[21] They found riboflavin deficiency in 34.2 and 27.7% of the institutionalized and noninstitutionalized subjects, respectively. A significant decline in both vitamins occurred with aging, and the status of both vitamins was worse in the institutionalized elderly. This study points to a widespread problem in maintenance of adequate micronutrient as well as protein and calorie nutrition among the elderly. This problem is exacerbated among the institutionalized, perhaps due to the presence of diseases causing them to be institutionalized or related to the drugs commonly given to the patients.[21]

The most commonly used measurements which predict adipose tissue or muscle mass stores are percent ideal body weight, triceps skinfold, mid-arm muscle circumference, and creatinine/height index (CHI). Visceral protein stores are usually evaluated by measuring serum albumin and serum transferrin. Iron deficiency is determined by hemoglobin and hematocrit levels. In addition, total lymphocyte count and measurements of cell-mediated immunity are used as indicators of cell function.

The norms for most of these measurements have been developed from data obtained from children and young or middle-aged adults.[22,23] It is clear that these standards are not suitable for application to the elderly. For instance, in the healthy elderly, CHI has been shown to be significantly lower than in the young.[24] This is because increasing age in the adult brings atrophy of skeletal muscle and declining fat-free body mass. The decline of creatinine excretion with age may reflect both dietary change (decreased meat ingestion) and decreased muscle mass status.[25] Wide overlaps between malnourished and well-nourished elderly groups have been seen. Anemia has been distinguished from malnutrition in elderly only when a lower limit of normal hemoglobin has been used. When creatinine excretion was related to total arm length rather than to height, and when a lower limit of normal was used in a recent study, improved prediction of malnutrition was seen with elderly males.[24] Percent of ideal body weight and arm muscle circumference were limited in their ability to predict malnutrition in elderly, and triceps skinfold thickness was not useful in elderly males in this same study.[24] The need for standards to be developed specifically for the elderly population is apparent.

This paper will attempt to review the measurements, uses, and limitations of clinical, biochemical, and anthropometric assessment in the elderly. Dietary assessment, while important, will not be discussed.

PHYSIOLOGICAL CHANGES

There are a number of changes that may significantly influence the nutritional requirements of the elderly person. As seen in Table 1, body composition changes throughout life, with fat increasing, lean body mass decreasing, and metabolically active tissues being slowly reduced. This reduction accounts for the fall in basal energy metabolism, which is often accompanied with an even greater reduction in physical activity.[26] Another example is that

Table 1
BODY COMPOSITION AND AGE

	Age	
	25 years	**70 years**
Fat (%)	20	36
Cell mass (%)	47	36
Bone mineral (%)	6	4
Other (%)	27	24

From Gregerman, R. I. and Bierman, E. I., *Textbook of Endocrinology*, Williams, R. H., Ed., W. B. Saunders, Philadelphia, 1974. With permission.

the nerve-conduction velocity of the older person is reduced by 10 to 15%. Kidney function is reduced and glomerular filtration rates may be as low as 60% of values in earlier years. The renal plasma flow is decreased, sometimes by as much as 50%. Breathing capacity is usually affected in an elderly person. Maximal breathing capacity may be 70 to 80% of normal and vital capacity may be 60% of normal values. Bone deposition declines, resulting in a net loss which is manifested through a progressive thinning of the walls and a concomitant decrease in overall skeletal mass. Furthermore, in the GI tract, there is a reduction in gastric acidity and a slowing of intestinal motility.[27] However, some absorptive functions in the elderly, such as those in the bowel, may be almost as efficient as in the young. Apparent digestibility of neutral detergent fiber has been shown to be comparable between elderly and young adults.[28]

Deteriorative changes may account for decreased efficiency of nutrient absorption and in some instances justify giving increased levels of nutrients or supplemental minerals and vitamins in the diet. Healthy elderly may also experience greater metabolic function changes compared to a young adult population. For instance, reduced glucose tolerance has been seen in the aged. This phenomenon has been associated with peripheral tissue glucose-uptake reduction. Insulin efficiency in the periphery may be impaired in the healthy elderly person.[29] On the other hand, hepatic glucose production mechanisms remain intact in the aging.[29] The peripheral insensitivity to insulin points up the importance of exercise in the elderly to aid in moving blood glucose into the periphery for use.

Other factors which may affect an elderly person's intake or produce GI dysfunction include poor dentition, constipation, diarrhea, urinary or fecal incontinence, decreased olfactory acuity, and impaired vision or hearing.[30]

NONPHYSIOLOGICAL CONSIDERATIONS

In addition to physiological factors which influence nutritional status in the elderly, psychological, social, and economic factors must be taken into account. All of these can be adversely affected, especially if an elderly person has a chronic disease(s) producing an undesirable decline in their nutritional status. Administrators of nutritional programs for the elderly have recognized the importance of the social component in such programs. Nutriture for vitamins A and C was found to be better for elderly Missourians who participated in conjugate meals under Title VIIs Nutrition Program compared to those elderly who did not participate. In addition, those who regularly participated in the program, which has a social and recreational component along with the nutritional component, had a larger proportion of more adequate intakes of riboflavin and thiamin.[31]

Lifelong patterns of eating affect specific nutrient levels and stores. In a recent review, it was revealed that diets of the elderly assessed by intake studies have been shown to be most deficient in energy calcium, iron, and B complex vitamins. Biochemical studies have identified folacin and B complex vitamins as being most often deficient.[32] Specific micronutrient problems are often exacerbated by unbalanced or generally inadequate diets eaten in some developing countries and among some minority groups in developed countries. For instance, a group of Indian women living in South Africa, with a 38% anemia incidence, had significantly more iron deficiency than a sample of women studied in Washington, U.S.[33] Similarly, Spanish Americans may have a greater risk of developing vitamin A deficiency than the general American population.[34] Elderly, urban, black Americans have a high (39%) incidence of low serum zinc values compared to other Americans.[35]

CLINICAL ASSESSMENT

The purpose of a clinical assessment is to detect possible nutritional deficiencies through physical signs and symptoms of malnutrition. It includes a medical history, physical exam, and if necessary, X-rays. This information should be obtained by the physician and the results found in the patient's medical record. However, a nutritionist should be able to recognize the physical signs and symptoms of malnutrition.

In general, the medical history is designed to obtain prevalent disease patterns such as infections or parasitic infestations, which influence absorption and metabolism of nutrients independent of dietary intake.

The physical examination should reveal the outward signs that will provide clues to the patient's nutritional history. The physician and nutritionist should notice any physical signs that are considered symptoms of nutrient deficiency; however, many of these signs and symptoms can be indicative of another problem or a specific disease and therefore, may not be considered as a deficiency of a particular nutrient. It must be emphasized that (1) signs of malnutrition may be related to non-nutritional factors (i.e., environmental or socioeconomic factors) and (2) they may not correlate with information obtained from dietary data or biochemical values in the individual.[36] These two factors should be kept in mind, especially when dealing with an elderly person, particularly if the person has a chronic disease which may manifest signs or symptoms of malnutrition; also, the physiological changes that have occurred in the person must be kept in mind. Minds should also be kept open to new ways of assessing nutrient status in these chronic disease states. For example, many elderly, due to loneliness and depression, are prey to alcoholism. Alcoholism has a progressive effect in producing malnutrition. Buccal smears from alcoholic patients have recently been shown to have good associations between selected standard indexes of nutritional status and oral cytologic characteristics.[37] The use of these oral cytologic characteristics to assess malnutrition in alcoholics may be of value in the future. Hair pluckability as an index of protein depletion has been recently used on Nigerians aged 1 month to 75 years, and epilation force was shown to be significantly lower in those suffering from protein malnutrition with serum albumin less than or equal to 3 g/dℓ. The use of a trichotillometer, which measures epilation force, has been suggested for field use.[38] It too may prove to be of use in future nutritional assessment.

The physical signs of a clinical deficiency, with considerations of a deficiency or excess of a nutrient are listed in Table 2.[39]

ANTHROPOMETRIC MEASUREMENTS

Anthropometry is the measurement of the variations of the physical dimensions and the gross composition of the human body at different age levels and degrees of nutrition.[40]

Table 2
CLINICAL NUTRITION EXAMINATION

Clinical findings	Consider deficiency of	Consider excess of
Hair, nails		
Flag sign (transverse depigmentation of hair)	Protein, copper	
Hair easily pluckable	Protein	
Hair thin, sparse	Protein, biotin, zinc	Vitamin A
Nails spoon-shaped	Iron	
Nails lackluster, transverse ridging	Protein-caloria	
Skin		
Dry, scaling	Vitamin A, zinc, essential fatty acids	Vitamin A
Erythematous erutpion (sunburn-like)		Vitamin A
Flaky paint dermatosis	Protein	
Follicular hyperkeratosis	Vitamins A,C, essential fatty acids	
Nasolabial seborrhea	Niacin, pyridoxine, riboflavin	
Petechiae, purpura	Ascorbic acid, vitamin K	
Pigmentation, desquamation (sun-exposed area)	Niacin (pellagra)	
Subcutaneous fat loss	Calorie	
Yellow pigmentation spring sclerae (benign)		Carotene
Eyes		
Angular palpebritis	Riboflavin	
Band keratitis		Vitamin D
Corneal vascularization	Riboflavin	
Dull, dry conjunctiva	Vitamin A	
Fundal capillary microaneurysms	Ascorbic acid	
Papilledema		Vitamin A
Scleral icterus, mild	Pyridoxine	
Perioral		
Angular stomatitis	Riboflavin	
Cheilosis	Riboflavin	
Oral		
Atrophic lingual papillae	Niacin, iron, riboflavin, folate vitamin B_{12}	
Glossitis (scarlet, raw)	Niacin, pyridoxine, riboflavin, vitamin B_{12}, folate	
Hypogeusesthesia (also hyposmia)	Zinc, vitamin A	
Magenta tongue	Riboflavin	
Swollen, bleeding gums (if teeth present)	Ascorbic acid	
Tongue fissuring, edema	Niacin	
Glands		
Parotid enlargement	Protein	
"Sicca" syndrome	Ascorbic acid	
Thyroid enlargement	Iodine	
Heart		
Enlargement, tachycardia high output failure	Thiamine ("wet beriberi")	
Small heart, decreased output	Caloria	
Sudden failure, death	Ascorbic acid	
Abdomen		
Heptaomegaly	Protein	Vitamin A
Muscles, extremities		
Calf tenderness	Thiamine, ascorbic acid (hemorrhage into muscle)	
Edema	Protein, thiamine	
Muscle wastage (especially temporal area, dorsum of hand, spine)	Calorie	
Bones, joints		
Beading of ribs (child)	Vitamins C, D	
Bone and joint tenderness	Ascorbic acid (subperiosteal hemorrhage) Vitamin A (child)	

Table 2 (continued)
CLINICAL NUTRITION EXAMINATION

Clinical findings	Consider deficiency of	Consider excess of
Bone tenderness (adult)	Vitamin D, calcium, phosphorus (osteomalacia)	
Bulging fontanelle (child)		Vitamin A
Craniotabes, bossing (child)		Vitamin D
Nerologic		
Confabulation, disorientation	Thiamine (Korsakoff's psychosis)	
Decreased position and vibratory senses, ataxia	Vitamin B$_{12}$, thiamine	
Decreased tendon reflexes, slowed relaxation phase	Thiamine	
Drowsiness, lethargy		Vitamins A,D
Ophthalmoplegia	Thiamine, phosphorus	
Weakness, paresthesias, decreased fine tactile sensation	Vitamin B$_{12}$, pyridoxine, thiamin	
Other		
Delayed, healing and tissue repair (e.g., wound, infarct, abscess)	Ascorbic acid, zinc, protein	
Fever (low-grade)		Vitamin A

From Weinsier, R. L. and Butterworth, C. E., *Handbook of Clinical Nutrition*, C. V. Mosby, St. Louis, 1981, 30.

Therefore, repeated anthropometric measurements over a period of time can provide an index of an individual's response to changes in the environment, i.e., changes of weight and lean body mass. At the 1969 White House Conference on Food, Nutrition, and Health, the following recommendations on anthropometric measurements were recommended for elderly adults: weight, height, arm circumference, and triceps skinfold.[41]

When gathering and evaluating anthropometric measurements, it is very important that standardized equipment and procedures are used.

Height

When measuring the height of an adult, the subject must stand erect, without shoes, with his feet parallel, and with heels, buttocks, head, and shoulders touching the wall. However, obtaining the correct height of an elderly individual may be complicated due the physical and anatomical changes that occur with aging.

The most obvious physical alteration is a change in body stature, which is primarily caused by a progressive decline in absolute height. This decrease in height that occurs with age has been documented by several cross-sectional studies.[42-48] One study using 855 cadavers documented a reduction in height of 1.2 cm/20 years postmaturity for whites and blacks of both sexes,[48] whereas another study found a decrease of 7 cm for older white subjects and black females, but not for black males over 50 years of age.[47] Fisher et al.[49] reported declines in mean heights of elderly Utahans to be 2 in. for men and 8 in. for women between the ages of 41 to 64 years and over 74 years. The mean heights of older age groups are less than the mean heights of younger age groups.[49] Shortening of the spinal column is primarily responsible for the loss of stature that occurs with aging since it has shown that the long bones do not undergo a significant reduction in size with age.[50,51]

The changes in the vertebral column result from narrowing of the discs as well as from loss of height of the individual vertebrae; the changes are responsible for a progressive decrease in height thereafter.[46] Another factor affecting height is the kyphosis, or curvature of the spine which gives an individual the hunchbacked appearance that is present in many elderly persons. It is believed that this condition is due to generalized osteoporosis.[46] Severe osteoporosis may cause the bones in the legs to bow. This bowing, in combination with a curvature of the spine in some elderly subjects makes it difficult to obtain the correct height. An assessment of 30 elderly subjects sought to find if there was a discrepancy between heights expressed by the subjects and the actual measured heights. The difference between them was found to be 2.4 in.[18] Therefore, it is advised that referring to elderly subjects for their heights may bring misleading results. Bony chest breadth measurement by roentgenogram has been used in 45- to 65-year-old Scots and has been suggested as a useful indicator of frame size, fat-free mass, and relative fatness. Unlike stature, bony chest breadth (BCB) remains stable with aging. The measurement can be expressed as weight to bony chest breadth ratio.[52] Perhaps BCB may come to replace height in some cases, particularly where the nutritional assessment of the elderly is concerned.

Weight

The weight of a person is the most commonly used indicator of the body size which reflects adequate nutriture; however, it has limited use because it does not distinguish between body size and composition.

The table of standards that is most commonly used to compare height/weight observations is the 1959 Metropolitan Life Insurance tables, as seen in Table 3.[53] The data from which the Metropolitan Life Insurance tables were derived were collected over a wide span of time and are different from more recent surveys in the U.S. This suggests that, on the whole, the normal healthy American tends to be bigger and heavier than the tables suggest.[54] The techniques used to obtain the measurements for these tables were not standardized.

In addition, they were based on a potentially erroneous assumption that weight at age 25 is ideal. Their acceptance and overinterpretation reflected a cultural bias. That is, they were used in excess of what the data would support. They were used as an ideal medical classification instead of an insured population index only.[55] The Metropolitan Life Insurance tables have recently been revised, and the weights for shorter men and women have been increased to more realistic levels. The tables, based upon the ''Build Study 1979/Society of Actuaries and Association of Life Insurance Medical Directors of America (Chicago)'', have been only briefly described in news releases and require further evaluation (Table 4).

The standards recommended by the Committee on Dietary Allowances, Food and Nutrition Board is based on recent data.[26] Table 5 presents desirable weights for heights for adult males and females.

Master et al.[56] reported data on heights and weights of Americans aged 65 to 94 years and found that the percentage of overweight males declined steadily over the 30-year span to constitute only 10% of the group surveyed. Simultaneously, the number of underweight men increased from 20 to 50%. For the same age groups, the percentage of overweight females decreased from 40 to 10%, with underweight females increasing from 20 to 55%. From their data, standard tables for heights and weights have been developed for this older population (Table 6). These tables may be more accurate than the widely used Metropolitan Life Insurance tables from 1959 for calculating ideal body weight of persons over 65. However, these data are representative of an elderly white population. Thus, a need to develop more comprehensive standards which take into account both varied ethnic and socioeconomic status, currently exists.

Most males continue to gain weight until about age 42, while women do not attain their maximum weight until approximately age 50. Thereafter, weight usually remains fairly constant up to about age 65 or 70, whereupon it will progressively decline.[57,58] Most cross-sectional data, as well as the few published longitudinal studies, document a cumulative decrease in lean body mass with age accompanied by an increase in adipose tissue.[43,59,60]

Table 3
HEIGHT AND WEIGHT STANDARDS FOR ADULTS

Men (in indoor clothing)[a]

Feet	Inches	Centimeters	Small frame	Medium frame	Large Frame
5	1	154.9	112—120	118—129	126—141
5	2	157.5	115—123	121—133	129—144
5	3	160.0	118—126	124—136	132—148
5	4	162.6	121—129	127—139	135—152
5	5	165.1	124—133	130—143	138—156
5	6	167.6	128—137	134—147	142—161
5	7	170.2	132—141	138—152	147—166
5	8	172.7	136—145	142—156	151—170
5	9	175.3	140—150	146—160	155—174
5	10	177.8	144—154	150—165	159—179
5	11	180.3	148—158	154—170	164—184
6	0	182.9	152—162	158—175	168—189
6	1	185.4	156—167	162—180	173—194
6	2	188.0	160—171	167—185	178—199
6	3	190.5	164—175	172—190	182—204

Women (in indoor clothing)[a]

4	8	142.2	92—98	96—107	104—119
4	9	144.8	94—101	98—110	106—122
4	10	147.3	96—104	101—113	109—125
4	11	149.9	99—107	104—116	112—128
5	0	152.4	102—110	107—119	115—131
5	1	154.9	105—113	110—122	118—134
5	2	157.5	108—116	113—126	121—138
5	3	160.0	111—119	116—130	125—142
5	4	162.6	114—123	120—135	129—146
5	5	165.1	118—127	124—139	133—150
5	6	167.6	122—131	128—143	137—154
5	7	170.2	126—135	132—147	141—158
5	8	172.7	130—140	136—151	145—163
5	9	175.3	134—144	140—155	149—168
5	10	177.8	138—148	144—159	153—173

Note: These tables correct the 1959 Metropolitan standards to height without shoe heels.

[a] Allow 5 to 7 lb for men and 2 to 4 lb for women.

From the *Build and Blood Pressure Study*, Chicago Society of Actuaries, Chicago, 1959. With permission.

Current evidence suggests that the increased amounts of fat are deposited primarily around internal organs, especially in females, while the amount of subcutaneous fat increases only slightly.[43,58]

The end result is a reduced ratio of subcutaneous fat to total body fat. After age 65 to 70, total body weight decreases without an increase in fat tissue.[60] Nonetheless, obesity is common among elderly Americans. Kohrs[61] reported on 400 Missourians 59 years and older. When obesity was defined as greater than 119% of desirable weight for height, 22% of the males and 59% of the females were obese. Fisher et al.,[49] reporting on elderly Utahans,

Table 4
HEIGHT AND WEIGHT STANDARDS FOR ADULTS

Men (in indoor clothing)[a]

Feet	Inches	Centimeters	Small frame	Medium frame	Large frame
5	1	154.9	128—134	131—141	138—150
5	2	157.5	130—136	133—143	140—153
5	3	160.0	132—138	135—145	142—156
5	4	162.6	134—140	137—148	144—160
5	5	165.1	136—142	139—151	146—164
5	6	167.6	136—145	142—154	149—168
5	7	170.2	140—148	145—157	152—172
5	8	172.7	142—151	148—160	155—176
5	9	175.3	144—154	151—163	158—180
5	10	177.8	146—159	154—166	161—184
5	11	180.3	149—160	157—170	164—188
6	0	182.9	152—164	160—174	168—192
6	1	185.4	155—165	164—178	172—197
6	2	188.0	158—172	167—182	176—202
6	3	190.5	162—176	171—187	181—207

Women (in indoor clothing)[a]

Feet	Inches	Centimeters	Small frame	Medium frame	Large frame
4	8	142.2	102—111	109—121	118—131
4	9	144.8	103—113	111—123	120—134
4	10	147.3	104—115	113—126	122—137
4	11	149.9	106—118	115—129	125—140
5	0	152.4	108—121	118—132	128—143
5	1	154.9	111—124	121—135	131—147
5	2	157.5	114—127	124—138	134—150
5	3	160.0	117—130	127—141	137—155
5	4	162.6	120—133	130—144	140—159
5	5	165.1	120—133	133—147	143—163
5	6	167.6	123—136	136—150	146—167
5	7	170.2	126—139	139—153	149—170
5	8	172.7	129—142	142—156	152—173
5	9	175.3	135—148	145—159	155—176
5	10	177.8	138—151	148—162	158—179

Note: These tables correct the 1979 Metropolitan Standards to height without shoe heels.

[a]   Allow 5 to 7 lb for men and 2 to 4 lb for women.

From the *Build Study 1979*, Society of Actuaries and Association of Life Insurance Medical Directors of America (Chicago), 1980, 359. With permission.

used the weight for height of 25-year-olds as a standard. Approximately 50% of the women and 30% of the men had weights exceeding 120% of the standard.

Arm Measurements

The data documenting alterations in lean body mass and fat content with age have been obtained by utilizing sophisticated techniques such as measurement of total body potassium,[62,63] total body water,[64] radiography,[65,66] and calculations of total body density.[67] These procedures, however, are not widely available and are too time consuming and expensive

Table 5
SUGGESTED DESIRABLE WEIGHTS FOR HEIGHTS AND
RANGES FOR ADULT MALES AND FEMALES

Height[b]		Weight[a]			
		Men		Women	
in.	cm	lb	kg	lb	kg
58	147	—	—	102 (92—119)	46 (42—54)
60	152	—	—	107 (96—125)	49 (44—57)
62	158	123 (112—141)	56 (51—64)	113 (102—131)	51 (46—59)
64	163	130 (118—148)	59 (54—67)	120 (108—138)	55 (49—63)
66	168	136 (124—156)	62 (56—71)	128 (114—146)	58 (52—66)
68	173	145 (132—166)	66 (60—75)	136 (122—154)	62 (55—70)
70	178	154 (140—174)	70 (64—79)	144 (130—163)	65 (59—74)
72	183	162 (148—184)	74 (67—84)	152 (138—173)	69 (63—79)
74	188	171 (156—194)	78 (71—88)	—	—
76	193	181 (164—204)	82 (74—93)	—	—

[a] Without clothes, average weight ranges in parentheses. Adapted from RDA.[26]
[b] Without shoes.

for routine clinical use. Thus, several attempts have been made to correlate the results of these various tests with the more widely used anthropometrics, such as triceps skinfold and arm circumference.[67-69]

Triceps skinfold is an indirect measurement of adipose tissue stores or body fat. It consists of measuring a double-layer of skin and subcutaneous fat at the middle point of the upper arm with skinfold calipers. The most common reliable calipers used are Lange and Harpenden.

The status of somatic protein stores or lean body mass can be estimated by calculating arm muscle circumference and arm muscle area from triceps skinfold and arm circumference measurements.

Durnin and Womberley[67] found that body density could be reliably estimated from skinfold thickness in men and women aged 16 to 72 years. A combination of biceps, triceps, and subscapular measurements were found to be the most reliable measurements for females 50 to 68 years, whereas the combined measurement of triceps, subscapular, and suprailiac skinfold thickness was more reliable in males of the same age group. A more recent study of body composition in 70-year-old males and females in Gothenburg, Sweden compared the reliability of measuring body composition using total body potassium and total body water to that of the more simple anthropometric measurements.[68] They found it was possible to estimate body fat from body weight and subscapular skinfold thickness in males, and from body weight, thigh, and triceps skinfold thickness in females. From the above studies, it seems that skinfold measurements on the trunk (subscapular and suprailiac) may be more reliable predictors of body fat in males, whereas extremity skinfold measurements (triceps, biceps, and thigh skinfolds) seem to be more accurate in females. The utilization of multiple skinfold measurements may avoid some of the pitfalls that occur in relying on triceps skinfold alone in the elderly population. However, Seltzer[70] found that the subscapular skinfold was not a satisfactory indicator of body fat, especially for obese subjects.

In addition to precise measurement procedure, the usefulness and validity of anthropometric evaluation depend upon the use of appropriate standards for data interpretation. Data from the Ten State Nutrition Survey are generally accepted as the most representative standards of the U.S. population currently available.[22] These data are based on a cross-sectional sample of approximately 12,400 subjects aged 1 to 44 years measured between

Table 6
AVERAGE WEIGHT FOR HEIGHT: PERSONS AGE 65 TO 94 YEARS

Height (in.)	Age (years)					
	65—69	70—74	75—79	80—84	85—89	90—94
			Men			
61	156—128	153—125	151—123			
62	158—130	155—127	153—125	148—122		
63	161—131	157—129	155—127	150—122	146—120	
64	164—134	161—131	157—129	152—124	148—122	
65	166—136	164—134	160—130	155—127	153—125	143—117
66	169—139	167—137	163—133	158—130	156—128	146—120
67	172—140	170—140	166—136	162—132	160—130	150—122
68	175—143	174—142	169—139	165—136	163—133	154—126
69	179—145	178—146	174—142	169—139	167—137	158—130
70	184—150	182—148	178—146	175—143	172—140	164—134
71	189—155	186—152	183—149	180—148	176—144	169—139
72	195—159	190—156	188—154	187—153	182—148	
73	200—164	196—160	192—158			
			Women			
58	146—120	138—112	135—111			
59	147—121	140—114	136—112	122—100		121—99
60	148—122	142—116	139—113	130—106	124—102	
61	151—123	144—118	141—115	133—109	128—104	
62	153—125	147—121	144—118	136—112	132—108	131—107
63	155—127	151—123	147—121	141—115	136—112	131—107
64	158—130	154—126	151—123	145—119	141—115	132—108
65	162—132	158—130	154—126	150—122	146—120	136—112
66	166—136	162—132	157—128	154—126	152—124	142—116
67	170—143	166—136	161—131	158—130	156—128	
68	175—143	170—140				
69	180—148	176—144				

Note: All subjects in this survey were healthy noninstitutionalized individuals.

From Master, A. M., Lasser, R. P., and Beckman, G., Tables of average weight and height of Americans aged 65 to 94 years, *JAMA*, 712, 115/659, 1960. With permission.

1968 and 1970. Because Caucasian individuals from low, middle, and upper income levels were included in this stratified probability sample, the standards do not account for possible variation due to racial and ethnic background. Percentiles are reported for upper arm circumference and triceps skinfold; these can be seen in Table 7. From these measurements, age- and sex-specific standards and percentiles were determined. Although this survey may not be representative of the entire U.S. population, most investigators agree that they are superior to those published by Jeliffe.[23] These measurements probably underestimate average measurements of Americans because they were developed for nutrition evaluation in underdeveloped countries (Table 8). Unfortunately, both of these standards are based on data obtained from young and middle-aged adults and, therefore, may not be advisable to use for elderly subjects.[16,17,23]

Table 7

PERCENTILES FOR UPPER AND MID-ARM CIRCUMFERENCES AND TRICEP SKINFOLD

Age group (years)	Males					Females				
	5th	15th	50th	85th	95th	5th	15th	50th	85th	95th

Percentiles for Upper Arm Circumference (cm)

17.5—24.4	25.0	26.4	29.2	33.9	35.4	21.5	23.3	26.0	29.7	32.9
24.5—34.4	26.0	28.0	31.0	34.4	36.6	23.0	24.3	27.5	32.4	36.1
34.5—44.4	25.9	28.0	31.2	34.5	37.1	23.2	25.0	28.6	34.0	37.4

Percentiles for Tricep Skinfold (mm)

17.5—24.4	4	5	10	18	25	9	12	17	25	31
24.5—34.4	4	6	11	21	28	9	12	19	29	36
34.5—44.4	4	6	12	22	28	20	14	22	32	39

Percentiles for Mid-Arm Muscle Circumference (cm)

17.5—24.4	21.7	23.2	25.8	28.6	30.5	17.0	18.3	20.5	22.9	35.3
24.5—34.4	22.0	24.1	27.0	29.5	31.5	17.7	18.9	21.3	24.5	27.2
34.5—44.4	22.2	23.9	27.0	30.0	31.8	18.0	19.2	21.6	25.0	27.9

From Frisancho, A. R., *Am. J. Clin. Nutr.*, 27, 1052, 1974. With permission.

Table 8

TRICEPS SKINFOLD, MID-ARM CIRCUMFERENCE, AND MID-ARM MUSCLE CIRCUMFERENCE STANDARDS FOR ADULTS

	Standard	90% Standard	90 to 60% Standard	60% Standard

Triceps Skinfold (mm)

Male	12.5	11.3	11.3—7.5	7.5
Female	16.5	14.9	14.9—9.9	9.9

Mid-Arm Circumference (cm)

Male	29.3	26.3	26.3—17.6	17.6
Female	28.5	25.7	25.7—17.1	17.1

Mid-Arm Muscle Circumference (cm)

Male	25.3	22.8	22.8—15.2	15.2
Female	23.2	20.9	20.9—13.9	13.9

From Jelliffe, D. B., Monograph Series No. 53, World Health Organization, Geneva, 1966. With permission.

BIOCHEMICAL ASSESSMENT

Evaluating nutritional status by biochemical measurements is a more objective and precise approach than using other components. The interpretation of laboratory data is often difficult to accomplish and may not correlate with other findings, i.e., physical signs. An advantage of using biochemical measurements is that changes in nutritional status are usually reflected sooner than in other measurements. Many biochemical parameters have been shown to be affected by age. The alterations that occur, however, are not as easily delineated as are the physical changes associated with the aging process.

The creatinine/height index (CHI) is defined as the total 24-hr creatinine excretion for an individual and, in the absence of renal dysfunction, is used as a predictor of muscle protein status.[71,72] The aging process has little effect on serum creatinine measurements although a decline in renal function and in creatinine production has been well described.[30,73-75] Glomerular filtration decreases, resulting in a gradually decreasing creatinine clearance, but this is usually offset by a decrease in creatinine production due to a parallel loss of muscle mass.[74] The total amount of creatinine excreted every 24 hr will, however, be reduced in the elderly. The average creatinine excretion rate has been shown to decline from 23.6 mg/kg/24 hr in 18- to 29-year-old subjects to 12.1 mg/kg/24 hr in 80- to 92-year-old subjects.[73] Thus, if norms based on standards developed for young adults are being used to calculate CHI, the severity of body mass depletion for the geriatric age group may be highly overestimated.

Serum albumin levels have been shown to accurately reflect a decline in protein status due to a decreased rate of synthesis in malnourished children[76] and adult surgical and medical patients.[6-11] The various effects of disease on plasma proteins must be considered when assessing the elderly since there is an increased incidence of chronic disease in this age group.[76] Several nutrition surveys have reported the frequent occurrence of depressed serum levels in the elderly, yet no correlation with the protein intake of these patients was made.[49,54,61] However, other studies have indicated that total serum protein and albumin are not significantly depressed in the aged.[77,78] Another indicator of visceral protein status is serum transferrin, which has a shorter half-life than serum albumin and thus can result in depressed levels of this before any changes in serum albumin can occur.[79]

Hemoglobin and hematocrit values are often used to assess nutritional status, as anemia is frequently associated with malnutrition. With increased age, a higher prevalence of decreased hemoglobin and hematocrit levels is noted in both males and females. Thus, the incidence of anemia in subjects between age 75 and 85 is higher than in those between 65 and 75, and the highest in subjects about 85 years of age.[80] It is possible that the fall in hemoglobin which occurs with age results from a direct effect of senescence, and thus may represent a normal physiological response to aging. A reduction in lean body mass and decreased requirement for oxygen which occurs in aging may result in this reduction of circulating hemoglobin.[59]

Many immunological abnormalities described in severe malnutrition of children have also been documented in the elderly and have been ascribed to the effects of senescence.[81] It is well recognized that malnutrition results in a profound compromise of host defense mechanisms.[82] Lymphocytopenia and anergy to skin test evaluation have been commonly used as indicators of PCM; however, many studies have reported that the incidence of anergy and lymphocytopenia also increase with advancing age.[81] Reduction of T cell number, altered response to mitogens, and an increase in the number of T suppressor cells have been described in the elderly population.[83] Thus, the similarity of the effects of the aging process and of PCM on immune function places the usefulness of routine immunological testing in this population in question.

Inasmuch as malnutrition can affect social and physiological functions, their impairment may also be an index of malnutrition. In 1976, the National Institutes of Science identified five areas of functioning that malnutrition likely affects. They are social/behavioral performance, disease response, physical activity and work performance, reproductive competence, and cognitive ability.[84] Current studies are underway to determine the extent to which these functional indicators may define moderate protein-energy malnutrition. This work is both timely and relevant, for the majority of populations in the developing as well as developed countries are more likely to experience moderate as opposed to severe malnutrition. Since clinical signs often do not appear until the degree of deficiency is severe, functional indexes of malnutrition could prove to be the earliest and most sensitive type of nutritional measurement. Sensitive and reliable methods of assessing nutrition by functional indexes, and realistic standards by which to do so remain to be developed and proven practical in the field.[84]

More varied groups, including minority subgroups of the U.S. elderly population need to be surveyed in order to establish current and more accurate standards for all segments of this age group. Such surveys and the evaluation of their results will be useful in nutritional assessment, particularly when a better understanding of the physiology of aging is achieved.

ACKNOWLEDGMENTS

Support for the preparation of this article was by National Livestock and Meat Board grant. Appreciation is expressed to Sandra S. Dunbar for some of the literature reviewed.

REFERENCES

1. **Peck, R. C.,** Psychological developments in the second half of life, in *Middle Age and Aging,* Neugarten, B. L., Ed., The University of Chicago Press, Chicago, 1973.
2. **Weiner, M. B., Brok, A. J., and Snadowsky, A. M.,** *Working with the Aged,* Prentice-Hall, Englewood Cliffs, N.J., 1978.
3. **Driver, A. G., McAlevy, M. T., and Smith, J. L.,** Nutritional assessment of patients with chronic obstructive pulmonary disease and acute respiratory failure, *Chest,* 82, 568, 1982.
4. **Blumenkrantz, M. J., Kopple, J. D., Gutman, R. A., Chan, Y. K., Barbour, G. L., Roberts, C., Shen, F. H., Gandhi, N. C., Tucker, C. T., Curtis, F. K., and Coburn, J. W.,** Methods for assessing nutritional status of patients with renal failure, *Am. J. Clin. Nutr.,* 33, 1567, 1980.
5. **Albiin, N., Asplund, K., and Bjermer, L.,** Nutritional status of medical patients on emergency admission to hospital, *Acta Med. Scand.,* 212, 151, 1982.
6. **Bistrain, B. R., Blackburn, G. L., Vitale, J., Cochran, D., and Naylor, J.,** Prevalence of malnutrition in general medical patients, *JAMA,* 235, 1567, 1976.
7. **Willcutts, H. D.,** Nutritional assessment of 1000 surgical patients in an affluent suburban community hospital, *J. Parent. Ent. Nutr.,* 1, 25, 1977.
8. **Bistrain, B. R., Blackburn, G. L., Hallowell, E., and Hadelle, R.,** Protein status of general surgical patients, *JAMA,* 230, 858, 1974.
9. **Faintuch, J., Waitzberg, D. L., Azevado, S. A., Gama-Rodrigues, J. J., and Gama, A. H.,** Influence of malnutrition on the length of hospitalization after moderate and severe surgical injury, *Nutr. Support Serv.,* 1, 29, 1981.
10. **Mullen, J. L., Gertner, M. H., Buzby, G. P., Goodhart, G. L., and Rosato, E. F.,** Implications of malnutrition in the surgical patient, *Arch. Surg.,* 114, 121, 1979.
11. **Weinsier, R. L., Hunker, E. M., Krumdieck, C. L., and Butterworth, C. E., Jr.,** Hospital malnutrition: a prospective evaluation of general medical patients during the course of hospitalization, *Am. J. Clin. Nutr.,* 23, 418, 1979.
12. **Blackburn, G. L. and Harvey, K. B.,** Nutritional assessment as a routine in clinical medicine, *Postgrad. Med.,* 71, 46, 1982.
13. **Bastow, M. D.,** Anthropometrics revisited, *Proc. Nutr. Soc.,* 41, 381, 1982.
14. **Winborn, A. L., Banaszek, N. K., Freed, B. A., and Kaminski, M. N., Jr.,** A protocol for nutritional assessment in a community hospital, *J. Am. Dent. Assoc.,* 78, 129, 1981.

15. **Nazari, S., Dionigi, R., Dionigi, P., and Bonoldi, A.,** A multivariate pattern for nutritional assessment, *J. Parent. Ent. Nutr.,* 4, 499, 1980.

16. **Gray, G. E. and Gray, L. K.,** Anthropometric measurements and their interpretations: principles, practices, and problems, *J. Am. Diet. Assoc.,* 77, 534, 1980.

17. **Burgert, S. L. and Anderson, C. F.,** An evaluation of upper arm measurements used in nutritional assessment, *Am. J. Clin. Nutr.,* 32, 2136, 1979.

18. **Mitchell, C. O. and Lipschitz, D. A.,** Detection of protein-calorie malnutrition in the elderly, *Am. J. Clin. Nutr.,* 35, 398, 1982.

19. **Bistrain, B.,** Anthropometric norms used in assessment of hospitalized patients, *Am. J. Clin. Nutr.,* 33, p. 2, 211, 1980.

20. **Young, E. A.,** Nutrition, aging, and the aged, *Med. Clin. N. Am.,* 67, 295, 1983.

21. **Chen, L. H. and Fan-Chiang, W. L.,** Biochemical evaluation of riboflavin and vitamin B_6 status of institutionalized and noninstitutionalized elderly in central Kentucky, *Int. J. Vitam. Nutr. Res.,* 51, 232, 1981.

22. **Frisancho, A. R.,** Triceps skinfold and upper arm muscle size norms for assessment of nutritional status, *Am. J. Clin. Nutr.,* 27, 1052, 1974.

23. **Jeliffe, D. B.,** The assessment of the nutritional status of the community, Monograph Series No. 53, World Health Organization, Geneva, 1966.

24. **Mitchell, C. O. and Lipschitz, D. A.,** The effect of age and sex on the routinely used measurements to assess the nutritional status of hospitalized patients, *Am. J. Clin. Nutr.,* 36, 340, 1982.

25. **Heymsfield, S. B., Arteaga, C., McManus, C., Smith, J., and Moffitt, S.,** Measurement of muscle mass in humans: validity of the 24-hour urinary creatinine method, *Am. J. Clin. Nutr.,* 37, 478, 1983.

26. *Recommended Dietary Allowances,* 9th ed., National Research Council, Food and Nutrition Board, National Academy of Science, Washington, D.C., 1980.

27. **Tobin, J. D.,** Normal aging — the inevitable syndrome, in *The Quality of Life, the Later Years,* Brown, L. E., Ed., Publishing Sciences Group, Acton, Mass., 1975.

28. **Brauer, P. M., Slavin, J. L., and Marlett, J. A.,** Apparent digestibility of neutral detergent fiber in elderly and young adults, *Am. J. Clin. Nutr.,* 34, 1061, 1981.

29. **Robert, J. J., Cummins, J. C., Wolfe, R. R., Durkot, M., Matthews, D. E., Zhao, X. H., Bier, D. M., and Young, V. R.,** Quantitative aspects of glucose production and metabolism in healthy elderly subjects, *Diabetes,* 41, 203, 1982.

30. **Masoro, E. J.,** Physiologic changes with aging, in *Nutrition and Aging,* Winick, M., Ed., John Wiley & Sons, N.Y., 1975.

31. **Kohrs, M. B., Nordstrom, J., Plowman, E. L., O'Hanlon, P., Moore, C., Davis, C., Abrahams, Ol., and Eklund, D.,** Association of participation in a nutritional program for the elderly with nutritional status, *Am. J. Clin. Nutr.,* 33, 2643, 1980.

32. **Bowman, B. B. and Rosenberg, I. H.,** Assessment of the nutritional status of the elderly, *Am. J. Clin. Nutr.,* 35, 1142, 1982.

33. **MacPhail, A. P., Bothwell, T. H., Torrance, J. D., Derman, D. P., Bezwoda, W. R., Charlton, R. W., and Mazet, F. G. H.,** Iron nutrition in Indian women at different ages, *S. Afr. Med. J.,* 59, 939, 1981.

34. **Shank, R. E.,** Nutrition and Aging, in Epidemiology of Aging, Osfeld, A. M. and Gibson, D. C., National Institutes of Health, Department of Health, Education and Welfare Publ. No. 77-711, Bethesda, Md., 1977, 199.

35. **Wagner, P. A., Kristu, M. L., Bailey, L. B. K. et al.,** Zinc status of elderly black Americans from urban low-income households, *Am. J. Clin. Nutr.,* 33, 1771, 1980.

36. **Wade, J. E.,** Role of a clinical dietitian specialist on a nutrition support service, *J. Am. Diet. Assoc.,* 70, 185, 1977.

37. **Hillman, R. W. and Kissin, B.,** Oral cytologic patterns and nutritional status: relationships in alcoholic subjects, *Oral Surg.,* 49, 34, 1980.

38. **Smelser, D. N., Smelser, N. B., Krumdieck, C. L., Schreeder, M. T., and Lavin, G. T.,** Field use of hair epilation force in nutrition status assessment, *Am. J. Clin. Nutr.,* 35, 342, 1982.

39. **Weinsier, R. L. and Butterworth, C. E.,** *Handbook of Clinical Nutrition,* C. V. Mosby, St. Louis, 1981.

40. **Roberts, S. L. W.,** *Nutritional Assessment Manual,* 1977.

41. **Christakis, G.,** Nutritional assessment of the elderly, in Nutritional assessment in health programs, *American Journal of Public Health, Part II,* American Public Health Association, Washington, D.C., 1973, 63.

42. **Hertzog, K. P., Garn, S. M., and Hempy, H. O.,** Partitioning the effects of secular trend and aging on adult stature, *Am. J. Phys. Anthropol.,* 31, 111, 1969.

43. **Nappa, H., Anderson, M., Bengtsson, C., Bruce, A., and Isaksson, B.,** Longitudinal studies of anthropometric data and body composition: the population study in Gothenburg, Sweden, *Am. J. Clin. Nutr.,* 33, 155, 1980.

44. **Posner, B. M.,** *Nutrition and the Elderly,* D.C. Heath & Co., Lexington, Mass., 1979.
45. **Riley, G.,** How aging influences drug therapy, *U.S. Pharm.,* 2, 28, 1977.
46. **Rossman, J.,** The anatomy of aging, in *Clinical Geriatrics,* J. B. Lippincott, Philadelphia, 1979.
47. **McNeely, G. R., Heyssell, R. M., and Ball, C. O.,** Analysis of factors affecting body composition determined from potassium content of 915 normal subjects, *Ann. N.Y. Acad. Sci.,* 110, 271, 1963.
48. **Trotter, M. and Gleser, G.,** The effect of aging on stature, *Am. J. Phys. Anthropol.,* 9, 311, 1951.
49. **Fisher, S., Hendricks, D. G., and Mahoney, A. W.,** Nutritional assessment of senior rural Utahans by biochemical and physical measurements, *Am. J. Clin. Nutr.,* 31, 667, 1978.
50. **Miall, W. E., Ashcroft, M. T., Lovell, H. G., and Moore, F.,** A longitudinal study of the decline of adult height with age in two Welsh communities, *Hum. Biol.,* 39, 445, 1967.
51. **Garn, S. M.,** Bone-loss and aging, in *The Physiology and Pathology of Human Aging,* Academic Press, N.Y., 1975.
52. **Garn, S. M., Pesick, S. D., Hawthorne, N. M., and Hawthorne, B. S.,** The bony chest breadth as a frame size standard in nutritional assessment, *Am. J. Clin. Nutr.,* 37, 315, 1983.
53. Build & Blood Pressure Study, Chicago Society of Actuaries, 1959.
54. Ten-State Nutrition Survey 1968-1980, U.S. Department of Health, Education and Welfare, Publ. No. (HSM) 72-8133, Center for Disease Control, Atlanta, 1972.
55. **Ritenbaugh, C.,** Obesity as a culture-bound syndrome, *Cult. Med. Psychiat.,* 6, 347, 1982.
56. **Master, A. M., Lasser, R. P., and Beckman, G.,** Tables of average weight and height of Americans aged 65 to 94 years, *JAMA,* 172, 658, 1960.
57. **Garth, S. and Young, R.,** Concurrent fat loss and fat gain, *Am. J. Phys. Anthropol.,* 14, 497, 1956.
58. **Hedja, S.,** Skinfold in old and long-lived individuals, *Gerontology,* 8, 201, 1963.
59. **Forbes, G. and Reina, J.,** Adult lean body mass declines with age: some longitudinal observations, *Metabolism,* 19, 653, 1970.
60. **Young, C. M., Blondin, J., Tensuan, R., and Fryer, J. H.,** Body composition of "older" women, *J. Am. Diet. Assoc.,* 43, 344, 1963.
61. **Kohrs, M. B., O'Neill, R. O., Preston, A., Edlun, D., and Abrahams, D.,** Nutritional status of elderly residents in Missouri, *Am. J. Clin. Nutr.,* 31, 2186, 1978.
62. **Anderson, E. and Langham, W.,** Average potassium concentrations of the human body as a function of age, *Science,* 130, 713, 1959.
63. **Novak, L.,** Aging, total body potassium, fat-free mass and cell mass in males and females between ages 18 and 85 years, *J. Gerontol.,* 27, 438, 1972.
64. **Fryer, J. H.,** Studies on body composition in men aged 60 and over, in *Biological Aspects of Aging,* Shock, N. W., Ed., Columbia University Press, N.Y., 1962, 75.
65. **Young, C. M., Tensuan, R. S., Sault, F., and Holmes, F.,** Estimating body fat of normal young women. Visualizing fat pads by soft tissue x-rays, *J. Am. Diet. Assoc.,* 42, 409, 1963.
66. **Heymsfield, S. B., Olafson, R. P., Kutner, M. H., and Nixon, D. W.,** A radiographic method of quantifying protein-calorie undernutrition, *Am. J. Clin. Nutr.,* 32, 693, 1979.
67. **Durnin, J. V. and Womberley, S.,** Body fat assessed from total body density and its estimation from skinfold thickness: measurements on 481 men and women aged from 16 to 72 years, *Br. J. Nutr.,* 32, 77, 1974.
68. **Steen, B., Broce, A., Isaksson, B., Lewin, I., and Swanborg, A.,** Body composition in 70-year-old males and females in Gothenburg, Sweden. A population study, *Acta Med. Scand. Suppl.,* 611, 87, 1977.
69. **Watson, P. E., Watson, J. D., and Batt, R. D.,** Total body water volumes for adult males and females estimated from simple anthropometric measurements, *Am. J. Clin. Nutr.,* 33, 27, 1980.
70. **Seltzer, C. C. and Mayer, J.,** Greater reliability of the triceps skinfold over the subscapular skinfold as an index of obesity, *Am. J. Clin. Nutr.,* 20, 950, 1967.
71. **Mendez, J. and Buskirk, E. R.,** Creatinine height index, *Am. J. Clin. Nutr.,* 24, 385, 1971.
72. **Bistrain, B. R., Blackburn, G. L., Sherman, M., and Scrimshaw, N. W.,** Therapeutic index of nutritional depletion in hospitalized patients, *Surg. Gynecol. Obstet.,* 141, 512, 1975.
73. **Cockroft, D. W. and Gault, M. H.,** Prediction of creatinine clearance from serum creatinine, *Nephron,* 16, 31, 1976.
74. **Paper, S.,** The effects of age in reducing renal function, *Geriatrics,* 28, 83, 1973.
75. **Lott, R. S. and Hayton, W. L.,** Estimation of creatinine clearance from serum creatinine concentrations: a review, *Drug. Intell. Clin. Pharm.,* 12, 140, 1978.
76. **Rothschild, M. A., Oratz, M., and Schreiber, S. S.,** Albumin synthesis, *N. Engl. J. Med.,* 286, 748, 1972.
77. **Mitchell, C. O. and Lipschitz, D. A.,** A nutritional evaluation of healthy elderly, *J. Parent. Ent. Nutr.,* 4, 603, 1980.
78. **Goldman, R.,** Decline in organ function, in *Clinical Geriatrics,* Rossman, J., Ed., J. B. Lippincott, Philadelphia, 1979.

79. **Ingenbleek, Y., Van Den Schriek, H. G., De Nayer, P., and DeVesscher, M.,** Albumin, transferrin and the thyroxine-binding prealbumin/retinol-binding protein complex in nutritional assessment, *Clin. Chim. Acta,* 63, 61, 1975.

80. **McLennan, W. J., Andrew, G. R., Macleod, C., and Caird, F. L.,** Anaemia in the elderly, *Q. J. Med.,* 52, 1, 1973.

81. **Weksler, M. E. and Hutteroth, T. H.,** Impaired lymphocyte function in aged humans, *J. Clin. Invest.,* 53, 99, 1974.

82. **Bell, R. G., Hazell, L. A., and Price, P.,** Influence of dietary protein restriction on immune competence. II. Effect on lymphoid tissue, *Clin. Exp. Immunol.,* 26, 314, 1976.

83. **Chandra, R. K. and Schrimshaw, N. S.,** Immunocompetence in nutritional assessment, *Am. J. Clin. Nutr.,* 33, 2694, 1980.

84. **Solomons, N. W. and Allen, L. H.,** The functional assessment of nutritional status; principles, practice, and potential, *Nutr. Rev.,* 41, 33, 1983.

NUTRITION AND MENTAL FUNCTIONING IN THE ELDERLY

Clifford H. Swensen and Emma A. Montgomery

If we are what we eat, then it follows that nutrition affects every aspect of our being, including mental functioning. This is especially true of the elderly, who have a much smaller margin for deficiency in their functioning, and whose general functioning would be expected to be much more vulnerable to inadequate nutrition than that of a young, healthy adult. Therefore, one would expect that concern with the elderly would be a focus of considerable attention and research. This would be expected to be especially true concerning the area of mental functioning because the elderly appear to be somewhat more at risk for deficits in mental functioning than is the population in general. The elderly appear to be especially prone to deficits in memory, learning, and problem-solving. The elderly are also more likely to become depressed and develop disorders in their sleeping patterns.

The mental functioning of the elderly is of concern not only for humanitarian reasons, but also for economic reasons. The very rapid increase in the number of the elderly has been accompanied by a concomitant rise in the necessity of increasing the number of nursing homes, and with it the expense of nursing home care. Elderly people are admitted into nursing homes primarily because their intellectual functioning has declined to the point that they are judged to be no longer capable of caring for themselves, and have placed an intolerable burden of care upon their families. Therefore, if nutritional deficiencies could be found as one of the bases for the development of mental problems of older people, it would provide the basis for treatment that would prevent or ameliorate the development of some forms of the mental decline that is perhaps the greatest fear of old people themselves and the greatest single cause of older people becoming a burden upon their families.

The kinds of mental problems suffered by older people are often insidious in their onset. Problems of memory, learning, and problem-solving may develop over a period of years, but only be revealed when a specific stressful incident occurs that reveals the incapacity of the older person. Other common problems such as depression and insomnia may only become apparent to members of the family when a situation arises that reveals their presence. Frequently these problems are attributed to emotional distress, but the effect of emotional stress may be either exacerbated or ameliorated by the person's physical condition, which in turn, will be a function of the state of nutrition of the person. A well-nourished older person should be more capable of coping with stress adaptively than a poorly nourished older person.

Unfortunately, the literature reveals that there has been no definitive research on the specific nutritional needs of the elderly, nor has research revealed any particular nutritional lack which when remedied, produces a change for the better in mental functioning. A review of the literature reveals many admonitions concerning the need for a good diet, surveys that seem to indicate that most old people eat an adequate diet, and a growing body of research that points to the importance of the neurotransmitters in the CNS, but no apparent dietary solution to disorders in CNS functioning.

DIET OF OLDER PEOPLE

Although the calories needed by older people decline steadily so that at age 70 they need only about 75% of the calories needed at age 25, there is no decline in the need for essential nutrients.[1-5] This means that older people need a well-balanced diet in which they show more concern for the quality of food. Research suggests that fat in the diet may be more likely to lead to an increase in weight than carbohydrates or protein.[6] This would suggest that as one ages the consumption of fats should be reduced. A diet of 1650 to 1825 cal/day has been recommended, with less than 1474 calories providing little margin for altered nutritional needs due to stress.[3]

It has been asserted that the elderly tend to eat a carbohydrate-rich, protein-poor diet,[7] that their diets lack sufficient iron, potassium, and calcium,[4] and that the diets of women in particular are deficient in calcium, thiamin, and riboflavin.[5] However, surveys of the diets of older people generally indicate that they obtain an adequate diet, with the possible exception of a deficiency in calcium.[5]

One reviewer has concluded that ''Overt nutritional disease is rare today among the elderly in the United States.''[3] In any case, there does not seem to be any clear pattern of nutritional deficiency in the diet of older people that could consistently be related to the intellectual and emotional problems that are commonly found among the elderly.

POSSIBLE LINKS OF MENTAL FUNCTIONING TO DIETARY DEFICIENCIES

Although there has been no clear link established between dietary deficiencies and specific mental problems, there has been research that indicates a relationship between the neurotransmitters, acetylcholine, and serotonin and specific mental and emotional problems. Acetylcholine has been linked to problems related to learning and memory, and serotonin linked to depression and insomnia.

Cholinergic System and Disorders of Learning and Memory

Difficulties in learning new material and in memory for recent events are common complaints among the elderly. The cholinergic system appears to be involved with learning, memory, and cognitive processes in general.[8,9] Acetylcholine is the neurotransmitter that seems to be most directly involved in learning and memory. Postmortem examination of the brains of elderly people suffering defects in learning and memory reveal reduced levels of acetylcholine.[10] Senile dementia and Alzheimer's disease are disorders among the elderly that are characterized by loss of memory, difficulty in learning, disorientation, personality change, and apraxias and aphasias. Alzheimer's disease, in particular, is characterized by degeneration of neural tissue, including senile plaques and neurofibrillary tangles, and a reduced level of choline acetyltransferase (CAT), the enzyme that catalyzes the conversion of choline to acetylcholine. The extent of neural degeneration and reduced CAT correlates with the extent of intellectual deterioration in such patients.

Furthermore, research on younger patients in which the cholinergic system is blocked with scopolamine and other centrally acting cholinergic agents shows defects in memory and a reduction in performance IQ — intellectual deficits that are characteristic of aged people with organic brain syndromes.[9,10]

These data suggest, therefore, that the cholinergic system is involved in cognitive processes and methods by which choline could be increased within the brain might alleviate defects in memory and cognition.

Research found that administering choline, which is the precursor to acetylcholine, or lecithin, which is the most common dietary source of choline, increases the amount of choline in the cerebrospinal fluid[11] and blood.[12,13] Studies of the effect of administering choline or lecithin in a variety of ways have not produced any evidence of improved functioning as measured by behavioral rating scales or psychological tests.[11-18] However, members of the families of patients and ward personnel report that patients appear to be somewhat improved in orientation to their surroundings and in their learning and memory.[11,13-16] One study with younger patients below age 65 in the early stages of the disease did find that their memory and learning was improved.[15] The general conclusion suggested by the results of these studies is that the administration of choline or lecithin does not "cure" the decline in mental deterioration, but that it may be of some value in the early stages of the disease, either alleviating the symptoms in the early stages, or slowing down the degenerative process.[11,13]

Dietary sources of choline and lecithin are lean meats, liver, eggs, wheat germ, and legumes.

It has been suggested that the reason increasing the levels of acetylcholine in the brain does not improve learning and memory is that increasing the general level of acetylcholine increases the functioning of all neurones indiscriminately, but that to improve cognitive functioning it may, in fact, be necessary to differentially alter the function of the neurones, with some transmitting a stronger signal, and others a weaker signal, so that the desired signal may more easily be identified.[9] This process has been likened to increasing the signal to noise ratio in a radio receiver, in which it is necessary, in order to receive a clear signal, to reduce the level of static relative to the level of the signal that is being tuned in. Thus, any attempt to alleviate difficulties in mental functioning with a general dietary supplement is not likely to be successful.

In any case, increasing the intake of choline or lecithin does not seem to have any clear, positive effect on memory and learning in the aged, except for the possible exception of those whose deficits are minimal.

Relationship of Serotonin Deficiency to Depression and Sleep Disturbances

Serotonin is a neurotransmitter related to both depression and sleep.[10,17] The precursor to serotonin is tryptophan. Administration of tryptophan, to normal subjects at least, reduces the latency of sleep and reduces waking time during the sleeping hours.[17] Low tryptophan levels have been associated with depression, and postmortem examination of the brains of depressed patients have found low levels of serotonin.[10]

The treatment of depression with tryptophan administration has had mixed results. On the other hand, treatment that increased levels of serotonin in the brain, chiefly with tryptophan, has been of some success in treating insomnia.[10] The exact manner in which tryptophan works is not clear because carbohydrate consumption also increases the serotonin level. However, this does suggest that a meal high in protein and carbohydrate should make sleeping easier.

Other Possible Dietary Links to Mental Functioning

It is thought that reversible brain syndromes may be due, among other things, to malnutrition.[4] Deficiencies in the B vitamins, in particular, are thought to be related to the impaired cognition and the confusion that is sometimes observed in older people.[2,18,19] Vitamin C is also thought to be related to impaired cognition.[19] However, definitive research is lacking.

Research on treatment with megavitamin therapy has failed to provide evidence in support of their efficacy.[20]

CONCLUSIONS

The most obvious conclusion is that research into the relationship between nutrition and mental functioning in older people has barely begun. The gerontological literature, almost as a required gesture, states that a balanced diet is desirable, but of course that could be said about all people, young and old. It seems likely that the aged, being somewhat more vulnerable and having less physical reserve to cope with physical and psychological stress, probably have a greater need for an adequate diet and are more vulnerable to dietary dificiencies than are young, healthy adults. However, evidence for any special dietary needs or vulnerabilities is lacking.

The growing body of research on the cholinergic system points to it as a key element in learning and memory, but the evidence does not identify dietary deficiency as a source of degeneration in the cholinergic system. Rather, the research seems to suggest that a degeneration in the brain cells can be compensated for in the early stages of degeneration by increasing the supply of acetylcholine precursors, much as glasses may alleviate a decline in vision or a hearing aid may alleviate decrements in hearing.

The role of serontonin in depression is not quite as well established, and treating depression through increasing supplies of serotonin in the brain through increasing the availability of its precursor, tryptophan, is even less well established than the treatment of cognitive difficulties with choline or lecithin. Perhaps the clearest relationship found thus far is the effect of serotonin on sleep. However, the treatment of insomnia with tryptophan supplementation, although promising, requires much more evidence.

In any case, eating well is a good thing, and even though research in this area has scarcely scratched the surface, it seems safe to conclude that ''good food is a real morale builder''.[21]

REFERENCES

1. **Stare, F. J.,** Three score and ten plus more, *J. Am. Geriatr., Soc.,* 25, 529, 1977.
2. **Mayer, J.,** Aging and nutrition, *Geriatrics,* 29, 57, 1974.
3. **Kart, C. S., Metress, E. S., and Metress, J. F.,** *Aging and Health: Biologic and Social Perspectives,* Addison-Wesley, Menlo Park, Calif., 1978, 104.
4. **Kart, C. S.,** *The Realities of Aging,* Allyn & Bacon, Boston, 1981, 128.
5. **Barrows, C. H. and Roeder, L. M.,** Nutrition, in *Handbook of the Biology of Aging,* Finch, C. E. and Hayflick, L., Eds., Van Nostrand Reinhold, N.Y., 1977, chap. 23.
6. **Caster, W. O.,** The role of nutrition in human aging, in *Nutrition, Longevity, and Aging,* Rockstein, M. and Sussman, M. L., Eds., Academic Press, N.Y., 1976, 29.
7. **Wurtmann, J. J.,** Neurotransmitter regulation of protein and carbohydrate consumption, in *Nutrition and Behavior,* Miller, S. A., Ed., Franklin Institute, Philadelphia, 1981, 69.
8. **Berger, P. A., Davis, K. L., and Hollister, L. E.,** Cholinomimetics in mania, schizophrenia, and memory disorders, in *Nutrition and the Brain,* Vol. 5, Barbeau, A., Growdon, J. H., and Wurtmann, R. J., Eds., Raven Press, N.Y., 1979, 425.
9. **Drachman, D. A. and Sahakian, B. J.,** Effects of cholinergic agents on human learning and memory, in *Nutrition and the Brain,* Vol. 5, Barbeau, A., Growdon, J. H., and Wurtmann, R. J., Eds., Raven Press, N.Y., 1979, 351.
10. **Growdon, J. H.,** Neurotransmitter precursors in the diet: their use in the treatment of brain diseases, in *Nutrition and the Brain,* Vol. 3, Wurtmann, R. J. and Wurtmann, J. J., Eds., Raven Press, N.Y., 1979, 117.
11. **Christie, J. E., Blackburn, I. M., Glen, A. I. M., Zeisel, S., Shering, A., and Yates, C. M.,** Effects of choline and lecithin in CSF choline levels and on cognitive function in patients with presenile dementia of the Alzheimer type, in *Nutrition and the Brain,* Vol. 5, Barbeau, A., Growdon, J. H., and Wurtmann, R. J., Eds., Raven Press, N.Y., 1979, 377.
12. **Etienne, P., Gauthier, S., Johnson, G., Collier, B., Mendis, T., Dastoor, D., Cole, M., and Muller, H. F.,** Clinical effects of choline in Alzheimer's disease, *Lancet,* 1, 508, 1978.
13. **Etienne, P., Gauthier, S., Dastoor, D., Collier, B., and Ratner, J.,** Alzheimer's disease: clinical effect of lecithin treatment, in *Nutrition and the Brain,* Vol. 5, Barbeau, A., Growdon, J. H., and Wurtmann, R. J., Eds., Raven Press, N.Y., 1979, 389.
14. **Boyd, W. D., Graham-White, J., Blackwood, G., Glen, I., and McQueen, J.,** Clinical effects of choline in Alzheimer's senile dementia, *Lancet,* 2, 711, 1977.
15. **Signoret, J. L., Whiteley, A., and Lhermitte, F.,** Influence of choline on amnesia in early Alzheimer's disease, *Lancet,* 2, 837, 1978.
16. **Growdon, J. H.,** Ways to predict clinical responses to lecithin administration, in *Nutrition and the Brain,* Vol. 5, Barbeau, A., Growdon, J. H., and Wurtmann, R. J., Eds., Raven Press, N.Y., 1979, 253.
17. **Karczmar, A.,** Overview: cholinergic drugs and behavior — what effects may be expected from a ''cholinergic diet?'', in *Nutrition and the Brain,* Vol. 5, Barbeau, A., Growdon, J. H., and Wurtmann, R. J., Eds., Raven Press, N.Y., 1979, 141.

18. **Lipton, M. A. and Wheless, J. C.,** Diet as therapy, in *Nutrition and Behavior*, Miller, S. A., Ed., Franklin Institute, Philadelphia, 1981, 213.
19. **Eisdorfer, C. and Cohen, D.,** The cognitively impaired elderly: differential diagnosis, in *The Clinical Psychology of Aging*, Storandt, M., Siegler, I. C., and Elias, M. F., Eds., Plenum Press, N.Y., 1978, 7.
20. **Gelenberg, A. J.,** Nutrition in psychiatry; we are what we eat, *J. Clin. Psych.*, 41, 328, 1980.
21. **Lipton, M. A., Mailman, R. B., and Nemeroff, C. B.,** Vitamins, megavitamin therapy, and the nervous system, in *Nutrition and the Brain*, Vol. 3, Wurtmann, R. J. and Wurtmann, J. J., Eds., Raven Press, N.Y., 1979, 183.
22. **Stare, F. J.,** Three score and ten plus more. *J. Am. Geriatr. Soc.*, 25, 533, 1977.

THE EFFECT OF AGE AND NUTRITION ON PROTEIN SYNTHESIS BY CELLS AND TISSUES FROM MAMMALS

Arlan Richardson

INTRODUCTION

Because proteins are essential for cell function and viability, the effect of aging on protein synthesis has been studied in detail in living organisms ranging from fungi and invertebrates to mammals, including man.[1,2] In this review, only studies that have focused on the synthesis of proteins by cells and tissues from mammals will be presented. The problems that confront an investigator measuring protein synthesis as a function of age by the incorporation of radioactively labeled amino acids has been discussed in detail in a previous review.[1]

CHANGES IN PROTEIN SYNTHESIS AS A FUNCTION OF AGE

The effect of aging on protein synthesis has been studied most extensively in tissues from rodents. Approximately 80% of the studies with rodents indicate that protein synthesis declines with increasing age.[1,2] Research by my laboratory over the past 7 years demonstrates that protein synthesis by a variety of tissues from Fischer F344 rats declines significantly with increasing age (Figure 1A). All the data given in Figure 1A were obtained with cell-free systems except for that of liver and spleen, which were obtained with isolated cells. Previous studies have shown the rate of protein synthesis by cell-free systems is only 1% of the rate observed in vivo;[1,3] one must be concerned with whether the age-related changes observed with cell-free systems reflect the situation in vivo. Using flooding amounts of [³H]-lysine, Fando et al.[9] accurately measured the rates of in vivo protein synthesis by the forebrain and cerebellum of Fischer F344 rats (Figure 1B). The age-related decline in brain protein synthesis observed using a cell-free system (Reference 5, Figure 1A) was similar to that observed by Fando et al.[9] Therefore, age-related changes in cell-free protein synthesis appear to accurately reflect the situation that occurs in the whole animal.

The effect of aging on the protein synthetic activities of a variety of tissues from Sprague-Dawley rats also has been studied in detail by a variety of investigators, and the results of these studies are summarized in Figure 2. Again, a general decline in protein synthesis was observed with increasing age. Recently, Starnes et al.[14] and Ibrahim et al.[15] measured the incorporation of [¹⁴C]-leucine into protein by mitochondria from the heart and liver, respectively. These are the first studies to describe the effect of aging on protein synthesis by the mitochondria and are of interest because the protein synthetic machinery of the mitochondria is different from that found in the cytoplasm of eukaryotic cells. The incorporation of [¹⁴C]-leucine into protein declined 45% with increasing age for mitochondria from these two tissues.

It should be noted that an increase in liver protein synthesis was observed in both Fischer F344 and Sprague-Dawley rats after 24 months of age (Figures 1A and 2). This phenomenon was first described by van Bezooijen et al.[16] and has been described in detail in a previous review.[1] Recently, van Bezooijen et al.[17] showed that the increase in liver protein synthesis in very old rodents was independent of the sex or strain of the animal. The molecular reason for the age-related increase in liver protein synthesis in very old rodents is unknown.

Although the effect of aging on total protein synthesis has been studied in detail, very little information is currently available on the effect of age on the synthesis of specific proteins. Only the synthesis of albumin by the liver has been studied in detail as a function

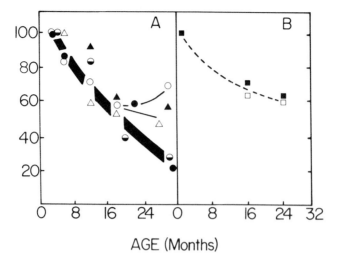

FIGURE 1.   Effect of aging on the protein synthetic activity of various tissues from Fischer F344 rats. Graph A: The protein synthetic activities of liver[4] (○), brain[5] (△), kidney[6] (●), testes[7] (▲), and mitogen-induced spleen lymphocytes[8] (◗) is expressed as a percentage of the 3- to 6-month-old rats. Graph B: The in vivo protein synthesis by forebrain (□) or cerebrum (■) is expressed as a percentage of the 1.5-month-old rats.[9]

of age. The synthesis of albumin by rodents has been observed to increase significantly with age using in vivo[18-20] and cell-free systems[21] and isolated hepatocytes.[22] However, an increase in the synthesis of protein by the liver is not always observed in very old rodents. Bolla and Greenblatt[10] recently reported that the cell-free synthesis of transferrin by the liver declined continuously between 6 and 30 months of age in Sprague-Dawley rats even though total protein synthesis increased after 24 months of age.

Only a limited amount of information is available on the effect of aging on protein synthesis by cells or tissues from humans. Using diploid fibroblasts obtained from the skin of human subjects, several investigators have measured the incorporation of radioactively labeled amino acids into protein as a function of aging in vitro (passage number), and a summary of these studies is presented in Table 1. In two thirds of these studies, a decline in amino acid incorporation was observed with increasing passage number. Recently, McCoy et al.[28] found that collagen synthesis by skin fibroblasts from human subjects decreased (30%) continuously with aging in vitro. Using fibroblasts from the foreskin of human donors of various ages, Chen et al.[29] measured the incorporation of [³H]-leucine into protein by fibroblast cultures permeabilized with a nonionic detergent. A 30 to 40% decrease in the incorporation of [³H]-leucine into protein was observed between 20 and 94 years of age. Neither the uptake nor the equilibration of the [³H]-leucine with the intracellular amino acid pool was observed to change significantly with increasing age.

Young's laboratory[30,31] has compared the rates of total body protein synthesis in human subjects of various ages by measuring the amount of [¹⁵N]-urea when constant [¹⁵N]-enrichment has been achieved after the administration of [¹⁵N]-glycine. The results of these studies are presented in Table 2. In the initial study,[30] in which data from males and females were pooled, the rate of total body protein synthesis by elderly subjects was 37% less than that of adult subjects. Subsequently, the rate of total body protein synthesis was measured in males and females separately.[31] The rate of total body protein synthesis by elderly females was significantly lower than adult females. Although the rate of total body protein synthesis by elderly males was lower than that of adult males, the difference was not statistically significant.

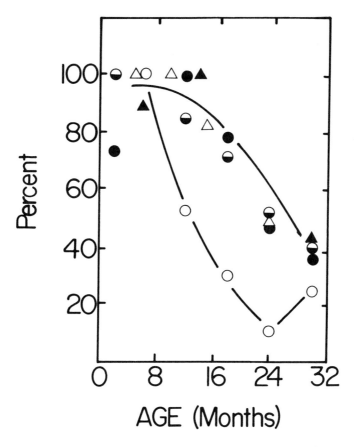

FIGURE 2.   Effect of aging on the protein synthetic activity of various tissues from Sprague Dawley rats. The protein synthetic activity of liver[10] (○), parotid gland[11] (◓), pancreas[12] (●), testes[13] (▲), and heart mitochondria[13] (△) is expressed as a percentage of the age that gave the highest value.

Table 1
EFFECT OF AGING IN VITRO ON PROTEIN SYNTHESIS BY HUMAN FIBROBLASTS

Source of fibroblasts	Methodology employed	Change	Ref.
Embryonic lung			
MRC	[³H]-Methionine incorporation into acid-insoluble material	89% decrease	23
WI-38	Autoradiography employing [³H]-algal protein hydrolysate	29—49% decrease	25
WI-38	[³H]-Proline incorporation into collagen	47% decrease	26
Newborn foreskin			
CF-3	[³H]-Leucine incorporation into acid-insoluble material	No change	24
HF-7 and HF-J	[³⁵S]-Methionine incorporation into proteins separated by 2-dimensional gel electrophoresis	No change	27
Skin biopsies from adult females	[³H]-Proline incorporation into collagen	30% decrease	28

Table 2
THE RATE OF WHOLE BODY PROTEIN SYNTHESIS IN
ADULT AND ELDERLY HUMAN SUBJECTS

	Age-range (years)	Total body protein synthesis ($g\ kg^{-1}\ day^{-1}$)	Ref.
Study I			
Male and female subjects combined			30
Adult	20—23	3.0 ± 0.2[a]	
Elderly	69—91	1.9 ± 0.2	
Study II			
Male subjects			31
Adult	20—25	3.33 ± 0.30	
Elderly	65—72	3.18 ± 0.71	
Female subjects			
Adult	18—23	2.63 ± 0.20	
Elderly	67—91	2.25 ± 0.37	

[a]   Mean ± standard deviation.

PHYSIOLOGICAL IMPORTANCE OF THE AGE-RELATED CHANGE IN PROTEIN SYNTHESIS

Although the current research evidence indicates that a general decline in protein synthesis occurs with increasing age in most tissues of living organisms,[1-3] it is not known if the decline in protein synthesis is involved in the molecular mechanism of the aging process or how the age-related decline in protein synthesis would physiologically affect the whole organism. It is clear that the decline in the protein synthetic activity of a tissue does not automatically lead to decline in the protein content of the tissue. A recent study by my laboratory showed that the protein synthetic activity of testicular tissue from three strains of rodents (Fischer F344 and Sprague Dawley rats and C57BL/6J mice) declined to a similar extent with senescence (Figure 3A). However, the age-related changes in the protein content of the testicular tissue was quite different for the three strains of rodents. In Fischer F344 rats, the protein content increased with senescence, while in the C57BL/6J mice a decrease in the protein content occurred. No significant change in the protein content of the testes from the Sprague-Dawley rats was observed. In the study shown in Figure 1A, in which a decrease in protein synthesis was observed with increasing age in a variety of tissues, no age-related decline in the protein content of any of the tissues was observed. Finch[32] concluded from a review of the literature that the activities of most liver and kidney enzymes of rodents do not change significantly with age.

If the levels of most proteins remain constant throughout the lifespan of an organism while the synthesis of proteins declines with age, one would predict that protein degradation would also decline with age, which would result in an age-related increase in the half-lives of proteins. Unfortunately, only a limited amount of information is currently available on the effect of aging on protein degradation and turnover. Several investigators have measured protein degradation during aging in vitro. Using cultured fibroblasts from human subjects, Bradley et al.,[33] Dell'Orco and Guthrie,[24] and Shakespeare and Buchanan[34] reported that the rate of protein degradation was higher in senescent cultures. However, Goldstein et al.[35] reported that ''true'' protein degradation declined during in vitro aging of cultured fibroblasts from human subjects. Kaftory et al.[36] also observed a decline in protein degradation with aging in vitro using cultures of chick fibroblasts. Gershon's laboratory has reported that the turnover of two liver enzymes declined significantly with increasing age in C57BL/6J mice.

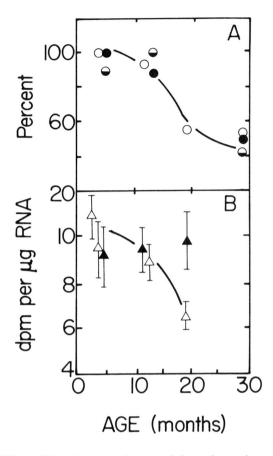

FIGURE 3.   Effect of aging and dietary restriction on the protein synthetic activity of testicular tissue. A comparison of the cell-free protein synthetic activities of testicular tissue from Fischer F344[7] (○) and Sprague-Dawley [13] (●) rats and from C57BL/6J mice[13] (◓) is shown in graph A. The cell-free protein synthetic activities of testicular tissue from Fischer F344 rats fed *ad libitum* (△) or 60% of *ad libitum* (▲) is shown in graph B. Each value represents the mean ± SEM of data obtained from 4 to 8 animals.

The half-life of ornithine decarboxylase increased 100% between 3 and 24 months of age,[37] while the half-life of aldolase increased over 50% between 5 and 28 months of age.[38] Young's laboratory measured the rates of protein degradation in human subjects in addition to total body protein synthesis. In an initial study, Winterer et al.[31] reported that the rate of total body protein degradation in elderly women (67 to 91 years) was significantly lower than that of adult women (18 to 23 years). However, no significant difference in the rates of total body protein degradation was observed for adult and elderly men. Subsequently, Uauy et al.[39] showed that the rates of total body and muscle protein degradation declined significantly with increasing age in both men and women (Table 3).

Several lines of evidence suggest that the age-related decline in protein turnover, which arises from a decline in protein synthesis and degradation, is the molecular basis for the changes that could be physiologically important to an organism. Age-related changes in protein turnover appear to play a role in the age-related appearance of altered enzymes and the decreased responsiveness of an organism to stimuli.

In 1970, Gershon and Gershon[40] reported the appearance of altered isocitrate lyase in old nematodes. Subsequently, several laboratories showed that altered enzymes, which have reduced specific activities, accumulate in older organisms. Although several enzymes become

Table 3
THE RATE OF PROTEIN DEGRADATION IN ADULT AND
ELDERLY HUMAN SUBJECTS[a]

| | Age range (years) | N[b] | Rate of protein degradation $(g\ kg^{-1}\ day^{-1})$ | |
			Total body protein	Muscle protein
Males				
Adult	20—25	4	2.94 ± 0.24[c]	0.76 ± 0.10
Elderly	68—77	6	2.64 ± 0.69	0.53 ± 0.17
Females				
Adult	18—23	4	2.35 ± 0.14	0.64 ± 0.17
Elderly	67—91	5	1.94 ± 0.36	0.31 ± 0.04

[a] The data were obtained from Reference 39.
[b] Number of subjects.
[c] Mean ± standard deviation.

altered with age, many enzymes remain unaltered throughout the lifespan of an organism. Rothstein[41] recently reviewed this area of research, and a list of the studies in this area since 1979 is given in Table 4.

Initially, the observation of altered enzymes with age was used as proof of the error catastrophe theory of aging proposed by Orgel.[42] However, data from research conducted over the past 5 years show that altered enzymes do not arise from errors in translation.[41] Recently, Sharma and Rothstein[43] demonstrated that the altered enolase in old nematodes arises from conformational changes in protein structure. Unaltered and altered enolase, which have different conformations and enzymatic properties, were unfolded in guanidine hydrochloride and subsequently refolded. The conformation and enzymatic properties of the two forms were identical after refolding. Therefore, changes in the amino acid sequence, which would arise from errors in protein synthesis, could not be responsible for the appearance of the altered enolase in old nematodes. In addition, cell-free studies indicate that the fidelity of translation does not decline with increasing age.[2]

What is responsible for the age-related changes in enzyme structure and activity if changes in the amino acid sequence of proteins do not occur with increasing age? In 1979, Rothstein[44] proposed that the altered enzymes arise from an increase in the ''dwell time'' of proteins in a cell, and the increase in the ''dwell time'' would lead to an increased probability of the subtle denaturation of the proteins. Those enzymes that are inherently stable would be unaltered with age while enzymes that are unstable would appear as altered enzymes in old organisms. The age-related increase in the ''dwell time'' of proteins in a cell would arise from an age-related decline in protein turnover, which has been observed to occur in several organisms.[45] Recently, Reznick et al.[38] suggested that the age-related decline in the turnover of aldolase in mouse liver may account for approximately 40% of the inactive aldolase molecules in the old mouse liver.

A general characteristic of all aging populations is the progressive impairment of the ability of an organism to respond to environmental challenges.[46] The biochemical mechanism responsible for the decreased ability to respond to challenges and stimuli is not known. Based on theoretical and experimental evidence, Richardson and Cheung[45] proposed that the decline in protein turnover was the molecular basis for the age-related decline in the responsiveness and adaptation of organisms. Because the rate of protein turnover is lower in senescent organisms, a longer time period would be required for the senescent organism to respond to stimuli that involve enzyme induction.

Table 4
EFFECT OF AGING ON THE APPEARANCE OF ALTERED ENZYMES IN MAMMALS

Enzyme	Tissue	Source	Ages studied (months)	Ref.
Enzymes altered with increasing age				
Enolase	Heart	Sprague-Dawley rats	6—12 and 28—30	48
Aldolase	Liver	C57BL/6J mice	7.5 and 30	49
	Liver	Fischer F344 rats	1 through 31	50
	Liver	Sprague-Dawley rats	12 and 30	50
	Liver	Wistar rats	12 and 30	50
	Liver	C57BL/6J mice	6 and 28	38
	Muscle	Rabbits	6—7 and 52—70	51
	Muscle	Rabbits	6—7 and 52—70	52
Phosphoglycerate kinase	Brain, muscle, liver	Sprague-Dawley rats	10, 18—20, and 28—30	53
Glyceraldehyde-3-phosphate dehydrogenase	Muscle	WF rats	5—6 and 27—38	54—56
Albumin	Blood serum	C57BL/6J mice	8 and 21	57
Enzymes unaltered with increasing age				
Enolase	Liver, muscle	Sprague-Dawley rats	6—12 and 28—30	48
Aldolase	Liver	C57BL/6J mice	3 and 30	58
	Lymphocytes	Humans	18—41 and 56—84[a]	59
	Liver	Dog	1—2 and 10—12[a]	60
	Liver	CBF₁ mice	3, 11, and 24	60
Phosphoglycerate kinase	Lung, kidney, heart	Sprague-Dawley rats	10, 18—20 and 28—30	53
Superoxide	Liver	Dogs	1—2 and 10—12[a]	60
Dismutase	Liver	CBF₁ mice	3, 11, and 24	60
Tyrosine aminotransferase	Liver	Wistar rats	3—6 and 27—31	61
β-D-Galactosidase	Liver cell lines	Human	3 through 17 passages	62

[a] This number represents years.

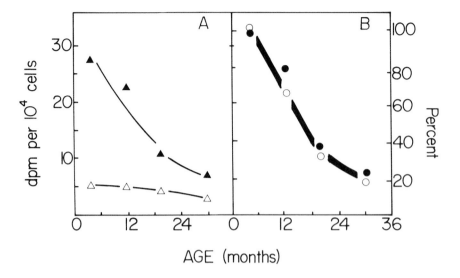

FIGURE 4.   The effect of aging on lymphocyte function and protein synthesis. Spleen lymphocytes from male Fischer F344 rats of various ages were incubated in the presence or absence of mitogens. Graph A shows the amount of [³H]-valine incorporated into protein in the presence (▲) or absence (△) of concanavalin A. Graph B compares the phytohemagglutinin-induced lymphocyte proliferaton (○) and concanavalin A-induced protein synthesis (●) by spleen lymphocytes.[8]

The immunosurveillance system is an example of a physiological process that must respond rapidly to environmental challenges. It is well documented that the responsiveness of the immune system declines dramatically with increasing age,[47] and this decline results in an increased incidence of a variety of diseases. Recently, Cheung et al.[8] measured the protein synthetic activity of lymphocytes obtained from the spleens of rats of various ages. The data in Figure 4A show that protein synthesis declines only slightly in resting lymphocytes; however, when the lymphocytes are stimulated with a mitogen, a dramatic age-related decline in protein synthesis is observed. The data in Figure 4B show that the decline in protein synthesis after mitogen stimulation parallels the decline in mitogen-induced lymphocyte proliferation, which is a measure of the immunological status of the organism. The data in Figure 4 suggest that the age-related decline in the responsiveness of the immune system to environmental challenges is due at least partially to the general decline in protein synthesis. This is the first report to relate the age-related decline in protein synthesis to a decline in a specific physiological function.

EFFECT OF DIETARY RESTRICTION ON AGE-RELATED CHANGES IN PROTEIN SYNTHESIS

The nutritional status of an organism is of importance in gerontological research because food restriction or undernutrition (but not malnutrition) is the most effective and reproducible method for increasing the longevity of an organism, and it is the only strategy known to retard the aging process in homeotherms.[63,64] The classic experiments by McCay[65] showed that a drastic restriction of the diet increased the survival of Osborn-Mandel rats. The restricted rats were allowed access to just enough food to maintain their weight, and at varying times, the rats were given additional diet to permit a slight weight gain. In McCay's study, the mean survival was increased from 483 to 820 days by dietary restriction. More importantly, the maximum survival was increased from 927 to 1321 days, which suggests

Table 5

STUDIES SHOWING AN INCREASED LIFESPAN OF RODENTS BY DIETARY RESTRICTION

Rodent strain	Sex	Dietary retriction regimen	Ref.
Rats			
Osborn-Mendel	Male and female	Severe restriction	65
Sprague-Dawley	Male and female	Severe restriction	68
Wistar	Male and female	54 and 66% of *ad libitum*	66,67
	Male	40—50% of *ad libitum*	69
	Male	Reduction of time with access to diet	70
Norway	Male	Reduction of time with access to diet	71
Wistar	Male	50% of *ad libitum*	72
	Male	60% of *ad libitum*	73
		Intermittent fasting	74,75
Simonsen	Male and female	60 and 80% of *ad libitum*	76
Fischer F344	Male	60% of *ad libitum*	77
Mice			
Unspecified	Male	50% of *ad libitum*	72
DBA/2f	Male and female	50% of *ad libitum*	78
NZB × NZW F₁ hybrid	Male and female	50% of *ad libitum*	78
A/J	Male	Low dietary protein	79
C57BL/6J	Male and female	Low dietary protein	79,80
	Male and female	Intermittent fasting	81
A/J × C57BL/6J F₁ hybrid	Male	Low dietary protein	79
B10C3F₁	Male	Intermittent fasting	82
Golden hamsters	Male	50% of *ad libitum*	72

that the increase in longevity arose from a decline in the aging process. However, the growth and sexual development of the rats was severely retarded by McCay's dietary restriction regimen. Experiments by Berg[66,67] in the 1960s showed that longevity could be increased by dietary restriction less severe than that used by McCay.[65] Although the rats on the restricted diets employed by Berg[66] were smaller than *ad libitum*-fed rats, the retardation of growth and sexual maturity was minimal compared to the rats on the restricted diet employed by McCay.[65]

The effect of dietary restriction on longevity has been studied extensively in rodents and a list of these studies is given in Table 5. The lifespan of rodents has been increased by dietary restriction using three types of regimens:

1. Reducing the dietary intake of a nutritionally adequate diet
2. Intermittent feeding of a nutritionally adequate diet (intermittent fasting)
3. Feeding *ad libitum* a diet containing an insufficient amount of protein necessary to support maximal growth (low dietary restriction)

In general, it appears that maximum survival is achieved when rodents are fed 50 to 60% of the amount of diet consumed by rodents fed *ad libitum*. Masoro's laboratory has just completed one of the most carefully controlled studies in which the effect of dietary restriction on longevity has been evaluated.[77] Male Fischer F344 rats, which were maintained under specific pathogen-free conditions, were fed a semisynthetic diet (21% casein, 15% sucrose, 43.65% dextrin, 3% Solka-Floc, 10% corn oil, 0.15% methionine, 0.2% choline chloride, and a vitamin and mineral mix). The restricted rats were fed 60% of the amount of the diet

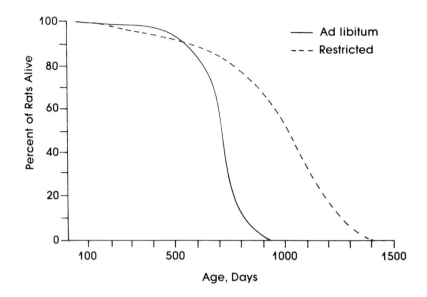

FIGURE 5.   The survival curves for male Fischer F344 rats fed *ad libitum* or 60% of *ad libitum* (restricted). (From Yu, B. P., Masoro, E. J., Murata, I., Bertrand, H. A., and Lynd, F. T., *J. Gerontol.*, 37, 130, 1982. With permission.)

consumed by rats fed *ad libitum*. Figure 5 shows the survival curves for the male Fischer F344 rats fed *ad libitum* or the restricted diet. It is obvious that dietary restriction has a dramatic effect on the survival of the rats. The medium and maximum length of life for the rats fed *ad libitum* was 714 and 963 days, respectively. The medium and maximum length of life for the rats fed the restricted diet was 1047 and 1435 days, respectively.

Dietary restriction has been shown to delay or decrease the incidence of a variety of diseases. The onset and severity of myocardial degeneration,[67,77] myocardial fibrosis,[77] periareritis,[67] and bile duct hyperplasia[77] were delayed and reduced in rats fed restricted diets. Several investigators have also shown that dietary restriction reduces the incidence of most spontaneous tumors in rodents.[67-69,71,82,83] In addition, dietary restriction has been shown to retard the age-related decline in the immune response.[79,83-86] and delay the development of life-shortening autoimmune disease[78,87] and the appearance of brain-reactive antibodies in the sera.[88] The effect of dietary restriction on renal disease has been one of the most studied areas of pathology because renal disease is one of the major diseases in most strains of rodents (e.g., in male Fischer F344 rats, Coleman et al.[89] found a highly significant correlation between aging and the severity of renal disease, and death in male Fischer F344 rats was often related to renal failure). Several investigators have shown that dietary restriction retards the onset and reduces the severity of renal disease.[67,71,73,77] The data in Table 6 show the severity of nephropathy in male Fischer F344 rats fed *ad libitum* and a restricted diet. Dietary restriction retarded the development of nephropathy and reduced its severity. At 26.5 to 27.5 months of age, 50% of the rats fed *ad libitum* showed a widespread glomerulosclerosis; however, none of the 35.5- to 36.5-month-old rats examined showed any sign of widespread glomerulosclerosis. Berg and Simms[67] made a similar observation with male Sprague-Dawley rats. At 802 days of age, 58% of the rats fed *ad libitum* showed the presence of severe glomerulonephritis, while none of the rats fed the restricted diet showed any presence of severe glomerulonephritis. Recently, Johnson and Barrows[90] showed that marked morphological differences in kidney glomeruli were observed in female C57BL/6J mice after feeding a restricted diet. Several investigators have measured the effect of aging on kidney function by the levels of protein excreted in the urine. An age-related increase in proteinuria is observed,[18,91,94] and dietary restriction has been observed to decrease proteinuria.[73,94]

Table 6
INCIDENCE OF SPONTANEOUS RENAL DISEASE IN MALE FISCHER F344 RATS FED *AD LIBITUM* OR A RESTRICTED DIET[a]

	Age range (months)	Percent of rats with following grade of nephropathy					
		0	1	2	3	4	E
Ad libitum							
	5.5—6.5	44	56	0	0	0	0
	11.5—12.5	0	10	80	10	0	0
	17.5—18.5	0	0	0	90	10	0
	23.5—24.5	0	0	0	20	80	0
	26.5—27.5	0	0	0	25	25	50
Restricted							
	5.5—6.5	80	20	0	0	0	0
	11.5—12.5	30	70	0	0	0	0
	17.5—18.5	0	30	70	0	0	0
	23.5—24.5	0	0	100	0	0	0
	29.5—30.5	0	0	89	11	0	0
	35.5—36.5	0	0	20	80	0	0

[a]   Male Fischer F344 rats were fed a semisynthetic diet either *ad libitum* or 60% of *ad libitum* (restricted) and killed at the ages shown above. The histopathological grading of chronic nephropathy was determined in 8 to 10 rats in each age group. The grading ranged from no lesions (grade 0) to widespread glomerulosclerosis (grade E). The data was taken from the study by Yu et al.[77]

Although dietary restriction has been shown to increase the longevity of a variety of living organisms,[64] the biochemical basis for the effect of dietary restriction on the aging process is unknown. Several investigators have suggested that dietary restriction exerts its mechanism of action at the level of gene expression. In 1972, Barrows[95] proposed that dietary restriction increased longevity by reducing protein synthesis and thereby reducing the use of the genetic code. Recently, Lindell[96] suggested that dietary restriction is a physiological ''stress'' that enhances gene expression, and this enhanced gene expression is a significant factor in the maintenance of cellular homeostasis in an organism as it ages. It is attractive to propose that dietary restriction acts at the level of gene expression because an age-related decline in gene expression appears to be a universal phenomenon in higher organisms.[1,45] However, at the present time, there is a dearth of information on the effect of long-term dietary restriction on gene expression. In 1959, Ross[97] reported that the age-related changes in the activities of four liver enzymes (ATPase, histidinase, catalase, and alkaline phosphatase) were delayed by dietary restriction regimens, which increased the longevity of the animals. Barrows[95] also reported that dietary restriction delayed the age-related changes in the activities of lactate dehydrogenase and malate dehydrogenase in rotifers. In a later study with C57BL/6J mice, Barrows[98,99] found that dietary restriction appeared to have no significant effect on the activities of a variety of enzymes in liver, kidney, and heart tissue. However, the activities of most of the enzymes studied by Barrows changed very little with increasing age. Recently, Roth reported that dietary restriction reversed the age-related decline in the number of striatal dopaminergic receptors in the brain of male rats,[75] and Bertrand et al.[100] found that dietary restriction increased the responsiveness of adipocytes to glucagon-promoted lipolysis, which was paralleled by an increase in glucagon-promoted adenylate cyclase activity.[101] Although the studies on enzyme levels would suggest that dietary restriction might affect gene expression, one cannot deduce what happens to gene expression and protein synthesis from data only on enzyme activity because the activities and levels of enzymes are dependent on both synthesis and degradation or transport.[32]

Although a large amount of information on the effect of short-term alterations in amino acid supply or fasting is currently available for a variety of tissues from rodents,[102-105] no one has directly measured how long-term dietary restriction, which increases longevity, affects gene expression. In a series of preliminary experiments, my laboratory has compared the protein synthetic activities of several tissues from male Fischer F344 rats fed *ad libitum* and a restricted diet. At 1.5 months of age, the rats were placed on dietary regimens identical to that described by Yu et al.,[77] i.e., the rats were fed a semisynthetic diet either *ad libitum* or 60% of *ad libitum*. This dietary restriction regimen has been shown to increase the mean and maximum survival of male Fischer F344 rats substantially (Figure 5). As indicated previously, a decline in testicular protein synthesis has been observed in a variety of strains of rodents (Figure 3A). The cell-free protein synthetic activity of testicular tissue from male Fischer F344 rats fed *ad libitum* and the restricted diet is shown in Figure 3B. Between 2 and 12 months of age, no significant age-related change in protein synthesis was observed and the protein synthetic activity of the testicular tissue from the rats fed the two dietary regimens was not significantly different. Between 12 and 19 months of age, testicular protein synthesis by the rats fed *ad libitum* decreased significantly, which is in agreement with our previous study.[7] However, no significant decline in the protein synthetic activity of testicular tissue was observed between 12 and 19 months of age for rats fed the restricted diet, and the cell-free protein synthetic activity of testicular tissue from the restricted rats was significantly greater than that observed for the rats fed *ad libitum*.

The effect of dietary restriction on the protein synthetic activity of the kidney was also studied because the physiological and pathological changes in the kidney appear to be correlated with a dramatic decline in the cell-free protein synthetic activity of kidney tissue.[6] Renal function and protein synthesis by 19-month-old male Fischer F344 rats fed *ad libitum* and the restricted diet is shown in Figure 6. Figure 6A shows that a significant increase in the amount of protein excreted in the urine occurs with increasing age for male Fischer F344 rats fed *ad libitum* a commercial lab-chow diet. The amount of protein excreted in the urine by 19-month-old rats fed the restricted diet was significantly lower than 19-month-old rats fed *ad libitum*. This observation is in agreement with previous studies, which showed that dietary restriction decreased proteinuria.[73,94] Using suspensions of freshly isolated kidney cells, the rate of protein synthesis by kidney cells isolated from male Fischer F344 rats fed *ad libitum* and the restricted diet was measured. A continuous decline in the rate of protein synthesis by kidney cells isolated from rats fed a commercial lab-chow diet was observed between 6 and 23 months of age. The rate of protein synthesis by kidney cells isolated from 19-month-old rats fed the restricted diet was 65% higher than that of kidney cells isolated from 19-month-old rats fed *ad libitum*. Therefore, dietary restriction, which has been shown to delay the age-related decline in renal function and the increase in renal disease, appears to retard the age-related decline in the protein synthetic activity of kidney tissue.

Dietary restriction has also been shown to delay the age-related decline in the immune system.[79,83-86] The age-related decline in the responsiveness of spleen lymphocytes to mitogens can be correlated to a decline in protein synthesis (Figure 4); therefore, the protein synthetic activity of spleen lymphocytes isolated from a limited number of 19-month-old rats maintained on the two dietary regimens was determined. Mitogen-induced lymphocyte proliferation and IL-2 production by spleen lymphocytes from 19-month-old rats fed the restricted diet was 50 to 90% greater than that of rats fed *ad libitum*, and the protein synthetic activity of the lymphocytes from the restricted rats was 40% greater than that of lymphocytes from rats fed *ad libitum*.[45]

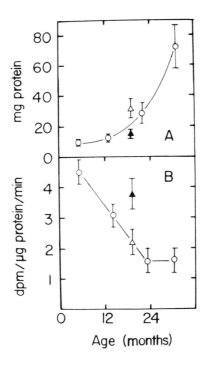

FIGURE 6. The effect of aging and dietary restriction on renal function and protein synthesis. Male Fischer F344 rats of various ages were fed a commercial lab-chow diet *ad libitum* (○) or a semisynthetic diet *ad libitum* (△) or 60% of *ad libitum* (▲). The amount of protein excreted in the urine over a 24-hr period is given in graph A. The rate of [³H]-valine incorporation into protein by suspensions of isolated kidney cells is shown in graph B. Each point represents the mean ± SEM of 4 (for rats fed the commercial lab-chow diet) or 6 (for rats fed the semisynthetic diet) animals.[107]

Based on the series of experiments described above, it appears that a life-long dietary restriction regimen, which has been shown to increase the longevity of rodents, results in an increase in the protein synthetic activities of several tissues from old male Fischer F344 rats. The data from these experiments do not support the hypothesis advanced by Barrows[95] that dietary restriction reduces the use of the genetic code by decreasing protein synthesis. The data do support Lindell's[96] hypothesis that gene expression is enhanced by dietary restriction. In addition, these data suggest that the age-related decline in protein synthesis plays a role in the aging process because dietary restriction retards the age-related decrease in protein synthesis.

SUMMARY

A decline in protein synthesis appears to be a common feature of the aging process.[1,2,45] Although the effect of aging on the total protein synthetic activity of a variety of tissues has been studied extensively, very little information is currently available on the effect of aging on the synthesis of individual proteins, except for albumin. The synthesis of albumin appears to initially decline with increasing age in rodents; however, after 2 years of age, an increase in albumin synthesis is observed.[22]

At the present time, the physiological importance of the age-related decline in protein synthesis is unknown. Recent studies have shown that the decline in the protein synthetic activity of the kidney is correlated to the increase in the incidence of renal disease,[6] and the age-related decline in the response of spleen lymphocytes to mitogens is paralleled by a decline in protein synthesis.[8] In 1982, Richardson and Cheung[45] proposed that the age-related decline in protein synthesis results in a decline in protein turnover, and the decline in protein turnover is the molecular basis for many of the physiological changes that are associated with senescence.

The age-related decline in protein turnover appears to be an important factor in the age-related appearance of altered proteins.[41,44] In addition, the age-related decline in protein turnover has been proposed to be responsible for the decline in the ability of an organism to respond to change with increasing age. The decline in the responsiveness of an organism to stimuli is a general characteristic of aging populations. Because proteins with a high turnover rate respond more quickly to stimuli, a decline in protein turnover would be expected to result in a decrease in the induction of enzymes in response to stimuli.[45]

Dietary restriction (not malnutrition) is currently the only experimental method that can extend the lifespan of mammals. Restriction regimens employing diets ranging from 50 to 60% of *ad libitum* result in a substantial increase in both the mean and maximum survival of rodents (Table 5). The incidence and severity of a variety of diseases in rodents have been observed to decrease when rodents are fed restricted diets, which enhance the immunosurveillance system. Preliminary experiments indicate that dietary restriction retards the age-related decline in protein synthesis in several tissues from male Fischer F344 rats, which can be correlated to an improvement in the functional status of the tissue. These preliminary observations support the hypothesis advanced by Lindell,[96] i.e., dietary restriction enhances gene expression by acting as a physiological ''stress''.

REFERENCES

1. **Richardson, A.,** The relationship between aging and protein synthesis, in *CRC Handbook Series in Biochemistry in Aging,* Florini, J. R., Ed., CRC Press, Boca Raton, Fla., 1981, 79.
2. **Richardson, A. and Birchenell-Sparks, M. C.,** Age-related changes in the synthesis and turnover of proteins, in *Review of Biological Research in Aging,* Vol. 1, Rothstein, M., Ed., Alan R. Liss, N.Y., in press.
3. **Waterlow, J. C.,** Protein turnover in the whole body, *Nature (London),* 253, 157, 1975.
4. **Coniglio, J. J., Liu, D. S. H., and Richardson, A.,** A comparison of protein synthesis by liver parenchymal cells isolated from Fischer F344 rats of various ages, *Mech. Ageing Dev.,* 11, 77, 1979.
5. **Ekstrom, R., Liu, D. S. H., and Richardson, A.,** Changes in brain protein synthesis during the lifespan of male Fischer rats, *Gerontology,* 26, 121, 1980.
6. **Hardwick, J., Hsieh, W. H., Liu, D. S. H., and Richardson, A.,** Cell-free protein synthesis by rat kidney: the effect of age on kidney protein synthesis, *Biochim. Biophys. Acta,* 652, 204, 1981.
7. **Liu, D. S. H., Ekstrom, R., Spicer, J. W., and Richardson, A.,** Age-related changes in protein, RNA, and DNA content and protein synthesis in rat testes, *Exp. Gerontol.,* 13, 197, 1978.
8. **Cheung, T. H., Twu, J. S., and Richardson, A.,** The role of IL-2-production and protein synthesis in the age-related decline in proliferation by spleen lymphocytes, *Exp. Gerontol.,* 18, 451, 1983.
9. **Fando, J. L., Salinas, M., and Wasterlain, C. G.,** Age-dependent changes in brain protein synthesis in the rat, *Neurochem. Res.,* 5, 373, 1980.
10. **Bolla, R. I. and Greenblatt, C.,** Age-related changes in rat liver total protein and transferrin synthesis, *Age,* 5, 72, 1982.
11. **Kim, S. K., Weinhold, P. A., Han, S. S., and Wagners, D. J.,** Age related decline in protein synthesis in the rat parotid gland, *Exp. Gerontol.,* 15, 77, 1980.
12. **Kim, S. K., Weinhold, P. A., Calkins, D. W., and Hartog, V. W.,** Comparative studies of the age-related changes in protein synthesis in the rat pancreas and parotid gland, *Exp. Gerontol.,* 16, 91, 1981.
13. **Richardson, A. and Meyers, J.,** A comparison of the cell-free protein synthetic activities of testicular tissue obtained from rats and mice of various ages, *Comp. Biochem. Physiol.,* 71B, 709, 1982.
14. **Starnes, J. W., Beyer, R. E., and Edington, D. W.,** Effects of age and cardiac work *in vitro* on mitochondrial oxidative phosphorylation and [^{3}H]-leucine incorporation, *J. Gerontol.,* 36, 130, 1981.

15. **Ibrahim, N. G., Marcus, D. L., and Freedman, M. L.,** Maintenance of cytochrome P-450 content in old rat livers in spite of decreased mitochondrial protein synthesis, *J. Clin. Exp. Gerontol.,* 3, 327, 1981.
16. **van Bezooijen, C. F. A., Grell, T., and Knook, D. L.,** The effect of age on protein synthesis by isolated liver parenchymal cells, *Mech. Ageing Dev.,* 6, 293, 1977.
17. **van Bezooijen, C. F. A., Sakkee, A. N., and Knook, D. L.,** Sex and strain dependency of age-related changes in protein synthesis of isolated rat hepatocytes, *Mech. Ageing Dev.,* 17, 11, 1981.
18. **Beauchene, R. E., Roeder, L. M., and Barrows, C. H.,** The interrelationship of age, tissue protein synthesis, and proteinurea, *J. Gerontol.,* 25, 359, 1970.
19. **Ove, P., Obenrader, M., and Lansing, A.,** Synthesis and degradation of liver proteins in young and old rats, *Biocim. Biophys. Acta,* 277, 211, 1972.
20. **Salatka, K., Kresge, D., Harris, L., Jr., Ebelstein, D., and Ove, P.,** Rat serum protein changes with age, *Exp. Gerontol.,* 6, 25, 1971.
21. **Chen, J. C., Ove, P., and Lansing, A. I.,** *In vitro* synthesis of microsomal protein and albumin in young and old rats, *Biochim. Biophys. Acta,* 312, 598, 1973.
22. **van Bezooijen, C. F. A., Grell, T., and Knook, D. L.,** Albumin synthesis by liver parenchymal cells isolated from young, adult, and old rats, *Biochim. Biophys. Res. Commun.,* 71, 513, 1976.
23. **Lewis, C. M. and Tarrant, G. M.,** Error theory and aging in human diploid fibroblasts, *Nature (London),* 239, 316, 1972.
24. **Dell'Orco, R. T. and Guthrie, P. L.,** Altered protein metabolism in arrested populations of aging diploid fibroblasts, *Mech. Ageing Dev.,* 5, 399, 1976.
25. **Razin, S., Pfendt, E. A., Matsumura, T., and Hayflick, L.,** Comparison by autoradiography of macromolecular biosynthesis in ''young'' and ''old'' human diploid fibroblast cultures: a brief note, *Mech. Ageing Dev.,* 6, 379, 1977.
26. **Houck, J. C., Sharma, V. K. and Hayflick, L.,** Functional failures of cultured human diploid fibroblasts after continued population doublings, *Proc. Soc. Exp. Biol. Med.,* 137, 331, 1971.
27. **Engelhardt, D. L., Lee, G. T-Y., and Moley, J. F.,** Patterns of peptide synthesis in senescent and presenescent human fibroblasts, *J. Cell. Physiol.,* 98, 193, 1979.
28. **McCoy, B. J., Galdun, J., and Cohen, K.,** Effects of density and cellular aging on collagen synthesis and growth kinetics in keloid and normal skin fibroblasts, *In Vitro,* 18, 79, 1982.
29. **Chen, J. J., Brot, N., and Weissbach, H.,** RNA and protein synthesis in cultured human fibroblasts derived from donors of various ages, *Mech. Ageing Dev.,* 13, 285, 1980.
30. **Young, V. R., Steffee, W. P., Pencharz, P. B., Winterer, J. G., and Scrimshaw, N. W.,** Total human body protein synthesis in relation to protein requirements at various ages, *Nature (London),* 253, 192, 1975.
31. **Winterer, H. C., Steffee, W. P., Davy, W., Perera, A., Uauy, R., Scrimshaw, N. S., and Young, V. R.,** Whole body protein turnover in aging man, *Exp. Gerontol.,* 11, 79, 1976.
32. **Finch, C. E.,** Enzyme activities, gene function, and aging in mammals, *Exp. Gerontol.,* 7, 153, 1972.
33. **Bradley, M. O., Hayflick, L., and Schimke, R. T.,** Protein degradation in human fibroblasts (WI 38). Effects of aging, viral transformation, and amino acid analogs, *J. Biol. Chem.,* 251, 3521, 1976.
34. **Shakespeare, V. and Buchanan, J. H.,** Increased degradation rates of protein in aging human fibroblasts in cells treated with an amino acid analog, *Exp. Cell Res.,* 100, 1, 1976.
35. **Goldstein, D., Stotland, D., and Cordeiro, B. A. J.,** Decreased proteolysis and increased amino acid efflux in aging human fibroblasts, *Mech. Ageing Dev.,* 5, 221, 1976.
36. **Kaftory, A., Hershko, A., and Fry, M.,** Protein turnover in senescent cultured chick embryo fibroblasts, *J. Cell. Physiol.,* 94, 147, 1978.
37. **Jacobus, S. and Gershon, D.,** Age-related changes in inducible mouse liver enzymes: ornithine decarboxylase and tyrosine aminotransferase, *Mech. Ageing Dev.,* 12, 311, 1980.
38. **Reznick, A. Z., Lavie, L., Gershon, H. E., and Gershon, D.,** Age-associated accumulation of altered FDP aldolase B in mice. Conditions of detection and determination of aldolase half life in young and old animals, *FEBS Lett.,* 128, 221, 1981.
39. **Uauy, R., Winterer, J. C., Bilmazes, C., Haverberg, J. N., Scrimshaw, N. S., Munro, H. N., and Young, V. R.,** The changing pattern of whole body protein metabolism in aging humans, *J. Gerontol.,* 33, 663, 1978.
40. **Gershon, H. and Gershon, D.,** Detection of inactive enzyme molecules in ageing organisms, *Nature (London),* 227, 1214, 1970.
41. **Rothstein, M.,** Posttranslational alteration of proteins, in *CRC Handbook Series in Biochemistry in Aging,* Florini, J. R., Ed., CRC Press, Boca Raton, Fla., 1981, 103.
42. **Orgel, L.,** The maintenance of the accuracy of protein synthesis and its relevance to ageing, *Proc. Natl. Acad. Sci. U.S.A.,* 49, 517, 1963.
43. **Sharma, H. K. and Rothstein, M.,** Altered enolase in aged *Turbatrix aceti* results from conformational changes in the enzyme, *Proc. Natl. Acad. Sci. U.S.A.,* 77, 5865, 1980.
44. **Rothstein, M.,** The formation of altered enzymes in ageing animals, *Mech. Ageing Dev.,* 9, 197, 1979.

45. **Richardson, A. and Cheung, H. T.,** The relationship between age-related changes in gene expression, protein turnover, and the responsiveness of an organism to stimuli, *Life Sci.,* 31, 605, 1982.
46. **Sartin, J., Chaudhur, M., Obenrader, M., and Adelman, R. C.,** The role of hormones in changing adaptive mechanisms during aging, *Fed. Proc.,* 39, 3163, 1980.
47. **Walford, R. L.,** Immunologic theory of aging: current status, *Fed. Proc.,* 33, 2020, 1974.
48. **Rothstein, M., Coppens, M., and Sharma, H. K.,** Effect of ageing on enolase from rat muscle, liver and heart, *Biochim. Biophys. Acta,* 614, 591, 1980.
49. **Barrows, C. H. and Kokkonen, G. C.,** Inconsistencies in the presence of age-associated altered aldolase in livers of mice, *Age,* 4, 123, 1981.
50. **Barrows, C. H. and Kokkonen, G. C.,** The effect of age on aldolase in the liver of rats, *Age,* 3, 53, 1980.
51. **Orlovaska, N. N., Demchenko, A. P., and Veselovnka, L. D.,** Age-dependent changes of protein structure. I. Tissue specific, electrophoretic and catalytical properties of muscle aldolase of old rabbits, *Exp. Gerontol.,* 15, 611, 1980.
52. **Demchenko, A. P. and Orlovaska, N. M.,** Age dependent changes of protein structure. II. Conformational differences of aldolase of young and old rabbits, *Exp. Gerontol.,* 15, 619, 1980.
53. **Sharma, H. K., Prasanna, H. G., and Rothstein, M.,** Altered phosphoglycerate kinase in aging rats, *J. Biol. Chem.,* 255, 5043, 1980.
54. **Gafni, A.,** Location of age-related modifications in rat muscle glyceraldehyde-3-phosphate dehydrogenase, *J. Biol. Chem.,* 256, 8875, 1981.
55. **Gafni, A.,** Purification and comparative study of glyceraldehyde-3-phosphate dehydrogenase from the muscles of young and old rats, *Biochemistry,* 30, 6035, 1981.
56. **Gafni, A.,** Age-related effects in coenzyme binding patterns of rat muscle glyceraldehyde-3-phosphate dehydrogenase, *Biochemistry,* 20, 6041, 1981.
57. **Schofield, J. D.,** Altered proteins in aging organisms-application of circular dichroism spectroscopy to the thermal denaturation of purified serum albumin from adult and ageing C57BL mice, *Exp. Gerontol.,* 15, 533, 1980.
58. **Petell, J. K. and Lebherz, H. G.,** Properties and metabolism of fructose diphosphate aldolase in livers of ''old'' and ''young'' mice, *J. Biol. Chem.,* 254, 8179, 1979.
59. **Steinhaugen-Thiessen, E. and Hilz, H.,** Aldolase activity and cross-reacting material in lymphocytes of ageing individuals, *Gerontology,* 25, 132, 1979.
60. **Burrows, R. B. and Davison, I. F.,** Comparison of specific activities of enzymes from young and old dogs and mice, *Mech. Ageing Dev.,* 13, 307, 1980.
61. **Weber, A., Guguen-Guillouzo, C., Szajnert, M. F., Beck, G., and Schapira, F.,** Tyrosine aminotransferase in senescent rat liver, *Gerontology,* 26, 9, 1980.
62. **Leray, G., Guenet, L., LeTrent, A., and LeGall, J.,** Age-related decrease in lysomal β-D-galactosidase activity of human liver cell lines: argument against post-traductional modifications, *Biochem. Biophys. Res. Commun.,* 100, 1491, 1980.
63. **Ross, M. H.,** Dietary behavior and longevity, *Nutr. Rev.,* 35, 257, 1977.
64. **Barrows, C. H. and Kokkonen, G. C.,** Diet and life extension in animal model systems, *Age,* 1, 131, 1978.
65. **McCay, C. M., Crowell, M. F., and Maynard, L. M.,** The effect of retarded growth upon the length of life span and upon the ultimate body site, *J. Nutr.,* 10, 63, 1935.
66. **Berg, B. N.,** Nutrition and longevity in the rat. I. Food intake in relation to size, health and fertility, *J. Nutr.,* 71, 242, 1960.
67. **Berg, B. N. and Simms, H. S.,** Nutrition and longevity in the rat. II. Longevity and onset of disease with different levels of food intake, *J. Nutr.,* 71, 255, 1960.
68. **Riesen, W. H., Herbst, E. J., Walliker, C., and Elvehjem, C. A.,** The effect of restricted caloric intake on the longevity of rats, *Am. J. Physiol.,* 148, 614, 1947.
69. **Ross, M. H. and Bras, G.,** Influence of protein under- and overnutrition on spontaneous tumor prevalence in the rat, *J. Nutr.,* 103, 944, 1973.
70. **Leveille, G. A.,** The long-term effects of meal-eating on lipogenesis, enzyme activity, and longevity in the rat, *J. Nutr.,* 102, 549, 1972.
71. **Drori, D. and Folman, Y.,** Environmental effects on longevity in the male rat: exercise, mating, castration and restricted feeding, *Exp. Gerontol.,* 11, 25, 1976.
72. **Stuchlikova, E., Juricova-Horakova, M., and Deyl, Z.,** New aspects of the dietary effect of life prolongation in rodents. What is the role of obesity in aging?, *Exp. Gerontol.,* 10, 141, 1975.
73. **Everitt, A. V., Seedsman, N. J., and Jones, F.,** The effects of hypophysectomy and continuous food restriction, begun at ages 70 and 400 days, on collagen aging, proteinuria, incidence of pathology and longevity in the male rat, *Mech. Ageing Dev.,* 12, 161, 1980.
74. **Carlson, A. J. and Hoelzel, F.,** Apparent prolongation of the life span of rats by intermittent fasting, *J. Nutr.,* 31, 363, 1946.

75. **Levin, P., Janda, J. K., Joseph, J. A., Ingram, D. K., and Roth, G. S.,** Dietary restriction retards the age-associated loss of rat striatal dopaminergic receptors, *Science,* 214, 561, 1981.

76. **Nolen, G. A.,** Effect of various restricted dietary regimens on the growth, health and longevity of albino rats, *J. Nutr.,* 102, 1477, 1972.

77. **Yu, B. P., Masoro, E. J., Murata, I., Bertrand, H. A., and Lynd, F. T.,** Life span study of SPF Fischer 344 male rats fed *ad libitum* or restricted diets: longevity, growth, lean body mass and disease, *J. Gerontol.,* 37, 130, 1982.

78. **Fernandes, G., Yunis, E. J., and Good, R. A.,** Influence of diet on survival of mice, *Proc. Natl. Acad. Sci. U.S.A.,* 75, 1279, 1976.

79. **Goodrick, C. L.,** Body weight increment and length of life: the effect of genetic constitution and dietary protein, *J. Gerontol.,* 33, 184, 1978.

80. **Leto, S., Kokkonen, G. C., and Barrows, C. H.,** Dietary protein, life-span, and physiological variables in female mice, *J. Gerontol.,* 31, 149, 1976.

81. **Gerbase-DeLima, M., Liu, R. K., Cheney, K. E., Michey, R., and Walford, R. L.,** Immune function and survival in a long-lived mouse strain subjected to undernutrition, *Gerontologia,* 21, 184, 1975.

82. **Weindruch, R. and Walford, R. L.,** Dietary restriction in mice beginning at 1 year of age: effect on life-span and spontaneous cancer incidence, *Science,* 215, 1415, 1982.

83. **Walford, R. L., Liu, R. K., Gerbase-DeLima, M., Mathies, M., and Smith, G. S.,** Long term dietary restriction and immune function in mice: response to sheep red blood cells and to mitogenic agents, *Mech. Ageing Dev.,* 2, 447, 1973/74.

84. **Weindruch, R., Gottesman, R. S., and Walford, R. L.,** Modification of age-related immune decline in mice dietarily restricted from or after midadulthood, *Proc. Natl. Acad. Sci. U.S.A.,* 79, 898, 1982.

85. **Weindruch, R. H., Kristie, J. A., Naeim, F., Mullen, B. G., and Walford, R. L.,** Influence of weaning-initiated dietary restriction on responses to T cell mitogens and on splenic T cell levels in a long-lived F_1-hybrid mouse strain, *Exp. Gerontol.,* 17, 49, 1982.

86. **Watson, R. R. and Safransk, D. V.,** Dietary restrictions and immune responses in the aged, in *CRC Handbook Series in Immunology in Aging,* Kay, M. M. B. and Makinodan, T., Eds., CRC Press, Boca Raton, Fla., 1981, 125.

87. **Friend, P. S., Fernandes, G., Good, R. A., Michael, A. F., and Yunis, E. J.,** Dietary restrictions early and late. Effects on the nephropathy of the NZB × NZW mouse, *Lab. Invest.,* 38, 629, 1978.

88. **Nandy, K.,** Effects of caloric restriction on brain-reactive antibodies in sera of old mice, *Age,* 4, 117, 1981.

89. **Coleman, G. L., Barthold, S. W., Osbaldeston, G. W., Foster, S. J., and Jonas, A. M.,** Pathological changes during aging in barrier-reared Fischer F344 male rats, *J. Gerontol.,* 32, 258, 1977.

90. **Johnson, J. E. and Barrows, C. H.,** Effects of age and dietary restriction on the kidney glomeruli of mice: observations by scanning electron microscopy, *Anat. Rec.,* 196, 145, 1980.

91. **Everitt, A. V.,** The urinary excretion of protein, nonprotein nitrogen, and creatine and uric acid in aging male rats, *Gerontologia,* 2, 33, 1958.

92. **Berg, B. N.,** Spontaneous nephrosis with proteinuria hyperglobulinemia and hypercholesterolemia in the rat, *Proc. Soc. Exp. Biol. Med.,* 119, 417, 1965.

93. **Perry, S. W.,** Proteinuria in the Wistar rat, *J. Pathol. Bacteriol.,* 80, 729, 1965.

94. **Tucker, S. M., Mason, R. L., and Beauchene, R. E.,** Influence of diet and food restriction on kidney function of aging male rats, *J. Gerontol.,* 31, 264, 1976.

95. **Barrows, C. H.,** Nutrition, aging, and genetic program, *Am. J. Clin. Nutr.,* 25, 829, 1972.

96. **Lindell, T. J.,** Molecular aspects of dietary modulation of transcription and enhanced longevity, *Life Sci.,* 31, 625, 1982.

97. **Ross, M. R.,** Protein, calories, and life expectancy, *Fed. Proc.,* 18, 1190, 1959.

98. **Leto, S., Kokkonen, G. C., and Barrows, C. H.,** Dietary protein, life span, and biochemical variables in female mice, *J. Gerontol.,* 31, 144, 1976.

99. **Barrows, C. H. and Kokkonen, G. C.,** The effect of various dietary restricted regimens on biochemical variables in the mouse, *Growth,* 42, 71, 1978.

100. **Bertrand, H. A., Masoro, E. J., and Yu, B. P.,** Maintenance of glucagon-promoted lipolysis in adipocytes by food restriction, *Endocrinology,* 107, 591, 1980.

101. **Voss, K. H., Masoro, E. J., and Anderson, N.,** Modulation of age-related loss of glucogon-promoted lipolysis by food restriction, *Mech. Ageing Mech.,* 18, 135, 1982.

102. **Munro, H. N., Hubert, C., and Baliga, B. S.,** Regulation of protein synthesis in relation to amino acid supply — a review, in *Alcohol and Abnormal Protein Biosynthesis,* Rothschild, M. A., Oratz, M., and Schreiber, S. S., Eds., Pergamon Press, Elmsford, N.Y., 1975, 33.

103. **Waterlow, J. C. and Garlick, P. J.,** Metabolic adaptations to protein deficiency, in *Alcohol and Abnormal Protein Biosynthesis,* Rothschild, M. S., Oratz, M., and Schreiber, S. S., Eds., Pergamon Press, Elmsford, N.Y., 1975, 67.

104. **Sidransky, H.,** Nutritional disturbances of protein metabolism in the liver, *Am. J. Pathol.,* 84, 649, 1976.
105. **Ferro-Luzzi, A. and Spadoni, N. A.,** Protein-energy malnutrition, *Prog. Food Nutr. Sci.,* 2, 515, 1978.
106. **Richardson, A. and Rehwaldt, C.,** Effect of dietary restriction on testicular protein synthesis in Fischer F344 rats, *Exp. Gerontol.,* submitted.
107. **Ricketts, W., Birchenall-Sparks, M. C., Hardwick, J. P., and Richardson, A.,** Effect of long-term dietary restriction on proteinuria and the protein synthetic activity of the kidney, *J. Cell Physiol.,* submitted.

DIET AND EXERCISE: MODIFIERS OF STRESS WHICH CHANGE HOST DEFENSES, LONGEVITY, AND CANCER GROWTH

Ronald R. Watson and A. H. Ismail

STRESS, CORTICOSTEROIDS, AND EXERCISE

Many types of stress in humans, including exercise stress, result in increased glucocorticoid production in subjects classified as having a sedentary physical status.[1,2] Subjects classified as active do not exhibit an exercise-related increase in glucocorticoid production unless given higher work loads.[1,2] Dietary stress such as starvation has long been known to cause stress with adverse effects associated, in part, with increased serum cortisol. Some exciting recent work suggests that diet and exercise can independently reduce corticosteroids, and hence lower some of the adverse effects of stress. This research is still in its infancy, yet the combination of exercise and dietary supplementation may help us in the future to overcome some of the adverse effects of emotional or physical stress including suppressed disease resistance in the young and the adult. We will discuss only that work involving high intakes of vitamin E and exercise as a model system for other nutrients.

Cortisol, the chief glucocorticoid in humans, is produced by the adrenal cortex under the direction of adrenocorticotropin hormone of the anterior pituitary. The increased production of glucocorticoids during stress induces the production of enzymes that break down muscle protein to the various amino acids, inhibits the uptake of glucose by muscle and fat in adipose tissue, and is catabolized for energy. The individual amino acids are converted to glucose via a stimulated production of a rate-limiting gluconeogenic enzyme.[4] Therefore, under conditions of stress, plasma glucose concentration is maintained. For many years, it has also been noted that elevated glucocorticoids cause a lymphocytopenia and suppression of thymus-derived T-lymphocytes and bursa-equivalent B-lymphocytes in humans[5] and animals.[6] Furthermore, the sensitivity of lymphocytes to glucocorticoids seems to be dependent upon the cystolic gluocorticoid receptor concentration.[7] Stimulated lymphocytes seem to be more sensitive to glucocorticoids than resting lymphocytes, possibly due to an increase in receptor concentration during transformation.[7] Immature T-lymphocytes seem to be heterogeneous with respect to their sensitivity to glucocorticoids.

In experimental animals, plasma glucocorticoids are elevated soon after stress induction, followed by a gradual recovery to basal levels.[9] The number of circulating lymphocytes inversely parallels the stress-related increases in plasma glucocorticoids in experimental mice[9] followed by a recovery to normal levels. While exercise stress is correlated with increased glucocorticoids in humans, it is not known what effect this has upon lymphocytes or the entire immune system in humans. Furthermore, it is not known for how long or what lymphocytes are affected. The importance and function of the immune system in animals in the surveillance against invasion by foreign substances or organisms has been well characterized.[10,11] The lymphocytes can be divided into two major groups, the thymus-derived T-lymphocytes and bursa-equivalent-derived B-lymphocytes. Respective B clones of lymphocytes involved in humoral immunity, upon exposure to the respective foreign substance, differentiate into antibody-secreting plasma cells. The antigen-specific antibody, an immunoglobulin protein between 150 and 900 kdaltons in size, is responsible for or participates in the elimination of the invading substance or antigen from mucosal surfaces[12] or in blood, lysis of bacterial cells in conjunction with complement,[13] neutralization of toxins or viruses, and the enhancement of phagocytosis by monocytes and macrophages.[11] The T-lymphocyte component of the immune system is involved in cell-mediated immunity and is responsible for enhancing the phagocytic and microbicidal ability of macrophages,[11] as well as graft rejection,[14] cytotoxicity against tumor cells,[15] delayed type hypersensitivity,[16] suppression of the autoimmune phenomenon, and regulation of the immune system. Because of these important functions, maintenance of an adequate immune response in animals is imperative.

The stress of short-term exercise causes an increase in plasma corticosteroids. These changes have been correlated with a decreased percentage of T-lymphocytes in the peripheral blood, decreased levels of secretory immunoglobulin (Ig) A,[17] and slight decreases in lymphocyte mitogenesis.[18] Therefore, it is not surprising that strenuous exercise is correlated with *increased* incidence to polio virus[19] and Coxsacki virus[20] as well as a *decrease* in tumor growth rate.[21,22] It is not known what beneficial or detrimental effects exercise stress has upon the other aspects of the immune system or host resistance.

ROLE OF NUTRITIONAL SUPPLEMENTATION TO ENHANCE HOST DEFENSES AND MODIFY STRESS

While stress due to exercise accompanied by increased glucocorticoid production has certain adverse effects upon host resistance to infection and the immune system, high dietary levels of vitamin E or C augment humoral- and cell-mediated immune responses in experimental animals.[23] Use of high levels of vitamin E to modify immune functions and stress is just beginning to be investigated, but is a useful model system to represent the possibility. Two possible mechanisms have been theorized by which high dietary levels of vitamin E increases the immune response. We have shown that high dietary levels of vitamin E in mice have been shown to decrease basal serum glucocorticoid levels.[24] Also, during the course of polyclonal lymphocyte activation, prostaglandin (PG) E production was decreased if lymphocytes were taken from animals fed high levels of vitamin E.[25] PGE are responsible for inducing suppressor T-lymphocytes during the immune response.[26] Interestingly, prostaglandin synthetase inhibitors significantly inhibit the immunosuppressive action of hydrocortisone upon human lymphocytes.[27]

A major physiological role of vitamin E is generally considered to be its ability to function as an antioxidant and as a free radical scavenger. Therefore, vitamin E may play a significant role in altering cell aging. While most human or animal diets contain enough vitamin E to prevent deficiency, the optimum amount for enhanced longevity, optimal immune responses, and resistance to infection and tumor growth is less clear and poorly defined. In addition, the added effect of exercise on needs and functions is unclear. For example, high vitamin E intake inhibited carcinogenesis,[30,31] protected patients from adverse side effects of radiation therapy,[32] stimulated humoral immune responses to antigen or resistance to bacterial infection,[23,25] enhanced helper T cell activity,[33] and accelerated maturation of cellular immune functions[24] in young mice. Absence of vitamin E in the diet may eventually cause significant suppression of resistance to infection.[34-36] However, research is critically needed to define optimum intakes of vitamins and exercise to produce the most active immune functions and anticancer defenses. It is necessary to understand the effects of intakes above that routinely found in what is now considered a "reasonable and prudent" diet, because as the amount of fat and polyunsaturated fatty acids increase in the diet, the requirement for some vitamins like vitamin E should increase.

Severe vitamin E deficiency is associated with suppressed host defenses to infectious diseases.[34-41] Supplementation of routine diets with vitamin E in animals improved the humoral immune responses of mice, chickens, turkeys, and guinea pigs.[42] Supplementation also improved the resistance of chickens and turkeys to *E. coli* and of sheep to *Chlamydia*.[42]

Why do these limited studies with young and mature animals suggest that high intakes of vitamin E improve disease resistance? Based upon our work, we believe that enhanced disease resistance, and tumor resistance in mice[24,43,44] is due to altered cellular immune functions and secretory immune functions. For example, we found that there were increased amounts of S-IgA in the intestinal secretions of young mice fed a high vitamin E diet.[44] We also found upon supplementation of high dietary vitamin E in young mice, elevated numbers of cytotoxic lymphocytes which could be important in anticancer defenses, perhaps due to more rapid maturation of T cell functions.[24,43] However, it appeared that vitamin E was accelerating maturation of only certain T cell functions. This hypothesis is supported by Levinson et al.[19] and Corwin and Shloss[45] who showed high dietary vitamin E increased the PHA/Con A response ratio suggesting an effect of the vitamin on maturation of T cells. Stimulation of T cell responses may be the result of an effect on $T_a +$ T cells.[45] Vitamin E did not function as a mitogen in athymic mice nor enhance the LPS response of the spleen as it does in normal mice. These data suggest that vitamin E may act with thymic factors to produce a mature T helper cell. Thus, vitamin E may act in part by selectively stimulating certain populations of T cells and/or reducing hormonal control on these T cells' function.

In the area of hormonal regulation, we have shown consistently in mice that high intakes of vitamin E reduce serum glucocorticoids.[24,43,44] Glucocorticoids have been clearly shown to affect T cell function and thymus gland development. This may be a partial explanation for enhanced immune functions. Vitamin E may affect the lymphocyte cell itself directly by altering prostaglandin synthesis.[46] There is growing evidence that prostaglandins regulate immune responses by increasing suppressor lymphocyte activity and thus are immunosuppressive. Vitamin E as an antioxidant may prevent the oxidation of arachidonic acid in the biosynthetic pathway leading to prostaglandins.[47,48] Chickens fed supplemental vitamin E at six times the normal level had decreased prostaglandins E, E_2, and E_{2a} in immunopoietic organs (bursa and spleen). Aspirin, a known prostaglandin inhibitor acted synergistically with vitamin E in depressing endogenous prostaglandin levels and decreasing mortality from E. coli infection.[46]

While a number of investigations show either enhancement or at least little effect of supplementary vitamin E, there are two recent reports which suggest that high levels may be inhibitory to certain aspects of the immune system.[49,50] A very recent animal study showed that some levels of vitamin E supplementation enhance while very high levels suppressed immune functions. Vitamin E is generally considered to be relatively nontoxic at high dosages. In addition, Yasunaga et al.[50] indicate that they have data in mice confirming our preliminary study[43] which showed some suppression of cancer cell growth in vivo as a result of supplementation with vitamin E mice.

VITAMIN E AND C SUPPLEMENTATION ON FIT AND NONFIT ADULTS

In our initial study, we had 21 subjects,[59] 12 subjects were classified as physically fit while 9 were classified as physically nonfit. An initial group of 8 nonfit and 4 fit subjects were given 400 IU of vitamin E and 500 mg of vitamin C daily for 30 days. A second group of 4 nonfit subjects and 5 fit subjects were given placebos for vitamins E and C.

The fit subjects continued on their regular physical activity schedule as evaluated by Ismail, while the unfit subjects participated in a program that did not improve their fitness status. Venous blood was collected from each subject before and immediately after the program. Lymphocytes were isolated from a Ficoll-Hypaque gradient and plasma collected for the various biochemical assays and in vitro analyses. An analysis of variance was conducted with respect to time (pre to post), diet (vitamin or placebo), and exercise (fit or nonfit) status to ascertain which conditions exhibited an effect and determination of significant interactions among the conditions.

The percentage of T cells in the isolated lymphocyte populations was measured using the sheep erythrocyte rosetting assay. There was a significant effect as to the percent T cells attributed to vitamin supplementation. Subjects that were fit exhibited a significant increase whereas nonfit subjects given the vitamin did not exhibit an increase in this immunological parameter from pre- to the post-time. The number of lymphocytes per milliliter of blood was determined by microscopically counting the isolated lymphocyte population. There did not seem to be a significant effect upon this parameter in subjects consuming high levels of the vitamins compared to subjects consuming placebo. The value for the number of T cells per milliliter of whole blood was generated from the product of percent T cells and number of lymphocytes. These data were extremely interesting. There did seem to be a significant placebo effect that was abrogated by consumption of the high levels of vitamins.

While the previous assays measured T-lymphocyte numbers in blood, the present assay measured the mitogenic activity of the T-lymphocyte in the population after exposure to a polyclonal T-lymphocyte mitogen. There was not significant effect due to consumption of the vitamin upon this parameter. However, the fit individuals exhibited an increased mitogenesis from pre to post regardless of their dietary supplementation. Stimulation units are defined as the cpm of tritiated thymidine incorporated into lymphocyte DNA due to mitogen exposure divided by the cpm of tritiated thymidine incorporated spontaneously in unstimulated lymphocytes. In other words, it takes into account background incorporation. When this parameter was examined, there was a significant effect due to consumption of the vitamins. Nonfit subjects given vitamins exhibited an increase in their lymphocytic PHA-driven stimulation units while placebo-fed subjects exhibited no significant increase in this parameter.

Con-A also stimulates cell division in T-lymphocytes. As with PHA stimulation, the fit subjects exhibited an increase in T cells activity, but nonfit subjects demonstrated a decreased T cell activity with this measure. The consumption of the vitamin supplementation did not seem to affect the T-lymphocyte activity in either of the subject groups. As with responsiveness to PHA, supplementation with vitamins A and C in both fit and nonfit individuals increased their stimulation units while subjects fed placebo exhibited no effect upon this parameter. Since stimulation units take into consideration background mitogenesis of the lymphocyte population, the value may be a more precise estimation of polyclonal mitogenesis when comparing different populations. However, scientific reviewers still insist that both values (cpm and SU) be expressed.

One mechanism by which lymphocytes control the immune response, such as that to PHA, is by production of prostaglandins (PGE) that stimulate suppressor lymphocytes. Therefore, any modulation of an immune response may be at this level. Previous results in animals indicate this to be one mechanism by which vitamin E alters the immune response. Interestingly, we have shown that lymphocytes of subjects given the vitamin supplementation demonstrated a significant increase in PGE production after exposure to PHA while subjects given placebo showed no significant alteration in this parameter.

Glucocorticoids suppress immune responses in man and animals. Therefore, any alteration of an immune response by diet may be the result of an alteration of glucocorticoid production. The basal level of cortisol in plasma was not affected by vitamin supplementation or fitness status.

Clearly there is a need to accurately define the levels of vitamin E which produce the best host defenses in humans! Whether supplementation of normal diets with vitamin E is considered "overcoming a deficiency disease" or "producing optimum functions" is largely irrelevant. What is critical is defining and determining what is important for health in man.

DIETARY RESTRICTIONS, LONGEVITY, AND IMMUNE RESPONSES

For centuries man has recognized the association between malnutrition and disease. One compelling hypothesis that might explain, in part, the increased morbidity and mortality suffered by malnourished populations is that nutritional deficiencies impair immune responsiveness, thus reducing antimicrobial protection in the nutritionally deprived.[51] On the other hand, nutritional intakes can be reduced in young and aged animals to enhance longevity by, in part, changing development of immune responses.[52] It is well recognized that mild or moderate restrictions of calories or protein in the diet significantly enhance longevity via immunological modifications in animals.[52] What needs to be learned are (1) the effects of mild dietary stresses of proteins or calories on two other vital psychological systems — work capacity and mental functions, (2) the effects of exercise on longevity of mildly malnourished animals, and (3) the long-term effects of these dietary restrictions in humans on their immune systems, work capacity, and longevity. In short, do the hundreds of millions of moderately malnourished people in the world benefit from these stresses or are they damaged via reduced work capacity, mental functions, and suppressed development of immunological functions?

Worldwide PCM is the most common form of acquired immunodeficiency.[51] Most of the human studies to date have focused on severe PCM in hospitalized children. While such children represent most dramatically the effects of nutritional deficiency on disease suscep- tibility, they comprise only a small percentage (3 to 5%) of malnourished children in most developing countries.[51] In contrast, up to 60% of all preschool children in many preindustrial societies may suffer from the milder forms of malnutrition. Many aged humans who are undernourished have major contributing complications or causes of the nutritional stress including cancer, alcoholism, prolonged hospitalization, major illness, lack of sufficient exercise, etc.

Undernutrition is often associated with increased incidence and/or severity of many bac- terial, viral, and parasitic diseases.[51] It is well documented that the lifespan of children, young animals, as well as adults are decreased when severely stressed by low protein or severe vitamin deficiencies. This appears to be caused, in large part, by immunosuppression induced by diet. Much less is known about the effects of marginal undernutrition, which is extremely prevalent worldwide.

The addition of the aging parameter to the long-term interactions between undernutrition, disease, and immunity has been studied with difficulty. As the organism ages, there occurs an overall decline in immune capabilities, accompanied by an increased incidence of in- appropriate responses directed toward the self. Mouse strains genetically predetermined to develop autoimmune diseases provide an ideal subject as their lifespan is severely reduced and can be more conveniently studied. For example, Fernandes and co-workers[53] have used nutritionally stressed female C3H/UMC mice and reduced the incidence of spontaneous mammary carcinomas from 70% in the controls to 0%. The undernourished mice were aliquoted food, that if shared equally, would yield a 40% calorie reduction as compared to *ad libitum*-fed controls. The response of B cells to lipopolysaccharides and the number of plaque-forming cells in the spleen were the same in both groups, while T cell proliferative responses to PHA, and con-A were increased in the undernourished group. The caloric restrictions in this case did not significantly lengthen lifespan; the longest surviving mice were on the restricted diet.

Walford et al.[54] working with C57BL/6J female mice, provide some tangible evidence supporting the theory that dietary restriction causes a delay in maturation of the immune system and hence enhances longevity. Walford et al.[54] fed 3.5 g of food (21.5% casein, 14.5 calories) 4 days a week followed by a triple portion, while the restricted mice were fed 3.5 g of the same diet (enriched in vitamins and salts) twice a week with a double proportion on the weekend. Body weights and spleen indexes for mice restricted 18 weeks and 55 weeks old were significantly lower than controls. At 52 weeks old, the mildly malnourished mice had the greater number of plaque-forming cells and T cells responding to mitogens.

Weindrach and co-workers,[55] in an extremely comprehensive project, manipulated unique dietary regimes in an effort to more thoroughly understand nutritional effect on longevity and immunity. Generally, the *ad libitum* diets showed the shortest lifespan with all moderately restricted groups surviving longer. The data indicate lifelong dietary restrictions lead to more immunologically youthful animals.

Almost all studies of longevity as modified by diet have used young, immature animals. Recently, Walford studied the effects of mild undernutrition on aging mice. He found enhanced imune functions, fewer tumors, and enhanced longevity, confirming the applicability of procedures used with young animals in old ones.

Similar studies in humans are very difficult. However, there is some indication in humans that if the nutritional stress is early and severe enough, impairment of cell-mediated immune responses may persist for several years. Dutz and co-workers[56,57] studied the delayed hypersensitivity response to dinitrochlorobenzene in orphans who had been severely malnourished and infected during early infancy and found significant depression of cellular immunity up to 5 years later, when all were free of intercurrent disease and were well nourished. This may be preliminary evidence of retarded maturation of human immune system. Our study in Colombian children severely malnourished later in life showed a rapid restoration of cellular and secretory immunity a few weeks after initiation of a high protein diet.[50]

Clearly, immunological processes basic and vital to the living organism can be manipulated and controlled by nutrition with much less known about the effects of work output (exercise) as an additive component. Functions such as growth, health, malignancies and tumor development, enzyme activity, longevity, and the immune response have all shown a dependence on diet.[51] The interrelationship between longevity, immunity, and diet is very complex; changes have been indicated to be dependent on how much food is allowed, components of the diet, age the animal is started on a diet, and duration of dietary manipulation. Generally, the studies involving normal mice indicate an increase in lifespan and immune capabilities during long periods of undernutrition. Some of this may be due to increased bacterial antigens stimulating some cellular immune systems (phagocytosis) as we and others have shown.[51,52] In addition, the increase in immune response in undernourished animals may be due to a delay in maturation of the immune response mediated by the dietary restriction. Combining both theories, undernourishing the animal (and perhaps children or aging human adults) will cause a delay in the maturation of the immune system, thereby delaying the onset of the aging process, allowing the animal to continue functionally sound. An animal's or person's lifespan is defined in genetic terms, and excluding death from disease or accident, the animal will fulfill this potential. Considering the lifespan of an animal as finite, with immune system intrinsically involved from the point of maturation of the immune system, there is a predetermined amount of time before it begins to dysfunction, ultimately causing death. The length of life from the maturation of the immune system is limited by the immune system and depends on its own gradual loss of ability to recognize itself.[51]

The lifetime of an animal or person can be considered as a numberline with the immune system as a subset of this line. The end of the line (or the end of life) is determined by the midpoint of the subset (or the maturation of the immune system).[51] Dietary restrictions (and exercise, although much more research is needed to determine its role) cause the subset to shift down the numberline, and correspondingly extend its length, or if the dietary restrictions are severe enough, decrease it due to increased disease. No matter if delayed or not, once the immune system begins to malfunction, the aging process begins, and death due to any number of situations can occur.

ACKNOWLEDGMENTS

We recognize the support of the National Livestock and Meat Board, Phi Beta Psi Sorority and the Wallace Genetic Foundation for some of our research which stimulated this paper.

REFERENCES

1. **White, J. A., Ismail, A. H., and Bottoms, G. D.,** *Br. J. Sports Med.,* 9, 3, 1975.
2. **White, J. A., Ismail, A. H., and Bottoms, G. D.,** *Med. Sci. Sports,* 8, 113, 1976.
3. **Hellman, H. L., Nakad, F., Curti, J., Weitzman, E. D., Kream, J., Roffward, J., Ellman, S., Fukushima, D. ., and Gallagher, T. F.,** *J. Clin. Endocrinol.,* 30, 411, 1970.
4. **Daco, J. C., Lefer, A. M., and Berne, R. M.,** *Physiologist,* 10, 153, 1967.
5. **Fauci, A. S. and Dale, D. G.,** *J. Clin. Invest.,* 53, 240, 1974.
6. **Fauci, A. S.,** *Immunology,* 28, 669, 1975.
7. **Smith, K. A., Grabtree, G. R., Kennedy, S. H., and Munck, A. U.,** *Nature (London),* 267, 523, 1977.
8. **Distelhorst, C. W. and Benutto, B. M.,** *J. Immunol.,* 126, 1630, 1980.
9. **Hedman, L. A.,** *Lymphology,* 13, 34, 1980.
10. **Heidelberger, M.,** *J. Exp. Med.,* 83, 303, 1946.
11. **Mackaness, G. B.,** *Infectious Agents and Host Reactions,* Mudd, S., Ed., W. B. Saunders, Philadelphia, 1970, 61.
12. **Tomasi, T. B. and Beinenstock, J.,** *Adv. Immunol.,* 9, 2, 1968.
13. **Johnston, R. B.,** *Malnutrition and the Immune Response,* Siskind, R. M., Ed., Raven Press, N.Y., 1977, 295.
14. **Simonsen, M.,** *Progr. Allergy,* 6, 349, 1962.
15. **Berendt, M. J. and North, R. J.,** *J. Exp. Med.,* 151, 69, 1980.
16. **Majarian, J. S. and Feldman, J. D.,** *J. Exp. Med.,* 114, 779, 1961.
17. **Tomasi, T. B., Trudeau, F. B., Czerwinski, D., and Erredge, S.,** *J. Clin. Immunol.,* 2, 173, 1982.
18. **Hedfors, E., Holm, G., and Ohnell, B.,** *Clin. Exp. Immunol.,* 24, 328, 1976.
19. **Levinson, S. O., Milzer, A., and Lewin, P.,** *Am. J. Hyg.,* 42, 204, 1945.
20. **Reyes, M. P. and Lerner, A. M.,** *Proc. Soc. Exp. Biol. Med.,* 151, 333, 1976.
21. **Rusch, H. P. and Kline, B. E.,** *Cancer Res.,* 4, 116, 1944.
22. **Good, R. A. and Fernandes, G.,** *Fed. Proc.,* 40, 1040, 1981.
23. **Tengerdy, R. P., Heinzerling, R. H., Brown, G. L., and Mathias, M. M.,** *Int. Arch. Allergy Appl. Immunol.,* 44, 221, 1973.
24. **Lim, T. S., Putt, N., Safranski, D., Chung, C., and Watson, R. R.,** *Immunology,* 44, 289, 1981.
25. **Heinzerling, R. P., Nockels, C. F., Quarles, C. L., and Tengerdy, R. P.,** *Proc. Soc. Exp. Biol. Med.,* 146, 279, 1974.
26. **Fischer, A., Durandy, A., and Griscelli, C.,** *J. Immunol.,* 126, 1452, 1981.
27. **Staszak, C. and Goodwin, J. S.,** *Cell Immunol.,* 54, 351, 1980.
28. **Banerjee, S. and Singh, H. D.,** *Am. J. Physiol.,* 190, 265, 1957.
29. **Clayton, B. E., Hammont, J. E., and Armitage, P. J.,** *Endocrinology,* 15, 284, 1957.
30. **Haber, S. L. and Wissler, R. W.,** *Proc. Soc. Exp. Biol. Med.,* 3, 774, 1962.
31. **Wattenberg, L. W.,** *Fundamentals in Cancer Prevention,* Magee, P. N. et al., Eds., University Park Press, Baltimore, 1976, 153.
32. **Black, H. S. and Chan, J. T.,** *J. Invest. Dermatol.,* 65, 412, 1975.
33. **Tanaka, J., Fugiwara, H., and Torisu, M.,** *Immunology,* 38, 727, 1979.
34. **Van Vleet, J. F.,** *J. Am. Vet. Med. Assoc.,* 176, 321, 1980.
35. **Feige, J., Jr., Mordstoga, K., and Aursjo, J.,** *Acta Vet. Scand.,* 18, 384, 1977.
36. **Teige, J., Jr., Sacegaard, F., and Froslie, A.,** *Acta Vet. Scand.,* 19, 133, 1978.
37. **Shetty, B. E. and Schultz, R. D.,** *Fed. Proc.,* 38, 2139, 1979.
38. **Serfass, R. E. and Ganther, H. E.,** *Life Sci.,* 19, 1139, 1976.
39. **Serfass, R. E. and Ganther, H. E.,** *Nature (London),* 255, 640, 1975.
40. **Bayne, R. and Arthur, J. C.,** *J. Comp. Pathol.,* 89, 151, 1979.
41. **Holmes, B., Park, B. H., Malawista, S. E., et al.,** *N. Engl. J. Med.,* 283, 217, 1970.
42. **Nockels, C. F.,** *Fed. Proc.,* 38, 2134, 1979.
43. **Watson, R. R., Chung, C., and Petro, T. M.,** New York Academy of Science, in, Proc. Conf. Vitamin E: Biochemical, hematological and clinical aspects, 393, 205, 1982.
44. **Watson, R. R. and Messiha, N.,** submitted.
45. **Corwin, L. M. and Shloss, J.,** *J. Nutr.,* 110, 916, 1980.
46. **Likoff, R. O., Guptill, D. R., Lawrence, L. M., McKay, C. C., Mathias, M. M., Nockels, C. F., and Tengerdy, R. P.,** *Am. J. Clin. Nutr.,* 34, 245, 1981.
47. **Lands, W. E. M. and Rome, L. H.,** Prostaglandins: *Chemical and Biochemical Aspects,* Karen, S. M., Ed., University Park Press, Baltimore, 1976, 87.
48. **Machlin, J. L., deDure, C., and Hayaishi, O., Eds.,** *Tocopherol, Oxygen and Biomembranes,* Elsevier/North Holland, N.Y., 1978, 179.
49. **Prasad, J. S.,** *Am. J. Clin. Nutr.,* 33, 606, 1980.

50. **Yasunaga, T., Kato, H., Ohgaki, K., Inmoto, T., and Hikasa,Y.,** *J. Nutr.,* 112, 1075, 1982.
51. **Watson, R. R. and McMurray, D. N.,** *CRC Crit. Rev. Food Sci. Nutr.,* 12, 113, 1979.
52. **Watson, R. R. and Safranski, D. W.,** *CRC Handbook of Immunology,* CRC Press, Boca Raton, 1981, 125.
53. **Fernandes, G., Yunis, E. J., and Good, R. A.,** *Nature (London),* 263, 504, 1976.
54. **Walford, R. L., Liu, R. K., Gerbase-DeLima, M., Mathies, M., and Smith, G. S.,** *Mech. Ageing Dev.,* 2, 447, 1973.
55. **Weindrach, R. H., Kristie, A., Chevey, K. E., and Walford, R. L.,** *Fed. Proc.,* 38, 2007, 1979.
56. **Dutz, W., Rosipal, E., Ghavami, H., Vessal, K., Kohout, E., and Post, C.,** *Eur. J. Pediatr.,* 122, 117, 1976.
57. **Dutz, W., Kohout, E., Rossipal, E., and Vessal, K.,** *Pathology Annual,* Sommers, S. E., Ed., Appleton-Century-Croft, N.Y., 1976, 11.
58. **McMurray, D. N., Watson, R. R., and Reyes, M. A.,** *Fed. Proc.,* 38, 613, 1979.
59. **Watson, R. R., Petro, T. M., and Ismail, A. H.,** *Fed. Proc.,* 42, 2092, 1983.

ADULT PROTEIN-CALORIE MALNUTRITION IN DEVELOPING COUNTRIES

Ronald R. Watson and Mary E. Mohs

INTRODUCTION

Protein-calorie malnutrition is one of the major health problems in the developing world. Its effects are far reaching and extend to the general socioeconomic well-being of society. Malnourished adults are a less productive element of society because of low resistance to disease as well as decreased physical strength. Such functional effects can set up a vicious cycle which further worsens the health of the country. Adult malnutrition may reflect recent acute or chronic moderate dietary problems as well as long-term effects of earlier malnutrition and disease. Animal research suggests that nutritional stresses during the growth phase result in suppressed immune functions in the next generation indicating one long-term aspect of malnutrition.[1] Protein-calorie malnutrition (PCM) should not be viewed as strictly a nutritional deficiency disorder. The causes often surpass inadequate and improper food intake to include poor living conditions, disease, reduced maternal-child interactions, the detrimental aspects of aging, and/or lack of primary health care. PCM in many ways can also be seen as a disease of socioeconomic inequalities and of maldistribution of food and wealth.[2] The precise incidence of malnutrition in the adult population throughout the world is very difficult to determine. For example, it is not listed as a cause of death in the *Annual Epidemiological and Vital Statistics* publication issue of the WHO. Very few studies have actually centered on adult PCM in less-developed countries, while a great deal of research has been conducted on infants and children. This is understandable. Clearly, nutritional stresses affect developing systems more readily than mature ones. In addition, adults, particularly men, in many societies get preferential access to food. This decreases their risk of PCM and increases that of the children who are relegated the leftovers. However, the adults are the people who will ensure the family's steady income, shelter, and food supply, as well as emotional support and sense of direction. Therefore, due to the importance of adults and their work in supporting and nurturing the family and the effects of aging on them, as well as the paucity of research on their nutritional problems in developing countries, we are analyzing the current literature. We will summarize the effects of age in incidence of PCM, the studies of PCM in adults during war, the relationship between immunosuppression and parasitic diseases in PCM, PCM in women, the adaptation to PCM by the body, and the effects of renutrition and prevention of PCM in adults.

When looking at the future public health picture for a country it is quickly seen that the nutritional health of the adult population is crucial for the survivability of the population as a whole. However, most nutritional aid has been targeted to the young and/or pregnant or lactating mothers on the basis of their being at higher nutritional risk due to high growth and development needs. While this tact has much merit, nutritional problems of the rest of the adult population have been largely overlooked. Marginal PCM in adults in the less developed world with reduced functions in undernourished adults could also be aiding in perpetuating the problem in the young through less efficient adult reproduction and work functions and reduced income and food for families.

A recent presidential commission on world hunger stated that the average daily intake of protein per person is 54 g in the developing countries compared with 97 g in the developed or industrialized countries.[3] A 70-kg adult requires 45 g/day to remain in positive nitrogen balance.[4] This would be especially true if the body makes adaptive compensations to adjust to a lower protein intake.[5] Nevertheless, those adults consuming less than the average of

54 g in developing countries are nearing or surpassing a critical minimum need. The protein supply must also be judged in relation to caloric intake. When insufficient calories are ingested, dietary or tissue protein is metabolized to serve as a source of calories. Therefore, with combined marginal protein and calorie intake it is usually impossible to meet the body's protein requirement. Thus, calorie malnutrition can cause protein inadequacy. On the other hand, strict protein malnutrition can occur in high calorie diets such as in areas which use a staple food supply like cassava.[6]

In areas of the world which experience chronic mild to moderate PCM, it is difficult to identify the precipitating cause of clinical and biochemical symptoms in a person. An infection, pregnancy, a chronic intestinal disorder, a seasonal reduction in food, or emotional stresses could trigger severe PCM in an already nutritionally stressed body.[4,7] There is a sharp difference between a state of chronic malnutrition endemic to some areas and a frank acute starvation (severe PCM). The latter condition has been vividly seen in POWs, displaced persons and refugees, and famine victims. However, it is the chronic form which affects most developing countries, and this has been insufficiently investigated to date.

Physicians in part of India have seen chronic adult PCM frequently enough to be very familiar with its etiology, clinical and biochemical features, diagnosis, prevention, and management. Bhattacharyya notes that adult Indian PCM usually results from combinations of poor diet, malabsorption, and infections, with wasting and edema being the chief clinical features. After organic diseases are tested for and excluded from the case, high protein and energy diets with vitamin and mineral supplements have usually brought a slow but good response in the Indian cases.[8]

The importance of adequate food intake for both the young and old is shown in famines where old persons (greater than 45 years) and children die first. Women and adolescents tend to survive better than men.[9] This may be in part associated with reduced immune functions and host resistance in healthy very young and aged.[1,10]

AGING EFFECTS AND PCM

The likelihood of PCM occurring in an adult in any country becomes greater as the aging process continues. This is true for a variety of reasons, both psychological and physiological. Many physiological systems are adversely affected in the aging process, including the neuromuscular, GI, and renal systems. Decreasing motor function results in decreasing ability to procure and chew foods. Declining ability to smell and taste foods can affect the elderly person's desire and ability to eat well. Poor dentition also contributes to inability to chew the food. This may be particularly true in developing countries where a high fiber content, unrefined diet is the norm and strong teeth are essential for successful mastication. The decline of the GI system can result in poor food absorption, utilization, and storage. Decreased digestive ability with aging can be severe enough to result in PCM regardless of continued intake of a formerly adequate diet. Body composition changes such as decreased lean muscle mass and increased fat content may occur, as well as a decline in the mineral content of bone through life.[11] Socioeconomic factors militating toward PCM are also exacerbated for many elderly in all countries. Isolation, dietary ignorance, and poverty associated with loss of work function are notable among them.

Nutritional standards are in need of revision for the elderly. A controversy currently exists regarding the protein needs for the elderly. One school of thought holds that decreased protein intake should be recommended for the elderly, as they have a reduced renal ability to handle protein solute loads. The other school holds that the reduced ability to digest protein requires a higher protein intake, not only for the sake of protein itself, but also for the protein-associated iron and trace mineral intakes. Lower basal metabolic rates, associated with lower lean muscle mass, also occur with aging.[12] This affects the total daily calorie expenditure in the aged.

The assessment of PCM in the elderly is difficult at present, as some anthropometric and clinical measures normally done to pinpoint PCM have inadequate standards for the elderly. Triceps skinfold thickness, midarm muscle circumference, and 24-hr urinary creatinine excretion are in this category. The high degree of individual variation in anthropometric measurements of the elderly also makes interpretation of values more difficult. Nutritional parameters and standards need to be developed specifically for the elderly to increase the reliability of such measurements in the future.[13]

This is particularly true since some observers unfamiliar with aging may misinterpret changes normally seen with aging as evidence of nutritional deficiency. For instance, height is affected by spinal column shortening due to thinning of discs, loss of vertebrae height, and kyphosis believed to result from osteoporosis. Body composition changes in lean muscle mass and adipose tissue occur differentially between aging men and women. Several surveys have shown depressed serum albumin in the elderly, and it is unclear whether this meant high PCM incidence or simply aging effects. Both PCM and the aging process have been found to affect immune functioning, calling into question the usefulness of assessing immune function to indicate PCM in the aged. Similarly, Hb values, which are used to assess malnutrition-associated anemia, are also lowered with age.[14]

In addition to the previously mentioned changes of aging that increase risk for PCM in the elderly, the increased incidence of certain diseases also results in higher PCM occurrence. Increased susceptibility to infectious diseases is associated with decreased immune function in the elderly. Such diseases can push the marginally malnourished elderly person into a frank PCM state. Chronic pulmonary disease and acute respiratory failure are often seen in the aged. These conditions have been shown to occur accompanied by significant deficits in body protein and fat stores and by low serum transferrin, and retinol-binding protein, creatinine height index, and total lymphocyte counts are all evidences of PCM.[15]

GI problems such as Crohn's disease are often treated by surgical resection, and high prevalence of PCM has been seen, even in former surgical patients who had resumed full-time work and seemed subjectively well.[16] Decreased GI function in the elderly may be similar in effect to that experienced by patients with short-bowel following surgery. For all the above reasons, PCM in the developing countries' elderly may be exacerbated compared to that seen in the younger adult segment of developing country populations.

ADULTS AND PCM EFFECTS OF WAR

Some of the classic studies of acute adult malnutrition were done as a result of famines and captivity occuring during World War II.[1-10] These studies have shed light on the many post-World War II famines most recently occurring in the Sahel region of Africa and in Cambodia.[1-12] In a nonwar situation, social class shows a known and strong association with both nutrition and resulting mental performance of mature adults.[17] However, this association blurs under wartime conditions. In the early 1940s a group of Jewish physicians in the Warsaw ghetto performed a scientific study on the pervasive hunger and subsequent starvation of their people. Both to expose the horrors of the Nazi-dominated ghetto and to shed light on the physiological results of acute PCM, the physicians amassed a great deal of data. They observed constant thirst, persistent increase in urinary output, general weakness, weight loss, considerable lowering of resting basal metabolic rate, and lowered serum proteins.[18]

In male Dutch famine patients, common symptoms of PCM included lack of potency and diminished libido. Many of them had testicular atrophy and showed a low sperm count (below 60,000/cc), poor motility, and abnormal morphology of cells. They also had low 17-ketosteroids and low urinary gonadotropin levels. Observations of these victims of the 1944 to 1945 Dutch famine winter showed men experiencing malnutrition to have enlargement of the breasts and a ''feminine-type'' distribution of fat in the pelvic girdle. It was hypothesized that the liver was not breaking down the estrogen-type compounds present in the liver.[17] It should be noticed that these sexual abnormalities have an overall detrimental effect on the population fertility rate, and in prolonged periods of PCM there could be a potential reduction in size of future generations.

Semistarved adult patients in Leningrad during the famous 900-day seige of that city by the Germans showed an inability to adapt well to cool, let alone cold, temperatures. They displayed chronic hypothermia.[19] This body temperature-lowering effect may have represented an adaptive attempt of the body to conserve energy. However, in such a severely cold climate as Russia's, the trade-off between energy conservation and maintenance of body warmth invited the possibility of frostbite at the very least in these patients.

A number of investigations have showed the tremendous mortality effects from infectious diseases, particularly pulmonary tuberculosis and bronchopneumonia, in semistarved persons during World War II. The previously mentioned Leningrad population succumbed most frequently to bronchopneumonia.[19] Persons interned in various German concentration camps showed highest mortalities from tuberculosis. For instance, estimates of 40% of Buchenwald's, 30% of Porta's, and at least 40% of Belsen's total mortality among captive Danish resistance fighters and policemen were due to tuberculosis.[20] A direct correlation was seen between increasing percent of weight loss and increasing percent of TB as a cause of mortality among these prisoners.[21]

World War II experiences have also had long-range effects on the health of some unfortunates. The recurrence of life-long health problems in thousands of Britons subjected to severe malnutrition as adult POWs of the Japanese in the Far East (FEPOWs) during World War II has recently been studied.[22] Ex-FEPOWs suffered from peripheral neuropathy, nerve deafness, ataxia, vitamin deficiencies, amblyopia with optic atrophy, and occasional spasticity due to cord lesions at the time of repatriation. Most of these symptoms disappeared soon thereafter, but neurological symptoms reappeared years later. Ex-FEPOWs most commonly exhibited long-range neurological disease symptoms which were categorized into five classes. Of these, cord disease, progressive and nonresponsive spastic paraplegia, and Parkinson's disease were relatively common in the ex-FEPOWs, with Parkinson's disease incidence much greater than that seen in the general population. The progressive and nonresponsive nature of the spastic paraplegia and the cord disease prompted the investigators to postulate that the intense short period of protein deprivation in the slow protein turnover site of the CNS may have taken years to manifest itself. In addition, they suggested that if a slow virus were responsible for the progressive neurological symptoms, that the discrete period of malnutrition may have made the patient susceptible to the initial virus infection.[22]

Another study of ex-FEPOWs showed that 5.5% of them had persisting neurological disease, and an additional 4.2% had possible asymptomatic nutritional neurological damage including impaired sensation and tendon reflexes dating to the original malnutrition experience in captivity. Peripheral neuropathy, such as the "burning-foot syndrome," amblyopia and optic atrophy, and sensorineural deafness were the most common forms of persisting neuropathy. Most of these syndromes were similar in frequency and pattern to those experienced during imprisonment.[23] In addition, psychiatric syndromes were seen to be common. Overt and disabling depression and/or anxiety is seen in one third of the ex-FEPOWs.[24] Thus, long-range neurological and neuropsychological effects of severe malnutrition in these adults has been quite debilitating, regardless of subsequently adequate nutrition.

PARASITIC DISEASES, IMMUNITY, AND NUTRITION IN ADULTS

Many developing countries are in the tropics, where the high prevalence and severity of anemia has long been recognized. Adult Hb values obtained in the Gambia and in East Africa within the last 25 years have been consistently low, averaging 8.85 g/dℓ of blood for East African men, 7.86 g/dℓ for East African women,[25] and 10.38 g/dℓ for Gambian adults of both sexes.[26]

Anemia is a well-recognized feature of kwashiorkor (severe protein undernutrition). This type of anemia, which evidences lowered bone marrow erythroid activity, has been unresponsive to nutrients other than protein and is thought to result directly from impaired protein metabolism.[27]

Several causes of anemia are likely to be present in a given area, but each locale usually has one factor which predominates. Malaria or other infective agents, hookworm and Schistosoma infestations and their attendant blood losses, and nutrient deficiencies such as iron, various vitamins, and protein are common factors in the etiology of anemia.[28.] Dietary insufficiency of animal protein has been shown to result in a low absorption of iron from a high-iron-content vegetable diet. Thus, the type of dietary protein as well as the amount can affect the bioavailability of iron and the resulting incidence of anemia.[29]

Malaria is the most important parasitic disease worldwide, both in terms of economic issues and in terms of damage to human health. Anemia is only one of the adverse nutritional effects that malaria produces in the tropics. Because it can accentuate nutritional stresses, malaria also has a relationship to low birth weight infants, and in induction of protein-calorie malnutrition. Some researchers have reported higher death rates from malaria among malnourished children.[30] This may occur because PCM and the attendant reduction in immune function is related to increased susceptibility to malaria of increased severity. However, whether this relationship is significant or not remains currently open to question.[31] A relationship between immune defenses and nutritionally mediated immunosuppression has not been demonstrated.[32]

Starvation (energy deficiency) has actually reduced the severity of animal infections with malaria. Sharma[33] measured lipid metabolism during starvation and found a correlation between starvation and reduced malaria, although the mechanism is unclear. One possible factor, reduction of a key metabolite of parasites, para-amino benzoic acid, in the diet has long been known to reduce folic acid synthesis and malaria growth. This can result in animals in enhanced immunity to malaria later.[33] Supplementation with para-amino benzoic acid[32] greatly increases parasitemia. Alterations of iron levels in the host, as with some other parasite infections, can also result in greatly altered malaria growth rates.[34] For example, starving humans fed an increased supply of nutrients can experience increased attacks of malaria. This is due to the provision of increased free iron to the parasite[35] without sufficient recovery of transferrin levels. This is a risk associated with iron therapy in less developed countries where malaria is prevalent. Mild starvation has been recommended as a treatment for malaria.[36] However, the risk of infection with other types of pathogens associated with attending immunosuppression seems very undesirable.

In considering the effect that nutritional stresses in adults have upon resistance to other parasitic infections, it is necessary to determine what immune responses play a significant role in protection. Although much of the antiparasitic response is irrelevant to host defenses, some protective response does occur. Whether this protection is mediated by antibody or via dependent means involving killer cells, or both is often unclear.[32] The limited protection by the immune systems is often evidenced by lack of clearance of the parasite pathogen.

Intestinal protozoan infections are generally more severe in the malnourished host. A significant effect of malnutrition on such organisms is correlated with immunosuppression.[37] For example, experimental protein-energy deficient diets appear to result in more severe lesions in the gut wall after intracecal inoculation of *E. histolytica*.[37] A diet low in protein but containing an increased carbohydrate content leads to an increase in numbers infected but a decrease in the incidence of cecal wall damage.[32] Rats fed a diet deficient in both protein and vitamins showed very severe lesions after challenge with *E. histolytica*, while the routine laboratory diet resulted in a low infectivity rate.

E. histolytica has a high iron requirement for growth. In the gut lumen this is probably obtained from bacteria, while hemoglobin and ferritin are the most likely sources for invasive trophozoites. Zulu peoples living in South Africa have been seen to develop much more severe amoebic infections than neighboring white or Indian communities. The susceptibility of the Zulu was correlated to a diet which was mainly cereals combined with a high intake of inorganic iron.[38] It also appears that the requirement for iron is the reason why extraintestinal abscesses occur mainly in the liver. The host response to invasion of the gut is to induce a low iron state. Iron is transferred to the liver and thereby provides an environment in the liver in which the parasite subsequently can develop readily. Turkana nomads living in Kenya include populations whose major dietary constituent is milk.[39] They normally have a low infection rate with *E. histolytic*, but, when given iron orally to correct their iron deficiency, the infection rate increased dramatically.

There are few studies attempting to correlate nutritional state, parasitic disease, and functioning host defenses. These studies as well as those relating to bacterial diseases in both young and old have been reviewed elsewhere.[40] Basically, protein-energy malnutrition can affect the gut flora, and this may enhance invasion of the parasite. In addition, protein malnutrition makes the gut wall thinner and more susceptible to invasion.

PCM IN ADULTS IN DEVELOPING COUNTRIES

Some of the effects of PCM have been studied in the past 2 decades in chronically undernourished adults of developing countries. Some notable studies were made in Colombia,[41] Sri Lanka,[7] Uganda,[42] and Guatemala.[43] With minimal dietary deficiency, abnormalities included muscle wasting and loss of subcutaneous fat. Weakness and other clinical appearances of PCM were sometimes present. In adults with severe PCM, clinical features were more significant. Appearance sometimes showed marked muscle wasting, subcutaneous fat loss, edema, and weight loss. Skin, hair, and nail abnormalities were usually apparent; GI problems were common though causes were variable.[4]

The study in Sri Lanka involved 96 patients ranging from 17 to 62 years.[7] Protein intake was determined to be less than 30 g/day and caloric intake appeared to be low in the range of 1500 to 2000 kcal (75% carbohydrates). The majority of women dated their illness to their last pregnancy, demonstrating how increased nutritional needs and other stresses can precipitate PCM. The general appearance of the subjects was that of dull-looking adults, prematurely old, with sparse, unhealthy hair, dry skin, bloated moon face, and often edema of the ankles. They were mentally dull, apathetic, and depressed individuals. A common characteristic was hyperpigmentation of the skin which has been seen frequently in other studies. Patients were listless and complained of a lack of energy. Nutritional deficiencies of vitamins and minerals were often associated with PCM. This study showed all patients exhibited varying degrees of anemia. Biochemical indexes showed the presence of hypoalbuminemia, low serum cholesterol levels, decreased blood urea, flat glucose tolerance curves, liver biopsies showing varying degrees of fatty infiltration and fibrosis, and widespread disturbances of endocrine function.[7]

A 1956 study of malnutrition in autopsies of adults in Uganda showed that the muscles contained less protein than was reported for malnourished subjects in America. The researchers also showed that in 90% of the adult African subjects there was a negative potassium balance.[42]

In Cali, Colombia investigators found that all 41 adult patients with severe, well-defined primary protein malnutriton showed sporadic diarrhea and histological abnormalities of intestinal mucosa, though not all had detectable malabsorption.[41] Lactose intolerance is known to increase GI disorders and could become a very important detrimental factor in milk supplementation programs to people in developing countries. A recent study in Panama showed that intestinal parasite infestation (*Ascaris*) showed a positive correlation with lactose intolerance.[44] This might be due to actual physical abrasion of the intestinal villi which would cleave the tips, which are thought to produce the enzyme lactase.[45]

Other digestive secretions are reduced with PCM also. In the pancreas, apparent acinar cells have shown decreased numbers of zymogen granules.[4] A study of 14 adult Bengalese PCM patients studied in a Calcutta hospital were found to have significantly lowered gastric secretory activity. All of these patients showed achlorhydria or hypochlorhydria upon ametazole hydrochloride stimulation. Gastric biopsy revealed varying degrees of atrophy of gastric mucosa, distorted glands, and cellular infiltration of the lamina propria. The gastric secretory activity showed a fair correlation with gastric biopsy results and a high coliform count was found in 4 of 6 patients in whom the upper bowel bacterial count was measured.[46] Thus, decreased digestive secretions, altered absorptive surfaces, and increased bacterial loads may combine to alter absorptive capacity. GI problems of the protein-calorie malnourished may perpetuate nutritional inadequacy not only through malabsorption but also through less nutrient availability due to altered transit time of the food in the gut.

PCM IN WOMEN OF CHILD-BEARING AGE IN DEVELOPING COUNTRIES

PCM has an important adverse effect on the ability of a pregnant mother to reach successful full-term pregnancy. Maternal nutrition is one of the most important factors which affects both the successful completion of pregnancy and the growth of the fetus. The interaction between nutrition and other environmental and biological factors frequently interferes with proper interpretation of the results. These interaction factors include racial and genetic background, age, health, education, nutritional habits, and past nutritional status of the mother; parity, multiple pregnancies, and fertility pattern; socioeconomic conditions, especially as they relate to sanitation, infections, and the availability of health services, climate, and possibly the health and nutritional status of the father.

In the developing countries, the examples of risk indicators of maternal malnutrition include: family income, family size, sanitary conditions, availability of health care, pregnancy, weight, and height. Reports from various developing countries indicate that the values for energy requirement during pregnancy and lactation may be too liberal. Of course, adaptation by restriction of physical activity during pregnancy can reduce these requirements. Latham[47] stressed that nutritional needs of a woman during pregnancy are greater than at other times of her life because the tissues and organs of a new person are being built up in her body during that time. While in the developed countries, women during pregnancy live a relatively inactive life; pregnant women in Africa remain active even during their last few months of pregnancy. Thus, most women in Africa need comparatively more energy when they are pregnant. On the other hand, there is evidence from developed countries that higher requirements for protein may be necessary for optimal intrauterine growth. The whole issue of differential protein and energy requirements during pregnancy and lactation should be reevaluated, taking several factors into consideration. Research on optimal nutritional requirements is needed, especially in developing countries where the cost of intervention programs represents an economic burden.

Gabr[48] reviewed the various aspects of malnutrition during pregnancy and lactation of women in Egypt. In a study which was carried out in Cairo, a significant but slight difference was observed in birth weight, height, and skull circumference between infants of wellnourished and moderately malnourished mothers. It was concluded that marginal degrees of maternal malnutrition are more likely to be reflected in the breast-fed infant than in the fetus.[48]

Women in developing countries may need an additional 300 kcal/day while pregnant, and 500 kcal/day extra while they are breast feeding. As most women in the developing world do heavy manual work, a lactating woman may need a total of 3300 kcal or more each day. In addition, higher protein, vitamin, and mineral needs continue throughout a woman's lactation period. However, this extra nutrient intake pays great dividends in terms of her baby's nutritional adequacy. This is true not only because of actual nutrient content of human milk, but also because of immune factors transmitted through breast milk.

Concern over the decline of breast feeding has spread from industrial societies to developing countries where living conditions are chronically impoverished. Major health organizations as well as leading medical researchers promote breast milk as the optimum food of choice for the human infant.[49-52] A possible exception to this general observation may be an extraordinary circumstance such as prematurity. Even here there is evidence that milk from the mother may compensate in part for the premature birth and reduced development of the infant by increased protein and immunoglobulin levels.[53,54] Since breast milk is being so strongly promoted for developing countries, an examination needs to be made of dietary factors that may negatively or positively affect its host defense components and nutrients.

There are significant differences between protein and lactose content of breast milk of nonprivileged and privileged groups of Ethiopian mothers.[55] Hanafy et al.[55] found that the protein content of breast milk from well-nourished mothers was significantly higher than that of malnourished mothers. Khin et al.[56] studied lactation of Burmese mothers from low socioeconomic status and the effects of their nutritional state on the amount and proximate composition in their breast milk. Their results indicated that maternal undernutrition affects potential milk output while proximate composition is not affected. Malnourished humans often produce milk which is suboptimal in quantity, with some studies also showing deficiencies in protein, calcium, water soluble vitamins, and fat.[57] Dietary supplementation of these mothers appears to lead to increased milk output and quality, but not necessarily higher protein concentrations.[4] While the total protein content of milk produced by malnourished human mothers may not differ significantly from the variations seen among adequately nourished mothers, the composition of milk protein could vary significantly with respect to the concentrations of the various immune components. A recent review featured several studies which showed little change in secretory immunoglobulins in milk related to nutritional status.[57] Significant reductions in episodes of enteric infections and diarrheal diseases occur among infant breast- rather than bottle-fed is underdeveloped areas where living conditions are poor.[58] This has been attributed to the special protection afforded by lessened exposure to pathogens at the breast and to the specific SIgA antibodies to antigens in the infant's environment present in his mother's milk.[58]

Human studies comparing levels of immunoglobulins in the milk of mothers from well-nourished and undernourished populations indicate that the mammary gland is capable of synthesizing and/or secreting adequate levels of antibodies under a wide range of nutritional stresses,[57] but no examination has been made of the secretory immune response and vigor of the mammary gland's SIgA production under experimental antigenic challenge when undernutrition is present in the mother.[57]

Malnutrition affects other aspects of reproductive function also. In rural Guatemala with food supplementation in a chronically malnourished population there was a significant negative association between the nutritional status of the mother during the third trimester of pregnancy and the length of amenorrhea postpartum.[59] This phenomenon may represent a positive adaptation to malnutrition, resulting in more widely spaced births and a longer period of lactation for each infant.

A report[60] on pregnant women in Addis Ababa, Ethiopia who ate less than 60% of the energy and 70% of the protein recommended for pregnancy in the U.S. had the surprising finding that infants on the average were only 7% lighter than U.S. newborn infants. They showed that undernourished U.S. infants usually had little or no subcutaneous fat and had disproportionately small livers, spleens, adrenals and thymuses. The subcutaneous fat was more plentiful and organs more uniformly undergrown in undernourishd Ethiopian infants, who had a relatively undergrown cerebellum and significant delay in descent of testes. The investigators suggested that undernutrition had no effect on the weight of other parts of the brain. Also in Ethiopia they found that income was not related to signs of undernutrition, since the Ethiopian mothers with higher incomes had 40% and the lower incomes 8% of their protein from animal sources but the total energy intakes were similar in the two groups.[60]

Tafari et al.[61] demonstrated that term infants of mothers in Ethiopia who did heavy physical labor during pregnancy and had an energy intake below 70% of the WHO/FAO recommendation, had a mean birth weight of 3068 ± 355 g compared with 3270 ± 368 g for children born to less physically active mothers on similar diets. The weight of the mothers did have a significant influence on birth weight when the mothers were on low energy intake.[61]

Malnourished Indian women who had been treated with oral-type contraceptives containing 0.03 or 0.05 mg ethinyl estradiol and 0.15 mg D-norgesteral had significantly higher serum retinol-binding protein and vitamin A values after 2 or 5 cycles of treatment than had untreated control subjects.[62]

In Bangladesh, both the cultural emphasis that women must be the last to partake of the food prepared in the house and the custom that the children share the same food as regularly prepared for the adults are considered to contribute substantially to the overall picture of inadequate protein intake of women and children in the country. In view of survey findings it is suggested that education is needed, as an increased supply of protein food alone would not solve the protein problem of Bangladesh. The data showed that energy deficiency was found in 13.8% of households. The incidence of protein deficiency rose from 12.3 to 26.1% when energy deficiency was taken into consideration. Protein deficiency as such was present only in 0.8% of the households compared with 11.5% of the households which were deficient in both protein and energy. The magnitude of deficiency of protein as well as of energy which resulted from maldistribution of food within the household has not yet been quantitated.[63]

To have a lasting effect, the nutrition education that accompanies nutritional rehabilitation must be couched in local terms and new concepts must be presented in a manner in keeping with tradition. Acceptance by mothers of new methods of child feeding is crucial to long-term success. A recent article about a hospital-affiliated nutrition rehabilitation unit in South Africa describes how this is being done there. Mothers live with their malnourished children in mud huts in a little community near the hospital during the rehabilitation period. They work in the unit's gardens, cook the food communally, and are taught a traditional type of work chant centering on three types of food: body-builders, body-protectors, and energy-givers. Locally available and acceptable foods are delegated to these three groups. The emphasis on good child nutrition via use of the traditionally acceptable foods makes it more highly likely that the mothers will not only understand the new nutrition concepts, but will also continue to feed the improved dietary balance to their children on returning home.[64]

Endocrine disturbances in men and women are common in malnutrition. Women show disturbed ovarian function and atrophy of the breasts, uterus, and lower genital tract. Sterility is not present except in acute phases of the illness. Acute malnutrition reduces the number of conceptions and live births. As many as 60 to 60% of young women in a famine region may have amenorrhea, absence of or abnormal stoppage of menstruation.[65]

Fertility and physique in a malnourished population in Columbia was evaluated in 403 families.[66] About 30% of the households in a subsidiary farming community contained members who were moderately malnourished. The number of living children at the time was taken as the measure of reproductive performance of each person. There was a positive association of fertility with amount of soft tissue. Paralleling the association, there was a tendency for the wealthier to have larger families and heavier body build, and the fat fertility association to decrease when both the husband's and wive's fat components were taken together. Mueller[66] suggested that there was a curvilinear relation between fertility and stature such that subjects with very small or very large bones had reduced fertility.

Women and their children in developing countries which have great family instability may suffer even more disproportionately than those with stable family infrastructures where nutrition is concerned. In the Kingston/St. Andrew metropolitan area of Jamaica, 95% of the children admitted to two hospitals with PCM had unmarried mothers. Of these, 48% of the mothers could count on little or no financial support from the fathers, and in 68% of cases, multiple partners had fathered the woman's children. The mothers who were employed had very low wage jobs, and 57% of the mothers were unemployed. Decrease in their standard of living directly correlated with their childrens' decreased level of nutrition. The trends toward early weaning and bottle feeding, particularly among teenage mothers, hastened and exacerbated the malnutrition of their children.[67]

In view of the nutritional, biochemical, anti-infectious, contraceptive, economic, convenience, maternal, and infant health benefits, human milk should be given a top resource priority in national development planning and in international planning for health, food production, and fertility control.[68] Finally, in support of this, Berg[69] stated: "The economic and social implications of an unusual depletion in the supply of crude oil reserves of a non-oil producing country like Asia or Latin America is very obvious. A comparable crisis, involving a valuable natural resource and losses in the hundreds of millions of dollars is going virtually unnoticed in many of the poor countries of the world. The resource is human breast milk and the loss is caused by the dramatic and steady decline of maternal nursing in recent decades!"

ADAPTATION TO MALNUTRITION

In the developing countries, there are large numbers of adults subsisting on energy intakes far below the recommended dietary allowance (RDA) and yet leading apparently healthy lives.[70] This phenomenon prompts some important questioning. Are the RDAs too high? Do lower levels of energy intake necessarily mean poorer human performance? There does appear to be some adaptation which allows productive and physical activity at levels of energy intake much less than the RDA.

In a recent study, a high-energy-intake group of adult male Javanese farmers expended significantly more energy in performing standard work tasks than a low-energy-intake group.[70] There may be long-term genetic adaptations and short-term phenotypic adaptations such as a decrease in the basal metabolic rate which facilitates a higher level of metabolic efficiency for energy-stressed subjects. There are also differences in the rate of energy expenditure of resting activities between healthy European men and malnourished African men. European men increased their energy output significantly more than African men when changing from a lying to a standing position. This may explain why some body movements during work (hoeing or harvesting for instance) or leisure (certain dances) are more exhausting for Europeans than for Africans. On the other hand, there is an optimal height and weight shown for maximum work performance by farm as well as factory workers in India. Individuals with a small body mass as well as obese subjects have been shown to be at a disadvantage.[71]

On the basis of the few reports, one can speculate there may be adaptation in some adults which enables individuals with low energy and specific nutrient intakes to maintain adequate body size and work output. One research goal worth pursuing would be to determine the prevalence of such a phenomenon. Adaptation is defined as an adjustment of the organism that enables it to maintain normal structure and function under different environmental conditions. Demonstrated efficiency of the body's mechanisms of adaptation and economy could reinforce confidence in the "safe levels" of protein intake proposed by FAO and

WHO. These levels are currently considered by many as being too low. Adaptation control mechanisms of the body during shortage of protein are two-pronged; one involves protein metabolism and the other involves changing hormonal levels. Some women in developing countries may have the capacity to protect themselves and their babies during pregnancy and lactation when food is in short supply. They can produce almost as much milk as women in developed countries although they eat considerably less food.

For example, Guatemalan women responded to additional energy stress of lactation primarily by reducing their body fat stores rather than by either significantly increasing their energy intake (even when food supplement was available at no cost) or decreasing their energy level of expenditure.[72]

REPLETION OR RENUTRITION OF PCM ADULTS

A study in rural Colombia was done to investigate changes in adult body fat content and in body cell mass that occurred during nutritional rehabilitation. Investigators found that body fat content increased markedly with the provision of a diet with adequate protein content. However, the accumulation of fat in the body did not lead to anthropometric differences in thickness of the scapular or tricep skinfolds. During this period some decrease in cell hydration occurred. In contrast, body cell mass did not increase unless the diet had a higher protein content. Increases in erythrocyte mass and total body hemoglobin lagged behind the increases in body and muscle cell masses. The investigators summarized their work by stating that body fat (energy) depots could be replenished rapidly if a positive energy balance occurs in severely undernourished subjects, even when the protein content of the diet is maintained at a low 26 g/day.[73] There was an excellent response to a high protein diet by Colombian adults without therapeutic doses of folate, B_{12}, or antibiotics.[41] In contrast to children who sometimes are retarded developmentally, in these adults there seemed to be no long-term effects following rehabilitation.

There are nutritional stresses which can be increased during renutrition. For example, in patients who have a vitamin A deficiency along with PCM with a high protein diet given too rapidly the body can develop a more severe vitamin A deficiency state. This occurs when retinol-binding protein synthesis increases in the liver with an increase in binding of available body stores of vitamin.

Oxygen consumption in maximum exercise (VO_2 max) increased during nutritional rehabilitation of severely undernourished sedentary Colombian men only when dietary protein was increased from 27 to 100 g daily.[74] Adequate energy intake maintained for 45 days before the increase in protein content of the diet had no effect on VO_2 max. The VO_2 max values were best correlated with the muscle cell mass at the different stage of repletion. The VO_2 max/muscle cell mass ratio was correlated only with blood and plasma volumes at the different stages of repletion. After 2.5 months of protein repletion, the VO_2 max values were still lower than those in mildly undernourished or normal Colombian men. They found that the maximum endurance time at 80% aerobic load decreased significantly.[74]

Balanced growth, especially in the agricultural sector, has a direct relation with the pattern of malnutrition in the population. The economic implication of adult malnutrition in developing countries appears significant in that it leads to inferior labor output, lower productivity, and consequently lower economic growth and development of the countries.

CHANGES IN MALNOURISHED ADULTS IN LDC ON A CELLULAR OR MOLECULAR LEVEL

Many biochemical abnormalities associated with PCM were examined in a recent review.[4] Depending upon the severity of adult PCM, serum proteins can be substantially reduced.

These include serum albumin and the higher molecular weight transport proteins such as transferrin, ceruloplasmin, lipoproteins, thyroxin, and cortisol-binding proteins. Retinol-binding protein is also reduced, which might precipitate vitamin deficiency symptoms. Serum amino acids (i.e., leucine, isoleucine, valine, and methionine) are either normal or at depressed levels including nonessential amino acids (i.e., glycine, serine, glutamine). Urea, creatinine, and hydroxyproline apparent in urinary excretion may decrease. Many times, critical abnormalities in electrolytes develop although serum levels may be normal.[4]

Intestinal biopsies from Egyptian patients with severe protein deficiency showed that there were multiple large areas devoid of ribosome granules in the undifferentiated cells.[75] The nuclei of cell organelles were compressed. The changes were accompanied by atrophy of the absorptive cells with reduction in the numbers and variation in the length and density of their microvilli. They showed also that there was morphological evidence of delayed transport of fat in the epithelial cells.

A number of studies on PCM correlates have been done in Zaire. A study of children and adults of European and African descent failed to show an effect of PCM on transcobalamin-II levels in malnourished patients compared to those of healthy controls.[76] However, PCM has been shown to affect the absorption and utilization of certain amino acids and their specific enzyme levels. Young Zairian mothers suffering from PCM were given loads of phenylalanine and tyrosine and the results were compared with those seen in healthy controls. Phenylalanine hydroxylase deficiency was seen as causing the higher blood levels of phenylalanine with a phenylalanine load, as well as the lower blood tyrosine formed from phenylalanine. Simultaneous disturbances in tyrosine catabolism were seen as causing higher urinary excretion of secondary phenylalanine and tyrosine metabolites. Blood tyrosine levels were low, even with tyrosine loads, possibly as a result of tyrosine malabsorption and increased tissue utilization. Results also indicated deficiencies in the enzymes tyrosine-transaminase and *p*-hydroxyphenylalanine-oxidase.[77]

Studies in surgical[78] and hemodialysis[79] patients in developed countries have also shown a high correlation between lowered plasma valine levels and severity of PCM. Many other PCM-indicating variables also correlated well with plasma valine. On this basis, plasma valine levels have been suggested for use as an indicator of PCM.[70]

The findings of at least one study in a developing country tends to support the validity of this suggestion. Valine decreases in maternal and cord vein plasma have been observed in subclinically malnourished, underprivileged women and their newborn infants in Ethiopia. Moreover, an increased glycine/valine ratio was seen in the maternal and cord vein plasma and a reduced cord/maternal ratio of tyrosine and methionine was found. The latter was thought to indicate idiopathic placental dysfunction. During the immediate neonatal period a delayed decrease of branch-chain amino acids took place. *In utero* PCM malnutrition was deemed evidenced by all these amino acid level changes.[80]

Normally, vitamin E is not required as an erythropoietic factor for humans, but various studies have shown a reticulocytosis response and a limited hemoglobin concentration increase in PCM patients supplemented with vitamin E.[81,82] These effects have occurred prior to reversal of the patients' metabolic derangement and are the basis for a claim that vitamin E should be seen as a potential erythropoietic factor for humans. The role of PCM in this case would be to unmask the essential role of vitamin E by interfering with the metabolic processes which normally by-pass it.[83]

Since developing countries have a high incidence of PCM and since PCM affects every cell and tissue, it is encumbent to be aware of interactions of malnutrition and cancer.[84] Some effects of PCM have been found to interact negatively and others positively with the carcinogenic process. For example, depressed cellular proliferation by a prolonged DNA synthesis cycle occurs with PCM. However, cell-mediated immunity is depressed with defective mobilization of macrophages. This removes a barrier to cancer cell proliferation.

Because of the conflicting effects of PCM, it is not possible to say at the present time that PCM has any dominant modulating effect on carcinogenesis.[84]

There has long been concern expressed about life-long detrimental effects of severe infantile PCM on brain function of adults. Recent CT brain scans of adults in their early 20s in Cape Town, South Africa, showed no significant differences between a control group and a formerly severely undernourished group. However, a highly critical evaluation showed 5 out of the 20 previously malnourished adults to have minimal changes which correlated positively with tests for visuomotor organization disturbances.[85]

The secretive function of endocrine glands can also be altered in adult PCM. Brazilian men and women with PCM showed a low peripheral T_3 concentration. This was thought to be due to deficient iodination of T_4, since the thyroids were able to release substantial T_3 after endogenous TSH stimulation. Indeed, the basal levels of serum TSH were significantly higher than normal in the PCM patients and all but one of them showed a sustained, high response of TSH secretion to TRH administration. Similarly, fasting PRL (prolactin) levels were very low in PCM patients, but the peak PRL response to TRH was not significantly different from that of normal controls.[86] The exaggerated TSH response may be the adaptive mechanism which allows adult PCM patients to maintain normal serum T_4 levels.

PREVENTION OF PCM IN ADULTS

Prevention of PCM from the public health standpoint is critical. Possible partial prevention techniques include increased crop production, new types of food storage, improved sanitation, and direct targeted food aid. In countries with centralized food preparation or manufacturing, one of the most feasible methods of improving nutrition on a broad population basis is the use of food fortification programs. For those groups whose staple food is a cereal crop, fortification with amino acid or protein supplements is often relatively economical and technically feasible. It has the advantage of improving the protein adequacy of both adults and children and not just hitting a small target group.[6] Since cereal crops are generally much higher in protein content than tuber crops (e.g., cassava), efforts might be made to introduce more nutritious crops to those areas lacking them. Development and use of strains able to thrive in difficult soils and climates help make this possible.

A study in Nigeria during the famine of the mid-1970s showed the benefits of supplementation through leaf proteins.[87] In the famine region, supplementing home diets with approximately 10 g of leaf protein a day resulted in significant clinical improvement. Within 10 days edema disappeared (in kwashiorkor cases), appetite improved, and children became more mentally alert. There was a progressive increase in total serum proteins but most of the increase was due to increased globulin synthesis. Leaf protein appears to offer much promise for the large scale prevention and treatment of protein malnutrition in tropical areas such as Africa.[87]

The problems of eliminating PCM in the developing world are manifold. In Guatemala, studies showed a relationship between the amount of farmable land a family has access to and their nutritional status. Relative risk of having moderate malnutrition was 2.3 times greater in families with access to less than 1.4 ha than in those with access to more than 3.5 ha.[41] Thus, the social framework of a society plays an important role in the differential prevalence of malnutrition. In areas where hunger and starvation are rife, social unrest is common.[88] This can set up a vicious cycle, interrupting agriculture, transportation, and other factors essential to the people's adequate nutrition.

In a Sudanese farming community, the average protein intake was adequate, but the average per capita per day energy intake was found to be deficient.[89] It appears that food intake of the vulnerable groups was not related to food availability *per se* but was determined by inequitable intrafamily distribution. The investigators postulated that this was the main factor in the ecology of malnutrition which was more prevalent among the children in that area. Thus, long-standing familial customs also play a role in malnutrition patterns. The investigators recommended increase in the food crops production, together with an intensive nutrition education program to combat malnutrition on both macro- and microsocietal scales.[89]

A universal Declaration of Human Rights was adopted by the United Nations 25 years ago. It states: ''Everyone has the right to a standard of living adequate for the health and well-being of himself and his family''. High population growth rates continue to add to the problem even though some are now refuting this notion. As urban centers in the developing world increase in population (in Asia and Africa the population living in cities almost doubled in a single decade, 1950 to 1960, and has since continued to increase), nutritionally related disorders increase. Many urban factors, including increased infections and infestations, and cultural confusion, frequently are associated with an increased incidence and severity of malnutrition.[6]

The world has changed from the time in which cycles of starvation were judged both inevitable and unsolvable. With the systems of mass food production, communication, and transportation now possible, this attitude is no longer defensible. Developing countries are striving to achieve the modernization of their infrastructures of communication, transportation, and production, and the provision of increased health care, including nutrition, for their peoples. We now have the knowledge to install early warning systems on a worldwide basis based on economic and medical surveillance. Such diverse data as weather reports, crop forecasts, food reserve tallies and retail price trends, height-to-weight ratios, and skin fold thickness measurements can warn us in advance of nutritionally stressed areas and impending famine spots.[80] It is hoped that all the nations of the world will continue to tackle the endemic problem of chronic protein-calorie malnutrition in developing countries in a united effort. As shown in this paper, though the overall adult population of developing countries is not the most nutritionally stressed group, there are adult subgroups who do live with chronic moderate malnutrition. Since these groups contain the breadwinners and mothers of families of developing countries, more attention should be given to ameliorating their sufferings.

ACKNOWLEDGMENTS

Preparation of some of the literature review was done by Bruce Hamaker. Review supported in part by a National Livestock and Meat Board grant.

REFERENCES

1. **Watson, R. R. and McMurray, D. N.,** The effects of malnutrition on the secretory and cellular immune processs, *CRC Crit. Rev. Foods Nutr.,* 113, 1979.
2. **Jaya Rao, K. S.,** Protein-calorie malnutrition, *Indian J. Med. Res.,* 23(68), 17, 1978.
3. Overcoming World Hunger: The Challenge Ahead, Report of the Presidential Commission on World Hunger, 1980, p. 4.
4. **Freeman, H. J., Kim, Y. S., and Sleisenger, M. H.,** Protein digestion and absorption in man, *Am. J. Med.,* 67, 1030, 1979.
5. **McCance, R. A. and Widdowson, E. M.,** *Calorie Deficiencies and Protein Deficiencies,* Little, Brown & Co., Boston, 1968.
6. **Ershoff, B. H.,** Food fortification in the prevention and treatment of malnutrition in the urban areas of developing countries, *J. Appl. Nutr.,* 40, 114, 1978.
7. **Obeyesekere, I.,** Malnutrition among Ceylonese adults, *Am. J. Clin. Nutr.,* 6, 38, 1966.
8. **Bhattacharyya, A. K.,** Common deficiency diseases. I, *J. Indian Med. Assoc.,* 70, 57, 1978.

9. **Mayer, J.,** Management of famine relief, *Science,* 188, 571, 1975.
10. **Watson, R. R. and Safranski, D.,** Dietary restrictions and immune responses in the aged, *CRC Handbook of Immunology in Aging,* CRC Press, Boca Raton, 1981, 125.
11. **Gambert, S. R.,** Nutritional assessment in the elderly, *Wisc. Med. J.,* 81, 18, 1982.
12. **Gambert, S. R. and Guansing, A. R.,** Protein-calorie malnutrition in the elderly, *J. Am. Gerontol. Soc.,* 26(6), 272, 1980.
13. **Pifer, J. M.,** Nutritional aspects of aging, *Primary Care,* 9, 223, 1982.
14. **Mitchell, C. O. and Lipschitz, D. A.,** Detection of protein-calorie malnutrition in the elderly, *Am. J. Clin. Nutr.,* 35, 398, 1982.
15. **Driver, A. G., McAlevy, M. T., and Smith, J. L.,** Nutritional assessment of patients with chronic obstructive pulmonary disease and acute respiratory failure, *Chest,* 82, 568, 1982.
16. **Bambach, C. P. and Hill, G. L.,** Long-term nutritional effects of extensive resection of the small intestine, *Aust. N. Z. J. Surg.,* 52, 500, 1982.
17. **Stein, Z., Susser, M., Saenger, G., and Marolla, F.,** *Famine and Human Development,* Oxford University Press, N.Y., 1975.
18. **Winick, M.,** *Hunger and Disease: Studies by the Jewish Physicians in the Warsaw Ghetto,* John Wiley & Sons, N.Y., 1979.
19. **Brozek, J., Wells, S., and Keys, A.,** Medical aspects of semistarvation in Leningrad (seige 1941-1942), *Am. Rev. Sov. Med.,* 4, 7, 1946.
20. **Helwig-Larsen, P., Hoffmeyer, H., Kieler, J., et al.,** Famine disease in concentration camps. Complications and sequels, *Acta Med. Scand. Suppl.,* 274, 1, 1952.
21. **Barbosa-Saldivar, J. L. and Van-Itallie, T. B.,** Semistarvation: an overview of an old problem, *Bull. N.Y. Acad. Med.,* 55, 774, 1979.
22. **Gibbard, F. B. and Simmonds, J. P.,** Neurological disease in ex-Far-East prisoners of war, *Lancet,* 2(8186), 135, 1980.
23. **Gill, G. V. and Bell, D. R.,** Persisting nutritional neuropathy amongst former war prisoners, *J. Neurol. Neurosurg. Psychiat.,* 45, 861, 1982.
24. **Gill, G. V. and Bell, D. R.,** The health of former prisoners of war of the Japanese, *Practitioner,* 225, 531, 1981.
25. **Rowland, H. A. K.,** Anaemia in Dar-es-Salaam and methods for its investigation, *Trans. R. Soc. Trop. Med. Hyg.,* 60, 143, 1966.
26. **Woodruff, A. W. and Schofiedh, F. D.,** Hemoglobin values among Gambians, *Trans. R. Soc. Trop. Med. Hyg.,* 51, 217, 1957.
27. **Allen, D. M. and Dean, R. F. A.,** The anaemia of kwashiorkor in Uganda, *Trans. R. Soc. Trop. Med. Hyg.,* 59, 326, 1965.
28. **Woodruff, A. W.,** Recent work concerning anemia in the tropics, *Semin. Hematol.,* 19, 141, 1982.
29. **Martinez-Torres, C. and Layrisse, M.,** Effect of amino acids on iron absorption from a staple vegetable food, *Blood,* 35, 669, 1970.
30. **Gregor, I. A., Rahman, A. K., Thomson, A. M., Billewicz, W. Z., and Thompson, B.,** The health of young children in a West African (Gambian) village, *Trans. R. Soc. Trop. Med. Hyg.,* 64, 48, 1970.
31. **McGregor, I. A.,** Malaria: nutritional implications, *Rev. Infect. Dis.,* 4, 798, 1982.
32. **Langhorne, J. and Choen, S.,** Specific mechanisms of immunity to parasitic protozoa, in *The Impact of Malnutrition on Immune Defense in Parasitic Infection,* Isliker, H. and Schurch, B., Eds., 1981, 84.
33. **Sharma, O. P., Shukla, R. P., Singh, C., and Sen, A. B.,** Suppression of malaria by starvation: a biochemical study, *Indian J. Med. Res.,* 69, 251, 1979.
34. Jerusalem. Active immunization against malaria (Plasmodium berghei). I. Definition of anti malarial immunity, *A. Tropenmed. Parasit.,* 19, 171, 1968.
35. **Murray, M. J., Murray, A. B., Murray, N. J., and Murray, M. B.,** Refeeding malaria and hyperferraemia, *Lancet,* 1, 653, 1975.
36. **Weinberg, E. D.,** Iron and infection, *Microbiol. Rev.,* 42, 45, 1978.
37. **Targett, G. A. T.,** Malnutrition and immunity to protozoan parasites, in *The Impact of Malnutrition on Immune Defense in Parasitic Infection,* Isliker, H. and Schurch, B., Eds., 1981, 158.
38. **Diamond, L. S., Harlow, D. R., Phillips, B. P., and Keister, D. B.,** Entamoeba histolytica: iron and nutritional immunity, *Arch. Invest. Med.* (Mex.), 9, 329, 1978.
39. **Murray, M. J., Murray, A., and Murray, C. J.,** The salutary effect of milk on amoebiasis and its reversal by iron, *Br. Med. J.,* 1, 1351, 1980.
40. **Watson, R. R. and Petro, T. M.,** Resistance to bacterial and parasitic infections in the nutritionally compromised host, *CRC Crit. Rev. Microbiol.,* in press.
41. **Mayoral, L. G., Bolanos, O., Lotero, H., and Duque, E.,** Enteropathy in adult protein malnutrition: a review of the Cali experience, *J. Clin. Nutr.,* 28, 894, 1975.
42. **Holmes, E. G., Jones, E. R., Lyle, M. D., and Stainer, M. W.,** Malnutrition in African adults, *Br. J. Nutr.,* 10, 198, 1956.

43. **Valverde, V. et al.,** Relationship between family land availability and nutritional status, *Ecol. Food Nutr.,* 6, 1, 1977.

44. **Carrera, E., Crompton, D., and Nesheim, M.,** Lactose tolerance in *Ascaris* infected preschool children, *Fed. Proc.,* 41, (Abstr.), 751, 1982.

45. **Bayless, T. M., Rothfeld, B., Massa, C., et al.,** Lactose and milk intolerance: clinical implications, *N. Engl. J. Med.,* 292, 1156, 1975.

46. **Guha Mazumdar, D. N., Mitra, R. C., Mitra, N., Sen, N. N., and Chatterjee, B. D.,** Gastric secretory study in protein calorie malnutrition (PCM) in adults and correlation with gastric mucosal structure, small bowel microflora and absorption, *J. Indian Med. Assoc.,* 71, 25, 1978.

47. **Latham, M. C.,** Human nutrition in tropical Africa, *FAO Food Nutr. Stud.,* 1979.

48. **Gabr, M.,** Malnutrition during pregnancy and lactation, *World Rev. Nutr. Diet.,* 36, 90, 1981.

49. **A.A.P. Committee on Nutrition,** Commentary on breastfeeding and infant formulas, including proposed standards for formulas, *Pediatrics,* 57, 278, 1976.

50. **WHO,** Resolution of the 34th World Health Assembly: international code of marketing of breastmilk substitutes, *Nutr. Today,* 16, 13, 1981.

51. **Jelliffe, D. B. and Jelliffe, E. F. P.,** The volume and composition of human milk in poorly nourished communities. A review, *Am. J. Clin. Nutr.,* 31, 492, 1978.

52. **Fomon, S. J., Ziegler, E. E., and Zavquez, H. D.,** Human milk and the small premature infant, *Am. J. Dis. Child.,* 131, 463, 1977.

53. **Gross, S. J., David, R. J., Bauman, L., and Tomarelli, R. M.,** Nutritional composition of milk produced by mothers delivering preterm, *J. Pediatr.,* 96, 641, 1980.

54. **Gross, S. J., Buckley, R. H., Wakil, S. S., McAllister, D. C., David, R. J., and Faix, R. G.,** Elevated IgA concentration in milk produced by mothers delivered of preterm infants, *J. Pediatr.,* 99, 389, 1981.

55. **Hanafy, M., Morsey, M., Seddick, Y., Habile, Y., and Lozy, M.,** Maternal nutrition and lactation performance, *J. Trop. Pediat., Environ. Child Health,* 18, 187, 1972.

56. **Khin, M. N., Tin, O. T., Thein, K., and Hlaing, N. N.,** Study on lactation performance of Burmese mothers, *Am. J. Clin. Nutr.,* 33, 2665, 1980.

57. **Listman, J. A. and Watson, R. R.,** Effect of nutritional stresses on host defenses components of milk, in *Malnutrition, Disease and Immunity,* Marcel Dekker, N.Y., 1983, in press.

58. **Chandra, R. K.,** Prospective studies of the effect of breastfeeding on incidence of infection and allergy, *Acta Paediatr. Scand.,* 68, 691, 1979.

59. **Delgado, H. et al.,** Effect of maternal nutritional status and infant supplementation during lactation on postpartum amenorrhea, *Am. J. Obstet. Gynecol.,* 135(3), 303, 1979.

60. **Naeye, R. and Tafari, N.,** Effects of long-term maternal undernutrition on the human fetus, *Pediatric Res.,* 13 (Abstract), 392, 1979.

61. **Tafari, N., Naeye, R. L., and Gobergie, A.,** Effects of maternal undernutrition and heavy physical work during pregnancy on birth weight, *Br. J. Obstet Gynecol.,* 87, 222, 1980.

62. **Ram, M. M. and Bangaji, M. S.,** Serum vitamin A and retinol-binding protein in malnourished women, treated with oral contraceptives: effects of estrogen dose and duration, *Am. J. Obstet. Gynecol.,* 135, 470, 1979.

63. **Hussain, M. A. and Ahmad, K.,** Protein problems in Bangladesh, *Ecol. Food Nutr.,* 6, 31, 1977.

64. **Hamer-Hodges, M.,** Kwazup: nutritional rehabilitation in Transkei, *Br. Med. J.,* 281, 1708, 1980.

65. **Parham, E. S.,** Starvation, *J. Home Econ.,* 67, 7, 1975.

66. **Mueller, W. H.,** Fertility and physique in malnourished population, *Hum. Biol.,* 51, 153, 1979.

67. **Bailey, W.,** Clinical undernutrition in the Kingston/St. Andrew metropolitan area: 1967-1976, *Soc. Sci. Med.,* 15D, 471, 1981.

68. **Brozek, J.,** Energy-protein malnutrition and behavior, *Nutr. Rev.,* 38, 164, 1980.

69. **Bengoa, J. M.,** Prevention of protein-calorie malnutrition, in *Protein-Calorie Malnutrition,* Olson, R. E., Ed., Academic Press, N.Y., 1975.

70. **Edmundson, W.,** Individual variations in basal metabolic rate of mechanical work efficiency in East Java, *Ecol. Food Nutr.,* 8, 189, 1979.

71. **Satyanarayana, K., Naidu, A. N., and Narasinga Rao, B. S.,** Work output in undernourished adolescents: effect of early malnutrition, *Nutr. Rev.,* 38, 143, 1980.

72. **Shultz, Y., III, Lechtig, A., and Bradfield, R.,** Energy expenditures and food intakes of lactating women in Guatemala, *Am. J. Clin. Nutr.,* 33, 892, 1980.

73. **Barac-Nieto, M., Spurr, G. B., Lotero, H., Maksud, L., and Dahners, H. W.,** Body composition during nutritional repletion of severely undernourished men, *Am. J. Clin. Nutr.,* 28, 894, 1979.

74. **Barac-Nieto, M. et al.,** Aerobic work capacity and endurance during nutritional repletion of severely undernourished men, *Am. J. Clin. Nutr.,* 33, 2268, 1980.

75. **Nassar, A. M. et al.,** Ultrastructural changes in the mucosa of the small intestine due to protein calorie malnutrition, *J. Trop. Ped.,* 26, 62, 1980.

76. **Mozes, N.,** Transcobalamin II in protein-energy malnutrition among residents of the Kivu area, *Clin. Chim. Acta,* 124, 157, 1982.

77. **Antener, I., Verwilghen, A. M., van Geert, C., and Mauron, J.,** Biochemical study of malnutrition. V. Metabolism of phenylalanine and tyrosine, *Int. J. Vit. Nutr. Res.,* 51, 297, 1981.

78. **Young, G. A. and Hill, G. L.,** Evaluation of protein-energy malnutrition in surgical patients from plasma valine and other amino acids, proteins and anthropometric measurements, *Am. J. Clin. Nutr.,* 34, 166, 1981.

79. **Young, G. A., Swanepoel, C. R., Croft, M. R., Hobson, S. M., and Parsons, F. M.,** Anthropometry and plasma valine, amino acids and proteins in the nutritional assessment of hemodialysis patients, *Kidney Int.,* 21, 492, 1982.

80. **Gebre-Medhin, M., Larsson, U., Lindblad, B. S., and Zetterstrom, R.,** Subclinical protein-energy malnutrition in under-privileged Ethiopian mothers and their newborn infants, *Acta. Paediatr. Scand.,* 67, 213, 1978.

81. **Whitaker, J., Fort, E. G., Vimokesant, S., and Dinning, J. S.,** Hematologic response to Vitamin E in the anemia associated with protein-calorie malnutrition, *Am. J. Clin. Nutr.,* 20, 783, 1967.

82. **Kulapongs, P.,** The effect of Vitamin E on the anemia of protein-calorie malnutrition in northern Thai children, in *Protein-Calorie Malnutrition,* Olson, R. E., Ed., Academic Press, N. Y., 1975, 263.

83. **Drake, J. R. and Fitch, C. D.,** Status of Vitamin E as an erythropoietic factor, *Am. J. Clin. Nutr.,* 33, 2386, 1980.

84. **Deo, M. G.,** Implications of malnutrition in chemical carcinogenesis, *J. Cancer Res. Clin. Oncol.,* 99, 77, 1981.

85. **Handler, L. C., Stoch, M. B., and Smythe, P. M.,** CT brain scans: part of a 20-year development study following gross undernutrition during infancy, *Br. J. Radiol.,* 54, 953, 1981.

86. **Medeiros-Neto, G. A., Sucupria, M., Knobel, M., and Cintra, A. U.,** Prolactin, TSH and thyroid hormones, responses to TRH in adult protein calorie malnutrition, *Horm. Metab. Res.,* 9, 525, 1977.

87. **Olatunbosun, D. A.,** Leaf protein for human use in Africa, *Indian J. Nutr. Diet.,* 13, 168, 1976.

88. **Goldsmith, G. A.,** Food and population, *Am. J. Clin. Nutr.,* 28, 934, 1975.

89. **Lechtig, A., Klein, C., Daza, C., Read, M., and Kahn, S.,** Maternal malnutrition: a serious obstacle to 3rd world development, *Nutr. News,* 44, 13, 1981.

Mineral and Vitamin Nutrition in the Adult

ZINC AND COPPER NUTRITION IN AGING

Robert A. Jacob, Robert M. Russell, and Harold H. Sandstead

INTRODUCTION

Since human zinc (Zn) deficiency was first described in the early 1960s, research has elucidated many of its specific metabolic interactions. Moreover, the discovery of a variety of Zn-related clinical disorders, particularly in adults, has directly demonstrated the importance of Zn in human nutrition.

Zn is widely distributed throughout many foods. In the human diets, red and visceral meats, dark poultry, and shellfish provide most dietary requirements; oysters are particularly rich in Zn content. The recommended dietary allowance (RDA) for Zn in adults is 15 mg/day.[1] Whether or not Zn deficiency is a problem among the elderly is uncertain.[2] The factors thought to affect Zn metabolism and absorption and the evidence for Zn deficiency among the elderly will be discussed in this chapter.

Copper (Cu) deficiency manifested by leukopenia, anemia, and osteopenia rarely occurs among adults, including the elderly. However, data are insufficient to categorize the Cu status of the elderly. Some studies have shown an age-related increase in blood Cu levels. Speculations as to the possible significance of the increase in blood Cu with aging and current evidence that supports the importance of Cu in diets of the elderly will be presented.

ZINC NUTRITION AND THE ELDERLY

Biochemistry and Metabolism

The biochemical interactions of Zn are based largely on Zn metalloenzymes, for which over 70 have been identified in various living systems.[3,4] Zn metalloenzymes are involved in sundry metabolic processes including carbohydrate, lipid, protein, and nucleic acid metabolism. Some important Zn metalloenzymes studied in humans include carbonic anhydrase, alkaline phosphatase, thymidine kinase, carboxypeptidases, and alcohol dehydrogenase.

Zn is absorbed in both the proximal and distal small bowel and is normally transported in serum by a variety of protein ligands.[5] Almost 30 to 40% is firmly bound to an alpha-1 macroglobulin or transferrin and approximately 60% is bound to albumin.[6] Hence, serum albumin is a major determinant of serum or plasma Zn levels. About 1% of serum Zn is normally bound to amino acids, particularly cysteine and histidine.[7,8] Although the mechanism of Zn absorption has been extensively studied, it is still not conclusively established. The process is known to be active, energy dependent, and apparently mediated by specific Zn transport ligands. Evidence suggests that the Zn absorption mechanism can, like Fe, play a significant role in homeostatic regulation. Many substances which appear to facilitate intestinal absorption of Zn have been identified. Substances of possible physiological significance include prostaglandin E_2, citrate, picolinic acid, and a metallothionein-like protein.[9-15] Their respective roles in the physiologic process of Zn absorption are the subject of continuing research.

Since aging is accompanied by significant changes in hormonal activity, it is of interest to note reports of Zn interactions with various hormones. Insulin contains approximately 0.5% Zn. Certain studies have shown a decrement in glucose tolerance and an impaired release of insulin from the pancreas in Zn-deficient rats. Others, however, have not found evidence of impaired carbohydrate metabolism in Zn-deficient rats. Evidence that poor Zn nutriture contributes to impaired carbohydrate metabolism in adult humans is lacking. It has also been shown that Zn is essential for maturation of the testes and is necessary for synthesis of testosterones.[16] Other metabolic functions of concern in the elderly, which have been related to Zn nutriture include immune function and wound healing.[2]

The bioavailability of Zn among the elderly is a legitimate question to raise. Physicians often recommend increased dietary fiber for elderly patients to prevent constipation and diverticulitis. Studies by Ismail-Beigi, Reinhold, Kies, and Kelsay[17-21] have demonstrated that phytate and dietary fiber decrease the intestinal absorption of Zn by humans. Modest intakes of sources of dietary fiber (up to 26 g/day), however, appear to have no adverse effect on Zn nutriture.[22] Pharmacologic amounts of ferrous sulfate have been shown to inhibit Zn absorption when inorganic Zn and Fe are ingested together.[23] Health practitioners should be aware of this interactive effect when prescribing mineral supplements. Further, it has been observed that increased dietary polyunsaturated fat or phosphorus will impair Zn retention.[24-26]

Whether or not the aged intestine absorbs Zn less well is uncertain because few studies have been reported. In rats, radiolabeled Zn was more efficiently absorbed by young as compared to old animals.[27] In 44 men and 33 women aged 18 to 84, Aamodt[28] found that intestinal absorption of ^{65}Zn decreased with age in both men and women.

ZINC STATUS OF THE ELDERLY

Biochemical Measures

The difficulty of assessing Zn nutriture of any population rests with the uncertainty of any single tissue providing an accurate measure of Zn nutriture. Assays of hair Zn have been used in many surveys of the elderly. However, because of the poor correlation between hair and plasma Zn, the validity of this index alone is questionable.[29] In one study by Greger, 5% of 65 elderly, institutionalized individuals had hair Zn levels that were consistent with Zn deficiency (<75 µg/g).[30] Results of a separate study by Greger and Sciscoe[31] among 44 elderly subjects (age range 52 to 89 years) who participated in a congregate feeding program showed no hair Zn levels below 70 µg/g. Wagner et al.[32] studied an elderly black population in Florida for Zn status (age range 60 to 87 years). Within this population, 51% were using food stamps and 28% participated in a congregate meal program. It was found that 11% had hair Zn levels less than 70 µg/g and/or serum levels less than 70 µg/dℓ, findings consistent with Zn deficiency. The few studies reported described a range of 0 to 8% of elderly subjects with concentrations of Zn in hair below 70 µg/g.

Lindeman[33] studied 250 male subjects (age range 20 to 84 years) and 54 female subjects (age range 20 to 58 years) for plasma and red cell Zn concentrations in an attempt to determine if a change in these levels occurred with age. A significant decrease in plasma Zn concentration with increasing age in both males and females was found, although red cell levels of Zn did not decrease. One possible explanation for these findings is the known decrease in serum albumin which occurs with age. Because serum albumin is the primary Zn-binding protein in the circulation, the decrease in plasma Zn might not represent Zn deficiency *per se*. A number of other studies have not confirmed Lindeman's observations. Davies et al.[34] found no change with age in plasma Zn among normal subjects and patients without malignancy in the age range of 19 to 58 years. Similarly, Herring[35] found among 61 normal subjects ages 10 to 60 years, no age-related change in plasma or red blood cell levels of zinc. A study of 146 Caucasian subjects in Belfast (age range 65 to 95 years) revealed no relation between plasma or hair zinc levels and age, although, plasma Zn and albumin levels were related.[36] Deeming and Webber[37] found that hair Zn levels decreased for both males and females after 40 years of age, whereas serum levels did not among 11 male and 16 female elderly subjects (age range 16 to 84 years). In this small study, serum Zn levels did not change with age when divided into groups of 16 to 20, 21 to 30, 31 to 40, and 40 years and over. With the exception of Lindeman's work, these studies are

Table 1
**RELATION OF INCOME TO ENERGY
AND ZINC INTAKES OF PERSONS 65
YEARS AND OLDER WHO
PARTICIPATED IN THE TEN STATE
NUTRITION SURVEY**

Race/income ratio	Energy (kcal)	Zinc (mg)
White/low	1670	11
White/high	1795	13
Black/low	1295	9
Black/high	1484	10
Spanish American/low	1710	11
Spanish American/high	1563	10

From Sandstead, H. H., *Am. J. Clin. Nutr.*, 35, 801, 1982b.
With permission.

consistent with findings from the U.S. National Health and Nutrition Examination Survey II (Hanes II) for serum Zn that showed no decrease with age up to 75 years.[2]

These various studies suggest that low hair, serum or red blood cell levels of zinc are not usual findings among the elderly. When low levels are observed, they may reflect a decreased level of serum albumin and/or may reflect poor zinc nutriture, either caused by low intake or underlying disease. It seems likely that these indexes of zinc nutriture are more frequently low among persons from the lower socioeconomic groups because of the cost of foods rich in zinc, and because of inappropriate food selections.[2]

Dietary Intake of Zinc

Estimates of dietary Zn of elderly persons have been made by extrapolation from dietary intake data reported for the USDA Nationwide Food Consumption Survey of 1977 to 1978 (NFCS), the DHEW Ten State Nutrition Survey of 1968 to 1970 (TSNS), and Hanes II reported in 1981.[2,38] The findings suggest that an important factor influencing Zn intake is the energy intake. As people age and their energy needs decrease, intake of dietary Zn decreases. Another important factor appears to be food selection. Economic factors are also probably important, particularly for persons in the lowest 10% of households economically. Estimated mean Zn intakes of elderly persons who participated in these surveys ranged from 7.0 mg/day in women over 75 years of age who participated in the NFCS to 13 mg/day in high income white persons over 65 years of age in the TSNS (Tables 1,2). Mean Zn intakes of persons who participated in Hanes II ranged from 12.6 mg/day in men 55 to 64 years of age whose energy intakes were 2071 kcal to 7.2 mg/day in women 65 to 74 years of age whose energy intakes were 1295 kcal/day (Table 3). These retrospective estimates were supported by the few prospective studies that have been published. For example, Holden et al.[39] found that Zn intake ranged from 5.9 to 12.4 mg/day with a mean of 8.6 mg in 22 persons aged 14 to 64 years. Similarly, Abdulla et al.[40] found a mean Zn intake of 8.3 and 7.2 mg/day (range 3.7 to 20.4 mg) in self-selected diet composites of 17 men and 20 women pensioners, age 67 years. Calculated estimates of Zn intakes of institutionalized and free-living persons in other studies[30,31,41] revealed similar findings.

Zinc Requirement

Knowledge concerning human requirements is incomplete. Factors that influence requirements and experimental observations on human Zn requirements have been reviewed.[2] Possible influences of these factors on Zn requirements of elderly persons have also been

Table 2
RELATION OF AGE TO ENERGY AND ZINC
INTAKES OF PERSONS WHO PARTICIPATED
IN THE NATIONWIDE FOOD CONSUMPTION
SURVEY

Sex	Age (years)	Energy (kcal ± SD)	Zinc (mg)
Male	65—74	1970 ± 744	10.5
	75+	1808 ± 735	9.3
Female	65—74	1444 ± 565	7.6
	75+	1367 ± 492	7.0
All persons		1865	9.8

From Sandstead, H. H., *Am.J. Clin. Nutr.*, 35, 801, 1982b. With permission.

Table 3
RELATION OF AGE TO ENERGY AND
ZINC INTAKE OF PERSONS WHO
PARTICIPATED IN THE NATIONWIDE
HEALTH AND NUTRITION
EXAMINATION SURVEY II

Sex	Age (years)	Energy (kcal)	Zinc (mg)
Male	55—64	2071	12.6
	65—74	1828	10.6
Female	55—64	1401	8.2
	65—74	1295	7.2

From Sandstead, H. H., *Am. J. Clin. Nutr.*, 35, 801, 1982b. With permission.

reviewed.[2] Particularly important are substances in the diet referred to previously that can alter the bioavailability of Zn.

Zn requirements of healthy male adults assessed by multiple regression analysis of data from 157 man-months of balance studies suggest that dietary protein and P are major predictors of dietary Zn requirement.[26] The following equation explained 83% of the variance ($p < 0.0001$) in required Zn intake to maintain balance in equilibrium.

$$\text{Intake} = -1.466 + 0.23 \, (\text{Zn balance}) + 5.19 \, (\text{phosphorous intake})$$
$$+ 0.40 \, (\text{nitrogen intake}) - 0.30 \, (\text{phosphorous intake} - 1.389)$$
$$(\text{nitrogen intake} - 14.646)$$

Using this equation, dietary Zn requirements can be predicted for various levels of dietary P and protein. Thus, when dietary P is 1500 mg and protein is 100 g daily, the predicted Zn requirement is 12.57 mg with a 95% confidence of 9.78 to 15.36 mg/day.

If Zn requirements of elderly persons are similar to those of normal men, the Zn requirements of elderly persons can be predicted by the above equation. Table 4 shows the predicted Zn requirements of elderly persons who participated in the NFCS.[2]

On the basis of these studies, it appears that the RDA for Zn of 15 mg/day[1] is more than adequate for the elderly. Mean Zn intakes of persons who participated in the above cited national surveys (Tables 1 to 3) exceeded mean requirements estimated for persons who

Table 4
ZINC REQUIREMENT OF PERSONS WHO
PARTICIPATED IN THE NATIONWIDE FOOD
CONSUMPTION SURVEY

Sex	Age (years)	Protein intake (g)	Phosphorous intake (g)	Zinc requirement (mg)
Male	65—74	81.0	1.246	10.06
	75+	74.6	1.137	8.95
Female	65—74	60.4	0.930	6.49
	75+	54.1	0.880	5.61

From Sandstead, H. H., *Am. J. Clin. Nutr.*, 35, 801, 1982b. With permission.

participated in the NFCS (Table 4). The fact that the predicted requirements were only slightly lower than the mean intakes suggests that the intakes did not provide a large margin of safety. Thus, it is probable that intakes were less than adequate for persons whose intakes were in the lower half of the distribution. The percent involved is unknown.

Therapeutic Trials

A third method for assessing Zn status is a therapeutic trial to correct an abnormal state. In some patients with disease, taste and smell sensations have been related to Zn nutriture and have improved with Zn administration. However, as taste acuity decreases with age, it is uncertain if hypogeusia is related to Zn nutriture in the elderly. To evaluate this question, Greger and Geissler[42] studied 49 institutionalized elderly people (mean age 75 years) for taste changes before and after administration of 15 mg of Zn or a placebo daily for 95 days. After 95 days the concentrations of Zn in hair was increased ($p<0.002$) though taste acuity for sodium chloride and sucrose was not improved. In another study of 44 noninstitutionalized individuals, Greger and Sciscoe[31] found no relation between taste acuity for sodium chloride and levels of dietary or hair Zn. Similarly, among 65 institutionalized persons (mean age 75 years), sweet and salty taste were not related to dietary or hair Zn, and taste acuity was found to decrease with age.[30] Thus, taste acuity was not a good measure of Zn nutriture among healthy elderly persons.

In contrast to taste acuity, response of wound healing to Zn supplementation has been indicative of Zn deficiency in some elderly persons. In two double-blind human trials, supplementation improved healing of leg ulcers.[43,44] In one of these studies, Zn nutriture was not evaluated prior to therapy.[43] In the second study, improvement in healing occurred compared to placebo in persons with serum Zn concentrations less than 100 μg/dℓ but not in persons with higher levels of serum Zn.[44] These observations in patients are consistent with the essentiality of Zn for wound healing in experimental animals.[45] It thus appears evident that Zn deficiency is responsible for poor healing in some elderly persons. The frequency of this phenomenon is unknown.

Another clinical effect of Zn deficiency is impaired immunity.[46] Limited evidence suggests that Zn nutriture and immune response are related in elderly persons. For example, 36 women aged 66 to 96 displayed a correlation ($r = 0.325$, $p <0.03$) between serum Zn and postimmunization titers to an influenza antigen.[47] In a second study that did not include an assessment of Zn nutriture, 15 institutionalized elderly individuals (mean age 81) displayed an increased total number of lymphocytes in the circulation and improved delayed hypersensitivity skin reactions and IGG responses to tetanus toxoid subsequent to supplementation with 100 mg Zn daily. Similar changes did not occur in 15 control individuals who were given placebo.[48] In a third study, 173 individuals, ages 74 ± 7 years, displayed a weak relationship between serum Zn and their dermal response to four standard antigens. Reevaluation of 5 anergic subjects occurred 5 months later; after 4 weeks of Zn supplementation, the dermal responsiveness to antigens became positive.[2] These few observations support the hypothesis that immune responsiveness is a useful, though nonspecific functional index of Zn nutriture, and that the diagnosis of Zn deficiency should be considered in elderly persons with impaired immune responsiveness. Additional testing of this hypothesis is needed.

A new area of research that may be related to immune function is the role of Zn in resistance to neoplasia. Epidemiologic studies have suggested a relationship between Zn deficiency and esophageal carcinoma.[49,50] Experimental studies in rodents have shown a relation between Zn nutriture and the incidence of esophageal cancer subsequent to exposure to methylbenzylnitrosamine.[51,52] It is unknown if the occurrence of other neoplasms is facilitated by marginal or deficient Zn nutriture.

Finally, an intriguing report on the effect of Zn on gastric acid secretion in rats has shown that Zn may be necessary to maintain the acidity of gastric contents.[53] This may be of relevance in the elderly due to the increased prevalence of gastric atrophy and achlorhydria with aging. With achlorhydria, intestinal absorption of Fe is impaired. Gastric atrophy is also associated with an increased incidence of vitamin B_{12} malabsorption. Further, a high gastric pH, by influencing intraluminal small intestinal pH, might inhibit the intestinal absorption of folic acid and could be a factor in bacterial overgrowth of the proximal small intestine with attendant intestinal malabsorption.

Among some elderly persons, alcoholism may contribute to the occurrence of poor Zn nutriture. Alcoholism is, unfortunately, not rare among the elderly. Its relation to poor Zn nutriture is well established.[54] Therefore, it seems reasonable to presume that elderly persons who drink to excess and in association have a poor diet, have an increased risk of Zn deficiency.

In summary, decreased taste acuity with aging does not appear to be related to Zn nutriture nor does it appear to be correctable by administration of Zn. In some instances it seems highly likely that impaired wound healing and suppressed immune function are related to poor Zn nutriture in the elderly. The frequency of these phenomena are unknown. Further study of these problems will be necessary to define the incidence of Zn deficiency-related impaired healing and immunity. Other related questions in need of study concern the possible pharmacologic effects of zinc on these processes. Previous evidence suggests wound healing will not be improved in persons with adequate Zn nutriture by supplemental Zn.[44] The possibility that Zn deficiency is related to the formation of neoplasms is also in need of further epidemiologic study and basic research. Finally, studies of the activity of Zn-dependent enzymes (e.g., alkaline phosphatase, red cell carbonic anhydrase, and alcohol dehydrogenase) as possible indexes of Zn nutriture in the elderly remain to be done.

CONCLUSION

Findings to date suggest that various groups of elderly (diseased, institutionalized, poor, low income) have an increased risk of poor Zn nutriture. Among healthy elderly persons, indexes of Zn nutriture are apparently similar to those of younger persons. When low levels of serum or hair Zn are observed, they in most instances probably reflect poor Zn nutriture. The functional consequences of deficient Zn nutriture among the elderly are probably similar to those observed in younger persons. Because of the greater frequency of conditioning factors[26] such as disease, alcoholism, ingestion of chelating drugs, the voluntary increased consumption of sources of dietary fiber, the decreased selection of foods rich in readily bioavailable Zn such as red meat (either due to taste preference, decreased energy need, or economic deprivation), the occurrence of functional deficits caused by Zn deficiency probably occurs in much greater frequency among the elderly than among younger persons.

COPPER NUTRITION AND THE ELDERLY

Biochemistry and Metabolism

Like Zn, Cu is important as an integral component of certain metalloenzymes. Examples include cytochrome oxidase, superoxide dismutase, uricase, dopamine-B-hydroxylase, lysyl oxidase, ceruloplasmin, and tyrosinase. A variety of mammalian plasma and connective tissue oxidases (monoamine, spermine, diamine, and benzylamine oxidases) are Cu-dependent, if not Cu metalloproteins.

Studies of the amount of ingested Cu absorbed from the human GI tract show variable results (15 to 97%), but a review of the older literature suggests 40 to 60% as a reasonable range for adults.[55] Cu absorption in 7 elderly men aged 65 to 74 years has recently been determined by a stable isotope/mass spectrometry technique.[56] Absorption, corrected for estimated endogenous GI Cu secretion was 29 ± 3% (mean ± SD). The dose of tracer Cu given in this study was 3.33 mg, an amount more than 2 times the usual intake of many people. This might account for the somewhat lower absorption than that cited previously. An alternative explanation is that older persons absorb less Cu than younger persons.

Factors that affect Cu bioavailability and dietary requirements have been reviewed.[57] Bioavailability is influenced both by facilitators and inhibitors of absorption. Potential facilitators include certain L-amino acids. Potential inhibitors include Zn, Cd, phytate, hemicellulose, lignin, unabsorbed fat, bile, and ascorbic acid. The presence of large amounts of the latter substances in the intestinal lumen should, in theory, increase the amount of dietary Cu needed for maintenance of homeostasis.

After absorption, Cu is transported via albumin to the liver where it complexes with metallothionein and other proteins. It is then released into the peripheral circulation primarily as a component of ceruloplasmin (about 93% of circulatory Cu). The remainder is complexed with albumin and amino acids. Ceruloplasmin is a copperoxidase enzyme that is involved in the metabolism of Fe and transferrin.

The homeostatic regulation of Cu metabolism is complex. Intestinal absorption is apparently influenced by metallothionein in mucosal cells. This protein avidly binds Cu and apparently inhibits its movement across intestinal mucosal cells into the body.[58,59] Induction of synthesis of metallothionein by Zn or Cd appears to account in part for the impairment of Cu homeostasis caused by these metals. The principal excretory vehicle for Cu is bile where it is complexed with large molecular weight ligands and is for practical purposes, not reabsorbed.[60] Levels of Cu in blood are primarily related to the level of ceruloplasmin. Synthesis and release of ceruloplasmin by the liver is stimulated by estrogenic hormones[61] progesterone, and testosterone, while cortisol tends to decrease ceruloplasmin.[62] Ceruloplasmin is also released from the liver along with other acute phase proteins in infections. Endogenous pyrogen and leukocyte endogenous mediator (LEM) apparently influence this phenomenon.[46]

COPPER NUTRITURE OF THE ELDERLY

Biochemical Indexes

The relationship of serum Cu to age and sex has been reviewed by Yunice.[63] Observed increases in serum Cu and ceruloplasmin in females relative to males may well be due to hormonal effects on Cu metabolism. If so, it is not known whether these effects might be of endogenous or exogenous (e.g., oral contraceptives) origin.

Hormonal effects on Cu metabolism have also been mentioned as a mechanism for the apparent increase in serum Cu levels with age. In addition to the serum Cu-age correlation found by Yunice in males,[63] two other studies have shown a similar relationship. Harman's results showed a linear increase in serum Cu with age, with mean serum levels of 124 μg/

dℓ at age 24 years and 145 µg/dℓ at age 60 years.[64] Herring et al.[35] showed a positive correlation of serum Cu with age in subjects between 10 to 50 years of age. Data regarding this subject is particularly lacking, however, for elderly subjects older than 70 years of age. Studies showing a concomitant increase in tissue Cu in the aged vs. younger adults have not been reported. Cu in aorta and coronary arteries has been reported to decrease with age.[65,66] More data is needed to clarify the relationship between age, serum, and body Cu.

Requirements and Dietary Intake

The RDA for Cu has not been established because of limited data on the human requirements. However, estimated "safe and adequate" intakes have been published for a number of trace elements for which RDA values cannot be established on the basis of present knowledge. The estimated adequate and safe Cu intake for adults is 2.0 to 3.0 mg/day.[1] Recommendations for its intake, specifically for the elderly population, have not been proposed.

Metabolic balance studies provide information relevant to the Cu needs of the elderly, but cannot be relied on exclusively for defining Cu requirements. In a recent study, 6 elderly male subjects (aged 65 to 74 years) maintained overall positive Cu balances and normal levels of serum Cu and ceruloplasmin during a 12-week period with Cu intakes of 3.2 $\pm$ 0.1 mg/day (mean $\pm$ SEM).[67] In another recent study, 5 males and 6 females consuming 2.3 mg Cu/day over 30 days showed overall positive Cu balances and hair Cu levels indicative of adequate Cu nutriture.[68] Cu retention was decreased by a high Zn intake (23.3 mg/day), consistent with the well-known Zn-Cu antagonism. These studies indicate that the estimated safe and adequate Cu intake for adults of 2 to 3 mg/day[1] is adequate for elderly adults. A Swedish study found Cu intakes in an elderly population to average 1.1 mg/day (0.5 to 3.0 mg/day), with most individual intakes below the estimated adequate intake at 2 mg/day. The intakes, however, were judged minimally adequate since none of the traditional signs of Cu deficiency were recognized upon follow-up clinical examination.[40] Metabolic balance studies on young men have shown that dietary Zn and protein levels influence the Cu requirement. Using multiple regression to evaluate 161 balance studies, dietary Zn and dietary protein accounted for 59% of the variance in Cu requirements ($p < 0.0001$). As dietary Zn increased, the requirement for Cu increased. In contrast, increased dietary protein decreased the requirement for Cu. On this basis, estimated Cu requirement exclusive of sweat loss was 1.14 mg/day (95% confidence: 0.92 to 1.48 mg/day) when dietary Zn was 10 mg and dietary protein 80 g.[57]

Cu intake data has been reviewed[55,69] and considerable worldwide variability is apparent. Recent data suggests previous assumed levels of intake may be in error as far as western diets are concerned. Several studies have revealed mean intakes between 0.34 to 1.5 mg/day. Klevay[70] has suggested that a high proportion of U.S. diets provide much less than 2 mg/day, the estimated adequate intake for adults. Estimates of regular, vegetarian, and renal diets of a U.S. hospital were 0.90, 1.10, and 0.51 mg/day, respectively. Much higher intakes have been reported, 4.5 to 5.8 mg/day in India and 2.3 to 7.3 mg/day in the Ukraine.

While the great bulk of studies pertaining to Cu intake and human requirements has been carried out on young rather than elderly adults, there is currently no data to suggest that Cu intakes or requirements are appreciably different for the elderly than for younger adults. The adequacy of present-day intakes is the subject of current research.

COPPER NUTRITURE AND DISEASES OF AGING

Human Cu deficiency has been documented in infants and children and experimentally in young men,[71] but not in the elderly *per se*, excepting isolated cases such as nephrosis, intestinal disease, and during total parenteral nutrition. Evidence does not implicate Cu deficiency in the deterioration of physiological functions which occur with aging. A number of conditions or diseases associated with aging have been shown to include disturbances in Cu metabolism. These include neoplasms and cardiovascular and rheumatic diseases. Specific mechanisms for these changes in Cu homeostasis in these diseases have usually not been suggested, much less established. Ceruloplasmin is classified as an acute phase protein; its increased hepatic synthesis is generally believed to underlie the hypercupremia seen in hormonal, infectious, and inflammatory body disturbances. This response is stimulated by endogenous pyrogen and leukocyte endogenous mediator (LEM).[46]

Hypercupremia is commonly seen in cases of leukemias, Hodgkin's disease, malignant tumors, and myelomas. Besides the increase in ceruloplasmin in response to inflammation mediated by endogenous pyrogen and LEM, no other explanation for the hypercupremia is apparent. Some studies indicate that arthritis/rheumatism sufferers may have increased levels of Cu in serum and synovial fluid and that Cu-containing drug treatments may be beneficial. Other studies do not substantiate this. Therefore, more research is needed to clarify the relation between inflammatory joint disease and Cu homeostasis.

Cu has been implicated in a variety of cardiovascular diseases. In animals, Cu deficiency causes impaired synthesis of elastin and collagen due to impaired lysyl oxidase-mediated cross-linking. Arterial aneurisms and rupture occur. Animals also display degenerative changes in the nervous system affecting myelin. In lambs, congenital Cu deficiency results in enzootic ataxia. Cytochrome oxidase levels in brain tissue are low and myelination of the nervous system is grossly retarded. In rats, swine, and cattle, severe injury to myocardium occurs with sudden death from rupture or arrythmia. Rats, swine, and monkeys also display increased serum cholesterol levels (increased LDL cholesterol) in Cu deficiency. In humans, a similar increase in serum cholesterol has been produced in one individual who was experimentally depleted of Cu.[71] These observations support the hypothesis of Klevay[72] that relates impaired Cu nutriture in humans to the occurrence of atherosclerotic cardiovascular disease, and most particularly, to ischemic heart disease. Klevay notes that an absolute deficiency of Cu or relative increase in dietary Zn relative to Cu will induce the phenomenon. Other researchers[73,74] have not always confirmed Klevay's findings in rats. Reasons for these differences are not readily apparent, but may relate to duration of studies, numbers of animals observed, or differences in diet that are not readily apparent from the reports.

CONCLUSION

Evidence at this time is insufficient to establish a role for Cu nutriture in aging and associated diseases. Some investigators have speculated that the apparent increase in serum Cu levels with aging may be associated with increased tissue-free radicals. This hypothesis, that excess Cu ions in the elderly act as oxidation catalysts, thus promoting free radical cellular peroxidation damage, has not been experimentally verified. Others have noted an association between deficient Cu nutriture in animals and disease of the cardiovascular system. It is unknown if chronic mild Cu deficiency occurs in humans and contributes to the occurrence of atherosclerotic cardiovascular disease.

REFERENCES

1. Food and Nutrition Board National Research Council, Recommended Dietary Allowances, 9th ed., National Academy of Sciences, Washington, D.C., 1980.
2. **Sandstead, H. H., Henriksen, L. K., Greger, J. L., Prasad, A. J., and Good, R. A.,** Zinc nutriture in the elderly in relation to taste acuity, immune response, and wound healing, *Am. J. Clin. Nutr.,* 36, 1046, 1982.
3. **Mildvan, A. S.,** *The Enzymes,* Vol. 2, 3rd ed., Boyer, P. D., Ed., Academic Press, N.Y., 1970, 445.
4. **Riordan, J. F. and Vallee, B. L.,** Structure and function of zinc metalloenzymes, in *Trace Elements in Human Health and Disease,* Prasad, A. S., Ed., Academic Press, N.Y., 1976, 227.
5. **Schwartz, F. J. and Kirchgessner, M.,** Experimental studies on the absorption of zinc from different parts of the small intestine and various zinc compounds, *Nutr. Metab.,* 18, 157, 1975.
6. **Boyett, J. A. and Sullivan, J. F.,** Distribution of protein-bound zinc in normal and cirrhotic serum, *Metabolism,* 19, 148, 1970.
7. **Henkin, R. E. and Smith, F. R.,** Zinc and copper metabolism in acute viral hepatitis, *Am. J. Med. Sci.,* 264, 401, 1972.
8. **Giroux, E. L. and Henkin, R. I.,** Competition for zinc among serum albumin and amino acids, *Biochem. Biophys. Acta,* 273, 64, 1972.
9. **Aggett, P. J. and Harries, J. T.,** Current status of zinc in health and disease states, *Arch. Dis. Child.,* 54, 909, 1979.
10. **Becker, W. M. and Hoekstra, W. G.,** The intestinal absorption of zinc, in *Intestinal Absorption of Metal Ions, Trace Elements and Radionuclides,* Skoryma, S. C. and Waldron-Edward, D., Eds., Pergamon Press, N.Y., 1971.
11. **Cousins, R. J.,** Regulation of zinc absorption: role of intra-cellular ligands, *Am. J. Clin. Nutr.,* 32, 339, 1979.
12. **Evans, G. W. and Johnson, P. E.,** Characterization and quantitation of a zinc binding ligand in human milk, *Pediatr. Res.,* 14, 876, 1980.
13. **Evans, G. W. and Johnson, E. C.,** Zinc absorption in rats fed a low protein diet and a low protein diet supplemented with tryptophan or picolinic acid, *J. Nutr.,* 110, 1076, 1980.
14. **Lonnerdal, B., Keen, C. L., and Hurley, L. S.,** Iron, copper, zinc, and manganese in milk, *Annu. Rev. Nutr.,* 1, 149, 1981.
15. **Song, M. F. and Adham, N. D.,** Evidence of an important role of prostaglandins E_2 and F_2 in the regulation of zinc transport in the rat, *J. Nutr.,* 109, 2152, 1980.
16. **Abbasi, A. A., Prasad, A. S., Rabbani, P., and DuMouchelle, E.,** Experimental zinc deficiency in man. Effect on testicular function, *J. Lab. Clin. Med.,* 96, 544, 1980.
17. **Ismail-Beigi, F., Faraji, B., and Reinhold, J. G.,** Binding of zinc and iron to wheat bread, wheat bran and their components, *Am. J. Clin. Nutr.,* 30, 1721, 1977.
18. **Reinhold, J. G., Nase, K., Lahimgorzaheh, A., and Hedayati, H.,** Effects of purified phytate and phytate-rich bread upon metabolism of zinc, calcium, phosphorus, and nitrogen in man, *Lancet,* 283, 1973.
19. **Reinhold, J. G., Fardji, S., Abadki, P., and Ismail-Beigi, F.,** Decreased absorption of calcium, magnesium, zinc and phosphorus consumption as wheat bread, *J. Nutr.,* 106, 493, 1976.
20. **Kies, C., Fox, H. M., and Beshgetoor, B.,** Effect of various levels of dietary hemicellulose on zinc nutritional status of men, *Cereal Chem.,* 56, 133, 1979.
21. **Kelsay, J. L., Jacob, R. A., and Prather, E. S.,** Effect of fiber from fruits and vegetables on metabolic responses of human subjects. III. Zinc, copper, and phosphorus balances, *Am. J. Clin. Nutr.,* 32, 2307, 1979.
22. **Sandstead, H. H., Klevay, L. M., Jacob, R. A., Munoz, J. M., Logan, G. M., Jr., Reck, S. J., Dintzis, F. R., Inglett, G. E., and Shuey, W. C.,** Effects of dietary fiber and protein level on mineral element metabolism, in *Dietary Fibers: Chemistry and Nutrition,* Inglett, G. E. and Falkehad, I., Eds., Academic Press, N.Y., 1979, 147.
23. **Solomons, N. W., Pineda, O., Viteri, F. E., and Sandstead, H. H.,** Mechanisms of intestinal interaction of iron and zinc in man, *Fed. Proc.,* 40, 856, 1981.
24. **Likaski, H. C., Klevay, L. M., Bolonchuk, W. W., Mahalko, J. R., Milne, D. B., Johnson, L., and Sandstead, H. H.,** Influence of dietary lipids on iron, zinc, and copper retention in trained athletes, *Fed. Proc.,* 41, 275, 1982.
25. **Greger, J. L. and Snedeker, S. M.,** Effect of dietary protein and phosphorus levels on the utilization of zinc, copper, and manganese by adult males, *J. Nutr.,* 110, 2243, 1980.
26. **Sandstead, H. H.,** Availability of zinc and its requirements in human subjects, in *Clinical, Biochemical, and Nutritional Aspects of Trace Elements,* Prasad, A. S., Ed., Alan R. Liss, N.Y., 1982, 83.

27. **Strain, W. H., Pories, W. J., Michael, E., Peer, R. M., and Zaresky, S. A.,** Influence of age on absorption and retention of trace elements, in *The Biomedical Role of Trace Elements of Aging,* Hse, J. M., Davis, R. L., and Neithamer, R. W., Eds., Eckert College Gerontology Center, St. Petersburg, Fla., 1976, 161.

28. **Aamodt, R. L., Rumble, W. S., and Henkin, R. I.,** Zinc Absorption in Human Subjects: Effects of Age, Sex, and Food, 183rd National Meeting of the American Chemistry Society, Division of Agricultural and Food Chemistry, Las Vegas, Nev., March/April, 1982, (abstr.) p.1.

29. **Solomons, N. W.,** On the assessment of zinc and copper nutriture in man, *Am. J. Clin. Nutr.,* 32, 856, 1979.

30. **Greger, J. L.,** Dietary intake and nutritional status in regard to zinc of institutionalized aged, *J. Gerontol.,* 32, 549, 1977.

31. **Greger, J. L. and Sciscoe, B. S.,** Zinc nutriture of elderly participants in an urban feeding program, *J. Am. Diet. Assoc.,* 70, 37, 1977.

32. **Wagner, P. A., Krista, M. L., Bailey, L. B., Christakis, G. J., Ternigan, J. A., Aravjo, P. E., Appledorf, H., Davis, C. G., and Dinning, J. S.,** Zinc status of elderly black Americans from urban low-income households, *Am. J. Clin. Nutr.,* 33, 1771, 1980.

33. **Lindeman, R. D., Clark, M. L., and Colmore, J. P.,** Influence of age and sex on plasma and red-cell zinc concentrations, *J. Gerontol.,* 26, 358, 1971.

34. **Davies, I. J. T., Musa, M., and Dormandy, T. L.,** Measurements of plasma zinc, *J. Clin. Path.,* 21, 359, 1968.

35. **Herring, W. B., Leavell, B. S., Paixao, L. M., and Yoe, J. H.,** Trace metals in human plasma and red blood cells, *Am. J. Clin. Nutr.,* 8, 846, 1960.

36. **Vir, S. C. and Love, A. H. G.,** Zinc and copper status of the elderly, *Am. J. Clin. Nutr.,* 32, 1472, 1979.

37. **Deeming, S. B. and Weber, C. W.,** Hair analysis of trace minerals in human subjects as influenced by age, sex, and contraceptive drugs, *Am. J. Clin. Nutr.,* 31, 1175, 1978.

38. Food and Nutrient Intakes of Individuals in 1 day in the United States, Preliminary Report No. 2, Nationwide Food Consumption Survey 1977-1978, USDA Science and Education Administration, Spring, 1977, 1980.

39. **Holden, J. M., Wolf, W. R., and Mertz, W.,** Zinc and copper in self-selected diets, *J. Am. Diet. Assoc.,* 77, 23, 1979.

40. **Adbulla, M., Jagerstad, M., Norden, A., Quist, I., and Svensson, S.,** Dietary intake of electrolytes and trace elements in the elderly, *Nutr. Metab.,* 21 (Suppl. 1), 41, 1977.

41. **Flint, D. M., Wahlquist, M. L., Smith, T. J., and Parish, A. B.,** Zinc and protein status of the elderly, *J. Human Nutr.,* 35, 287, 1981.

42. **Greger, J. L. and Geissler, M. S.,** Effect of zinc supplementation on taste acuity of the aged, *Am. J. Clin. Nutr.,* 31, 633, 1978.

43. **Haeger, K., Lanner, E., and Magnusson, P. O.,** Oral zinc sulfate in the treatment of venous leg ulcer, in *Clinical Applications of Zinc Metabolism,* Pories, W. J., Strain, W. H., Hsu, J. M., and Woosley, R. L., Eds., Charles C Thomas, Springfield, Ill., 1974, 158.

44. **Hallbook, T. and Lanner, E.,** Serum-zinc and healing of venous leg ulcers, *Lancet,* 2, 780, 1972.

45. **Wacker, W. E. C.,** Role of zinc in wound healing: a critical review, in *Trace Elements in Human Health and Disease,* Prasad, A. S., Ed., Academic Press, N.Y., 1976, 107.

46. **Beisel, W. R.,** Single nutrients and immunity, *Am. J. Clin. Nutr.,* 35, 417, 1982.

47. **Stiedemann, M. and Harrell, I.,** Relation of immunocompetence to selected nutrients in elderly women, *Nutr. Rep. Int.,* 21, 931, 1980.

48. **Duchateau, J., Delepesse, G., Vrigens, R., and Collet, H.,** Beneficial effects of oral zinc supplementation on the immune response of old people, *Am. J. Med.,* 70, 1001, 1981.

49. **Kaplan, H. S. and Tsuchitani, P. J.,** *Cancer in China,* Alan R. Liss, N.Y., 1978.

50. **Lin, H. J., Chan, W. C., Fong, Y. Y., and Newburne, P. M.,** Zinc levels in serum, hair and tumors from patients with esophageal cancer, *Nutr. Rep. Int.,* 15, 635, 1977.

51. **Fong, Y. Y., Sivak, A., and Newberne, P. M.,** Zinc deficiency and methylbenzylnitrosamine-induced esophageal cancer in rats, *J. Natl. Cancer Inst.,* 61, 145, 1978.

52. **Rensburg, J. J., Bruya, D. B., and Van Schelkwy, D. J.,** Promotion of methylbenzylnitrosamine induced esophageal cancer in rats by subclinical zinc deficiency, *Nutr. Rep. Int.,* 22, 891, 1980.

53. **Yamaguchi, M., Yoshino, T., and Okada, S.,** Effect of zinc on the acuity of gastric secretion in rats, *Toxicol. Appl. Pharmacol.,* 54, 526, 1980.

54. **Sullivan, J. F. and Lankford, H. G.,** Zinc metabolism and chronic alcoholism, *Am. J. Clin. Nutr.,* 17, 57, 1965.

55. **Mason, K. E.,** Copper metabolism and requirements of man, *J. Nutr.,* 109, 1979, 1979.

56. **Turnlund, J. R. et. al.,** Copper absorption in elderly men determined by using stable ^{65}Cu, *Am. J. Clin. Nutr.,* 36, 587, 1982.

57. **Sandstead, H. H.,** Copper bioavailability and requirements, *Am. J. Clin. Nutr.,* 35, 801, 1982b.
58. **Hall, A. C., Youngs, B. W., and Bremner, I.,** Intestinal metallothionein and the mutual antagonism between copper and zinc in the rat, *J. Inorgan. Biochem.,* 11, 57, 1979.
59. **Bremner, I. and Campbell, J. K.,** The influence of dietary copper intake on the toxicity of cadmium, *Ann. N.Y. Acad. Sci.,* 355, 319, 1980.
60. **Gollan, J. L.,** Studies on the nature of complexes formed by copper with human alimentary secretions and their influence on copper absorption, *Clin. Sci. Mol. Med.,* 49, 237, 1975.
61. **Smith, J. C., Jr. and Brown, E. D.,** Effects of oral contraceptive agents on trace element metabolism — a review, in *Trace Elements in Human Health and Disease,* Vol. 2, *Essential and Toxic Elements,* Prasad, A. S. and Oberleas, D., Eds., Academic Press, N.Y., 1976, 315.
62. **Cartwright, G. E.,** Copper metabolism in human subjects, in *A Symposium on Copper Metabolism,* McEloy, W. D. and Glass, B., Eds., Johns Hopkins Press, Baltimore, 1950, 274.
63. **Yunice, A. A.,** Serum Copper in Relation to Age, in *Ultratrace Metal Analysis in Science and the Environment, Advances in Chemistry* Series No. 172, American Chemical Society, Washington, D.C., 1979, 230.
64. **Harman, D.,** The free radical theory of aging: effect of age on serum copper levels, *J. Gerontol.,* 20, 151, 1965.
65. **Schroeder, H. A., Nason, A. P., Tipton, I. H., and Balassa, J. J.,** Essential trace metals in man: copper, *J. Chron. Dis.,* 19, 1007, 1966.
66. **Taylor, G. O. and Williams, A. O.,** Lipid and trace metal content in coronary arteries of Nigerian Africans, *Exp. Mol. Pathol.,* 21, 371, 1974.
67. **Fernandes, G., West, A., and Good, R. A.,** Nutrition, immunity and cancer — a review. III. Effects of diet on the diseases of aging, *Clin. Biol.,* 9, 91, 1979.
68. **Burke, D. M., DeMicco, F. J., Taper, L. J., and Ritchey, S. J.,** Copper and zinc utilization in elderly adults, *J. Gerontol.,* 36, 558, 1981.
69. **Waslien, C. I.,** Human intake of trace elements, in *Trace Elements in Human Health and Disease,* Vol. 2, Prasad, A. S., and Oberleas, D., Eds., Academic Press, N.Y., 1976, 347.
70. **Klevay, L. M.,** Dietary copper and the copper requirement of man, in *Trace Element Metabolism in Man and Animals,* Kirch-Gessner, M., Ed., Technisch Universita & Munchen, GDR, 1978.
71. **Klevay, L. M., Inman, L., Johnson, K. L., et al.,** Effects of a diet low in copper on a healthy man, *Clin. Res.,* 28, 758A, 1980.
72. **Klevay, L. M.,** Coronary heart disease: the zinc/copper hypothesis, *Am. J. Clin. Nutr.,* 28, 764, 1975.
73. **Fischer, P. W., Girous, A., Belonje, B., and Shah, B. G.,** The effect of dietary copper and zinc on cholesterol metabolism, *Am. J. Clin. Nutr.,* 33, 1019, 1980.
74. **Koo, S. I. and Williams, D. A.,** Relationship between the nutritional status of zinc and cholesterol concentration of serum lipoproteins in adult male rats, *Am. J. Clin. Nutr.,* 34, 2376, 1981.

VITAMIN E IN THE AGED

Miroslav Ledvina

INTRODUCTION

Interest in vitamin E developed from the observation that rats fed exclusively cow's milk, even supplemented with vitamin B from yeast and iron, were incapable of bearing young. In 1936, a group of substances named tocopherols (in Greek tokos means childbirth) were isolated from wheatgerm oil.

In the article presented we intend to discuss some of the more recent research results which we feel have extended current knowledge about the importance of vitamin E in the aged.

Undoubtedly, vitamin E is one of the most versatile of all the vitamins. The manifestations of vitamin E deficiency in laboratory animals are very diverse; the symptoms, organ specificity, and pathology vary widely among different animal species. On the other hand, manifestations of deficiency in adult humans are doubtful. This vitamin also shows additional peculiarities, e.g., the dietary requirement for vitamin E strongly depends on the intake of other substances (unsaturated fatty acids, vitamin A, selenium, etc.).

Vitamin E includes several closely related compounds, chemical derivatives of 6-hydroxychromane with an isoprenoid side chain containing 16 carbons. These compounds include α-tocopherol as well as β, α, δ, ϵ, $\zeta_1 + \zeta_2$, and η-tocopherols. From the practical standpoint, α-tocopherol is of greatest importance. However, its effectiveness is not highest in all respects, e.g., as to the antioxidant properties (see below), the δ-derivative is more effective than α. Naturally occurring α-tocopherol is D-(+)α-tocopherol; however, the synthetically produced vitamin is D, L-α-tocopherol.

Pure tocopherols are fat-soluble oils. They are relatively heat stable; in the absence of oxygen they are very stable. The loss during cooking amounts to approximately 50%. Tocopherols withstand acids even at elevated temperatures, but oxidation and ultraviolet light destroy the vitamin activity.

The daily requirement of vitamin E in adults is 10 to 30 mg. According to the Food and Nutrition Board (USA), the recommended dose for Middle Europe was estimated at 12 mg/day for both sexes in all age categories.[1] In fact, the real minimal necessary intake varies considerably with the intake of other food constituents (see below).

Vegetable oils are the richest sources of vitamin E. From these, wheatgerm oil has the highest concentration, but also barley, bean, carrot, rye, and soybean oil all have several hundred mg/100 g. Marine animal oils, so rich in highly unsaturated fatty acids, are a relatively good source as well. On the other hand, a small content is found in coconut and olive oils. Vegetables, lettuce, mallow, and spinach are rich in vitamin E. On the other hand, the content is low in all kinds of meat, grain products, fruits, animal fats, and backed cereals. Of the animal organs, only the liver contains relatively large amounts. The tocopherol content in polished rice, bleached flour, milk, cheese and yeast is very low, almost negligible. The data on the content of vitamin E in various foods and about the distribution between the individual tocopherols have been reviewed by Dicks.[2] However, we must take into account that not only the content of tocopherol itself in the food is significant, but also its relation to other food constituents. Moreover, some margarines, edible fats, and oils are artificially supplemented with vitamin E (along with A and D vitamins). The usual Middle European intake was about 185% of the recommended daily intake in 1974.[1]

Intestinal absorption of fat-soluble vitamin E depends on bile acids. Absorbed tocopherol occurs in the chyle,[3] and it is preferentially incorporated into chylomicrons. If the chylomicron formation is inhibited by puromycin, radioactive tocopherol does not appear in the chyle. The absorption may be substantially reduced in various diseases of the GI tract, as in sprue or long-lasting steatorrhea.[4,6] Also, cystic fibrosis is connected with vitamin E deficiency.[5] The finding that the resorption capacity for tocopherol is markedly depressed in vitamin E deficiency seems to be of great importance.[7] Tocopherols are distributed slowly to all tissues,[8] and associated primarily with biomembranes and chromatin.[9]

The serum level of tocopherol depends on its oral intake, but this is modified by several factors such as dosage, duration of treatment, age, and species.[24]

The normal level of tocopherols in the blood plasma seems to correlate not only with the food intake both of vitamin E and polyunsaturated fatty acids. To explain altered tocopherol levels, intestinal absorption must be taken into consideration. Numerous studies have indicated that the plasma tocopherol levels appear to be correlated with total lipid values rather than with any individual lipid component.

Vitamin E is also present in the formed blood elements. The content is relatively low in red cells and platelets, but it is many times higher in lymphocytes and granulocytes — 6.13 and 7.79 $\mu g/10^9$ cells, respectively.[10] This increased content might be due to a specific role of tocopherol during phagocytic activity of cells (connected with generation of free radicals).

INFLUENCE OF VITAMIN E ON VITAL FUNCTIONS

Vitamin E Deficiencies

Typical deficiencies have been described only in animals. The main organs and systems affected under these circumstances are the gonads, muscle, brain, hematopoietic system, liver, kidney, and some endocrine glands.[12]

On the other hand, clinical syndromes due to vitamin E deficiency in man are still almost exclusively restricted to premature infants, though vitamin E depletion based either on low serum levels or on in vitro erythrocyte hemolysis can be demonstrated in some conditions even in adults and aged people. Moreover, less pronounced hypovitaminoses may play a certain role in man, as indicated by laboratory symptoms (increased hemolysis with dialuric acid and H_2O_2, increased excretion of vitamin C in the urine, high urinary excretion of creatine and 1-methylhistidine, and high generalized aminoaciduria, increased blood cholesterol, and a decreased level of vitamin E in the plasma). Nevertheless, the fundamental E vitamin depletion effect in animals — infertility — has never been observed in the pathology of human reproduction. Studies in primates reveal a similarity to man; vitamin E deficiency results only in hemolytic anemia.

Herbivorous animals such as the rabbit, guinea pig, calf, and lamb are more sensitive to vitamin E deficiency than is the rat.

Typical vitamin E deficiency syndromes in different species can be summarized as follows

Species	Syndrome	Organ altered
Rat	Infertility, resorption of the fetuses	Gonads (female)
	Testicular degeneration	Gonads (male)
	Pigmentation	Adipose tissue
	Muscular dystrophy (no encephalomalacia, no exudative diathesis)	Muscle (skeletal and myocardial)
Rabbit	Muscular dystrophy	Muscle
	Myocardial degeneration	Muscle (myodardial)
Chick	Encephalomalacia	Brain
	Exudative diathesis	Subcutaneous tissue etc.
Monkey	Macrocytic anemia	Hematopoietic system

Pathological changes in the organs affected during vitamin E deficiency can be described in the following manner:

1. Reproductive organs: In female rats hypertrophy and dystrophy of the uterine musculature occur, ceroid is deposited, and the color of the uterus changes to brown. Beginning on the 8th day, the embryos are damaged, they perish, and are absorbed. However, the ovarium and endometrium remain unaltered. In male rats testicular atrophy appears. There is extensive degeneration of germinal epithelium, but no hyperplasia of the interstitial or Leydig cells. Excretion of 17-oxosteroids from the interstitial cells is depressed.
2. Muscles: Changes lead to dyskinesia, hind legs weaken, (adductors are affected), and atrophy follows. Striations disappear and the muscle is replaced by connective tissue. Also, the myocardial muscle is altered; it shows necroses and fibrosis with abnormal electrocardiogram.
3. Liver glycogen disappears and the cells undergo fatty degeneration. The process results in necroses. Ceroid of lipoprotein character is deposited.
4. Kidney: Advanced glomerulonephritis and later renal failure develops in the rat.
5. Hyperplasia of the anterior lobe of the hypophysis and a weight increase of the adrenals occur.
6. Subcutaneous tissues: In chicks fed cod liver oil, a massive accumulation of fluid in the subcutaneous tissues appears. Together with edema of the muscles and connective tissue in general, it forms the localizations of exudative diathesis, caused primarily by a damage of the capillaries in the affected tissue.[13] In rats fed a vitamin E-deficient diet containing 20% cod liver oil, yellow-brown pigments accumulate in the adipose tissue; the color is due to two pigments (ceroid) occurring by polymerization of oxidation products of polyunsaturated fatty acids.[14] A similar pigment was found in the male rat germinal epithelium and in the rat uterus.
7. Teeth: When rats are reared from weaning on a vitamin E-deficient diet containing 20% lard, their normally brown-colored incisor teeth become depigmentated. The cessation of production of this enamel pigment proves that the diet contains polyunsaturated fatty acids.[13] On the other hand, vitamin E deficiency causes an opposite effect on adipose tissue (see 6).

As outlined by Baessler et al.,[15] some of the syndromes of vitamin E deficiency are due to increased lipid peroxidation (see below). For instance, encephalomalacia of the birds is a consequence of increased vessel permeability, and the resorptive sterility is based on the inability of the fetuses to build their vascular system. Some symptoms (testicular degeneration, muscular dystrophy) are due to the combined effect of lipoperoxidation with other influences. Certain symptoms are probably a consequence of mixed deficiencies both of vitamin E and Se.

The Effect of Vitamin E Deficiency on Individual Specified Functions

Numerous metabolic changes and laboratory biochemical symptoms have been observed in various forms of vitamin E deficiency. It can be assumed that almost all manifestations of vitamin E deficiency appear to be secondary to the basic function of this vitamin-decreased peroxidation of polyunsaturated fatty acids.

As mentioned below, α-tocopherol is oxidized to tocopherylhydroquinone and tocopherylquinone, but this may be converted back to tocopherol, thus representing a redox system. The metabolic changes due to vitamin E deficiency may be caused either by the lack of tocopherol itself or by the absence of its metabolic products mentioned above.

There is indisputable evidence on the participation of tocopherol in the respiratory chain and in succinate oxidation. Oxygen consumption of the tissues depends on vitamin E. Its depletion leads to an increase of oxygen consumption.[16] In the mitochondria of vitamin E-deficient rabbit hearts there was a close correlation between the decline in respiration and the rate of membrane peroxidation;[17] following addition of vitamin E, the respiration turned back to normal conditions. As to the activity of individual redox systems of the respiratory chain, tocopherol inhibits cytochrome oxidase, cytochrome reductase, and succinate dehydrogenase. The ubiquinone redox system has also been studied. A higher saturation of the body with vitamin E is associated with increased ubiquinone concentrations, and the vitamin E-depleted diets induce a combined deficiency of vitamin E and ubiquinone, providing that the diets do not contain larger amounts of ubiquinone, as in corn oil.[18]

Enzyme superoxide dismutase plays an important role in protection against oxygen (see below). In tocopherol-depleted chicks, its activity is enhanced in the red blood cells and liver. This proves an adaptive role of superoxide dismutase against oxygen damage.[19]

Vitamin E deficiency substantially alters the metabolism of lipids. Apart from the effect on membrane lipids that belongs to the essential properties of vitamin E (see below), biosynthesis of cholesterol is increased in vitamin E deficiency while the rate of cholesterol catabolism is decreased.[21] Consequently, the concentrations of cholesterol in the blood plasma, skeletal muscle, liver, and lungs in the rabbit are increased.[22] In man, vitamin E may lead to a marked decrease of cholesterolemia although there is some lack of agreement on this.[23] Yang and Desai[24] report that vitamin E has no significant effect on total cholesterol concentration in normal subjects. However, in hyperlipidemic subjects, α-tocopherol ingestion has a dose-dependent effect by lowering serum cholesterol.

As for the saccharide turnover, vitamin E deficiency decreases the amount of liver glycogen. Tocopherols activate phosphoglucomutase, phosphohexoisomerase, and gluconeogenesis, and inhibit the activity of lactate dehydrogenase.

Vitamin E deficiency influences the immune responses of the body. Vitamin E enhances humoral immunity and stimulates phagocytosis, increases the number of in vivo and in vitro antibody-forming cells, and elevates IgG and IgM antibody titers. Vitamin E supplemented to diets increases markedly the resistance of mice and chicks to bacterial infection with *E. coli* and *Diplococcus pneumoniae*.[25] The immune response is dose-dependent; it has been shown to be inhibited in very high doses of vitamin E.[26]

Vitamin E is important for permeability of the vessel wall and capillaries. Above-mentioned exudative diathesis in the chicks is a consequence of the apparent damage of the capillaries. Vitamin E deficiency is associated with histological changes in the intima of the chick aorta (thickening of endothelium, formation of plaques, fibrosis of subendothelial tissue).[27] Vitamin E increases the capillary resistance.[28]

The integrity of red blood cells depends significantly on the presence of vitamin E. Thus, there is a deficiency symptom observed both in the man and animal — a decreased resistance to hemolysis when exposed to oxidants. For inducing hemolysis, dialuric acid and hydrogen peroxide are employed. Tocopherol prevents this hemolysis.[29] The hemolysis leads to release of K^+ from human erythrocytes.[30] The hydrogen peroxide hemolysis test has been commonly used to estimate the tocopherol blood concentration.

Vitamin E and its deficiency influence platelets and hemocoagulation. This vitamin is widely used in the treatment of various blood coagulation disturbances, and hemorrhages are involved in the clinical feature of vitamin E deficiencies. α-Tocopherol inhibits platelet aggregation; the sharp rise in lipid peroxides normally associated with platelet aggregation is markedly diminished.[31] Vitamin E inhibits substantially the synthesis of prostaglandins PGE_2 and PGE_2 in the platelets; thus, prostaglandin inhibitors (aspirin) may have a tocopherol-like effect. Synthesis of the highly potent aggregation agent thromboxane A_2 is inhibited, too, and the platelets from vitamin E-deficient rats aggregate much more readily than those from control animals.[32] This is a consequence of increased prostaglandin synthesis.

Experiments dealing with the influence of vitamin E on lifespan are very interesting. In full accordance with the generally accepted theory of the antioxidant properties of this vitamin, we may suppose that reduced lipoperoxidation may prolong the lifespan of animals. (Similar human experiments have not been reported.) Not only artificial antioxidants, but also tocopherol was capable of influencing the lifespan of some experimental animals. Some authors have succeeded in proving a small, but significant prolongation of life in some strains of mice,[33] Drosophila,[34] and in mice fed a diet with additional α-tocopherol.[35,36] The problem of the role of free-radical reactions in regard to aging has been reviewed recently.[37]

Much more must be done in this field to prove or deny the direct relationship of longevity to vitamin E. Of course, the explanation even of the positive experiments may be diverse, e.g., through a decrease of plasma cholesterol, thus based on preventing atherosclerosis.

Vitamin E as an antioxidant agent has effects on parameters relevant for the development of postirradiation changes (such as catalase, glutathione peroxidase, and superoxide dismutase). Nevertheless, the practical effect of vitamin E as a radioprotective agent is doubtful. Many authors have tried to demonstrate this effect. Most have found no protective action, and sometimes even an increase of lethality following ionizing radiation is found. From the more recent results, doses of 100 mg/100 g body weight of mice have been proven to be radioprotectively efficient.[38] In other experiments[39] even the postirradiation administration of vitamin E reduced lethality of mice exposed to 8 Gy (= 800 R). Such a protective effect of vitamin E was demonstrated in the erythrocyte membrane.[40] On the contrary, a recent paper[41] reports even about sensibilization against ionizing radiation in the presence of vitamin E. We have to take into account that higher exposures to ionizing radiation destroy also tissue vitamin E itself; some changes in serum tocopherol levels and its tissue distribution following irradiation have been reported.[42]

MECHANISMS OF THE EFFECT OF VITAMIN E

Vitamin E has been recognized as a biological lipid antioxidant for over 40 years since the discovery of its antioxidant property in 1941. This antioxidant ability is derived from its function as a free radical chain breaker. α-Tocopherol and other tocopherols are capable of efficiently quenching singlet oxygen $^1O_2$. Tocopherol quenches up to 120 molecules of $^1O_2$ per molecule of α-tocopherol before the latter is oxidized.[43] Besides this, tocopherols are reasonably good free-radical scavengers. These properties, in addition to the liposoluble character, make tocopherol an excellent candidate for being a protective agent in biological membranes that contain polyunsaturated fatty acids as well as oxidoreductases that produce reactive free radicals during their catalytic acticity.[44] A number of biological systems can convert oxygen to singlet oxygen $^1O_2$ to superoxide λO_2 and hydroxyl radicals (·OH). All these radicals are capable of either directly or indirectly initiating peroxidation reactions.[45]

Tocopherol interferes with the step of propagation of the lipid peroxidation process. In this step hydroperoxides are formed.

$$L· \text{ (lipid radical) } + O_2 \rightarrow LO_2^· \text{ (lipid hydroperoxy radical)}$$

$$LO_2^· \text{ is then converted to LOOH (hydroperoxide)}$$

α-Tocopherol reacts with $LO_2^·$; thus the amount of lipid radicals is diminished. Tocopherol prevents autoxidation mainly of polyunsaturated fatty acids. The process requiring NADPH and iron chelates[46] is as follows:

—CH=CH—CH$_2$—CH=CH—   (divinyl group in polyunsaturated
                                fatty acids)

          H

—CH=CH—ĊH— CH=CH—   (lipid radical)

      O$_2$

—CH=CH—CH— CH=CH—   (peroxy radical)

          O—O·

      H

—CH=CH—CH— CH=CH—   (hydroxyperoxide of fatty acid)

          O—OH

Peroxy radicals and hydroperoxides produce damage by oxidation of biologically important substances, thus damaging mainly biological membranes. Amino acids, proteins, enzymes, vitamin A, and other substrates are oxidized.

Free radicals are permanently generated in the body, and they are neutralized by numerous systems. Superoxide radical is the most important; its deteriorating influence is prevented by superoxide dismutase. When we compare the effect of tocopherol with that of other scavengers, the tocopherol effect seems to be less pronounced.[47]

The antioxidant function of tocopherols is obvious in vitro. In vivo, a number of symptoms due to vitamin E deficiency can be cured by the administration of nonphysiological antioxidants. This finding has been used as a strong argument that vitamin E itself acts as a physiological antioxidant.[16] Moreover, the vitamin E deficiency symptoms have been interpreted as secondary toxic effects of lipid hydroperoxides formed in the tissue.

However, the clearly documented theory of antioxidant function of tocopherol is not sufficient to explain all the effects of vitamin E in vivo (and its depletion as well).

Biological membranes, especially these of microsomes and to a less degree in mitochondria, are the essential loci where peroxidation of polyunsaturated fatty acids occurs. These fatty acids are found mainly in phospholipids. The normal state of these phospholipids maintains the integrity of the membranes.

Nowadays we know that tocopherol acts not only as an antioxidant, but also as a membrane stabilizer. The side chain of α-tocopherol *per se* can physically stabilize the biological membranes by specific physicochemical interactions between the phytyl side chain and the fatty acid chains of polyunsaturated phospholipids.[48] This is not surprising, since α-tocopherol is known to be distributed in biomembranes.[49,50] There is considerable evidence that free radicals attack polyunsaturated fatty acids in the membranes in vivo and that the degradation results in cell damage.

The detection and especially quantitative determination of lipid peroxidation in biological material is not an easy task.[51] Three procedures which have gained practical use in measuring lipid peroxidation are (1) the thiobarbituric acid (TBA) reaction, (2) the determination of conjugated dienes, and (3) the measurement of fluorescent products of lipoperoxidation.

The TBA reaction is the most widely used, although it has been criticized on several points, especially on its specificity.[52] In principle, this reaction is based on malondialdehyde, the main degradation product of peroxidized lipids. A better term is "TBA-reacting substances". The method cannot be applied to tissue extracts because of rapid metabolizing of malondialdehyde.

The second method fit to measure lipid peroxidation is the determination of conjugated dienes by ultraviolet spectrophotometry (absorption at 233 nm) under strictly anaerobic conditions.

Malondialdehyde forms secondary products that interact with other tissue constituents. Thus, fluorescent substances, both water- and lipid-soluble are synthesized, and may be determined with high sensitivity. The fluorescent substances are commonly named lipofuscins or age pigments.

There is a huge number of reports dealing both with lipoperoxidation in microsomes and the consequences caused by the peroxidation products in the microsomal fraction.[51]

Increased peroxidation also causes functional changes in the membranes. For example, increased permeability to Na^+ and K^+ in the erythrocyte membranes[53] and/or swelling and lysis of the mitochondria was observed. Plenty of enzymic changes were observed. For example, 10^{-5} molar linoleic acid hydroperoxide has been reported to cause complete inhibition of 6-phosphogluconate dehydrogenase, glucose-6-phosphate dehydrogenase, and glyceraldehyde-3-phosphate dehydrogenase.[54]

Numerous in vitro and in vivo experiments have proved that α-tocopherol inhibits the TBA-reactive material formation. For example, in microsomes of vitamin E-depleted rabbits, the intensity of the TBA reaction was 5.7 times higher than in control animals.[55] Such an effect was also shown[30] in the mitochondria of several animal species. The peroxidation of microsomal lipids was shown to be coupled with the formation of singlet oxygen accompanied by oxidation of NADPH by liver microsomes.[56] Recently, Lippman[57] demonstrated on human mitochondria that the scavenging effect of tocopherol against $O_2^{-\cdot}$ may be potentiated when used in a mixture of antioxidants and geroprotectors (α-tocopherol, butylated hydroxytoluene, "ACF 233", sulfur-containing amino acids).

Lipid peroxidation may be induced also by ionizing radiation[53] due to radiolysis of water with formation of free radicals. This increase was efficiently reduced by vitamin E as well.[30] It is well known that lipid peroxidation is promoted by phagocytosis. Besides this, the degree of lipoperoxidation (LPD) is influenced by the maturation of tissue. The intensity of ascorbate- and NADPH-induced lipoperoxidation[58] were shown to be dependent also on the age of rats (a decrease from 1 to 12 months). It must be emphasized that the activity of superoxide dismutase decreases with age as well.[59]

In addition, vitamin E changes the conditions for LPD. It was found that phospholipids of the testes of vitamin E-depleted rats contained considerably more arachidonic acid than those of animals with a normal vitamin E supply.[60] In vitamin E-deficient rats, an increased formation of arachidonic from linoleic acid could be shown; thus the degree of LPD could be increased.

Lipid peroxidation is closely bound with enzymic systems that ensure destruction of the oxidation products. Glutathione peroxidase (GSH-Px) and glutathione reductase (GSH-reductase) are the main factors representing the principal enzymes which remove harmful lipid hydroperoxides formed from polyunsaturated fatty acids and H_2O_2 (Figure 1).

Reduced glutathione (GSH) is a hydrogen donor for the GSH-Px reaction and is itself regenerated through GSH-reductase using NADPH.[61] The activity of GSH-Px, present mainly in cytosol, seems to be of great importance for removing hydroperoxides. On the other hand, catalase responsible for removing hydrogen peroxide does not react with lipid hydroperoxides.[62]

The activity of GSH-Px depends on numerous factors. First, the chain indicated in Figure 1 is active only if the supply of NADPH is sufficient, i.e., if the rate of phosphogluconate cycle of glucose catabolism operates normally. In addition, there are many drugs capable of modifying the GSH-Px activity (ethoxyquin, cholesterol, tri-o-cresyl phosphate, lipid peroxides, nucleosides, menadione, gentissic acid, potassium chlorate, and several metals such as silver). Moreover, genetically determined deficiency states of GSH-Px are known in humans.[63]

There is an extraordinarily important factor necessary for the activity of GSH-Px, namely selenium. The essential trace element selenium is an integral part of GSH-Px.[64] The activity of the enzyme in some tissues has been shown to depend on the dietary selenium intake,[65,66]

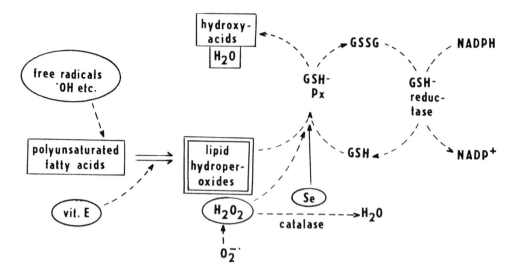

FIGURE 1.   Degradation of lipid hydroperoxides.

although high doses (>2 ppm) are toxic. Selenium acts as a water-soluble antioxidant. It is bound to proteins. In the rat and swine, diets deficient in tocopherol and sulfur-containing amino acids result in liver necrosis. In the sheep, calf, rabbit, and chick, muscular dystrophy occurs, while in birds exudative diathesis is observed. All these syndromes may be prevented or cured not only by administration of tocopherol, but also by selenium (in form of SeO_3^{2-}). Thus, the majority of the antioxidant activity associated with selenium in early dietary studies appears to be mediated through the GSH-Px activity.[45] There is an inverse relation between the level of GSH-Px in the liver and logarithm of dietary vitamin E ($r = -0.8$), excess of the intake depresses significantly the GSH-Px activity in the liver and other tissues.[61]

Aerobic organisms also possess other mechanisms which provide protection against uncontrolled free radical reactions and especially against lipid peroxidation.

Superoxide dismutase (SOD) is one of the most important factors in all aerobes. This enzyme catalyzes the following dismutation:

$$O^{-\cdot} + O^{-\cdot} + 2\,H^+ \rightarrow O_2 + H_2O_2$$

Hydrogen peroxide formed is removed by catalase and GSH-Px (see above). Although $O^{-\cdot}$ itself cannot initiate lipid peroxidation, it may be converted to $\cdot OH$ and $^1O_2$ radicals which are highly unstable and harmful toward biomembranes. Thus, SOD indirectly represents a major agent of oxygen toxicity. SODs are metalloproteins containing copper and zinc, or manganese. SOD has been shown[67] to directly reduce lipid peroxidation. An increase in SOD activity was observed when the diet was low in tocopherol.[19] The SOD activity is also relevant for the course of acute radiation syndrome.[68] The SOD activity may be correlated with lifespan in mammals. There exists a linear increase in the ratio of SOD to reciprocal specific metabolic rate [1/(cal/g/day)] as a function of maximum lifespan,[69] but this fact cannot support the role of SOD in aging.

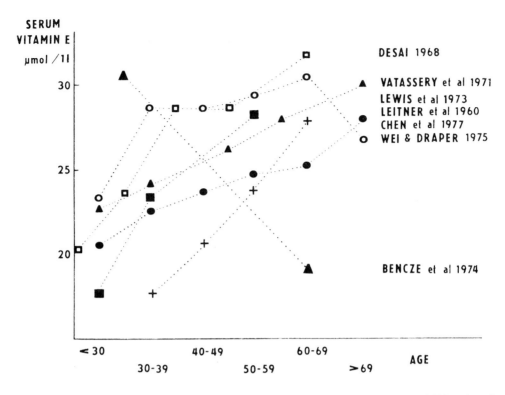

FIGURE 2. Mean serum vitamin E levels at different ages in control populations. (From Kelleher, J. and Losowsky, M. S., *Tocopherol, Oxygen, and Biomembranes*, de Duve, C. and Hayaishi, O., Eds., Elsevier North Holland, Amsterdam, 1978, 311. With permission.)

REQUIREMENT AND INTAKE OF VITAMIN E IN DIFFERENT AGE CATEGORIES

Adult Persons

I find no real scientific basis at present to estimate the optimum dietary requirement of vitamin E in man. The successful clinical use of vitamin E in man, though rather of empirical character, represents a sufficiently clear evidence for the importance of vitamin E. As summarized in the discussion about the functions of vitamin E and its role in redox processes, it is reasonable to assume that the intake of vitamin E is essential for adults despite the fact that the symptoms of deficiency are expressed neither in generative organs, nor in muscles and cardiovascular system in a typical form.

Numerous data are available as to the daily requirement of vitamin E in man. Substantially differing data have been published in the literature ranging from 12 mg/day in Middle Europe to 30 mg/day in the U.S. But Ketz[70] outlines that even the lowest numbers given (12 mg/day) are a little exaggerated with respect to the intake of polyunsaturated fatty acids.

Numerous reports have been published about the normal blood levels of vitamin E from various parts of the world. A satisfactory summary on this data is given in Figure 2. It includes studies from many countries (Canada, U.S.A., Great Britain, Hungary, Eskimo population, and China).

With one exception (Bencze), all the studies show a clear tendency to increase serum tocopherol with increasing age. The results of Wei and Draper are somewhat different — above the age of 70 the serum level falls again. However, the social positions of the populations investigated were not equal, therefore, the results must be accepted with caution.

It is useful to characterize the symptoms of long-lasting dietary deprivation of vitamin E. Above all, the consequences would be evaluated on the basis of changed blood values of tocopherol. Of course, in any subject with a low serum level, a diminished dietary intake of vitamin E must be considered, but to produce low blood levels, long-term dietary deprivation is required. A small percentage from the population reveals dangerously low blood levels, as reported from the U.S.[72] and Great Britain.[73]

Dietary restriction of vitamin E intake may also be evaluated according to the content of tocopherol in tissues of experimental animals. However, a decrease in tissue tocopherol was observed not only after a vitamin E-deficient diet, but also in animals fed various oils, since their consumption led to increasing need of vitamin E.

In the muscle (calves), normal tocopherol value of 0.95mg/100 g decreased if supplemented both with saturated and unsaturated fat.[2] Cod liver oil caused a decrease in the heart muscle from 0.44 to 0.32 mg/100 g.[74]

Aged Persons

The requirements of vitamin E in older individuals is assumed to be similar to younger adults. However, in rats this requirement decreases with age.[75]

Judging the requirement and daily intake of vitamin E in older individuals, consideration of their altered caloric intake is necessary. If the total caloric intake in control adults equals 100% then in age categories:[76]

65—79 years	Male	94.2%	80 years and more	Male	82.1%
	Female	95.3%		Female	78.2%

Such a decrease is related to reduced consumption of foods rich in vitamin E.

This decrease goes roughly parallel with diminishing basal metabolic rate with age. For Middle Europe, the mean basal metabolic rate in kJ/1.75 m^2/24 hr in both sexes is as follows:[15]

Age (years)	Men	Women	
18	7480	6630	(kJ = 4.16 × kcal)
24	7060	6220	
42	6630	6220	
66	6220	5810	
75	5810	5500	

Besides various pathological factors, some additional physiological influences alter the mean values (such as fasting, tropical climate). The daily intake of vitamins other than tocopherol in aged population is of significance. If the intake in the adult equals 100%, then the real consumption of vitamins in aged as a percent of the adult value is[76]

		65—79 years	80 years and above
Vitamin	Male	59.7	54.6
A, β-carotene	Female	50.8	39.7
Thiamine	Male	80.2	67.0
	Female	82.5	67.5
Riboflavin	Male	83.1	71.0
	Female	82.5	74.1
Niacin	Male	119.0	98.3
	Female	120.7	88.5
Ascorbic acid	Male	91.0	53.7
	Female	67.8	54.3

Certainly, there are numerous factors of geographic, social, or seasonal character which influence the allowance. This study has shown that the intake of vitamins by old populations inhabiting old people's homes has been higher than the recommended allowance, excepting lower values of vitamins A and C.

Now we will discuss the intake of vitamin E in the elderly in relation to their blood levels. As measured by Hejda,[76] daily consumption of vitamin E in aged population was 8.65 mg and this intake was roughly parallel to the decreased caloric consumption. Although the recommended allowance of vitamin E in adults is higher, some sources give 30mg/day[77] for man; the minimal requirement of vitamin E is probably not higher than 3 to 6 mg/day (Food and Nutrition Board, Washington 1974). The minimum requirement raises with increasing body weight.

How does one evaluate the changes of blood vitamin E concentration in aged population? The curves published here indicate their prevailing tendency to increased with age (Figure 2). There is no evidence that vitamin E deficiency is more frequent in elderly populations. Also, hemolysis test results and tissue storage data do not indicate any increasing tendency toward vitamin E deficiency with increasing age.[71] Other reports bring differing results. In 65% of the people over 65 years, the serum vitamin E level was lower than the lower limit found in a control population (under 630 g/100 mℓ), and 38% had levels less than 450 g/100 mℓ, the value proposed by Leonard et al.[78]) below which positive in vitro erythrocyte hemolysis occurs. It could be concluded in general, however, that the aged individuals have mostly blood concentrations of vitamin E not different from those found in the adults.

Factors Influencing the Requirement of Vitamin E

As explained in the foregoing sections, the requirement for tocopherol varies according to the accumulation of polyunsaturated fatty acids in the biomembranes. Thus, the requirement of vitamin E in man and animals is higher when large quantities of polyunsaturated fatty acids are ingested. The ratio tocopherol: polyunsaturated fatty acids (mg/g) in the food should be not lower than 0.4. If the consumption of vegetable oils rich in polyunsaturated fatty acids is extremely high (if it exceeds 20 g/day over the mean consumption), a sudden substitution of this diet for a diet containing a relatively small proportion of vegetable oils may result in relative vitamin E deficiency, since tissues are deprived of tocopherol more readily than of polyunsaturated fatty acids.[79] Foods rich in polyunsaturated fatty acids such as soy bean oil and other vegetable oils are also very rich in α– and γ–tocopherols.

The recommended daily intake seems to be not higher than about 5 mg/day if the individual has a slight reserve of linoleic acid and his diet is poor in this acid. Each gram of ingested polyunsaturated fatty acids raises the recommended intake of vitamin E by 0.5 to 0.8 mg.[70]

Vitamin A (retinol) is another factor of eminent significance in this sense. It was found as early as in 1939 that ingested vitamin A was deposited in the liver in larger amounts when the diet contained vitamin E than when the diet was vitamin E-deficient. Vitamin A reserves of rats were depleted more rapidly when both vitamin A and E were deficient. Nowadays it is known that retinol in vitro participates in radical reactions;[80] it quenches H^1O_2. High retinol intake in vivo depresses tissue tocopherol levels, indicating an oxidative role of excess vitamin A in tocopherol destruction.[81] Vitamin A contained in fish oils is partially destroyed by oxidation in vitamin E-deficient rats and chicks.

Dietary intake of selenium is a factor closely associated with the function of vitamin E. There is a striking similarity between the nutritional effects of vitamin E and selenium. Small amounts of selenium- and sulfur-containing amino acids were found to prevent many vitamin E deficiency symptoms. The explanation lies in the activity of glutathione peroxidase. A clear sparing effect of dietary vitamin E on selenium requirement and vice versa was demonstrated.[82] The following chain of causes and consequences is valid:

High level of vitamin E $\rightarrow$ ↓ peroxides in the membranes

= sparing the need for glutathione peroxidase (decomposing

the peroxides formed) = sparing of selenium

(all the steps vice versa as well)

Vitamin E supplementation was shown to reduce the selenium content of the liver and myocardium and to raise its amount in the kidney.[83] These effects are probably due to an elevated excretion of this element.

IMPORTANCE OF VITAMIN E IN THE OLD POPULATION

Possible Relationship between Vitamin E and the Lifespan

The question arises of how longevity may be altered by a higher or lower generation of free radicals. The observed changes of membrane peroxidation with aging may be retarded by antioxidants, especially by naturally occurring vitamin E. It has therefore been suggested that aging may also be retarded on the basis of the "free radical theory of aging".

The probable cause of the prolongation of the lifespan evidenced by some authors in animals treated with vitamin E need not be due exclusively to the antioxidant effect of tocopherol toward polyunsaturated lipids. An additional mechanism is the formation or inhibition of tissue protein crosslinks. This has been designated as the "crosslinking theory of aging."[84] There are numerous potential crosslinking agents reviewed recently by Bjorn-sten.[85] The crosslinking influence most thoroughly studied is that on crosslinking elements of macromolecules of connective tissue such as collagen and elastin. These crosslinkages are represented mostly by lysine-derived bridges and also crosslinks from tyrosine residues. The experiments with lathyrogen are in full agreement with the crosslinking theory. β-Aminopropionitrile was shown to inhibit not only the formation of crosslinks, but also to prolong the lifespan of animals.[86] Analogous changes may be observed in many types of proteins and enzymes; this represents the theoretical basis for the application of the cross-linking theory of aging. We may briefly say "proteins are step-by-step deteriorated in their functions".

There are many pathological processes determining the mean length of human life—diseases attacking mostly the old population. The relationship of these diseases to vitamin E is probably the most significant practical aspect discussed in this article. In animals, vitamin E deficiency can cause symptoms of many of these pathological states, but only exceptionally in man. Nevertheless, vitamin E is also highly important in these human diseases, namely as a therapeutic mean, naturally administered in large doses. It does not matter that the principle of the therapeutical use is mostly just empirical. The decisive factor is whether the treatment is successful or not.

Basic data about the influence of vitamin E on the blood vessels and atherosclerosis were obtained in experimental animals. Several papers on the influence of vitamin E in chicks may be summarized by saying that high doses of vitamin reduced atheromatous processes, and total lipids of the aorta wall are diminished.[87] In the rabbits fed excess of cholesterol, cholesterolemia increases and the accumulation of lipids in all three layers of the aorta occurs, mainly in the endothelium. Administration of a mixture of vitamin A and E had no effect on cholesterolemia, but the lipid content of the aorta (again mainly in endothelium) was markedly lower.[88]

The study of the relationship between vitamin E and atherosclerosis in man is incomparably more complicated. As to the individual signs, a decrease of serum cholesterol has been reported after vitamin E treatment[89] that turns, however, back to initial values again.[90] Other authors deny such an influence.[23]

Large-scale clinical studies are extraordinarily valuable. A complex study from seven German clinics[91] on 269 patients with arteriosclerosis gave rather negative results. Only subjective symptoms were improved, while blood lipids, peripheral vascular signs, symptoms of angina pectoris, and lifespan remained like in controls. The improvement of stenocardia was found also in another study.[92] On the other hand, a decrease of cholesterolemia was reported elsewhere.[93]

Arteriosclerotic retinopathy was shown to be substantially improved after a combined vitamin A and E therapy.[94]

Early stages of myocardial infarction reveal a moderate decrease of serum levels of tocopherol and increase of malondialdehyde within 48 hr of the acute event. The same change was found in patients after stroke.[95] Experimental results with stroke-prone strains of hypertensive rats evidenced that chronic administration of vitamin E neither markedly lowered hypertension nor decreased the incidence of stroke. However, the lifespan was significantly prolonged; this beneficial effect of tocopherol may be related to the improvement of biomembrane characteristics.[96] A recent finding of Farnworth et al.[97] is very interesting. The authors observed a dramatic lowering of the myocardial lesion incidence in rats fed diets enriched in saturated fatty acids.

Much has been done in this respect in the study of arteriosclerosis of leg arteries. In 88% of patients suffering from arteriosclerosis of leg arteries treated with α-tocopherol (average 4.6 years) improved walking distance and better arterial flow were demonstrated.[98] A large group of patients (1476) with intermittent claudication was treated with tocopherol, and a positive shift was found; not only the walking distance was lengthened, but also a certain prolongation of the lifespan of the cured patients was observed.[99] Besides this, a lowered content of tocopherol in the skeletal muscles of elderly men suffering from occlusive arterial disease was found.[100]

In other (nonarteriosclerotic) peripheral vascular diseases, a beneficial therapeutic effect of large doses of vitamin E was demonstrated as well. The effect seems to be due not to vasodilation, but is more likely associated with the improvement of tissue hypoxemia by tocopherol. Tissue hypoxemia is known to damage phospholipids.[101]

There are numerous indications to a successful therapy of various vascular diseases such as:

Diseases of veins: Acute and noncomplicated chronic
                Thrombophlebitis (treatment usually together with dicumarol)
                Postphlebitic syndrome
                Varices
                Indolent ulcus cruris (varicose and ischemic cases)
Diseases of arteries: Occlusive arteriosclerosis (mainly in diabetics)
                Buerger disease
                Raynaud disease

Malnutrition is a widely occurring phenomenon in the elderly. As mentioned above, the states associated with malnutrition have much to do with vitamin E. Absorption may generally be reduced in the elderly. Some authors[95-102] report that malabsorption syndrome is connected with low serum levels of vitamin E, but according to Kelleher and Losowsky,[71] no significant reduction of the absorption of vitamin E can be demonstrated. The absorption is also negatively altered in sprue and long-lasting steatorrhea[4]. Leonard et al.[103] reported that a high proportion of patients after gastric surgery (mainly after vagotomy and antrectomy) developed low plasma levels of vitamin E and abnormal in vitro red cell hemolysis. However, other authors[71] saw only insignificant differences in all age groups. Naturally, intestinal resection is a further cause of malnutrition.

In old populations, especially in men, relatively many persons suffer from ethanol-induced fatty liver. This pathological process in animals can be modified by treatment with antioxidants. Therefore, there is a hypothesis that the enhanced lipid peroxidation is an important pathogenetic factor. Numerous papers dealing with this problem have been reviewed by Plaa and Witschi.[51]

The finding that dietary supplementation with vitamin E prevents carcinogenesis after methylcholanthrene is of questionable significance for man.[104]

Changes Conditioned by the Transport System

The saturation of the body with vitamin E may be conditioned by the system competent to transport fat-soluble tocopherols. The concentrations of blood and tissue lipids increase with age. Blood lipids are transported by the help of lipoproteins. Low-density lipoproteins (LDL) and the LDL:HDL ratio are the main factors which induce atherogenesis, as thoroughly reviewed by Lewis and Naito.[105] The LDL fraction does not increase linearly with age. After 55 years in men and after 60 in women, the values do not raise any more, and after 70 years the average value is significantly lower than during the culmination.

α-Tocopherol was demonstrated to be bound to low-density lipoproteins (β-lipoproteins).[11] About 41 to 58% of total serum tocopherol was shown to be carried by LDL (S_f 3-9 fractions), whereas HDL (high-density lipoproteins) transfer only 29 to 37%. It was shown that II and IIb hyperlipidemias were associated with an increase in serum tocopherol.[95] On the contrary, abetalipoproteinemia reveals absolutely zero level of serum tocopherol. In some papers,[106,107] a very close positive correlation between serum tocopherol and β-lipoproteins (LDH) and with other serum lipids was found, but other reports deny it.[71]

The clearly pronounced transport function of β-lipoproteins is supported by the finding that oral administration of vitamin E is not able to increase the serum tocopherol level in abetalipoproteinemia.[108] The studies in humans and animals suggest that supplementation of the diet with α-tocopherol can cause a reduction in the serum cholesterol concentrations, especially in subjects with higher than normal levels of cholesterol. After tocopherol administration, high-density lipoproteins increase as well.[108] This is a beneficial effect.

One may conclude that low tocopherol levels may be due to diminished circulating transport system, but much more work must be carried out to prove or deny it.

Influencing the Degree of Lipoperoxidation by Nutritional Changes

It is essentially possible to change the composition of biomembranes and thus to alter peroxidation of lipids. There is much evidence that the diet is capable of changing the composition of the membranes. For example, during 3 months, the relation of long-chain polyunsaturated fatty acids (with 22 carbons) of heart muscle phospholipids increased substantially by feeding with cod liver oil,[109] and supplementation of the diet with corn oil resulted in an increase of malondialdehyde in microsomes and mitochondria.[110] Our experience with mice fed highly unsaturated oil from halibut agrees with the above-mentioned experiments.

Influence of Vitamin A on Vitamin E

Carotenoid pigments have been demonstrated to be quenchers of $^1O_2$ in vitro and in vivo. High vitamin A intakes depress tissue vitamin E levels, indicating the oxidative role of excess vitamin A in tocopherol destruction. Vitamin E improves the state after minimal vitamin A intake. Tocopherol has a pronounced protective (antioxidant) action toward dietary vitamin A.

Relation of Vitamin E to Oxidant Gases and Toxic Drugs

Vitamin E has a clean-cut relationship to oxygen and some noxious gases and drugs in the environment which contribute to pollution of air and water or used in industry and agriculture. These substances, acting, for example, in toxic concentrations in smog, are a cause of relevant changes in human beings, whose secondary complications shorten the lifespan. Accelerated free radical reactions and accompanying peroxidation processes are the cause of the changes in certain intoxications.

Air oxygen can be considered a potentially toxic gas for the lung. Oxygen was shown to be an element initiating lipid peroxidation through formation of singlet oxygen and superoxide and hydroxyl radicals. These active forms of oxygen which are produced from endogeneous oxidation reactions as well as from oxidation of xenobiotics are capable of either directly or indirectly initiating peroxidation reactions.[45] Lipid peroxidation is an important consequence of exposure to normobaric or hyperbaric oxygen. However, as reviewed by Plaa and Witschi,[51] in vivo α-tocopherol failed to raise the concentrations of lipid peroxides in the lung after hyperbaric oxygen, although in vitro the formation was enhanced.

On the other hand, the influence of ozone and nitrogen dioxide on the in vivo-produced lipid peroxides in the lung is unambiguous; the degree of peroxidation is increased considerably.[51] However, the presence of vitamin E does not protect against in vitro lipid peroxidation by ozone.[111] More recent studies indicate that the lung has other effective mechanisms which may protect against this peroxidation induced by oxidant gases, e.g., superoxide dismutases.[112]

Lipid peroxidation in the lung need not necessarily be produced by oxidant gases alone. Extensive lung injury will be produced by the herbicide paraquat. Its toxicity is greatly enhanced by oxygen,[113] and also it is significantly increased in selenium- and tocopherol-deficient mice.[114] Additional superoxide dismutase prevents the paraquat toxic effects.

Facultative Beneficial Effect of Vitamin E on Bacterial Resistance

Vitamin E, when supplemented to diets in 150 to 300 mg/kg doses, significantly increases the resistance of mice and chicks to bacterial infections (*E. coli, D. pneumoniae*). Vitamin E enhances humoral immunity and the number of antibody-forming cells both in vitro and in vivo, and elevates IgG and IgM antibody titers.[25]

Importance of Vitamin E for the Formation of "Age Pigments"

An appearance of various aberrant forms of lipids is the ultimate product of peroxidation. These forms are no longer capable of fulfilling the specific role in the cell. These waste products are named age pigments — lipofuscins, and they are accumulated in various organs not only in man,[115] but also in animals. In cases of vitamin E deficiency in animals, inordinate rate of accumulation of lipofuscin in heart muscle in mice depends on age; the levels of lipofuscin in 28-month-old mice supplemented with vitamin E are shown to be comparable with the levels in the controls 5 or 6 months earlier in life.[116]

In rat experiments, an accumulation of brown "ceroids", two pigments occurring by polymerization of autoxidation products of polyunsaturated fatty acids, was found in depot fat of vitamin E-deficient rats.[14]

All these unmetabolizable fluorescent pigments are known to be synthesized from malodialdehyde.[117] Their formation in the brain, adrenals, and muscle may be effectively prevented by antioxidants.[118]

As yet no deleterious consequences of these age pigments have been reported. There is also lack of any concomitant effects on longevity. Moreover, an analogous accumulation of lipofuscin is caused by γ-irradiation and ultraviolet light.

Pharmacological Use of Vitamin E in Other Pathological Processes Occurring in the Aged

At first, the effect of vitamin E in higher doses is to be understood not as a vitaminous one, but it has pharmacodynamical character.

Vitamin E has to do also with several groups of diseases attacking all age categories.

Some Liver Injuries

In some tocopherol-deficient rats with protein malnutrition, massive acute hepatic necrosis develops, but this alteration occurs only if the diet contains appreciable quantities of polyunsaturated fats. This observation is an evidence of the relationship between vitamin E and unsaturated fatty acids. Liver necrosis in the rat belongs to the syndromes due most likely to a combined deficiency of tocopherol and selenium. However, dietary liver necrosis in the rat may be prevented even without selenium, when diets very rich in proteins or sulfur-containing amino acids (cystine) together with tocopherol are administered. This fact may be helpful to explain the decreased levels of vitamin E in the liver of elderly humans.

It is known that lipid peroxidation could be one of the principles of the hepatotoxicity of CCl_4 and $CBrCl_3$. A great increase of conjugated dienes, and a little later, of malondialdehyde has been reported. Vitamin E has a protective action not only toward nutritional liver necrosis, but also against necrosis following CCl_4. The amount of accumulated liver fats decreases considerably after tocopherol administration.[119] Selenium can significantly reduce the amount of malondialdehyde in the liver of CCl_4-treated animals,[120] and this is consistent with the well-known antioxidant properties of this element. The role of selenium has been elucidated in the above paragraph dealing with glutathione peroxidase.

Neurological Disorders

A concise description of indications of a variety of neurological disorders, where large doses of vitamin E have been successfully used in clinical practice of the gerontologists follows:

1. Disturbances of blood circulation in the brain (functional cerebrovascular insufficiency)
2. Neuralgias connected with joint pains
3. Secondary neurotic syndromes of asthenic type (with early onset of senility)
4. Secondary myopathic syndromes of the aged (not progressive muscular atrophy), muscular atrophy ex-inactivitate
5. Bechterew's disease, fibrositis, and tendinitis
6. Postclimacteric vasomotory disorders

Various Rheumatological and Dermatological Diseases

With a certain degree of success, disseminated lupus erythematosus, dermatomyositis, periarteritis nodosa, progressive scleroderma, rheumatoid arthritis, psoriasis, keloid formation, xanthomatosis, early phase of Dupuytren's contraction, kraurosis of the vulva, and induratio penis plastica were all treated with vitamin E.

CONCLUSIVE RECOMMENDATIONS FOR THE NUTRITION OF OLD PEOPLE

A Relatively High Intake of Vitamin E

The elderly should have a level of vitamin E intake which will maintain the serum level of α-tocopherol in the normal range (above 0.5 mg/100 mℓ), and ensure the proper relation of vitamin E to polyunsaturated fatty acids in tissues. An excess daily intake is to be avoided excepting the cases with substantially higher body weight, or cases with impaired intestinal absorption. One must be aware of a possible vitamin E-like effect of aspirin during its long-lasting administration. Basic laboratory tests of low dietary intakes are: determination of serum α-tocopherol, estimation of osmotic resistance of blood red cells after dialuric acid or hydrogen peroxide, and excretion of creatine and 1-methylhistidine.

Balance Between the Dietary Intake of Vitamin E and Polyunsaturated Fatty Acids

The practical conclusion is not to increase too much consumption of the foods rich in polyunsaturated fatty acids (vegetable oils not more than 20 g/day). Of course, this recommendation contradicts the generally accepted instructions and practice of the "antiatherogenic" diet with a high proportion of unsaturated fats and a pronounced reduction of animal fats. A control of the lipid transporting system is advisable (lipoproteins).

There must be a guaranteed supply of suitable quantities of vitamin A, (β-carotene), selenium, and sulfur-containing amino acids (cystine). Also, ascorbic acid must be supplied in proper amounts, mainly during the winter and spring months.

Choice of Proper Foods as to their Technological Treatment

It is useful to choose those vegetable oils that contain more vitamin E and less polyunsaturated fatty acids, to take proper types of flour (to reject bleached flour and polished rice), and to prefer margarines and other processed fats artificially vitaminized. As a preventive factor, vitamin E may be considered to be used in old population rather in mixtures with other antioxidants and geroprotectors than as a sole drug administered in large doses.

REFERENCES

1. **Katz, H. A.,** *Grundriss der Ernährungslehre,* G. Fischer, Jena, 1978, 242.
2. **Dicks, M. W.,** Vitamin E content of foods and feeds for human and animal consumption, Bulletin 435, University of Wyoming, Laramie.
3. **Weber, F.,** Resorption und Stoffwechsel der Vitamine E und K und Verwandter Verbindungen, in *Vitamin A, E, and K. Clinical and Physiological Problems,* Kress, H. and Blum, K. U., Eds., K. K. Schattauer Verl., Stuttgart, 1969.
4. **Binder, H. J., Herting, D. C., Hurst, V., Finch, S. C., and Spiro, H. M.,** Tocopherol deficiency in man, *N. Engl. J. Med.,* 273, 1289, 1965.
5. **Farrell, P. M., Bieri, J. G., Fratantoni, J. F., Wood, R. E., and Sant'Agnese, P. A.,** The occurrence and effects of human vitamin E deficiency. A study in patients with cystic fibrosis, *J. Clin. Invest.,* 60, 233, 1977.
6. **Kelleher, J. and Losowsky, M. S.,** Absorption of α-tocopherol in man, *Br. J. Nutr.,* 24, 1033, 1970.
7. **Gassmann, B., Proll, J., Schmandke, H., and Rodel, W.,** Vergleichende Untersuchungen über die Vitaminresorption, *Int. Z. Vitaminforsch.,* 36, 117, 1966.
8. **Lussier, D. M. and Roy, R. M.,** Distribution of vitamin E among tissues and subcellular fractions in sublethally irradiated mice, *Radiat. Res.,* 70, 236, 1977.
9. **Hauswirth, J. W. and Nair, P. P.,** Aspects of vitamin E in the expression of biological information, *Ann. N.Y. Acad. Sci.,* 203, 111, 1972.
10. **Kayden, H. J.,** The transport and distribution of α-tocopherol in serum lipoproteins and the formed elements of the blood, in *Tocopherol, Oxygen and Biomembranes* de Duve, C. and O. Hayaishi, O., Eds., Elsevier/North Holland, Amsterdam, 1978, p. 131.
12. **Bieri, J. G. and Farrell, P. M.,** Vitamin E, *Vitam. Horm. (Leipzig),* 34, 31, 1976.
13. **Dam, H. and Sondergaard, E.,** The relationship between vitamin E and dietary polyunsaturated fatty acids, in *Vitamin A, E, and K. Clinical and Physiological Problems,* Kress, H. and Blum, K. U., Eds., K. K. Schattauer Verl., Stuttgart, 1969, 277.
14. **Porta, E. A. and Hartroft, W. S.,** Lipid pigments in relation to aging and dietary factors (lipofuscins), in *Pigments in Pathology,* Wolman, M., Ed., Academic Press, N.Y., 1969, 191.
15. **Baessler, K. H., Fekl, W., and Lang, K.,** *Grundbegriffe der Ernährungslehre,* Springer-Verlag, Berlin, 1973.
16. **Boguth, W.,** Action of vitamin E, *Vitam Horm. (New York),* 27, 1, 1969.
17. **Guarnieri, C., Flamingi, F., Ferrari, R., and Caldera, C. M.,** Studio della Funzionalia Mitocondriale nel Muscolo Cardiaco di Conigli Alimentari in Cadenza di α-Tocoferolo, *Acta Vitaminol. Enzymol.,* 3, 86, 1981.
18. **Page, A. C., Jr., Gale, P. H., Koninszy, F., and Folkers, K.,** Coenzyme Q. IX. Coenzyme Q_9 and Q_{10} content of dietary components, *Arch. Biochem. Biophys.,* 85, 474, 1959.
19. **Sklan, D., Rabinowitch, H. D., and Donogue, S.,** Superoxide dismutase: effect of vitamin A and E, *Nutr. Rep. Int.,* 24, 551, 1981.
20. **March, B. E., Wong, E., Seier, L., Sim, S. J., and Biely, J.,** Hypervitaminosis E in the chick, *J. Nutr.,* 103, 371, 1973.

21. **Eskelson, C. D. and Jacobi, H. P.,** Effects of vitamin A, K, and E on in vitro cholesterol biosynthesis, in *Vitamin A, E, and K. Clinical and Physiological Problems,* Kress, H. and Blum, K. U., Eds., K. K. Schattauer Verl., Stuttgart, 1969, 321.

22. **Vessby, B., Lithell, H., and Boberg, J.,** Supplementation with vitamin E in hyperlipidemic patients treated with diet and clotfibrate, *Am. J. Clin. Nutr.,* 30, 517, 1977.

23. **Nikitin, J. P.,** Influence of vitamin E on lipids and blood coagulation in patients suffering from atherosclerosis, *Vopr. Pitaniya (Moscow),* 21, 6, 22, 1962.

24. **Yang, N.-Y. J. and Desai, I. D.,** Effect of high levels of dietary vitamin E on liver and plasma lipids and fat soluble vitamins in rats, *J. Nutr.,* 107, 1418, 1977.

25. **Tengerdy, R. P., Heinzerling, R. H., and Mathias, M. M.,** Effect of vitamin E on disease resistance and immune responses, in *Tocopherol, Oxygen, and Biomembranes,* de Duve, C. and Hayaishi, O., Eds., Elsevier/North Holland, Amsterdam, 1978, 191.

26. **Yusunaga, T., Kato, H., Ohgaki, K., Inamoto, T., and Hikasa, Y.,** Effect of vitamin E as an immunopotentiation agent for mice, *J. Nutr.,* 112, 1075, 1982.

27. **Sulkin, N. M. and Sulkin, D. F.,** Intimal lesions in arteries of vitamin E deficient rats, *Proc. Soc. Exp. Biol. Med.,* 103, 111, 1960.

28. **Kanimura, M.,** Effect of vitamin E on skin microcirculation, *Vitamins (Japan),* 33, 166, 1966.

29. **Tudhope, G. R. and Hopkins, J.,** Lipid peroxidation in human erythrocytes in tocopherol deficiency, *Acta Haematol.,* 53, 98, 1975.

30. **Inouye, B., Aono, K., Iida, S., and Utsumi, K.,** Influence of superoxide generating system, vitamin E, and superoxide dismutase on radiation consequences, *Physiol. Chem. Phys.,* 11, 151, 1979.

31. **Steiner, M.,** Inhibition of platelet aggregation by alphatocopherol, in *Tocopherol, Oxygen, and Biomembranes,* de Duve, C. and Hayaishi, O., Eds., Elsevier/North Holland, Amsterdam, 1978, 143.

32. **Machlin, L.,** Vitamin E and prostaglandins, in *Tocopherol, Oxygen, and Biomembranes,* de Duve, C. and Hayaishi, O., Eds., Elsevier/North Holland, Amsterdam, 1978, 179.

33. **Harman, D.,** Free radical theory of aging: effect of amount and degree of unsaturation of dietary fat on mortality rate, *J. Gerontol.,* 26, 451, 1971.

34. **Miquel, J. and Johnson, J. E.,** Effect of various antioxidants on the life span and lipofuscin of Drosophila and of C57BL/6J mice, *Gerontologist,* 15, 25, 1975.

35. **Ledvina, M. and Hodanova, M.,** The effect of simultaneous administration of tocopherol and sunflower oil on the lifespan of female mice, *Exp. Gerontol.,* 15, 67, 1980.

36. **Blackett, A. D. and Hall, D. A.,** The effect of vitamin E on mouse fitness and survival, *Gerontology,* 27, 133, 1981.

37. **Leibovitz, B. and Siegel, B. V.,** Aspects of free radical reactions in biological systems: aging, *J. Gerontol.,* 35, 45, 1980.

38. **Sakamoto, K. and Sakka, M.,** Reduced effect of irradiation on normal and malignant cells irradiated in vivo in mice pre-treated with vitamin E, *Br. J. Radiol.,* 46, 538, 1973.

39. **Malick, M., Roy, R., and Sternberg, J.,** Effect of vitamin E on post-irradiation death in mice, *Experientia,* 34, 1216, 1978.

40. **Prince, E. and Little, J.,** The effect of dietary fatty acids and tocopherol on the radiosensitivity of mammalian erythrocytes, *Radiat. Res.,* 53, 49, 1973.

41. **Kagerud, A. and Peterson, H.-I.,** Tocopherol in irradiation of experimental neoplasms, *Acta Radiol. Oncol.,* 20, 97, 1981.

42. **Newmark, H. L., Pool, W. R., Bauernfeind, J. C., and De Ritter, E.,** Biopharmacological factors in parenteral administration of vitamin E, *J. Pharm. Sci.,* 64, 665, 1975.

43. **Fahrenholtz, S. R., Doleiden, F. H., Trozzolo, A. M., and Lamola, A. A.,** Quenching of singlet oxygen by α-tocopherol, *Photochem. Photobiol.,* 20, 505, 1974.

44. **McCay, P. B., Fong, K.-L., Lai, E. K., and King, M. M.,** Possible role of vitamin E as a free radical scavenger and singlet oxygen quencher in biological systems which indicate radical-mediated reactions, in *Tocopherol, Oxygen, and Biomembranes,* de Duve, C. and Hayaishi, O., Eds., Elsevier/North Holland, Amsterdam, 1978, 41.

45. **Bus, J. S., and Gibson, J. E.,** Lipid peroxidation and its role in toxicology, in *Review of Biochemistry and Toxicology,* Hodgson et al., Eds., Elsevier/North Holland, Amsterdam, 1979, 125.

46. **Svingen, B. A., Buege, J. A., O'Neil, F. O., and Aust, S. D.,** The mechanism of NADPH-dependent lipid peroxidation, *J. Biol. Chem.,* 254, 5892, 1979.

47. **Wills, E. D.,** Effect of antioxidants on lipid peroxide formation in irradiated synthetic diets, *Int. J. Radiat. Biol.,* 37, 403, 1980.

48. **Lucy, J. A.,** Functional and structural aspects of biological membranes. Suggested structural role for vitamin E in the control of membrane permeability and stability, *Ann. N.Y. Acad. Sci.,* 203, 4, 1972.

49. **Nakamura, T. and Hishinuma, I.,** Protective effect of tocopherol on the formation of TBA-reactive material in rat liver, in *Tocopherol, Oxygen, and Biomembranes,* de Duve, C. and Hayaishi, O., Eds., Elsevier/North Holland, Amsterdam, 1978, 95.

50. **Taylor, S. L. and Tappel, A. L.,** Effect of dietary antioxidants and phenobarbital pretreatment on microsomal lipid peroxidation and activation by carbon tetrachloride, *Life Sci.,* 18, 1151, 1976.

51. **Plaa, G. L. and Witschi, H.,** Chemicals, drugs, and lipid peroxidation, *Annu. Rev. Pharmacol. Toxicol.,* 16, 125, 1976.

52. **Gray, J. I.,** Measurement of lipid oxidation: a review, *J. Am. Oil Chem. Soc.,* 55, 539, 1978.

53. **Purohit, S. C., Bisby, R. H., and Cundall, R. B.,** Chemical damage in gamma-irradiated human erythrocyte membranes, *Int. J. Radiat. Biol.,* 38, 159, 1980.

54. **Khandwala, A. and Gee, J. B. L.,** Linoleic acid hydroperoxide: impaired bronchial uptake by alveolar macrophages, a mechanism of oxidant gas injury, *Science,* 182, 1364, 1973.

55. **Tappel, A. L. and Zalkin, H.,** Inhibition of lipid peroxidation in microsomes by vitamin E, *Nature (London),* 185, 35, 1960.

56. **King, M. M., Lai, E. K., and McCay, P. B.,** Singlet oxygen production associated with enzyme-catalyzed lipid peroxidation in liver microsomes, *J. Biol. Chem.,* 250, 6496, 1975.

57. **Lippman, R. D.,** The prolongation of life: a comparison of antioxidants and geroprotectors versus superoxide in human mitochondria, *J. Gerontol.,* 36, 550, 1981.

58. **Lemechko, V. V. and Nikitchenko, j. V.,** Peroxidation of lipids of rat liver mitochondria during aging and thyroid hyperfunction, *Biokhimiya (Moscow),* 47, 752, 1982.

59. **Massie, H. R., Aiello, V. R., and Iodice, A. A.,** Changes with age in copper and superoxide dismutase levels in brains of C57BL/6J mice, *Mech. Ageing Dev.,* 10, 93, 1979.

60. **Demopoulos, H. B., Flamm, E. S., Pietronigro, D. D., and Seligman, M. L.,** The free radical pathology and the microcirculation in the major central nervous system disorders, *Acta Physiol. Scand. Suppl.* 492, 91, 1980.

61. **Yang, N.-Y. and Desai, I. D.,** Glutathione peroxidase and vitamin E interrelationship, in *Tocopherol, Oxygen, and Biomembranes,* de Duve, C. and Hayaishi, O., Eds., Elsevier/North Holland, Amsterdam, 1978, 233.

62. **Sies, H. and Summer, H.-H.,** Hydroperoxide-metabolising system in rat liver, *Eur. J. Biochem.,* 57, 503, 1975.

63. **Steinberg, M. H. and Necheles, T. F.,** Erythrocyte glutathione peroxidase deficiency, *Am. J. Med.,* 50, 542, 1971.

64. **Rotruck, J. T., Pope, A. L., Ganther, H. E., Swanson, A. B., Hafeman, D. G., and Hoekstra, W. G.,** Selenium: biochemical role as a component of glutathione peroxidase, *Science,* 179, 588, 1973.

65. **Hafeman, D. G., Sunde, R. A., and Hoekstra, W. G.,** Effect of dietary selenium on erythrocyte and liver glutathione peroxidase in the rat, *J. Nutr.,* 104, 580, 1974.

66. **Serfass, R. E. and Ganther, H. E.,** Effect of dietary selenium and tocopherol on glutathione peroxidase and superoxide dismutase activities in rat phagocytes, *Life Sci.,* 19, 1139, 1976.

67. **Mazeaud, F. and Michelson, A. M.,** Inhibition de l'autoxydation des acides gras polyunsatures par le superoxide dismutase, *Ann. Nutr. Aliment.,* 34, 351, 1980.

68. **Krizala, J., Kovarova, H., Kratochvilova, V., and Ledvina, M.,** Importance of superoxide dismutase for the acute radiation syndrome, in *Biological and Clinical Aspects of Superoxide and Superoxide Dismutase,* Bannister, W. H. and Bannister, J. V., Eds., Elsevier/North Holland, N.Y., 1980, 327.

69. **Sullivan, J. L.,** Superoxide dismutase, longevity and specific metabolic rate, *Gerontology.,* 28, 242, 1982.

70. **Ketz, H.-A.,** *Grundriss der Ernährungslehre,* G. Fischer, Jena, 1978, 242.

71. **Kelleher, J. and Losowsky, M. S.,** Vitamin E in the elderly, in *Tocopherol, Oxygen, and Biomembranes,* de Duve, C. and Hayaishi, O., Eds., Elsevier/North Holland, Amsterdam, 1978, 311.

72. **Bieri, J. C., Teets, L., Belavady, B., and Andrews, E. L.,** Serum vitamin E levels in a normal adult population in the Washington, D.C. area, *Proc. Soc. Exp. Biol. Med.,* 117, 131, 1964.

73. **Leitner, Z. A., Moore, T., and Sharman, I. M.,** Vitamin A and vitamin E in human blood. II. Levels of vitamin E in the blood of British men and women, 1952-7, *Br. J. Nutr.,* 14, 281, 1960.

74. **Chatterton, R. T., Jr., Hazzard, D. G., Eaton, H. D., Dehority, B. A., Grifs, A. P. J., and Gosslee, D. G.,** Tissue storage and apparent absorption of α- and γ-tocopherols by Holstein calves fed milk replacer, *J. Dairy Sci.,* 44, 1061, 1961.

75. **Bieri, J. G.,** Kinetics of tissue α-tocopherol depletion and repletion, *Ann. N.Y. Acad. Sci.,* 203, 181, 1972.

76. **Hejda, S.,** Nutrition and Nutritional State of Aged Population, Prague, 1967, 56.

77. **Robinson, C.,** *Normal and Therapeutic Nutrition,* 14th ed., McMillan, N.Y., 1972.

78. **Leonard, P. J. and Losowsky, M. S.,** Relation between plasma vitamin E level and peroxide hemolysis test in human subjects, *Am. J. Clin. Nutr.,* 20, 795, 1967.

79. **Glatzel, H.,** *Ernährung, Ernährungskrankheiten, Appetitlosigkeit,* Urban-Schwarzenberg Verl., Munich, 1976, 67.

80. **Lichti, F. U. and Lucy, J. A.,** Reactions of vitamin A with acceptors of electrons: Formation of radical ions from 7,7;8,8-tetracyanoquinodimethane and tetrachloro-1,4-dibenzoquinone, *Biochem. J.,* 112, 221, 1969.

81. **Combs, G. F.,** Differential effects of high dietary levels of vitamin A on the vitamin E-selenium nutrition of young and adult chicks, *J. Nutr.,* 106, 967, 1976.

82. **Thompson, J. N. and Scott, M. L.,** Role of the selenium in the nutriton of the chicks, *J. Nutr.,* 106, 967, 1976.

83. **Yang, N. -Y. J. and Desai, I. D.,** Glutathione peroxidase and vitamin E interrelationship, in *Tocopherol, Oxygen, and Biomembranes,* de Duve, C. and Hayaishi, C., Eds., Elsevier/North Holland, Amsterdam, 1978, 233.

84. **LaBella, F. S.,** Pharmacological retardation of aging, *Gerontologist,* 6, 46, 1966.

85. **Bjorksten, J.,** A unifying concept for degenerative diseases, *Comp. Ther.,* 4, 44, 1978.

86. **LaBella, F. S. and Vivian, S.,** Beta-aminopropionitrile promotes longevity in mice, *Exp. Gerontol.,* 13, 251, 1978.

87. **Weitzel, G.,** Influencing atherosclerosis with fat-soluble vitamins, *Bull. Schweiz. Akad. Med. Wiss.,* 13, 356, 1957.

88. **Horn, Z., Palkovits, M., and Scher, A.,** Preventive action of vitamins A and E on the development of cholesterol-induced atherosclerosis, *Z. Vitamin. Horm. Fermentforsch.,* (Vienna), 13, 8, 1963.

89. **Chen, L. H., Liao, S., and Packett, L. V.,** Interaction of dietary vitamin E and protein level of lipid sources with serum cholesterol level in rats, *J. Nutr.,* 102, 729, 1972.

90. **Greenblatt, I. J.,** Use of massive doses of vitamin E in humans and rabbits to reduce blood lipids, *Circulation,* 16, 508, 1957.

91. **Schettler, G.,** *Arteriosklerose,* Thieme Verl., Stuttgart, 1961, 592.

92. **Simic, B. S., Sibalic, M., Naumovic, M., Simic, A., Markovic, R., and Rakovic, R.,** Relation among vitamin A, tocopherol, and cholesterol serum levels in elderly, *Int. Z. Vitaminforsch.,* 33, 48, 1963.

93. **Hammerl, H. and Pichler, O.,** Die Bedeutung der Vitamine A, E und B_6 fur die Genese und Therapie Arteriosklerotischer Gefässveranderungen, *Wien. Klin. Wschr.,* 72, 468, 1960.

94. **Sautter, H.,** Vitamine in der Therapie Arteriosklerotischer Chorioretinopathien, *Dtsch. Med. Wschr.,* 83, 1514, 1958.

95. **Kibata, M., Shimizu, Y., Miyake, K., Stoji, K., Niyahara, K., Nasu, Y., and Kimura, I.,** Studies on vitamin E in lipid metabolism, in *Tocopherol, Oxygen, and Biomembranes,* de Duve, C. and Hayaishi, O., Eds., Elsevier/North Holland, Amsterdam, 1978, 283.

96. **Yamori, Y., Nara, Y., Horie, R., and Ohtaka, M.,** Biomembrane characteristics and chronic effect of tocopherol in models with hypertension and stroke, in *Tocopherol, Oxygen, and Biomembranes,* deDuve, C. and Hayaishi, O., Eds., Elsevier/North Holland, Amsterdam, 1978, 247.

97. **Farnworth, E. R., Kramer, J. K. G., Thompson, B. K., and Corner, A. H.,** Role of dietary saturated fatty acids on lowering the incidence of heart leasions in male rats, *J. Nutr.,* 112, 231, 1982.

98. **Haeger, K.,** Long-term observation of patients with atherosclerotic dysbasia on alpha-tocopherol treatment, in *Tocopherol, Oxygen, and Biomembranes,* de Duve, C. and Hayaishi, O., Eds., Elsevier/North Holland, Amsterdam, 1978, 329.

99. **Boyd, A. M. and Marks, J.,** Treatment of intermittent claudication. A reappraisal of the value of α-tocopherol, *Angiology,* 14, 198, 1963.

100. **Larsson, H. and Haeger, K.,** Plasma and muscle tocopherol content during vitamin E therapy in arterial disease, *Pharmacol. Clin.,* 1, 72, 1968.

101. **Rehncrona, S., Siesjo, B. K., and Smith, D. W.,** Reversible ischemia of the brain: biochemical factors influencing restitution, *Acta Physiol. Scand. Suppl.,* 492, 135, 1980.

102. **Bieri, J. G. and Farrell, P. M.,** Vitamin E., *Vitam. Horm.* (New York), 34, 31, 1976.

103. **Leonard, P. J., Kelleher, J., Pulvertaft, C. N., and Losowsky, M. S.,** Vitamin E deficiency following partial gastrectomy, in *Vitamin A, E, and K. Clinical and Physiological Problems,* Kress, H. and Blum, K. U., Eds., K.K. Schattauer Verl., Stuttgart, 1969, 334.

104. **Haber, S. L. and Wissler, R. W.,** Effect of vitamin E on carcinogenicity of methylcholanthrene, *Proc. Soc. Exp. Biol. Med.,* 111, 774, 1962.

105. **Lewis, L. A. and Naito, H. K.,** Relation of hypertension lipids and lipoproteins to atherosclerosis, *Clin. Chem.,* 24, 2081, 1978.

106. **Rubinstein, H. M., Dietz, A. A., and Srinavasan, R.,** Relation of vitamin E and serum lipids, *Clin. Chim. Acta,* 23, 1, 1969.

107. **Kayden, H. J.,** Vitamin E deficiency in patients with abetalipoproteinemia, in *Vitamin A, E, and K. Clinical and Physiological Problems,* Kress, H. and Blum, K. U., Eds., K.K. Schattauer Verl., Stuttgart, 1969, 334.

108. **Sunderam, G. S., London, R., Manimekalai, S., Nair, P. P., and Goldstein, P.,** α-Tocopherol and serum lipoproteins, *Lipids,* 16, 223, 1981.

109. **Gudbjarnason, S., Doell, B., Oscarsdottir, G., and Hallgrimsson, J.,** Modification of cardiac phospholipids and catecholamine stress tolerance, in *Tocopherol, Oxygen, and Biomembranes,* de Duve, C. and Hayaishi, O., Eds., Elsevier/North Holland, Amsterdam, 1978, 297.

110. **Takeuchi, N., Iritani, N., Fukuda, E., and Tanaka, F.,** Effect of long-term administration and deficiency of alpha-tocopherol on lipid metabolism of rats, in *Tocopherol, Oxygen, and Biomembranes,* de Duve, C. and Hayaishi, O., Eds., Elsevier/North Holland, Amsterdam, 1978, 257.

111. **Kyei-Aboagye, K., Hazucha, M., Wyszogrodski, I., Rubinstein, D., and Avery, M. E.,** The effect of ozone exposure in vivo on the appearance of lung tissue lipids in the endobronchial lavage of rabbits, *Biochem. Biophys. Res Commun.,* 54, 907, 1973.

112. **Frank, L., Bucher, J. R., and Roberts, R. J.,** Oxygen toxicity in neonatal and adult animals of various species, *J. Appl. Physiol.,* 45, 699, 1979.

113. **Fisher, H. K., Clemens, J. A., and Wright, R. R.,** Pulmonary affects of the herbicide paraquat studied 3 days after injection in rats, *J. Appl. Physiol.,* 35, 268, 1973.

114. **Bus, J. S., Aust, S. D., and Gibson, J. E.,** Paraquat toxicity: proposed mechanism of action involving lipid peroxidation, *Res. Commun. Chem. Pathol. Pharmacol.,* 11, 31, 1975.

115. **Fletcher, B. L., Dillard, J., and Tappel, A. L.,** Measurement of fluorescent lipid peroxidation products in biological systems and tissues, *Anal. Biochem.,* 52, 1, 1973.

116. **Blackett, A. D., and Hall, A. D.,** Tissue vitamin E levels and lipofuscin accumulation with age in the mouse, *J. Gerontol.,* 36, 529, 1981.

117. **Trombly, R., Tappel, A. L., Coniglio, J. G., Grogan, W. M., Jr., and Rhamy, R. K.,** Fluorescent products and polyunsaturated fatty acids of human testes, *Lipids,* 10, 591, 1975.

118. **Weglicki, W. B., Reichel, W., and Nair, P. P.,** Accumulation of lipofuscin-like pigment in the rat adrenal gland as a function of vitamin E deficiency, *J. Gerontol.,* 23, 469, 1968.

119. **Canzler, H. and Hartmann, F.,** Vitamin E und experimentelle Leberschädigung, in *Vitamin A, E, and K. Clinical and Physiological Problems,* Kress, H. and Blum, K. U., Eds., K.K. Schattauer Verl., Stuttgart, 1969, 291.

120. **Benedetti, A., Ferrali, M., and Chieli, A.,** A study of the relationship between carbon tetrachloride-induced lipid peroxidation and liver damage in rats pretreated with vitamin E, *Chem. Biol. Interact.,* 9, 117, 1974.

SELENIUM IN THE AGED

Seetha N. Ganapathy and Saro Thimaya

INTRODUCTION

Selenium was discovered by the Swedish chemist Berzelius in 1817 in a lead chamber while making sulfuric acid.[1] It is surrounded by arsenic, bromine, sulfur, and tellurium in the periodic table. The atomic number of the element is 34, and it is intermediate in its properties with sulfur and tellurium.[2] The element is named after the Greek word *selene*, or moon, since selenium was closely associated with tellurium, named for the earth.

Selenium (Se) is found in the soil and the whole soil/plant cycle affects the eventual selenium content of human foods. The soil content of Se varies according to the geographical location in the U.S.[3,4] It was shown that there are extensive areas in the Pacific Northwest, the Northeast, and the Southeast, where crops are generally low in Se. Se-Responsive diseases in livestock are most likely to occur in these areas. In the West Central states, the Se content of crops is predominantly in the protective, but nontoxic range of concentration. Other parts of the country have variable Se concentrations in plants.

Se occurs in the cells and tissues of the body in a concentration that is proportionate to the level of intake from the diet. The kidney, particularly the kidney cortex, is the highest in concentration followed by the glandular tissues, in particular the pancreas, the pituitary, and the liver, respectively. The kidney and the liver are the most sensitive indicators of the Se level in the animal, and the Se level present in these organs provides valuable diagnostic criteria. At toxic intake levels, considered to be 10 to 100 times or more greater than those ingested, the tissue concentrations rise steadily until levels as high as 5 to 7 ppm in the liver and kidneys and 1 to 2 ppm in the muscles are reached. Beyond these tissue levels, excretion begins to keep pace with absorption.[5]

Information about the metabolism of Se is fragmentary and little is known about its participation in biochemical reactions, but much speculation surrounds its biochemistry. The main functional form of Se that has been identified is the enzyme glutathione peroxidase (GSHPx). This enzyme catalyzes the reduction of hydrogen peroxide to water and of fatty acid hydroperoxides to hydroxy acids in the tissue, and thus protects the lipid in the cell membrane from peroxidation.[6] Glutathione peroxidase is present in a wide variety of tissues, and about 90% of the Se in red blood cells is in this form. In animals, GSHPx levels depend dramatically on dietary Se intakes. There is a strong correlation between blood GSHPx levels and selenium in man.[7] Pinto and Bartley[8] discussed the effect of age and sex on GSHPx and reductase. It has been reported that the discovery of Se as a component of GSHPx offers a convenient explanation for many relationships of Se, vitamin E, and sulfur amino acids.[9] Further, it was stated that it, as a part of GSHPx, catalyzes the destruction of peroxide; the sulfur amino acids act as precursors of glutathione, while vitamin E decreases the formation of peroxides. These strong interrelationships between vitamin E and Se suggest that Se might function as an antioxidant.

There are many theories of aging and many still stand. Aging may be considered as a sum total of most of those processes or changes elucidated in those theories. Lipid peroxidation is one of the several theories attributed to aging. Lipid peroxidation involves the formation of semistable peroxides from free-radical intermediates by the reaction of oxygen and unsaturated lipid. Lipid peroxidation damage occurs mainly in the membranes of subcellular organelles including microsomal, mitochondrial membranes which contain large amounts of polyunsaturated fatty acids. These membranes are key sites for lipid peroxidation because they contain, or are in proximity to, powerful catalysts of the reaction. Free radicals which are generated in lipid peroxidation are important in the development of degenerative diseases associated with aging.[10] Tissues most vulnerable to damage through free-radical lipid peroxidation are the lungs, heart, and brain.[11]

Oxidation of lipids in the cell membrane destroys its integrity and leads to destruction of the cells and accumulation of oxidized lipoproteins in the cells. The ceroid pigment which accumulates with aging appears to be a complex of peroxidized lipids and proteins.[12] The reasons for accumulation of these pigments are not known, but lipid peroxidation has been suggested as a major cause. The free-radical theory relates aging to the formation of damaged cell components; this can also be enhanced by radiation. These free radicals disrupt enzyme activity by causing defective DNA and RNA and destroying cell membranes. It has been proposed that antioxidants such as vitamin E and Se may prevent cell damage and inhibit the aging effect of the formed free radicals. Numerous studies on Se and vitamin E deficiency suggested that these might be responsible for protecting cells against oxidative damage.[13,14]

Further, the enzyme GSHPx appears to be an important factor in controlling endogenous formation of peroxides from unsaturated fatty acids in cell membranes. A lower activity of this enzyme in selenium deficiency could explain why some symptoms of Se deficiency can be relieved by α-tocopheral and other antioxidants.

Since GSHPx acts to prevent endogenous formation of peroxide from unsaturated fatty acids in cell membranes and according to the lipid peroxidation theory, oxidation of lipids in the cell membranes destroys its integrity and leads to the destruction of cells, it is justifiable to say that selenium plays an important part in the aging process.[14]

SELENIUM IN NUTRITION

Se toxicity was a problem in agriculture long before the nutritional significance was recognized. Se is a dietary essential which is independent of vitamin E, though it can be substituted for the others.

It is now well documented that Se toxicity or deficiency can cause severe health problems in domestic animals and humans in many parts of the world. The beneficial use of Se for animals was first recognized in a study by Schwarz and Foltz[15] in 1957 which showed that small amounts of Se prevented liver necrosis in rats fed on a torula yeast diet.[15] Several other studies showed that Se prevents exudative diathesis in chicks and young ruminants against white muscle disease.[16-19] Thompson and Scott[20] reported that Se was essential in a purified diet in which protein was entirely replaced by crystalline amino acids. All these experiments revealed that Se is an essential nutrient for domestic animals and birds.

A deficiency of this element causes many pathological changes in the tissues of animals. The role of Se in the fertility and growth of many animals has been noted. Recent work carried out by Segerson and Ganapathy[21] showed that supplementation of Se to adequately fed ewes in geographical regions low in Se can result in an increased rate of ova fertility. In primates, Se deficiency is characterized by a loss of weight, listlessness, and alopecia.[22]

Acute selenium poisoning in animals grazing on alkali (high salt) pastures is characterized by blindness, abdominal pain, salivation, muscle paralysis, and death from respiratory failure. Chronic toxicity (alkali disease) in cattle produces dullness and roughness of coat, loss of hair, sore hoofs, erosion of the joints of long bones, cardiac atrophy, cirrhosis of the liver, and anemia.[23] Poisoning occurs owing to the ingestion of certain plant species which accumulate Se from the soil.

Although Se deficiency disorders in man are not well established, its toxic effects were seen in industrial workers and the symptoms observed were depression, languor, nervousness, occasional dermatitis, GI disturbances, giddiness, garlic odor of the breath, sweating, and high urinary Se.[24] Besides food, man is susceptible to Se contamination from the atmosphere, and it usually enters the body in the form of gaseous matter from the dust. The blood and urinary Se are considered to be reliable indicators of total Se ingestion and absorption.[25]

The first account of a possible negative Se deficiency in man involved the use of New Zealand dried milk to treat Jamaican children suffering from Kwashiorkor.[26,27] It was later found that administering Se resulted in stimulated growth in two children with Kwashiorkor. McKeehan and co-workers[28] demonstrated the essentiality of Se for the optimum growth of human fetal tissue in a culture media. On the other hand, it was shown that the consumption of small amounts of Se during the development of the teeth increases the incidence of dental caries in children and laboratory animals.[29,30]

Although a Se-responsive clinical syndrome comparable to that of animals has not so far been described in man, low blood Se has been observed in pathological conditions such as cirrhosis and colonic, gastric, and pancreatic carcinoma.[31-34] Low blood Se has also been observed in two groups of infants with kwashiorkor.[25,35] Recent work carried out by Awasthi and co-workers[36] demonstrated that Se is an essential nutrient for man.

A study conducted by van Rij and co-workers[37] on a New Zealand patient on long-term parenteral nutrition with muscle pain tenderness and weakness associated with almost un-detectable Se concentration that responded to replacement of Se suggests that this is probably the first case of clinical deficiency noted in man. The extremely low Se level in New Zealand soils results in a low Se content of foods, low dietary intake, low urinary excretions, and low blood Se concentrations and GSHPx.[38] Of these, plasma Se gives a short-term index of nutritional status; whole, erythrocyte, and glutathione peroxidase give a long-term index.

Even though there is a wide variation in human Se intake, as Se concentration varies in different locations, the Se-responsive diseases are not as well characterized in humans as in animals. The World Health Organization (WHO) has considered Se, but dietary recommendations have not yet been made.[39] The U.S. National Research Council (Food and Nutrition Board) has recently established a "safe-to-adequate" level for this nutrient at 50 to 200 μg/day.[40] Except for the people in low Se areas, patients under total parenteral nutrition, and vulnerable population groups such as children and elderly consuming poor diets, the supply of Se seems to be quite adequate in most areas of the U.S. To date, there have not been many studies on the role of Se in the aged and even fewer on the effect of age on Se metabolism.

SELENIUM IN THE AGED

The body tolerance for Se has been observed to be 14.6 mg, with 100 to 150 μg/day eliminated by normal humans via urine, feces, skin, and lungs.[41] In a study conducted by Palmquist and co-workers[42] it was noted that urinary Se excretion in 5 females was 36.3 ± 1.9 and in 6 males it was 52.9 ± 3.85 μg/day, respectively.[42] Ganapathy and co-workers[43] observed the mean daily urinary and fecal excretions in males (26 to 30 years old) receiving 75 to 84 μg Se per day from all plant protein diets, to be 50 and 27 μg/day, respectively. The absorption of dietary Se was found to be 60 to 68% depending upon the diets.

Se composition of human organs and fluids is presented in Table 1. It can be seen from the data that a wide variation in blood, plasma, and serum exists. The levels in preadolescents seem to be higher than in infants or adults. The lowest level of Se among the fluids was in the saliva (3 ppb). While studying the distribution pattern of Se in hair, decreasing levels can be noted in the elderly. Among the organs, liver and testes seem to have high levels. A higher concentration can be noted in tooth enamels and the highest seems to be in nails (1.15 ppm). Similar results are found in many fluids and tissues of animals, and lower levels in the fluids and higher in tissues and organs.

Of the many reported values for human tissues, fluids, and diets, many do not show consistent values for any one sample. The method of handling, preparation, and analysis of the samples cause these variations. It is often noted that the experimental details and methods are not adequately discussed. It can be further revealed that there are wide variations in

Table 1
SELENIUM IN HUMAN TISSUES AND FLUIDS

Fluid or tissue	Case no.	Characteristics of subjects	Selenium (μg) mean	Ref.
Blood/mℓ	—	Adults	0.068 ± 0.013[a]	44
	—	60 years	0.047 ± 0.010[a]	45
	—	Young and middle age	0.060 ± 0.012[a]	45
	5	Cancer	0.03[a]	45
	1	5 months	0.036[a]	46
	8	1—12 months	0.035 ± 0.006[b]	46
	6	1—5 years	0.057 ± 0.015[a]	46
	24	5—10 years	0.054 ± 0.009[a]	46
	24	1—5 years	0.048 ± 0.012[b]	46
	29	5—10 years	0.057 ± 0.017[b]	46
	—	Children	0.813[b]	46
	626	—	0.265 ± 0.056[b]	48
	—	Infants	0.049 ± 0.012	51
	14	1—6 years	0.059 ± 0.011	51
	13	7—12 years	0.060 ± 0.014[c]	51
	13	Premature infants	0.040 ± 0.012[c,d]	51
	6	Infants	0.036 ± 0.017[c]	51
	—	—	0.043[b,c]	54
	—	Adults	0.059[b]	54
	52	Adults	0.136 ± 0.033	54
	210	Males	0.10 ± 0.34	61
	8	Men	0.081 ± 0.015[a]	51
	17	Women	0.09 ± 0.017[a]	51
Plasma/mℓ	25	—	0.098 ± 0.021	54
	253	—	0.081 ± 0.016	55
	—	Adults	0.096 ± 0.026	57
	—	Children	0.15	5
Serum/mℓ	9	1—12 months	0.051	47
	10	1—5 years	0.081	47
	7	5—12 years	0.097	47
	5	Adults	0.14 ± 0.006	50
	8	Cancer	0.11 ± 0.009	50
	11	Cardiovascular disease	0.10 ± 0.008	50
	6	Other disease	0.12 ± 0.006	50
	9	4—15 years	0.073 ± 0.020	51
	—	Newborn	0.050	7
	—	Infants	0.034	7
	—	Adults	0.102	7
Serum/mℓ	37	20—80 year male	0.12 ± 0.001	56
	20	Cancer	0.11 ± 0.03	56
	—	Cirrhosis	0.09 ± 0.02	56
	6	Adults	0.118	33
	—	Children	0.14	5
RBC/HB[d]	100	Normal adults	0.73 ± 0.38	57
RBC/mℓ blood	—	Children	0.36	5
Urine/24 hr	4	Females	6.5 ± 14.7[a]	58
	5	Females	36.3 ± 1.9	42
	6	Males	52.9 ± 3.85	42
	12	Males (black)	50.0	43
Saliva[a] (ppb)	26	11—12 years	3.0 ± 0.3	60
Milk	241	—	0.018 (0.007 − 0.033)	62
Hair/g	6	1—5 year normal	0.66 ± 0.07	50
	12	6—10 year normal	0.64 ± 0.04	50

Table 1 (continued)
SELENIUM IN HUMAN TISSUES AND FLUIDS

Fluid or tissue	Case no.	Characteristics of subjects	Selenium (μg) mean	Ref.
	13	11—15 year normal	0.76 ± 0.04	50
	21	16—20 year normal	0.66 ± 0.04	50
	26	21—30 year normal	0.62 ± 0.03	50
	9	31—40 year normal	0.59 ± 0.04	50
	7	41—50 year normal	0.64 ± 0.04	50
	9	51—60 year normal	0.64 ± 0.06	50
	8	61—70 year normal	0.55 ± 0.03	50
	4	70 year	0.62 ± 0.03	50
	73	11—15 year dyslexic	1.1	52
	44	11—15 year normal	0.6	52
	42	Adults	0.42	53
	40	Multiple sclerosis	0.50	53
Nails/g	16	21—53 years	1.14 ± 0.06	59
Enamel/g	45	—	0.43 ± 0.03	5
Muscle	6	Accident victims	0.11 ± 0.01[d]	41
Liver	11	Accident victims	0.30 ± 0.10[d]	41
Kidney	8	Accident victims	0.10 ± 0.02[d]	41
Lung	11	Accident victims	0.10 ± 0.02[d]	41
	5	Accident victims	0.02 ± 0.04[d]	41
Ovary	6	Accident victims	0.09 ± 0.03[d]	41
Lymph nodes	6	Accident victims	0.05 ± 0.01[d]	41

[a]   Low Se area.
[b]   Seleniferous area.
[c]   Hospitalized subjects (S).
[d]   ppm wet weight basis.

blood plasma and serum of the population group which can be attributed to their living location (geographical location), age, and pathological condition.

Tissue Se content of autopsy samples of two special case studies is shown in Table 2. Out of the 17 tissues analyzed, the Se content of 10 of those infant tissues are lower than the adult tissues. This might be a reflection of lower stores of minerals in the infants. Data collected from the same study on liver, skin, and muscle tissue of 10 individuals with ages ranging from 29 to 77 years gave variable values, but when grouped into under and over 60 years, the level in both liver and muscle seems to be lower in the elderly (Table 3).

Further, 18.2, 23.6, and 14.4 μg of Se was found in 100 mℓ of blood, blood cells, and plasma, respectively, in 253 subjects in Canada. They also noted a gradual decline in both plasma and cell Se with age. Kasperek and co-workers[63] noted a gradual rise of serum Se up to age 35 and then a decline in older individuals.[63] Both groups followed the neutron activation technique. Allaway and co-workers conducted a study to determine Se content of blood from blood banks across the U.S. using fluorimetric methods. Higher values were found in seleniferous areas with a mean of 20.6 μg/100 mℓ of blood (range: 10 to 34 μg/ 100 mℓ of blood).[61]

The findings of a study conducted in our laboratory to determine the Se content of human hair in relation to age, diet, pathological conditions, and serum levels seem to indicate that age is a factor in the level of Se in hair.[50] The Se level in hair varied from a high 0.76 μg/ g of hair in the adolescents to 0.55 μg/g in the elderly. After reaching the peak level, a decline was noted in the age groups of 16 to 40 years, the levels being 0.64 to 0.59 μg/g. The lowest level was in the 61- to 70-year group. These differences were statistically

Table 2
TISSUE SELENIUM
CONTENT OF ORGANS IN
TWO SPECIAL CASE
STUDIES

	Se content[a]	
Tissue	Infant[b]	Adult[c]
Stomach	0.19	0.17
Liver	0.34	0.39
Pancreas	0.05	0.13
Spleen	0.37	0.27
Kidney	0.92	0.63
Intestine	0.31	0.22
Heart	0.55	0.22
Lung	0.17	0.21
Artery	0.27	0.27
Muscle	0.31	0.40
Fat	0.09	0.12
Trachea	0.14	0.24
Gonad	0.46	0.47
Thyroid gland	0.64	1.24
Brain	0.16	0.27
Adrenal gland	0.21	0.36
Lymph node	0.26	0.10

[a] μg/g whole tissue.
[b] Aged $1^{1}/_{2}$ years and died of pneumonia.
[c] Aged 52 years and died of ruptured aortic aneurism.

From Dickson, R. C. and Tomlinson, R. H., *Clin. Chim., Acta*, 16, 311, 1967. With permission.

Table 3
SE CONTENT (μg/g) OF TISSUES IN TWO AGE GROUPS

Age group (years)	Number of cases	Skin	Liver	Muscle
>60	8	0.286 ± 0.147	0.431 ± 0.135	0.345 ± 0.081
<60	2	0.205 ± 0.085	0.44 ± 0.01	0.47 ± 0.12

significant ($p < 0.05$). This is graphically represented in Figure 1 and given in Table 4 along with dietary data. No significant correlation was noted between dietary and hair Se in all groups. The mean daily intake of Se in the aged seems to be within the recommended "safe" level for the element. The mean daily intake of Se was found to be 87 ± 3 μg; the lowest intake was noted in children and the elderly. The highest consumption was observed in the 16- to 20-year group. The dietary Se intake in all subjects ranged from a low of 25 to a high of 204 μg/day. The low value was due to poor dietary habits in one of the subjects. The Se supplied by the university cafeteria omnivorous diet ranged from 70 to 142.5 μg/ day, while cooked lacto-ovo-vegetarian diet provided 65.1 to 108.2 μg/day. Se content of an all-plant protein diet varied from 79.7 to 100.5 μg/day. These findings reveal that the supply of Se by an average diet irrespective of omnivorous or vegetarian diet falls in the range of 65 to 143 μg/day (Table 5).

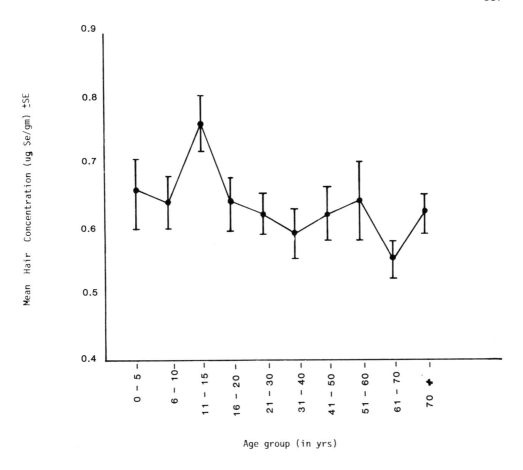

FIGURE 1. Variation of selenium concentration in human hair with age. (From Thimaya, S. and Ganapathy, S. N., *Sci. Total Environ.*, 24, 41, 1982. With permission.)

There are several other reports on Se content of foods and diets.[43,48,49,64,65] Se content of a cross-section of representative foods of the major food groups is shown in Table 6. It is clear according to many of these reports that meat, cereal, and legumes are rich sources of Se. Dairy products and eggs are moderate, and fruits and vegetables are poor sources. In addition to the geographical variations, Se is volatile and can be affected by the type, intensity, and/or duration of heat applications involved in a given cooking method. It has been shown that broiling and frying of certain meat cuts can destroy this element up to 31%, and the cooking method did not influence the frozen and fresh vegetables and legumes.[66]

Dietary supply of Se can vary from as low as 6 to 200 µg/day or more from one population group to another, depending upon the food habits in addition to the other factors discussed earlier.[66-71] It appears at this time that a well-balanced diet consisting of foods from various food groups should meet the dietary requirements as established by the National Research Council in 1980.[40] Our recent study[72] on the food consumption patterns of the healthy elderly revealed that 80% had diets meeting the requirements for all food groups except for dairy products, fruits, and vegetables. Their meal contained all foods rich in Se such as meat, cereals, fish, and beans. It can be assumed that the elderly consuming diets containing a wide variety of foods from all food groups would receive adequate amounts of Se.

In summary, it can be stated that there is a decline of Se level in a number of tissues, especially in plasma and hair with increasing age. Similar observations were made by others for Zn and Cu in hair and serum with age. Since a majority of the aged have co-existing

Table 4
SELENIUM LEVELS IN HAIR AND DIET

Age group	Hair[a]	Diet[b]	Recommended daily dietary allowances (μg/day)
5 and below	0.66 ± 0.07	61 ± 22[c]	30—120
6—10	0.64 ± 0.04	87 ± 25	50—200
11—15	0.76 ± 0.04	93 ± 28	50—200
16—20	0.64 ± 0.04	109 ± 40	50—200
21—30	0.62 ± 0.03	87 ± 38	50—200
31—40	0.59 ± 0.04	78 ± 22[d]	50—200
41—50	0.62 ± 0.04	73 ± 20[d]	50—200
51—60	0.64 ± 0.06	75 ± 18	50—200
61—70	0.55 ± 0.03	63 ± 21	50—200
above 70	0.62 ± 0.03	65 ± 13	50—200

[a]   Mean levels ± SEM (μg/g).
[b]   Mean levels ± SD (μg intake/day).
[c]   Two third of the serving was used for calculation of Se for this group.
[d]   Correlations found between the levels were significant.

From Thimaya, and Ganapathy, *Sci. Total Environ.*, 24, 41, 1982. With permission.

Table 5
SELENIUM CONTENT OF VARIOUS DIETS (μg Se/day)

Diets	Days						
	1	2	3	4	5	6	7
University omnivorous cafeteria diet[a] Range: 70—142.5 Mean: 92.7	99.6	142.5	79.0	102.0	70.0	80.0	76.0
Home cooked lacto-ovo-vegetarian diet[b] Range: 65.1—108.2 Mean: 84.8	71.2	86.7	83.6	65.1	99.3	97.4	108.2
All plant experimental diet[c] Range: 79.7—100.5 Mean: 85.9	80.6	79.7	100.5	95.9	82.5	79.9	83.7

[a]   Protein (g/day) = 80—100.
[b]   Protein (g/day) = 40—50.
[c]   Protein (g/day) = 46.

From Ganapathy, S. N. and Dhanda, R., *Ind. J. Nutr. Diet.*, 17, 53, 1980. With permission.

disease states, it is important to control these associated conditions in the aged prior to the study on mineral status in relation to the aged. It remains to be seen if the observed Se status reflects on various impaired health conditions, bioavailability, and/or dietary deficiency of the element. Evaluation of foods consumed by the healthy elderly for Se content revealed that they are receiving adequate amounts of Se. It will be interesting to evaluate the diets consumed by the aged for both Se and vitamin E to understand the dietary role of

Table 6
SELENIUM CONTENT OF SELECTED FOODS[48,49] (μg/g dry wt)

Food groups	Selenium	Food groups	Selenium
Meat group		Northern beans	0.193
Ground beef	0.346	Navy beans	0.185
Beef T-bone steak	0.340	Navy beans and pork	0.287
Beef liver	1.592	Lentils (dry)	0.225
Beef kidney	5.977	Cereals	
Bacon	0.395	Rice krispies	0.330
Pork chops	0.515	Ralston	1.197
Pork kidney	5.135	Golden grahams	0.197
Pork liver	1.577	Apple jacks	0.270
Chicken breasts	0.207	Bran buds	0.353
Chicken liver	1.977	Wheat germ (1 tbsp.)	0.663
Seafood		Special K	0.664
Croaker	2.874	Oats	0.413
Blackbass	1.929	Cornflakes	0.210
Trout	1.391	Shredded wheat	0.200
Flounder	2.128	Fresh and frozen vegetables	
Egg (raw)	0.184	Green beans and peas, cauliflower, green vegetables, (leafy and nonleafy), tomatoes, cucumber, lettuce, and carrots	
Dairy products			
Whole milk	0.010		
Non-fat dry milk	0.199		
Skim milk	0.011		
Buttermilk	0.011		
Yogurt (S. F.)	0.037	Range	0.002—0.003
Ice Cream (V. F.)	0.041	Mean	0.002
Cottage cheese	0.043	Fresh and frozen fruits	
Cheese	0.103	Peaches, grapes, strawberries, apples, bananas, and grapefruit	
Legumes (canned)			
Field beans	0.195		
Lima beans	0.233		
Black eyed peas	0.255	Range	0.002—0.002
Kidney beans	0.265	Mean	0.002
Pinto beans	0.300	Miscellaneous	
Pork and bean	0.230	Rice flour	0.254—0.361
Garbanzo chick peas	0.270	Wheat flour	0.652—0.717
October beans	0.193	Corn	0.187
		Potato (fried, baked, broiled)	0.002—0.003

these nutrients in the aged. Further, the interrelationships between Se, GSHPx, and vitamin E in aging should be explored in controlled animal model experimentation. If studies are carried out through the lifespan of the animals, it may throw more light on the relationships of these in relation to aging.

REFERENCES

1. **Scott, M. L.,** The selenium dilemma, *J. Nutr.*, 103, 803, 1973.
2. **Partington, J. R.,** Selenium, in *General Organic Chemistry*, 4th ed., St. Martin's, N.Y., 1967, 729.
3. **Carter, D. L., Brown, M. J., Allaway, W. H., and Cary, E. E.,** Selenium content of forage and hay crops in the pacific northwest, *Agric. J.*, 60, 532, 1968.
4. **Kubota, J., Allaway, W. H., Carter, D. L., Cary, E. E., and Lazar, V. A.,** Selenium in crops in the United States in relation to selenium-responsive diseases of animals, *J. Agric. Food Chem.*, 15, 3:448, 1967.

5. **Underwood, E. J.,** Selenium, in *Trace Elements in Human and Animal Nutrition,* 4th ed., Academic Press, N.Y., 1977, chap. 12.
6. **Ganther, H. E., Hafeman, D. G., Lawrence, R. A., Serfass, R. E., and Hoekstra, W. G.,** Selenium and glutathione peroxidase in health and disease, in *Trace Elements in Human Health and Disease,* Vol. 2, Prasad, A. S., Ed., Academic Press, N.Y., 1976, 165.
7. **Lombeck, I., Kasperek, K., Feinendegen, L. E., and Bremer, H. J.,** The state and supply of selenium in healthy children and dietetically treated patients with inborn errors of metabolism, in *Trace Element Metabolism in Man and Animals,* Vol. 3, Kirchgessner, M., Ed., Freising-Weihenstephan, 1978, 312.
8. **Pinto, R. E. and Bartley, W.,** The effect of age and sex on glutathione reductase and glutathione peroxidase activities and on aerobic glutathione oxidation in rat liver homogenate, *Biochem. J.,* 112, 109, 1969.
9. **Rotruck, J. T.,** Discovery of the role of selenium in glutathione peroxidase, in *Selenium in Biology & Medicine,* 2nd ed., Spallholz, J. E., Martin, J. L., and Ganther, H. E., Eds., AVI, Westport, Ct, 1981, chap. 2.
10. **Roe, D. A.,** The physiology and pathology of aging, in *Geriatric Nutrition,* Prentice-Hall, N.J., 1983, chap. 2.
11. **Slater, T. F.,** *Free Radical Mechanisms in Tissue Injury,* Pion, London, 1972.
12. Lipid peroxidation in membrane lipids and action of glutathione peroxidase, *Nutr. Rev.,* 36, 23, 1978.
13. Selenium and human health, *Nutr. Rev.,* 34, 347, 1976.
14. Selenium: an essential element for glutathione peroxidase, *Nutr. Rev.,* 31, 289, 1973.
15. **Schwarz, K. and Foltz, C. M.,** Selenium as an integral part of factor 3 against dietary liver necrotic degeneration, *J. Am. Chem. Soc.,* 79, 3292, 1957.
16. **Patterson, E. L., Milstrey, R., and Stokstad, E. L. R.,** Effect of selenium in preventing exudative diathesis in chicks, *Proc. Soc. Exp. Biol. Med.,* 95, 617, 1957.
17. **Scott, M. L., Bierri, J. G., Briggs, G. M., and Schwarz, K.,** Prevention of exudative diathesis by factor 3 in chicks on vitamin E deficient torula yeast diet, *Poult. Sci.,* 36, 1155, 1951.
18. **Stokstad, E. L. R., Patterson, E. L., and Milstrey, R.,** Factors which prevent exudative diathesis in chicks on torula yeast diets, *Poult. Sci.,* 36, 1160, 1957.
19. **Muth, O. H., Oldfield, J. E., Remmert, L. F., and Schubert, J. R.,** Effect of selenium and vitamin E on white muscle disease, *Science,* 128, 1090, 1958.
20. **Thompson, J. N. and Scott, M. L.,** Selenium deficiency in chicks and quail, in *Proc. Cornell Nutr. Conf. Feed Manufacturers,* N.Y. 1967, 130.
21. **Segerson, E. C. and Ganapathy, S. N.,** Fertilization of ova in selenium/vitamin E treated ewes maintained on two planes of nutrition, *J. Anim. Sci.,* 51, 2, 1980.
22. **Muth, O. H., Weswig, P. H., Whanger, P. D., and Oldfield, J. E.,** Effect of feeding selenium deficient ration to the subhuman primate (saimiri sciureus), *Am. J. Vet. Res.,* 32, 1603, 1971.
23. **Muth, O. H., Oldfield, J. E., and Weswig, P. H.,** Selenium in biomedicine, 1st Int. Symp. Oregon State University, AVI, Connecticut, 1967, pp. 194, 237.
24. Preliminary Air Pollution Survey of Selenium and Its Compounds, National Air Pollution Control Administration, A Literature Review, Public Health Service, U.S. Department of Health, Education, and Welfare, Raleigh, N.C., 1969, 58.
25. **Burk, R. F., Pearson, W. N., Wood, R. P., and Viteri, F.,** 75 Blood selenium, levels and in vitro red blood cell uptake of Se in kwashiorkor, *Am. J. Clin. Nutr.,* 20, 723, 1967.
26. **Schwarz, K.,** Development and status of experimental work on Factor 3-selenium, *Fed. Am. Soc. Exp. Biol.,* 20, 666, 1961.
27. **Majaj, A. S. and Hopkins, L. I.,** Selenium and kwashiorkor, *Lancet* 2, 592, 1966.
28. **McKeehan, W. L., Hamilton, W. G., and Ham, R. G.,** Selenium is an essential trace element for the growth of WI-38 diploid human fibroblasts, *Proc. Natl. Acad. Sci.,* 73, 2023, 1976.
29. **Hadjimarkos, D. M.,** Selenium and dental caries, *Caries Res.,* 3, 14, 1969.
30. **Shearer, T. R.,** Developmental and post-developmental uptake of dietary organic and inorganic selenium into the molar teeth of rats, *J. Nutr.,* 105, 338, 1975.
31. **Shamberger, R. J., Rukovena, E., Longfield, A. K., Tytko, S. A., Deodhar, S., and Willis, C. E.,** Antioxidants and cancer. I. Selenium in blood of normals and cancer patients, *J. Natl. Cancer Inst.,* 50, 863, 1973.
32. **Broghamer, W. L., Jr., McConnell, K. P., and Blotcky, A. L.,** Relationships between serum selenium levels and patients with carcinoma, *Cancer,* 37, 1384, 1976.
33. **McConnell, K. P., Broghamer, W. L., Jr., Blotcky, A. J., and Hurt, O. J.,** Selenium levels in human blood and tissues in health and disease, *J. Nutr.,* 105, 1026, 1975.
34. **Shamberger, R. J. and Willis, C. E.,** Selenium distribution and human cancer mortality, *CRC Crit. Rev. Clin. Lab. Sci.,* 25, 211, 1971.
35. **Levine, R. J. and Olson, R. E.,** Blood selenium in Thai children with protein calorie malnutrition, *Proc. Soc. Exp. Biol. Med.,* 134, 1030, 1970.

36. **Awasthi, Y. C., Beutler, E., and Srivastava, S. K.,** Purification and properties of human erythrocyte glutathione peroxidase, *J. Biol. Chem.,* 250, 5144, 1975.

37. **van Rij, A. M., Thomson, C. D., McKenzie, J. M., and Robinson, M. F.,** Selenium deficiency in total parenteral nutrition, *Am. J. Clin. Nutr.,* 32, 2076, 1979.

38. **Thomson, C. D. and Robinson, M. E.,** Selenium in human health and disease with emphasis to those aspects peculiar to New Zealand, *Am. J. Clin. Nutr.,* 33, 303, 1980.

39. World Health Organization, Trace elements in human nutrition, WHO Tech. Rep. Ser., No. 532, 1973.

40. Recommended Dietary Allowances, National Academy of Sciences, Washington, D.C., 1980.

41. **Hamilton, E. I., Minski, J. J., and Cleary, J. J.,** Concentrations and distribution of some stable elements in healthy human tissues in the United Kingdom, *Sci. Total Environ.,* 1, 341, 1973.

42. **Palmquist, D. L., Moxon, A. L., and Cantor, A. H.,** Pattern of urinary selenium excretion in normal adults, *Fed. Proc.,* 38(3), 391, 1979.

43. **Ganapathy, S. N., Booker, L. K., Craven, R., and Edwards, C. H.,** Trace minerals amino acids and plasma proteins in adult men fed wheat diets, *J. Am. Diet. Assoc.,* 78, 490, 1981.

44. **Kay, R. G. and Knight, G. S.,** Blood selenium values in an Adult Auckland population group, *N. Z. Med. J.,* 11, 90(639), 1979.

45. **Robinson, M. F., Godfrey, P. J., Thomson, C. D., Rea, H. M., and van-Rij, A. M.,** Blood selenium and glutathione peroxidase activity in normal subjects and in surgical patients with or without cancer in New Zealand, *Am. J. Clin. Nutr.,* 32, 1477, 1979.

46. **McKenzie, R. L., Rea, H. M., Thomson, C. D., and Robinson, M. F.,** Selenium concentration and glutathione peroxidase activity in blood of New Zealand infants and children, *Am. J. Clin. Nutr.,* 31, 1413, 1978.

47. **Lombeck, I., Kasperek, K., Feinendengen, L. E., and Bremer, H. J.,** Serum selenium concentrations in patients with maple-syrup urine disease and phenylketonuria under diet therapy, *Clin. Chem. Acta,* 64, 57, 1975.

48. **Joyner, B. T.,** The Effect of Food Preparation Methods on the Selenium Content of Selected Meats and Dairy Products, M. S. thesis, N. C. Agricultural & Technical State University, Greensboro, N. C., 1975.

49. **Sawyerr, D. R.,** Analysis of Selenium from Convenience Foods, M. S. thesis, N. C. Agricultural & Technical State University, Greensboro, N.C., 1977.

50. **Thimaya, S. and Ganapathy, S. N.,** Selenium in human hair in relation to age, diet, pathological conditions and serum levels, *Sci. Total Environ.,* 24, 41, 1982.

51. **Westermarck, T., Raumu, P., Kirjarinta, M., Lappalainen, L.,** Selenium content of whole blood and serum in adults and children of different ages from different parts of Finland, *Acta Pharmacol. Toxicol.,* 40, 465, 1977.

52. **Capel, I. D., Pinnock, M. H., Darrell, H. M., Williams, D. C., and Grant, E. C. G.,** Comparison of some trace, bulk and toxic metals in the hair of normal and dyslexic children, *Clin. Chem.,* 27(6), 879, 1981.

53. **Ryan, D. E., Holzbecher, J., and Stuart, D. C.,** Trace elements in scalp hair of persons with multiple sclerosis and of normal individuals, *Clin. Chem.,* 24(11), 1996, 1978.

54. **Rea, H. M., Thomson, C. D., Campbell, D. R., and Robinson, M. F.,** Relation between erythrocyte selenium concentrations and glutathione peroxidase activities of New Zealand residents and visitors to New Zealand, *Br. J. Nutr.,* 42, 201, 1979.

55. **Dickson, R. C. and Tomlinson, R. H.,** Selenium in blood and human tissues, *Clin. Chem. Acta,* 16, 311, 1967.

56. **Sullivan, J. R., Blotcky, A. J., Jetton, M. M., Hahn, H. K. J., and Burch, R. E.,** Serum levels of selenium, calcium, copper, magnesium, manganese and zinc in various human diseases, *J. Nutr.,* 109, 1432, 1979.

57. **Lane, H. W., Dudrick, S., and Warren, D. C.,** Blood selenium levels and glutathione peroxidase activities in university and chronic intravenous hyperalimentation subjects., *Proc. Soc. Exp. Biol. Med.,* 167, 383, 1981.

58. **Thomson, C. D.,** Urinary excretion of selenium in some New Zealand women, *Proc. Univ. Otago Med. Sch.,* 50, 31, 1972.

59. **Hadjimarkos, D. M. and Sherer, T. R.,** Selenium content of human nails: a new index for epidemiologic studies of dental caries, *J. Dent. Res.,* 52(2), 389, 1973.

60. **Hadjimarkos, D. M. and Shearer, T. R.,** Selenium concentrations in human saliva, *Am. J. Clin. Nutr.,* 24, 1210, 1971.

61. **Allaway, W. H., Kubota, J., Losee, F., and Roth, M.,** Selenium molybdenum and vanadium in human blood, *Arch. Env. Health,* 16, 342, 1968.

62. **Shearer, T. R. and Hadjimarkos, D. M.,** Geographic distribution of selenium in human milk, *Arch. Env. Health,* 30, 230, 1975.

63. **Kasperek, K., Shich, H., Siller, V., and Feinendegen, L. E.,** Selenium in man, in *Trace Elements in Human Health and Disease,* Vol. 3, Prasad, A. S., Ed., 1976, chap. 30.

64. **Ganapathy, S. N. and Dhanda, R.,** Selenium content of omnivorous and cafeteria diets, *Ind. J. Nutr. Diet.,* 17, 53, 1980.

65. **Morris, V. C. and Levander, O. A.,** Selenium content of foods, *J. Nutr.,* 100, 1383, 1970.

66. **Ganapathy, S. N., Joyner, B. T., Sawyer, D. R., and Häfner, K.,** Selenium content of selected foods, in *Trace Element Metabolism in Man and Animals,* Vol. 3, Kirchgessner, M., Ed., Freising-Weihenstephan, Germany, 1978, 322.

67. **Schroeder, H. A., Frost, D. V., and Balassa, J. J.,** Essential trace metals in man: selenium, *J. Chron. Dis.,* 23, 227, 1970.

68. **Mondragon, M. C. and Jaffee, W. G.,** Selenium in foods and urine of school children: different regions of Venezuela, 21, 185, 1972.

69. **Griffiths, N. M.,** Daily dietary intake of Se by some New Zealand women, in *Trace Elements in Human Health and Disease,* Vol. 3, Proc. Univ. Otago Med. Sch., Prasad, A. S., Ed., 51, 8, 1973.

70. **Thompson, J. N., Erdody, P., and Smith, D. C.,** Selenium content of food consumed by Canadians, *J. Nutr.,* 105, 274, 1975.

71. **Thorn, J., Robertson, J. and Buss, D. H.,** Trace nutrients. Selenium in British food, *Br. J. Nutr.,* 39, 391, 1978.

72. **Ganapathy, S. N. et al.,** unpublished observations.

CALCIUM METABOLISM AND NEEDS IN THE ELDERLY

Ira Wolinsky and Shailaja D. Telang

INTRODUCTION

Senescence is said to commence with the termination of growth and maturity and is characterized by a tendency towards involution and functional impairment.[1] As a result, the expectation of aging is a progressive weakening of the capacity to adapt to an ever-varying environment. Much recent attention has been devoted to understanding the nutritional dynamics and dietary needs of the elderly. Calcium (Ca) metabolism and dietary Ca requirements in aging is of interest not only from a general gerontological viewpoint, but also because of its possible involvement in the etiology of age-related osteoporosis.[2-6]

ABSORPTION OF CALCIUM IN AGING

Animals

A striking feature of Ca metabolism in aging is an inverse relationship between age and the percent absorption of Ca in the small intestine.

The phenomenon of declining percent absorption of Ca in aging animals is well documented in several species using both in vivo and in vitro methodologies.[7-16] Bronner[7] reviewed earlier literature on the subject and summarized findings from cattle and rats; recently Kenny[8] discussed the phenomenon in rats in summary form. Some examples can serve to illustrate the effect of aging on the efficiency of Ca absorption in animals. In a study on aging in our laboratory,[17] Fischer F344 rats aged 2 months (young, growing) and 28 months (aged) were fed a diet containing 0.6% Ca, 0.6% phosphorus (P), and 16% protein for a period of about 1 month. Ca intake and fecal and urinary excretion were monitored in order to calculate percent apparent absorpton of Ca (Table 1). Absorption of Ca dropped markedly between 2 and 28 months of age. In other experiments with other age groups, intermediate drops in Ca absorption can be demonstrated. Others[13] have also used the metabolic balance technique coupled with measurements of the endogenous fecal Ca excretion in rats in order to come up with a measurement of the actual Ca absorption and have obtained similar results; at a fixed level of dietary Ca, the percent of Ca absorbed from the diet was greatest at the youngest age (1 month) and progressively diminished thereafter (up to 26 weeks). Using an in vitro everted gut sac technique which measures the active mucosal/serosal transport of ^{45}Ca by the proximal duodenum, Armbrecht et al.[9] measured Ca transport in rats of various ages fed 0.6% Ca and 0.6% P diets. Intestinal Ca transport decreased between 1.5 and 12 months of age, with no further decline between 12 and 18 months of age.

Humans

In 1964, Bronner[7] summarized data on Ca absorption in human males 11 to 72 years old and females 12 to 59 years old, but no clear effect of aging on Ca absorption could be discerned from this review. It must be pointed out though, that the data compiled by Bronner were composite data and no one report cited experiments with a sufficient number of age points to demonstrate a trend. Since then, however, a number of reports have described an age-related decline of Ca absorption in the normal, nonosteoporotic elderly;[18-23] these are summarized in Table 2. This change with increasing age may be considered to represent the normal physiological intestinal adjustment to senescence. With the exception of the perfusion study by Ireland and Fordtran,[20] the studies cited in Table 2 used a variation of

Table 1
EFFECT OF AGE ON ABSORPTION OF Ca IN RATS

Age (months)	Fecal excretion of Ca (mg)	Urinary excretion of Ca (mg)	Apparent absorption[a] (%)
2	608 ± 15	5.0 ± 0.6	67.3 ± 0.5
28	2230 ± 69	21.9 ± 2.1	3.3 ± 0.9

[a] $\dfrac{\text{Intake - fecal excretion}}{\text{Intake}} \times 100$

From Wolinsky, I. and Tso, N. T-B., unpublished data.

Table 2
EFFECT OF AGE ON INTESTINAL ABSORPTION IN NONOSTEOPOROTIC INDIVIDUALS

Age (years)	Sex	No. subjects	Method employed	Results	Ref.
28—87	M	16	Appearance of ^{45}Ca in venous blood after oral ingestion of ^{45}CaCl$_2$ by fasted subjects	^{45}Ca appearance time is later and plasma maximum radioactivity is less in 70+ years individuals	19
	F	10			
12—85	F	59	Measurement of ^{47}Ca content in serum after oral dose of ^{47}CaCl$_2$ by fasted subjects	Inverse correlation between 1-hr plasma ^{47}Ca levels and age	22
20—95	M	75	Measurement of fraction of dose of ^{47}Ca or ^{45}Ca circulating in plasma after oral ingestion of ^{47}CaCl$_2$ or ^{45}CaCl$_2$ by fasted subjects	Absorption of Ca fell with age over 60 years	18
	F	115			
22—31	M	8	Triple-lumen jejunal perfusion	Ca absorption fell over age 60	20
61—75	F	5			
12—81	M	52	Measurement of plasma for ^{47}Ca given orally, and ^{45}Ca injected i.v.	Negative linear correlation between log of true absorption coefficient of Ca and age	23
	F				
30—90	M	17	Measurement of plasma for ^{47}Ca given orally, and ^{45}Ca injected i.v.	Inverse correlation between fractional Ca absorption and age	21
	F	77			

the in vivo technique whereby an oral or intravenous dose of Ca isotope is administered to a subject and the time course and extent of appearance in the blood is monitored as an indirect parameter of Ca absorption. Whether this procedure adequately reflects Ca absorption may be questioned.[24] However, Bullamore et al.[18] claim that Ca absorption measured with the radiocalcium test correlates very closely to that of true absorption measured by a Ca balance technique (coefficient of correlation, 0.93), and Avioli et al.[25] present evidence for support of the radiocalcium procedure as a specific measure of Ca absorption.

ADAPTATION TO LOW CALCIUM INTAKES

In several animal species[8,9,12,15,17,26-29] and in man[20,21,28,30-33] it has been amply demonstrated that the organism can adapt to low Ca intakes by a facultative increase in percent

Table 3
Ca BALANCE DATA, RAPIDLY GROWING DOGS

Ca in diet	Ca intake (g/day)	Fecal Ca excretion (g/day)	Retention	
			(g/day)	(% intake)
0.114	0.63	0.06	0.57	90
0.634	1.62	0.87	0.75	46
1.234	2.67	1.96	0.71	27

From Gershoff, S. N., Legg, M. A., and Hegsted, D. M., *J. Nutr.*, 64, 303, 1958. With permission.

absorption and retention of Ca; raising Ca intake may result in reduced utilization. In man, the adaptative response is so striking that even a retrospective dietary history of hospital outpatients was sufficiently accurate to clearly demonstrate the inverse relationship between dietary Ca intake levels and the efficiency of Ca absorption.[33] By means of this adaptation, and depending on prior dietary Ca history,[34] a negative Ca balance due to inadequate dietary intake may often, but not necessarily[3,32,35] be avoided. Using young rats fed diets containing 0.3, 0.6, and 1.2% Ca, our laboratory observed decreasing apparent absorption values of 78.3 ± 1.4, 67.3 ± 0.5, and 17.5 ± 0.6%, respectively.[17] Gershoff et al.[28] demonstrated that whereas absolute retention of Ca was greater in rapidly growing dogs fed a high Ca diet, dogs fed a low Ca diet were much more efficient, retaining 90% of the Ca in the diet vs. only 27% in the high Ca group (Table 3). The difference in total retention between the groups was considerably less than the difference in intakes, and all dietary groups were in Ca balance. In aging animals the capacity to adapt to restricted intakes of Ca is sluggish or impaired.[9,15,16,26] Armbrecht's group[9,26] reported on the effect of age on adaptation to dietary Ca in the Fischer 344 rat 1.5 to 18 months of age. In 1.5 month old animals fed a low Ca diet (0.014%), there was an increase in Ca active transport as measured by the everted gut sac technique; however, the magnitude of this intestinal adaptation decreased with age until there was only marginal adaptation by 1 year of age. In nonosteoporotic humans, two studies demonstrate the blunting of adaptation in aging.[20,21] Employing six young (mean age, 28) and six old (mean age, 68) males and females, Ireland and Fordtran,[20] using a triple-lumen jejunal perfusion method, measured mean absorption rates of the volunteers after they had been on a 2000-mg high Ca diet for 4 to 8 weeks and then placed on a 300-mg low Ca diet for 4 to 8 weeks. The change in diet from high to low Ca intake resulted in an increase of about 66% in the mean absorption rate in the young subjects but only about 50% in the elderly ones. Using a double-isotope method to assess intestinal Ca absorption in normal subjects, Gallagher and co-workers[21] found that when dietary Ca intake was assessed for its effect on fractional Ca absorption there was a significant ($p<0.02$) inverse correlation in 54 nonelderly subjects, in 20 elderly (<65 years) subjects, and no significant correlation in 20 elderly individuals (>65 years) studied.

One may seek a possible explanation for the phenomenon of age-related diminished adaptation to low Ca intake by reference to recent work on the primary regulators of Ca metabolism viz., the active metabolite of vitamin D, 1,25-dihydroxycholecalciferol [1,25(OH)$_2$D$_3$][8,9,21,35-41] and parathyroid hormone (PTH).[35,42,43] It is well established that the major determinant of Ca absorption is the circulating level of 1,25 (OH)$_2$D$_3$; a consequence of decreased production is decreased fractional Ca absorption. Several laboratories have demonstrated an age-related drop in production or decreased serum levels of 1,23(OH)$_2$D$_3$ in the Japanese quail,[38] rat,[35] and humans.[21,40] The in vitro production by kidney homogenates of 1,25(OH)$_2$D$_3$ decreased markedly between 8 and 108 weeks of age in both sexes of quail regardless of the Ca content of their diets (0.2% or 2.3 to 3.3%).[38] In humans[21] the mean serum $1,25(OH)_2D_3$ was significantly decreased in normal elderly subjects (>65 years) when compared to subjects less than 65 years old, 20.2 ± 2.4 pg/mℓ and 34.0 ± 1.9 pg/mℓ, respectively ($p<0.001$). Further, in nonelderly, normal subjects, dietary Ca intake correlated inversely with both Ca absorption ($r = -0.39$, $p<0.01$) and with serum $1,25(OH)_2D_3$ ($r = -0.50$, $p<0.1$), whereas in elderly normals and osteoporotic patients (women, 54 to 75 years old), these correlations were absent. Slovik et al.[40] stimulated the secretion of $1,25(OH)_2D_3$ via an i.v. infusion of PTH in normal young (men and women, 22 to 44 years old) and elderly patients (men and women, 50 to 80 years old) with untreated osteoporosis. Whereas serum levels of $1,25(OH)_2D_3$ were similar in both groups before the 24-hr infusion, serum levels about doubled (92% increase) in the young normals, but did not change significantly in the older osteoporotics. This research group suggested that the decreased absorptive ability and diminished capacity to adapt to lowered Ca intakes may be the result of impaired production of the hormonal form of vitamin D_3 in response to PTH stimulation. Supporting evidence from animal studies to support this hypothesis have been forthcoming[36,37,39,42] It is logical to suppose that an age-related deficient renal production of $1,25(OH)_2D_3$ would result in Ca malabsorption which would in turn stimulate an increase in PTH production. Elevated PTH levels could then produce bone resorption. If so, this phenomenon may play a part in the pathogenesis of senile osteoporosis. Armbrecht and colleagues[35] and Gallagher et al.[43] reported increases in PTH levels with age both in the rat and in women, and PTH has been implicated in low Ca adaptation,[35,39,42,44] presumably via its stimulation of $1,25(OH)_2D_3$ biosynthesis.[35,39,42,44-46]

OSTEOPOROSIS

One of the major aspects of Ca metabolism in aging is the phenomenon of progressive reduction in bone mass. This starts in almost all human populations beyond about 40 years of age, resulting in a progressive simultaneous loss of both the organic and inorganic components of bone with no compositional changes in the residual tissue.[47] That is to say, the composition of the bone is normal but there is less of it. When bone mass is reduced to such an extent that it is below the norm of an individual of a given age, sex, and race, the skeleton may become vulnerable to fractures arising from mild trauma or the stress of daily activities. This condition is known as osteoporosis.

Various investigators[48-69] have studied the rate of bone loss and age of onset in compact and trabecular bone in men and women (Tables 4, 5). Compact bone decreases at the rate of 3 to 5% each decade starting at the age of 40 years in both sexes and continuing at the same rate in men. In women, however, an additional loss occurs after menopause, bringing their total rate of decrease to 9% per decade between the ages of 45 to 75. The evidence suggests that the decrease of trabecular density begins during young adulthood (20 to 40 years) in both sexes and proceeds at similar rates (6 to 8% per decade) thereafter.[70] Since trabecular bone fractures such as of the vertebrae and distal radius occur earliest in post-menopausal women compared to compact bone fractures such as fractures of the proximal femur and humerus, it was assumed that more rapid loss occurs in trabecular bone compared to compact bone. However, studies on the rate of bone loss on compact bone and trabecular bone suggest that the latter starts at an earlier age and as a result, at any particular age will be greater than compact bone loss. This may contribute to early susceptibility of trabecular bones to fracture.[70]

Although aging bone loss evidently occurs in all populations, irrespective of socioeconomic status, food culture, and lifestyle, there are unexplained differences in the rate of bone loss among individuals within, as well as between populations. Additionally, there are differences attributable to heredity, race, and sex. Males are less susceptible to osteoporosis than females where the incidence of postmenopausal osteoporosis is extremely large.[71-75]

Table 4
AGE-RELATED RATE OF BONE LOSS IN COMPACT BONES

Measurement technique	Site of bone studied	Age of onset		Rate of bone loss per decade (%)		Ref.
		Male	Female	Male	Female	
Compact bone area	Metacarpal	40	40—50	4	3	48
			50—70		9	
Single photon absorptiometry	Distal radius	45	45	5	9	49
Single photon absorptiometry	Midshaft radius	45	45	4	10	50
Single photon absorptiometry	Midshaft radius	45	45	3	9	51
Single photon absorptiometry	Midshaft radius	45	35—45	3	2	52
			45—75		8	
Single photon absorptiometry	Midshaft radius	55	35—45	4	2	53
			46—75		9	
		—	45	—	9	54
Single photon absorptiometry	Midshaft radius	40	36—45	3	4	55
			45—90	—	9	
Radiographic photodensitometry	Proximal radius	40	40	5	10	56
Single photon absorptiometry	Midshaft radius	40	45	5	9	57

Although the mechanism of bone loss in aging has been investigated extensively in the past few decades, there has not been scientific consensus and the etiology of osteoporosis is still obscure. Since postmenopausal women are more susceptible to osteoporosis, a direct link between ovarian function and osteoporosis was proposed by Albright and associates in 1940.[76] A loss of ovarian function at any age is followed by a loss of bone,[74,77,78] which is particularly severe after an early menopause.[79] Long-term administration of estrogens to postmenopausal women greatly reduces the incidence of classic osteoporotic fractures,[80-83] and the increased rate of bone loss after treatment is withdrawn.[71,80,84] However, all women apparently lose bone after loss of ovarian function, but not all develop a degree of osteoporosis leading to fractures. The aged skeleton is less responsive to external administration of estrogen and PTH than that of a young adult because of a differential loss of trabecular bone during aging.[85] Though fewer in number, men are also susceptible to osteoporosis, suggesting that there are other factors responsible for the etiology of osteoporosis.

Other factors which have been largely implied in the etiology of osteoporosis have been nutritional ones. Considering the osteolytic nature of osteoporosis and the fact that 99% of body Ca is in the skeleton, it is not at all surprising that Ca nutriture has long been a subject of conjecture in relation to the etiology of osteoporosis.[2-6] In 1961, Nordin[86] proposed that Ca deficiency was a major factor in senile osteoporosis based on the observation that postmenopausal osteoporotics had a long history of below average Ca intake and that Ca supplementation resulted in positive balance. Subsequent eipdemiological surveys using cross-sectional methods to study the relationship between bone loss and Ca intake showed no such relationship.[87,88] Some correlation was observed by Nordin[89] in Japan and Finland, but not in other countries such as Gambia and Jamaica where dietary Ca was low and osteoporosis rare. Crosscultural surveys carried out in several geographic areas are based on the assumption that Ca requirements of different populations are similar, but this may not be true since varying environmental, cultural, dietary, and other conditions may influence the skeletal status directly or through changes in hormones or vitamin D.[90] Also, for each

Table 5
AGE-RELATED RATE OF BONE LOSS IN TRABECULAR BONE

Measurement technique	Site of bone studied	Age of onset		Rate of bone loss per decade (%)		Ref.
		Male	Female	Male	Female	
Radiographic photodensitometry	Lumbar vertebrae	45	38	6	8	58
Radiographic photodensitometry	Third lumbar	20	20	9	10	59
Radiographic photodensitometry	Femur	20	20	6	7	60
Single photon absorptiometry	Calcaneous	20	20	2	2	61
Dual-photon absorptiometry	Second and fourth lumbar vertebrae	—	30	—	10	62
Dual-photon absorptiometry	Third lumbar vertebrae	—	20	—	10	63
Dual-photon absorptiometry	First to fourth lumbar vertebrae	20	20	2	6	64

Postmortem apparent density method

	Lumbar vertebrae	50	50	13	13	65
	Third lumbar vertebrae	40	40	<5%	9	66
	Lumbar vertebrae	30	30	6	10	67
	Third lumbar vertebrae	30	30	7	8	68
	Second and third lumbar vertebrae	25	25	10	10	69

population there must be some minimal intake necessary for Ca homeostasis or maximal supression of bone loss; this intake may vary significantly among individuals and populations.[47]

Studies of individuals within the same geographical areas are useful to consider. When 382 Ohio residents with Ca intakes varying from 300 to 2000 mg/day were examined, no relationship between Ca intake and metacarpal cortical thickness was observed; similar conclusions were derived from longitudinal studies.[87,91] On the other hand, Thanangkul et al.[92] reported phalangeal bone density to be 12% greater in individuals ingesting more than 1250 mg/day of Ca than in persons whose intakes were only 350 to 150 mg/day. Similarly, a recent study by Metcovic et al.[93] on bone mass and Ca intake shows a clear relationship between the two. Also, an increase in the incidence of femoral neck fractures was observed in the group ingesting low amounts of Ca. Thus, results of studies conducted within single geographical areas are not conclusive. In contrast to these conflicting results, several studies on osteoporotic patients show that the Ca consumption of these patients is lower than that of age-matched controls.[59,86,94-96]

If low Ca intake is the major factor in senile osteoporosis, Ca supplementation should have beneficial effects in subjects suffering from osteoporosis and should have preventive effects if supplied at the early menopausal stage in women. Various reports have demonstrated beneficial effects of Ca supplementation to monosteoporotic menopausal women.[77,97-100]

However, Mazess and associates[71,78] could not demonstrate such beneficial results. Ca supplementation of osteoporotic women who have already suffered considerable bone loss does not seem to be very beneficial; no changes were observed by Shapiro et al.[101] and only a very small insignificant change was reported by Nordin et al.[102] Thus, the evidence of beneficial effects of Ca supplements to the postmenopausal nonosteoporotic women is provocative in that it does not prove that dietary Ca exerts a major impact on skeletal mass. The lack of beneficial effect of Ca supplementation could be due to alteration in other factors such as efficiency of utilization viz., absorption, skeletal turnover, and renal handling. These factors are in turn influenced by dietary constituents other than Ca. The most important variable which determines Ca utilization is the efficiency of GI absorption, and as discussed earlier, this function (Table 2) declines with age.[18-23]

Two major dietary constituents, among others, which could influence Ca metabolism are protein and P. The ready availability of P in commonly used foodstuffs and its widespread use in food processing have focused attention on the possibly deleterious effects of excessive intake. Animal studies have shown that high intakes of P can initiate a series of homeostatic adjustments which may result in bone loss.[103] Excessive dietary P causes an increase of plasma concentration of P and a decline in serum Ca. The resulting hypocalcemia stimulates secretion of PTH which in turn increases the rate of bone resorption. This sequence of events has been induced experimentally in a number of animal species,[103-108] but studies on man are not conclusive. Bell et al.[109] found a deleterious effect of increased P consumption (Ca/P ratio was less than 0.5). However, these studies were carried out only for a period of 8 weeks. With a high P intake, Reiss et al.[110] showed only a transient increase in serum PTH levels. Van den Berg et al.[111] showed only a slight increase in serum immunoreactive PTH levels, whereas, in the studies of Goldsmith et al.,[112] a high P intake seemed to have improved bone status of the subjects studied. Whether the high P content of the modern processed diet affects bone loss is not known, but recent studies caution that when the Ca/P ratio is low and is associated with some other factors, the rate of bone loss could increase.

Recent studies have demonstrated that a high dietary protein diet increases the urinary excretion of Ca.[113-116] This effect can be ameliorated by increasing the intake of Ca[116-118] and when dietary P is allowed to increase with the protein. The increase in Ca excretion with high protein diet is attributed to a lower efficiency of renal reabsorption.[114] Some laboratories have ascribed this to an increased production of acids such as sulfate arising from the oxidation of excess sulfur amino acids.[119,120] Further work is necessary to elucidate the complex interactions among high dietary protein, P, and Ca, and the net effect of high protein diets normally consumed over a lifetime on bone metabolism and the etiology of osteoporosis.

ADULT CALCIUM REQUIREMENTS

A number of reviews on Ca requirements have been written.[45,121-128] Almost all the reviews point out the difficulties in determining the exact Ca requirements for humans. The difficulties are due to the interaction of Ca with other dietary components to either facilitate or inhibit availability or utilization of Ca, a lack of an indisputable Ca deficiency disease in humans which would enable one to assess how much Ca must be ingested to prevent or cure the disease, and the ability of the human organism to adapt to Ca deprivation and restore balance over time.[3,5,6,8,20,21,27,28,30-34,45,123,124,129] Thus we cannot, at present state with certainty the optimum Ca intake for man.

In 1962, the FAO/WHO Expert Group on Calcium Requirements proposed that: (1) high intakes of Ca were not necessary and (2) requirements could be met safely with less than half the customary intakes in most European countries and in North America.[123] They based this claim on the ability of man to be in Ca balance on low dietary intakes and the absence

Table 6

**A SUMMARY OF STUDIES ON Ca REQUIREMENTS OF THE
ELDERLY (60 TO 88 YEARS) BASED ON RETENTION DATA**

Subjects	No. of studies	Mean requirement	Ref.
95 men and women	8	290—1000 mg	133—140
7 men and women (winter)	1	7.0 mg/kg	136
8 men and women (summer)	1	8.0 mg/kg	136
8 men	1	18.5 mg/kg, or 1031 mg/day	133
8 women	1	16.7 mg/kg, or 923 mg/day	134

of nutritional disability in populations on Ca intakes of, or even below, 300 mg/day. This was done at a time when dietary Ca deficiency was felt to be widespread and fortification of cereals and other foods was advocated. The FAO/WHO suggested a practical allowance, defined as one likely to meet the needs of the great majority of persons ingesting adequate amounts of vitamin D. The allowance suggested for adults (19 +) was 400 to 500 mg/day. The recommendations appeared practical and realistic to many and a number of countries have adopted them with minor modifications including, among others, the industrialized nations of the United Kingdom, Canada, and Australia.[127]

There are those, however, who consider that higher intakes of Ca are desirable.[125] In 1941 the Food and Nutrition Board of the U.S. National Research Council recommended 800 mg Ca per day as a suitable daily allowance for nonpregnant, nonlactating adults. It has remained at that level in the U.S. since then. The recommendation has been based in part on balance studies conducted with groups of individuals accustomed to ample intakes of foods high in Ca. In 1980, however, it was recognized by the Food and Nutrition Board[125] that adults remain in Ca balance with intakes as low as 200 to 400 mg/day and that a higher proportion of Ca is utilized when intake is low than when it is generous. Thus, it might seem to some that a recommendation of 800 mg/day was an exaggerated figure. But there was a reluctance to lower the recommendation feeling that ''no advantage accrues from such low intakes, and for a variety of reasons it seems unwise to recommend such low calcium intake.'' These reasons include the possible long-range deleterious effects of high dietary intakes of protein[130] or P[131] on Ca retention and the possibility of reduced Ca absorption during aging.[18-23] A number of nations have found the arguments of the U.S. Food and Nutrition Board convincing and have, with modification, adopted their recommendations, including, among others, Sweden and the Netherlands.[127]

The situation with regards to recommendations for Ca intake in the aged presents even additional controversy.[4-6] It is unknown whether Ca requirements of the elderly are the same or similar to those already proposed for younger adults. It is thought by some that it might be prudent to add another age-sex category to the U.S. Recommended Dietary Allowance (RDA) for those over 60 years of age whose Ca requirements may be elevated. Advocates of higher intakes in the elderly stress the phenomenon of reduced Ca absorption in aging, hormonal changes which influence the metabolism of Ca and bone,[132] and reduced physical activity. Interpretation of the available evidence presents some difficulties in implementing this goal. First, much of the work on Ca requirements in the elderly has been performed on osteoporotics and/or involved balance studies where zero Ca balances achieved over short-term periods was taken as the Ca requirement but cannot, necessarily, be assumed to be sustained in the aged over longer durations; second, Ca balance studies performed on non-osteoporotic elderly aged 60 to 88 (Table 6) have revealed considerable variation in requirement from one study to another[133-140] as well as from one individual to another.[124]

Clearly more experimentation and exposition are required before the controversy between the minimalist viewpoint[123] and the maximalist stance[6,125] can be resolved. Wherever possible, inappropriate concerns about meeting artificially inflated Ca requirements should be avoided since they may divert attention and resources from perhaps more important nutritional problems; on the other hand, dismissing the possible need for more Ca in our diets would be a potential disservice to keeping and improving the health of our ever-increasing aged population. Perhaps if the accumulated data of the many studies available to us were reevaluated so that Ca requirements were expressed in terms of intake of another critical nutrient (e.g., protein or perhaps P) some of the controversy would evaporate.

ACKNOWLEDGMENTS

The portion of unpublished work reported carried out in the laboratory of Ira Wolinsky was supported by the National Institute of Aging grant AGO2396. S.D. Telang was on research leave at the Department of Neurobiology and Anatomy, University of Texas Medical Branch, Houston, Texas and at Psychiatric Research Division, University Hospital, University of Saskatchewan, Saskatoon, Canada.

REFERENCES

1. **Timiras, P. S.,** *Developmental Psychology and Aging,* Macmillan, N.Y., 1972.
2. **Albanese, A. A.,** *Bone Loss: Causes, Detection and Therapy.* Alan R. Liss, N.Y., 1977.
3. **Spencer, H., Kramer, L., and Osis, D.,** Factors contributing to calcium loss in aging, *Am. J. Clin. Nutr.,* 36, 776, 1982.
4. **Heaney, R. P.,** Calcium intake requirement and bone mass in the elderly, *J. Lab. Clin. Med.,* 100, 309, 1982.
5. **Marcus, R.,** The relationship of dietary calcium to the maintenance of skeletal integrity in man — an interface of endocrinology and nutrition, *Metabolism,* 31, 93, 1982.
6. **Heaney, R. P., Gallagher, J. C., Johnston, C. C., Neer, R., Parfitt, A. M., BChir, M. B., and Whedon, G. D.,** Calcium nutrition and bone health in the elderly, *Am. J. Clin. Nutr.,* 36, 986, 1982.
7. **Bronner, F.,** Dynamics and function of calcium, in *Mineral Metabolism: An Advanced Treatise,* Vol. 2, Comar, C. L. and Bronner, F., Eds., Academic Press, N.Y., 1964, 342.
8. **Kenny, A. D.,** *Intestinal Calcium Absorption and its Regulation,* CRC Press, Boca Raton, Fla., 1981.
9. **Armbrecht, H. J., Zenser, T. V., Bruns, M. E. H., and Davis, B. B.,** Effect of age on intestinal calcium absorption and adaptation to dietary calcium, *Am. J. Physiol.,* 236, E769, 1979.
10. **Schachter, D., Dowdle, E. B., and Schenker, H.,** Accumulation of ^{45}Ca by slices of the small intestine, *Am. J. Physiol.,* 198, 275, 1960.
11. **Milhaud, G., Cherian, A. G., and Moukhtar, M. S.,** Calcium metabolism in the rat studied with calcium^{45}: effect of age, *Proc. Soc. Exp. Biol. Med.,* 114, 382, 1963.
12. **Henry, K. M. and Kon, S. K.,** The relation between calcium retention and body stores of calcium in the rat: effect of age and of vitamin D, *Br. J. Nutr.,* 7, 147, 1953.
13. **Hansard, S. L. and Crowder, H. M.,** The physiological behavior of calcium in the rat, *J. Nutr.,* 62, 325, 1957.
14. **Taylor, D. M., Bligh, P. H., and Duggan, M. H.,** The absorption of calcium, strontium, barium and radium from the gastrointestinal tract of the rat, *Biochem. J.,* 83, 25, 1962.
15. **Braithwaite, G. D.,** Studies on the absorption and retention of calcium and phosphorus by young and mature Ca-deficient sheep, *Br. J. Nutr.,* 34, 311, 1975.
16. **Henry, K. M., Kon, S. K., Todd, P. E. E., Toothill, J., and Tomlin, D. H.,** Calcium and phosphorus metabolism in the rat: effect of age on rate of adaptation to a low calcium intake, *Acta Biochim. Pol.,* 7, 167, 1960.
17. **Wolinsky, I. and Tso, N. T-B.,** unpublished data.
18. **Bullamore, J. R., Gallagher C., Wilkinson, R., and Nordin, B. E. C.,** Effect of age on calcium absorption, *Lancet,* 2, 535, 1970.
19. **Caniggia, A., Gennari, C., Cesari, L., and Romano, S.,** Intestinal absorption of ^{45}Ca in adult and old human subjects, *Gerontologia,* 10, 193, 1964/65.
20. **Ireland, P. and Fordtran, J. S.,** Effect of dietary calcium and age on jejunal calcium absorption in humans studied by intestinal perfusion, *J. Clin. Invest.,* 52, 2672, 1973.

21. **Gallagher, J. C., Riggs, B. L., Eisman, J., Hamstra, A., Arnaud, S. B., and DeLuca, H. F.**, Intestinal calcium absorption and serum vitamin D metabolites in normal subjects and osteoporotic patients. Effect of age and dietary calcium, *J. Clin. Invest.*, 64, 729, 1979.

22. **Avioli, L. V., McDonald, J. E. and Lee, S. W.**, The influence of age on the intestinal absorption of [47]Ca in women and its relation to [47]Ca absorption in postmenopausal osteoporosis, *J. Clin. Invest.*, 44, 1960, 1965.

23. **Alevizaki, C. C., Ikkos, D. G., and Singhelakis, P.**, Progressive decrease of true intestinal calcium absorption with age in normal man, *J. Nucl. Med.*, 14, 760, 1973.

24. **Mautalen, C. A., Cabrejas, M. L., and Soto, R. J.**, Isotopic determination of intestinal calcium absorption in normal subjects, *Metabolism*, 18, 395, 1969.

25. **Avioli, L. V., McDonald, J. E., Singer, R. A., and Henneman, P. H.**, A new oral isotopic test of calcium absorption, *J. Clin. Invest.*, 44, 128, 1965.

26. **Armbrecht, H. J., Zenser, T. V., Gross, C. J., and Davis, B. B.**, Adaptation to dietary calcium and phosphorus restriction changes with age in the rat, *Am. J. Physiol.*, 239, E322, 1980.

27. **Kimberg, D. V., Schachter, D., and Schenker, H.**, Active transport of calcium by intestine: effects of dietary calcium, *Am. J. Physiol.*, 200, 1256, 1961.

28. **Gershoff, S. N., Legg, M. A., and Hegsted, D. M.**, Adaptation to different calcium intakes in dogs, *J. Nutr.*, 64, 303, 1958.

29. **Wolinsky, I. and Guggenheim, K.**, Effect of low calcium diet on bone and calcium metabolism in rats and mice — a differential species response, *Comp. Biochem. Physiol.*, 49A, 183, 1974.

30. **Heaney, R. P., Saville, P. D., and Recker, R. R.**, Calcium absorption as a function of calcium intake, *J. Lab. Clin. Med.*, 85, 881, 1975.

31. **Bhandarkar, S. D. and Nordin, B. E. C.**, Effect of low-calcium diet on urinary calcium in osteoporosis, *Br. Med. J.*, 1, 145, 1962.

32. **Spencer, H., Lewin, I., Fowler, J., and Samachson, J.**, Influence of dietary calcium intake on Ca[47] absorption in man, *Am. J. Med.*, 46, 197, 1969.

33. **Agnew, J. E. and Holdsworth, C. D.**, The effect of fat on calcium absorption from a mixed meal in normal subjects, patients with malabsorption disease and patients with a partial gastrectomy, *Gut*, 12, 973, 1971.

34. **Kemm, J. R.**, Effect of previous dietary history of calcium intake on the skeleton and calcium absorption in the rat, *J. Phisiol.*, 230, 643, 1973.

35. **Armbrecht, H. J., Gross, C. J., and Zenser, T. V.**, Effect of dietary calcium and phosphorus restriction on calcium and phosphorus balance in young and old rats, *Arch. Biochem. Biophys.*, 210, 179, 1981.

36. **Norman, A. W.**, *Vitamin D. The Calcium Homeostatic Steroid Hormone*, Academic Press, N. Y., 1979.

37. **Omdahl, J. L. and DeLuca, H. F.**, Mediation of calcium adaptation by 1,25-dihydroxycholecalciferol, *J. Nutr.*, 107, 1975, 1977.

38. **Baksi, S. and Kenny, A. D.**, Vitamin D metabolism in Japanese quail: dietary calcium and estrogen effects, *Am. J. Physiol.*, 241, E275, 1981.

39. **Pento, J. T.**, The influence of interrupted vitamin D metabolism on acute low calcium adaptation in the rat, *Nutr. Metab.*, 20, 321, 1976.

40. **Slovik, D. M., Adams, J. S., Neer, R. M., Holick, M. F., and Potts, J. T., Jr.**, Deficient production of 1,25-dihydroxyvitamin D in elderly osteoporotic patients, *N. Engl. J. Med.*, 305, 372, 1981.

41. **Armbrecht, J. H., Zenser, T. V., and Davis, B. B.**, Effect of age on the conversion of 25-hydroxyvitamin D_3 to 1,25-dihydroxyvitamin D_3 by kidney of rat, *J. Clin. Invest.*, 66, 1118, 1980.

42. **Ribovich, M. L. and DeLuca, H. F.**, Intestinal calcium transport: parathyroid hormone and adaptation to dietary calcium, *Arch. Biochem. Biophys.*, 175, 256, 1976.

43. **Gallagher, J. C., Riggs, B. L., Jerpbak, C. M., and Arnaud, C. D.**, The effect of age on serum immunoreactive parathyroid hormone in normal and osteoporotic women, *J. Lab. Clin. Med.*, 95, 373, 1980.

44. **Pento, J. T., Waite, L. C., Tracy, P. J., and Kenny, A. D.**, Adaptation to calcium deprivation in the rat: effects of parathyroidectomy, *Am. J. Physiol.*, 232, E336, 1977.

45. **Paterson, C. R.**, Calcium requirements in man: a critical review, *Postgrad. Med.*, 54, 244, 1978.

46. **Gallagher, J. C., Riggs, B. L., and DeLuca, H. F.**, Effect of estrogen on calcium absorption and serum vitamin D metabolites in postmenopausal osteoporosis, *J. Clin. Endocrinol. Metab.*, 51, 1359, 1980.

47. **Draper, H. H. and Bell, R. R.**, Nutrition and osteoporosis, in *Advances in Nutritional Research*, Draper, H. H., Ed., Plenum Press, N.Y., 1977, 79.

48. **Garn, S. M., Rohmann, C. G., Wagner, B., Davila, G. H., and Ascoli, W.**, Population similarities in the onset and rate of adult endosteal bone loss, *Clin. Orthop.*, 65, 51, 1969.

49. **Goldsmith, N. F., Johnston, J. O., Picetti, G., and Garcia, C.**, Bone mineral in the radius and vertebral osteoporosis in an insured population, *J. Bone Jt. Surg.*, 55A, 1276, 1973.

50. **Roh, Y. S.**, Evaluation of Bone Mass and Bone Remodeling In Vivo and its Clinical Application in Osteoarthrosis and Osteoporosis, Ph.D. thesis, Katholic University of Louvian, Belgium, 1973.

51. **Boyd, R. M., Cameron, E. C., McIntosh, H. W., and Walker, V. R.,** Measurement of bone mineral content in vivo using photon absorptiometry, *Can. Med. Assoc. J.,* 111, 1201, 1974.

52. **Donath, A., Indermuhle, P., and Baud, R.,** Mineralometric osseuse: measuree par l'absorption de photons d'une source d'^{125}I, *Radiol. Clin. Biol.,* 43, 393, 1974.

53. **Mazess, R. B. and Cameron, J. R.,** Bone Mineral Content in Normal U.S. Whites, in *International Conference on Bone Mineral Measurement,* Mazess, R., Ed., U.S. Department of Health, Education, and Welfare, Publ. NIH 75-683, Washington, D.C., 1974.

54. **Smith, D. M., Khairi, M. R. S., and Johnston, C. C., Jr.,** The loss of bone mineral with aging and its relationship to risk of fracture, *J. Clin. Invest.,* 56, 311, 1975.

55. **Ringe, J. D. and Von Rehpenning, W.,** Kuhlencordt F. Physiologische Anderung des Mineralgehalts von Radius und Ulna in Abhangigkeit von Lebensalter und Geschlecht, *Fortschr. Rontgenstr.,* 126, 376, 1977.

56. **Meema, H. E. and Meema, S.,** Compact bone mineral density of the normal human radius, *Acta Radiol. Oncol.,* 17, 342, 1978.

57. **Runge, H., Fengler, F., Franke, J., and Koall, W.,** Ermittlung des peripheren Knochenmineralgehaltes bei Normal Personen und Patienten mit verschiendenen Knochenerkrankungen, bestimmt mit Hilfe der Photonabsorptionstechnik am Radius, *Radiologe,* 20, 505, 1980.

58. **Haasner, E., Krakowski, E., and Bach, K.,** Normalwerte des Hydroxylapatitgehaltes im Skeletin Abhangigkeit von Lokalisation, Lebensalter und geschlecht, *Klin. Wochenschr.,* 45, 575, 1967.

59. **Hurxthal. L. M. and Vose G. P.,** The relationship of dietary calcium intake to radiographic bone density in normal and osteoporotic persons, *Calcif. Tissue Res.,* 4, 245, 1965.

60. **Heuck, F. H. W.,** Quantitative measurements of mineral content in bone diseases in *Symposium Ossium,* Jeffife, A. M. and Strickland, B., Eds., E.S. Livingston, Edinburgh, 1970.

61. **Banzer, D. H., Schneider, U., Risch, W. D., and Botsch, H.,** Roentgen signs of vertebral demineralization and mineral content of peripheral cancellous bone, *Am. J. Roentgenol.,* 126, 1306, 1976.

62. **Krolner, B. and Pors Nielsen, S.,** Measurement of bone mineral content (BMC) of the lumbar spine. I. Theory and application of a new two dimensional dual-photon attentuation method, *Scand. J. Clin. Lab. Invest.,* 40, 653, 1980.

63. **Hansson, T., Ross, B., and Nachemson, A.,** The bone mineral content and ultimate compressive strength of lumbar vertebrae, *Spine,* 5, 46, 1980.

64. **Riggs, B. L., Wahner, H. W., Dunn, W. L., Mazess, R. B., Offord, K. P., and Melton, L. J., III,** Differential changes in bone mineral density of the appendicular and axial skeleton with aging, *J. Clin. Invest.,* 67, 328, 1981.

65. **Muller, K. H., Trias, A., and Ray, R. D.,** Bone density and composition: age related and pathological changes in water and mineral content, *J. Bone Jt. Surg.,* 48A, 140, 1966.

66. **Weaver, J. K., Chaterji, S. K., and Jeffery, J. S.,** Cancellous bone: its strength and changes with aging and an evaluation of some methods for measuring its mineral content. I. Age changes in cancellous bone, *J. Bone Jt. Surg.,* 48A, 289, 1966.

67. **Ahuja, M.,** Normal variation in the density of selected human bones in North India, *J. Bone Jt. Surg.,* 51B, 719, 1969.

68. **Havivi, E., Reshef, A., Schwartz, A., Guggenheim, K., Bernstein, D. S., Hegsted, D. M., and Stare, F. J.,** Comparison of metacarpal bone loss with physical and chemical characteristics of vertebrae and ribs, *Isr. J. Med. Sci.,* 7, 1055, 1971.

69. **Arnold, J. S.,** Amount and quality of trabecular bone in osteoporotic vertebral fractures, *Clin. Endocr. Metab.,* 2, 221, 1973.

70. **Mazess, R. B.,** On aging bone loss, *Clin. Orthop.,* 165, 239, 1982.

71. **Christiansen, C., Mazess, R. B., Transbol, I., and Jensen, G. F.,** Factors in response to treatment of early postmenopausal bone loss, *Calcif. Tissue Int.,* 33, 575, 1981.

72. **Heer, K. R., Alexandrow, K., Lauffenburger, T., and Hass, H. G.,** Changes of bone mineral in healthy menopausal and premenopausal women: two year preliminary results of a longitudinal study, *Am. J. Roentgenol.,* 126, 1298, 1976.

73. **Johnston, C. C., Jr., Norton, D. A., Khairi, R. A., and Longcope, C.,** Age related bone loss, in *Osteoporosis,* Barzel, U., Ed., Grune & Stratton, N.Y., 1979.

74. **Lindsay, R., Aitken, J. M., Anderson, J. B., Hart, D. M., McDonald, E. B., and Clarke, A. C.,** Long term prevention of postmenopausal osteoporosis by estrogen, *Lancet,* 1, 1038, 1976.

75. **Newton-John, H. F. and Morgan, D. B.,** Osteoporosis: disease of senescence, *Lancet,* 1, 232, 1968.

76. **Albright, F., Bloomberg, E., and Smith, P. H.,** Postmenopausal osteoporosis, *Tran. Assoc. Am. Physicians,* 55, 298, 1940.

77. **Horsman, A., Gallagher, J. C., Simpson, M., and Nordin, B. E. C.,** Prospective trial of oestrogen and calcium in postmenopausal women, *Br. Med. J.,* 2, 789, 1971.

78. **Christiansen, C., Christiansen, M. S., McNair, P. I., Hagen, C., Stocklund, E., and Transbol, I.,** Prevention of early postmenopausal bone loss: controlled 2 year study in 315 normal females, *Eur. J. Clin. Invest.,* 10, 273, 1980.

79. **Aitken, J. M.,** Osteoporosis and its relation to oestrogen deficiency, in *The Management of the Menopause and Postmenopausal Years,* Campbell, S., Ed., MTP Press, Lancaster, Pa., 1976, 225.

80. **Lindsay, R., Hart, D. M., Mclean, A., Clarke, A. C., Kraszewski, A., and Garwood, J.,** Bone response to termination of oestrogen treatment, *Lancet,* 1, 1325, 1978.

81. **Hutchinson, T. A., Polansky, S. M., and Feinstein, A.,** Postmenopausal oestrogens protect against fractures of hip and distal radius, *Lancet,* 2, 706, 1979.

82. **Weiss, N. S., Ure, C. L., Ballard, J. H., Williams, A. R., and Daling, J. R.,** Decreased risk of fractures of the hip and lower forearm with postmenopausal use of estrogen, *N. Engl. J. Med.,* 303, 1195, 1980.

83. **Paganin-Hill, A., Ross, R. K., Gerkins, V. R., Henderson, B. G., Arthur, M., and Mack, T. M.,** A case control study of menopausal estrogen therapy and hip fractures, *Ann. Int. Med.,* 95, 28, 1981.

84. **Horsman, A., Nordin, B. E. C., and Crilly, R. G.,** Effects on bone of withdrawal of oestrogen therapy, *Lancet,* 2, 33, 1979.

85. **Jowsey, J.,** *Metabolic Disease of Bone,* Monographs in Clinical Orthopedics, Vol. 1, W. B. Saunders, Philadelphia, 1977.

86. **Nordin, B. E. C.,** The pathogenesis of osteoporosis, *Lancet,* 1, 1011, 1961.

87. **Garn, S. M.,** *The Earlier Gain and the Later Loss of Cortical Bone,* Charles C Thomas, Springfield, Ill., 1970.

88. **Chalmers, J. and Ho, K. C.,** Georgraphical variations in senile osteoporosis the association with physical activity, *J. Bone Jt. Surg.,* 52B, 667, 1970.

89. **Nordin, B. E. C.,** International patterns of osteoporosis, *Clin. Orthop.,* 45, 17, 1966.

90. **Nordin, B. E. C., Young, M. M., Bulusu, L., and Horsman, A.,** Osteoporosis reexamined, in *Osteoporosis,* Barzel, U., Ed., Grune & Stratton, N.Y., 1970, 47.

91. **Garn, S. M., Rohmann, C. G., and Wagner, B.,** Bone loss as a general phenomenon in man, *Fed. Proc.,* 26, 1729, 1967.

92. **Thanangkul, D., Johnston, F. A., Kime, N. S., and Clarke, S. J.,** Adaptation to a low calcium intake, *J. Am. Diet. Assoc.,* 35, 23, 1959.

93. **Metcovic, V., Kostial, K., and Simonovic, I.,** Bone status and fracture rates in two regions of Yugoslavia, *Am. J. Clin. Nutr.,* 32, 540, 1979.

94. **Riggs, B. L., Kelly, P. J., and Kinney, V. R.,** Calcium deficiency and osteoporosis: observations in one hundred and sixty patients and critical review of literature, *J. Bone Jt. Surg.,* 49A, 915, 1967.

95. **Birge, S. J., Jr., Ketumann, H. T., and Cuatrecasas, P.,** Osteoporosis, intestinal lactase deficiency and low dietary calcium intake, *N. Engl. J. Med.,* 276, 445, 1967.

96. **Newcomer, A. D., Hodgson, S. F., McGill, D. B., and Thomas, P. J.,** Lactase deficiency: prevalance in osteoporosis, *Ann. Int. Med.,* 89, 218, 1978.

97. **Recker, R. R., Saville, P. D., and Heaney, R. P.,** Effect of estrogens and calcium carbonate on bone loss in postmenopausal women, *Ann. Int. Med.,* 87, 649, 1977.

98. **Albanese, A. A., Edelson, A. H., Lorenze, E. J., Jr., Woodhull, M. L., and Wein, E. H.,** Problems of bone health in elderly, a ten year study, *N.Y. State J. Med.,* 75, 326, 1975.

99. **Albanese, A. A., Lorenze, E. J., Jr., Edelson, A. H., Wein, E. H., and Carroll, L.,** Effects of calcium supplements and estrogen replacement therapy on bone loss of postmenopausal women, *Nutr. Rep. Int.,* 24, 403, 1981.

100. **Smith, E. L., Jr., Reddan, W., and Smith, P. E.,** Physical activity and calcium modalities for bone mineral increase in aged women, *Med. Sci. Sports Excer.,* 13, 60, 1981.

101. **Shapiro, J. R., Moore, W. T., Jorgensen, H., Reid, J., Epps, C. H., and Whedon, D.,** Osteoporosis: evaluation of diagnosis and therapy, *Arch. Int. Med.,* 135, 563, 1975.

102. **Nordin, B. E. C., Horsman, A., Crilly, R. G., Marshall, D. H., and Simpson, M.,** Treatment of spinal osteoporosis in postmenopausal women, *Br. Med. J.,* 1, 451, 1980.

103. **Krook, L.,** Dietary calcium phosphorous and lameness in the horse, *Cornell Vet.,* Suppl. 58, 59, 1968.

104. **Krishna Rao, G. V. G. and Draper, H. H.,** Influence of dietary phosphate on bone resorption in senescent mice, *J. Nutr.,* 102, 1143, 1972.

105. **Draper, H. H., Sie, T. L., Bergan, J. G.,** Osteoporosis in aging rats induced by high phosphorous diets, *J. Nutr.,* 102, 1133, 1972.

106. **Laflamme, G. H. and Jowsey, J.,** Bone and soft tissue changes with oral phosphate supplements, *J. Clin. Invest.,* 51, 2834, 1972.

107. **Jowsey, J. and Balasubramaniam, P.,** Effect of phosphate supplements on soft tissue calcification and bone turnover, *Clin. Sci.,* 42, 289, 1972.

108. **DeLuca, H. F., Castillo, L., Jee, W., Oldham, S., and Grummer, R. H.,** Studies on high phosphate diets. Food Research Institute. Annual Report. University of Wisconsin, Madison, 1976, 394.

109. **Bell, R. R., Draper, H. H., Tzeng, D. Y. M., Shin, H. K., and Schmidt, G. R.,** Physiological responses of human adults to foods containing phosphate additives, *J. Nutr.* 107, 42, 1971.

110. **Reiss, E., Canterbury, J. M., Bercovitz, M. A., and Kaplan, E. L.,** The role of phosphate in the secretion of parathyroid hormone in man, *J. Clin. Invest.,* 49, 2146, 1970.

111. **Van den Berg, C. J., Kumar, R., Wilson, D. M., Heath, H., III, and Smith, L. H.,** Orthophosphate therapy decreases urinary calcium excretion and serum 1-25 dihydroxyvitamin D concentrations in idiopathic hypercalciuria, *J. Clin. Endocrinol. Metab.,* 51, 998, 1980.

112. **Goldsmith, R. S., Jowsey, J., Dube, W. J., Riggs, B. L., Arnaud, C. D., and Kelly, P. J.,** Effects of phosphorous supplementation on serum parathyroid hormone and bone morphology in osteoporosis, *J. Clin. Endocrinol. Metab.,* 43, 523, 1976.

113. **McCance, R. A., Widdowson, E. M., and Lehmann, H.,** The effect of protein intake on the absorption of calcium and magnesium, *Biochem. J.,* 36, 686, 1942.

114. **Allen, L. H., Bartlett, R. S., and Block. G. D.,** Reduction of renal calcium reabsorption in man by consumption of dietary protein, *J. Nutr.,* 109, 1345, 1979.

115. **Robertson, W. G., Heyburn, P. J., and Peacock, M.,** The effect of high animal protein intake on the risk of calcium stone formation in the urinary tract, *Clin. Sci.,* 57, 285, 1979.

116. **Hegsted, M. and Linkswiler, H. M.,** Long term effects of level of protein intake on calcium metabolism in young adult women, *J. Nutr.* 111, 244, 1981.

117. **Linkweiler, H. M., Zemel, M. B., Hegsted, M., and Schuette, S.,** Protein-induced hypercalciuria, *Fed. Proc.,* 40, 2429, 1981.

118. **Spencer, H., Kramer, L., Osis, D., and Norris, C.,** Effect of a high protein (meat) intake on calcium metabolism in man, *Am. J. Clin. Nutr.,* 31, 2167, 1978.

119. **Schuette, S. A., Zemel, M. B., and Linkswiler, H. M.,** Studies on the mechanism of protein-induced hypercalciuria in older men and women, *J. Nutr.,* 110, 305, 1980.

120. **Block, G. D., Wood, R. J., and Allen, L. H.,** A comparison of the effects of feeding sulfur amino acids and protein on urine calcium in man, *Am. J. Clin. Nutr.,* 33, 2128, 1980.

121. **Hegsted, D. M.,** Calcium and phosphorus, in *Modern Nutrition in Health and Disease,* 5th ed., Goodhart, R. S. and Shils, M. E.,Eds., Lea & Febiger, Philadelphia, 1973, 268.

122. **Avioli, L. V.,** Calcium and phosphorus, in *Modern Nutrition in Health and Disease,* 6th ed., Goodhart, R. S. and Shils, M. E., Eds., Lea & Febiger, Philadelphia, 1980, 294.

123. **FAO/WHO,** *Calcium Requirements,* WHO Technical Report, Ser. No. 230, FAO, Rome, 1962.

124. **Irwin, M. I. and Kienholz, E. W.,** A conspectus of research on calcium requirements of man, *J. Nutr.,* 103, 1019, 1973.

125. National Research Council, *Recommended Dietary Allowances,* 9th ed., Washington, D.C., p. 125, 1980.

126. **Widdowson, E. M.,** Nutritional requirements and its assessment, with special reference to energy, protein and calcium, *Bibl. Nutr. Dieta,* 28, 148, 1974.

127. International Union Nutritional Sciences, Report of the Committee on International Dietary Allowances of the International Union of Nutritional Sciences, *Nutr. Abstr. Rev.,* 45, 89, 1975.

128. **McBean, L. D. and Speckmann, E. W.,** A recognition of the interrelationship of calcium with various dietary components, *Am. J. Clin. Nutr.,* 27, 603, 1974.

129. **Allen, L. H.,** Calcium bioavailability and absorption: a review, *Am. J. Clin. Nutr.,* 35, 783, 1982.

130. **Altchuler, S. I.,** Dietary protein and calcium loss: a review, *Nutr. Res.,* 2, 193, 1982.

131. **Draper, H. H. and Scythes, C. A.,** Calcium, phosphorus, and osteoporosis, *Fed. Proc.,* 40, 2434, 1981.

132. **Parson, J. A.,** *Endocrinology of Calcium Metabolism,* Raven Press, N.Y., 1982.

133. **Ackermann, P. G. and Toro, G.,** Calcium and phosphorus balance in elderly men, *J. Gerontol.,* 8, 289, 1953.

134. **Ackermann, P. G. and Toro, G.,** Calcium balance in elderly women, *J. Gerontol.,* 9, 446, 1954.

135. **Bogdonoff, M. D., Shock, N. W., and Nichols, M. P.,** Calcium, phosphorus, nitrogen and potassium balance studies in the aged male, *J. Gerontol.,* 8, 272, 1953.

136. **Fujitani, M.,** Studies on calcium, phosphorus, sodium and chlorine balances in the aged, *J. Osaka City Med. Cent.,* 9, 2063, 1960.

137. **Ohlson, M. A., Brewer, W. D., Jackson, L., Swanson, P. P., Roberts, P. H., Mangel, M., Leverton, R. M., Chaloupka, M., Gram, M. R., Reynolds, M. S., and Lutz, R.,** Intakes and retentions of nitrogen, calcium and phosphorus by 136 women between 30 and 85 years of age, *Fed. Proc.,* 11, 775, 1952.

138. **Owen, E. C.,** The calcium requirements of older male subjects, *Biochem. J.,* 33, 22, 1939.

139. **Owen, E. C., Irving, J. T., and Lyall, A.,** The calcium requirements of older male subjects with special reference to the genesis of senile osteoporosis, *Acta Med. Scand.,* 103, 235, 1940.

140. **Roberts, P. H., Kerr, C. H., and Ohlson, M. A.,** Nutritional studies of older women. Nitrogen, calcium, phosphorus retentions of nine women, *J. Am. Diet. Assoc.,* 24, 292, 1948.

CHROMIUM REQUIREMENTS AND NEEDS IN THE ELDERLY

Richard A. Anderson

INTRODUCTION

In a general sense, aging begins at conception and progresses at different rates that may or may not be controlled, and ends at death. Proper nutrition, obviously, will not prevent aging but it may decrease the rate or kinetics of this inevitable process. If proper nutrition is to exert its primary effects on the aging process, good nutrition must be practiced throughout life and not left until the "twilight years". Proper nutrition should not be considered a cure for diseases and other processes associated with aging but rather a preventive measure.

This is particularly true for chromium nutrition. Chromium (Cr) deficiency symptoms often are slow to develop compared to other trace elements. For example, in rats, signs of Fe or Zn deficiency are obvious after 3 weeks or less of feeding the respective deficient diets. However, when Cr-deficient diets are fed to experimental animals, signs of Cr deficiency take months to develop.[1] In humans, the manifestations of marginal Cr deficiency may take several decades to become apparent.

Cr, an essential trace element for higher animals and man, is involved in carbohydrate, lipid, and nucleic acid metabolism. The role of Cr in carbohydrate and lipid metabolism is as a potentiator of insulin action, and in nucleic acid metabolism Cr is postulated to be involved in maintaining the structural integrity of the nuclear strands.[1,2] Cr functions by increasing the effectiveness of insulin but not as its substitute. Insulin-dependent processes which are potentiated by Cr include glucose uptake, oxidation of glucose to carbon dioxide, and incorporation of glucose into fat. The rate of incorporation of acetate into fat, which is not insulin-dependent, is not activated by Cr. With only nanogram quantities of Cr, optimal in vitro effects of insulin can be demonstrated in insulin-sensitive tissues. In the absence of Cr, much higher levels of insulin are required.[3]

CONSEQUENCES OF MARGINAL DIETARY INTAKE OF Cr

The signs of Cr deficiency listed in Table 1 are similar to many of the problems associated with aging. Some of the problems routinely associated with aging may more correctly be associated with poor nutrition that manifests itself with age. This may be one possible explanation for the grossly varying rates of aging in different individuals. Chronological age is independent of nutrition while the actual aging process may be critically dependent upon nutrition.

Overt signs of Cr deficiency in humans were first observed in a female patient receiving total parenteral nutrition.[4] She developed diabetic-like signs including glucose intolerance, unexpected weight loss, and impaired nerve conduction. These diabetic-like signs, that were refractory to 45 units of insulin per day, were reversed by daily infusion of 250 µg of inorganic Cr. Once body stores were replenished, a daily maintenance dose of 20 µg was sufficient. Similar overt signs of Cr deficiency that were refractory to insulin but reversed by supplemental chromium have also been observed in a second patient receiving total parenteral nutrition.[5]

These overt symptoms of Cr deficiency have not been reported in the general population, yet marginal signs of Cr deficiency have appeared. In three separate studies at least 50% of the subjects showed significant improvement in their impaired glucose tolerance after Cr supplementation.[6-8] Cr has also been reported to significantly improve high density lipoprotein cholesterol in normal male subjects.[9] We have demonstrated that subjects with near optimal

Table 1
SIGNS OF CHROMIUM DEFICIENCY

Impaired glucose tolerance
Elevated circulating insulin
Glycosuria
Fasting hyperglycemia
Impaired growth
Decreased longevity
Elevated serum cholesterol and triglycerides
Increased incidence of aortic plaques
Peripheral neuropathy
Brain disorders
Decreased fertility and sperm count

blood glucose are not affected by supplemental Cr. However, the glucose intolerance of subjects with marginally elevated or depressed serum glucose following a glucose challenge is improved significantly following Cr supplementation. We did not observe an increase in high density lipoprotein cholesterol following Cr supplementation.[10] Liu et al.[11] reported that more that two thirds of normal and hyperglycemic women displayed a decrease in blood glucose and circulating insulin after supplementation with yeast that was high in biologically active Cr. Cr-rich brewer's yeast also improved the glucose tolerance and total lipids in elderly subjects, while Cr-poor torula yeast had no effect.[12] Other investigators reported little difference between subjects treated with placebo and Cr.[13,14] However, a number of factors, of which Cr is only one, affect glucose removal rates. Therefore, Cr deficiency should not be construed as the sole criterion of impaired glucose tolerance. Furthermore, subjects who do not respond to Cr supplementation tend to display seriously impaired glucose tolerance and/or are overweight; subjects that respond are usually better able to tolerate a glucose load and are of normal body weight.[3]

EVALUATION OF REPORTED Cr CONTENT OF BIOLOGICAL MATERIALS

Evaluation of reported values pertaining to the Cr content of biological materials is difficult due to problems associated with total Cr analysis. Methods for the analysis of total Cr in biological materials have not been clearly defined nor generally accepted. A variety of methods are available, but none of these is suitable for all samples. Cr concentration in the parts per billion (ppb) range, extreme matrix effects, possible volatility, and the inherent property of Cr to bind nonspecifically to reaction vessels, graphite tubes, etc. make determination of Cr in biological materials an extremely difficult problem. Methods of Cr analysis have been summarized.[1]

Literature values for Cr concentration of biological fluids have decreased precipitously in the past 2 decades. For example, reported values for serum Cr range from several hundred ppb to the presently accepted values of less than 1 ppb for normal individuals. Similar decreases have been observed in reported values for urinary Cr concentration. Values greater than 1 ppb for the Cr content of normal human urine, serum, and milk samples may be in error and, unless verified by independent means, should not be considered accurate. Values for the Cr concentration of human tissues should be accepted with caution since most of the reported values have not been verified using sample collection techniques without stainless steel scalpels, trays, etc., and samples have not been analyzed using improved laboratory techniques and instrumentation. There appears to be reasonable agreement among laboratories for the Cr content of foods, water, and hair; therefore older and newer reported values usually appear to be indicative of the true Cr content of these materials.

Table 2
FOODS HIGH IN CHROMIUM

Mushrooms
Brewers yeast
Black pepper
Prunes
Raisins
Nuts
Asparagus
Wine

Standard Reference Materials, containing Cr concentrations similar to the samples to be analyzed, should be employed for all analytical studies involving Cr in biological samples. More than one Standard Reference Material should be utilized, since Cr from different sources behaves differently during digestion and analysis and total Cr content of individual Standard Reference Materials varies.

CHROMIUM REQUIREMENTS

Problems in analysis of Cr and dependence of amount required on form have prevented select committees from arriving at a recommended daily allowance (RDA) for Cr. However, a provisional recommended daily safe and adequate intake of 50 to 200 μg of Cr for adults and children 11 years or older has been suggested with no special reference to form other than the plus 3 valence state.[15]

DIETARY SOURCES OF CHROMIUM

Approximately one third of the diets designed by a nutritionist to provide the recommended dietary allowance (RDA) of vitamins and minerals contained less than the minimum safe and adequate dietary intake of Cr.[16] If these diets, which were designed to provide three well-balanced meals (total of 2800 cal) containing the RDA for vitamins and minerals other than Cr failed to provide the minimum safe and adequate intake of Cr, it is not surprising that a significant number of individuals, eating their respective normal diets, are consuming marginal levels of dietary Cr. The long-term consequences of eating a marginally Cr-deficient diet may lead to signs of Cr deficiency (Table 1), many of which are also associated with aging.

There have not been any comprehensive studies determining the Cr content of foods eaten in the U.S. However, Finnish workers completed a large study involving the nutrient content, including Cr, of numerous foods from each of the basic food groups.[17] Foods high in Cr are shown in Table 2. The approximate amount of Cr from different food groups consumed by Finnish people is shown in Table 3. The distribution of Cr is similar among fruits and vegetables, dairy products, beverages, and meat, with lesser amounts from cereal products and negligible dietary Cr from fish and seafood. Doisy et al.[18] also reported that diets rich in seafoods contain low levels of Cr. Beverages, including milk, account for almost one third of the daily intake of Cr. Cr in drinking water usually accounts for negligible quantities of dietary Cr. Mean Cr content of drinking water from 100 selected cities was 0.43 ng/mℓ.[19] Therefore, a representative liter of water would be expected to contain less than 2% of the minimum safe and adequate daily intake of Cr. However, several cases of Cr-contaminated public water supplies have been cited.

Diets of elderly subjects in the U.S. contained from 5 to 115 μg of Cr daily with an average daily intake of 52 μ.[20] It is apparent that diets low in Cr could inadvertently be

Table 3
DAILY INTAKE OF CHROMIUM FROM VARIOUS FOOD GROUPS

Food group	Average daily intake (μg)	Comments
Cereal products	3.7	55% from wheat
Meats	5.2	55% from pork; 25% from beef
Fish and seafoods	0.6	
Fruits, vegetables, nuts, and mushrooms	6.8	70% from fruits and berries
Dairy products, eggs, and margarine	6.2	85% from milk
Beverages, confectionaries, sugar, and condiments	6.6	45% from beer, wine, and soft drinks
Total	29.1	

selected. Several studies have reported much higher daily intakes of Cr with some of more than 800 μg/day (See Table 2, Kumpulainen, 1980).[21] However, these values appear to be much too high and should not be considered as an accurate estimate of the normal dietary intake of Cr. Problems of Cr analysis and contamination of foods using blenders or similar devices containing stainless steel (which is about 18% Cr) may have contributed to these erroneously high results.

Methods of food preparation also affect Cr intake since Cr tends to leach from stainless steel cookware.[22] The long-term consequences of food preparation using teflon-coated and other types of cookware needs to be evaluated. Anderson and Bryden[23] demonstrated that Cr in beer samples, a significant fraction of which appears to originate from contamination during preparation, was absorbed and therefore presumably utilized similar to Cr in foods.

CHROMIUM ABSORPTION AND AGING

Absorption of Cr by rats decreases with age[24] but this does not appear to explain the decreased levels of Cr in human tissues (see section on Aging Effects on the Cr Content of Tissues, Blood, Urine, and Hair) since Cr absorption by human subjects does not decrease with age.[25] Absorption of orally administered inorganic, radioactive Cr for elderly and younger subjects was similar and less than 0.5%.[25] Absorption of supplemental inorganic chromium has been shown by Anderson et al.[26] to be about 0.4% which was similar to that of Cr found in foods. However, while Cr absorption does not appear to significantly decrease with age, the ability to convert Cr to a useable physiological form may be age-dependent. Children responded to inorganic Cr supplementation by an improvement in glucose tolerance within 24 hr[27-29] while adult subjects required 1 to 3 months to respond.[6-8] This suggests that age may play a role in the ability to convert Cr to forms that function in vivo. Aging also affects food preferences. For example, older individuals often prefer sweeter foods[30] which not only tend to be low in Cr, but sugars stimulate urinary Cr excretion.[2,31] Therefore, these foods lead to a depletion of body Cr stores since more Cr is excreted than is absorbed.

AGING EFFECTS ON THE Cr CONTENT OF TISSUES, BLOOD, URINE, AND HAIR

Chromium content of human tissues, except lung, appears to decrease with age.[32] Mean hepatic Cr concentration in children less than 10 years old was 17.2 μg/g of ash, while for subjects over 30 it was less than 2 μg/g. A marked decline of Cr concentration with age

Table 4
DAILY URINARY Cr EXCRETION AND AGING

Age	Subjects	Cr excretion (μg/day) Mean ± SEM
20—29	7	0.22 ± 0.06
30—39	10	0.23 ± 0.05
40—49	16	0.14 ± 0.02
50 plus	9	0.18 ± 0.04

Note: Daily urinary Cr excretion of normal free-living subjects. There was no significant difference in the urinary Cr excretion of male and female subjects; therefore, sexes were grouped. Urinary chromium excretion determined by graphite furnace atomic absorption.[26]

was also reported for samples of kidney, aorta, heart, and spleen.[33] These results need to be verified using newer methods of sample collection not involving stainless steel scalpels, trays, etc., and samples need to be analyzed in a clean environment using updated methods and equipment.

Urinary Cr excretion has also been reported to decrease with age. Mean Cr concentration of 15 subjects aged 23 to 49 years was 32.9 ± 4.1 ng/mℓ and for 12 subjects aged 51 to 84 years was significantly lower, 8.0 ± 1.8 ng/mℓ, $(p<0.001)$.[34] Gurson et al.[35] reported a stepwise increase in urinary Cr excretion to age 35 followed by a slight decline. However, these Cr concentrations are more than 10-fold higher than the presently accepted values. Therefore, extreme caution must be exercised in interpretations involving values of this magnitude. Using validated modern methods for analysis of total Cr in urine, we did not find any significant difference in the urinary Cr excretion of subjects over for either male or female subjects older than 20. (Table 4). A specific form of volatile Cr that was postulated to be the form of Cr that functions in vivo, was also reported to decline with age.[34] However, these results are difficult to evaluate since volatile urinary Cr may be an artifact of older analytical methods that did not employ suitable background correction.

Effects of age on Cr in whole blood, RBC, and serum need to be evaluated. However, difficulties in analysis of these samples which are compounded by low levels (about 0.2 ppb) and therefore inherent problems of Cr contamination have prevented completion of verifiable studies involving the Cr content of blood and its components.

Hair Cr content has been suggested as a possible indicator of body Cr content.[36] Included among the potential advantages of hair Cr analysis is its relatively high concentration which permits analysis of total Cr by techniques that are less sensitive and therefore less affected by contamination and matrix effects.[37] Hair Cr concentrations can be correlated with age but usually only for children. For example, the Cr concentration distal from the scalp of an 18-month-old child whose hair had never been cut was 940 ppb, reflecting the high levels at birth, but the Cr concentration in the 2-cm region closest to the scalp was only 144 ppb, a value similar to that for older children and adults.[38] Hair Cr content of premature infants may be indicative of gestational age; presumably, the Cr status of premature infants is considerably less than full-term babies. Hair Cr content of babies at a gestational age of 32 weeks was 6 times lower than that of babies at a gestational age of 36 weeks.[38]

AGING, INSULIN, AND GLUCOSE INTOLERANCE — THE Cr CONNECTION

A number of studies have demonstrated an age-related decline of glucose tolerance in man. Depending on the specific criteria used, 10 to 50% of otherwise healthy elderly people

Table 5
POSTULATED CHROMIUM EFFECTS ON SUBJECTS WITH VARYING LEVELS OF SERUM GLUCOSE AND INSULIN

Glucose	Insulin	Postulated Cr effects
Normal	Normal	No effect
Normal	Elevated	Decrease insulin
Elevated	Elevated	Decrease glucose and insulin
Elevated	Normal or below	No effect or marginal lowering of glucose
Low	Normal or elevated	Normalize glucose and insulin

have glucose intolerance.[39-41] Rats also display decreased glucose tolerance with age.[40] Fasting plasma glucose increases from a mean of about 92 mg/dℓ at 25 years to 100 mg/dℓ at age 60.[42] Larger age-associated changes are observed following a glucose challenge. The magnitude of the increases in fasting plasma glucose with age is similar to the magnitude of the decrease in fasting glucose in Cr-responsive subjects following Cr supplementation.[26]

Insulin secretion with aging also varies. Some studies show increased insulin levels, others decreased, and yet others show no change with age.[40] A major confounding variable in the interpretation of insulin secretion studies is the presence of obesity.[42] Most obese subjects, at least initially, are able to compensate for insulin resistance by increasing insulin secretion so that glucose homeostasis is maintained, but at higher rates of insulin secretion.[43] Cr supplementation and supplementation with yeast containing high amounts of insulin-potentiating forms of Cr leads to a decrease in circulating insulin.[4,11,12,18]

The postulated effects of Cr supplementation on subjects with varying levels of blood glucose and insulin are shown in Table 5. Normal subjects with normal levels of glucose and insulin are not affected by supplementation with nutritional levels of Cr.[10,18] Subjects with normal glucose but elevated insulin would likely show a decrease in serum insulin (Table 5).[10,18] Subjects in the early stages of both elevated glucose and insulin would likely show decreases in glucose with no change in insulin (increased insulin sensitivity) followed by a decrease in insulin concentration. Subjects that are overtly diabetic with elevated glucose but normal or below normal levels of insulin would be in the advanced stages of glucose intolerance and may not respond to Cr. These postulated Cr responses would be expected to occur if the reason for the abnormal glucose and insulin levels were related to Cr nutrition. Obviously, glucose intolerance is related to many factors of which Cr is only one. Anderson et al.[44] have also demonstrated that Cr may have an effect on subjects with low blood sugar following a glucose challenge. This work needs to be verified on hypoglycemic subjects.

METHODS TO ASSESS Cr STATUS

Direct and indirect methods can often be used to assess trace element status. Direct methods to assess Cr status would include determinations of the Cr concentrations in biological samples such as blood, urine, hair, and body tissues. Indirect methods would include determinations of physiological parameters e.g., for selenium and iron, determination of the glutathione peroxidase and hemoglobin levels, respectively. However, Cr is not known to be quanitatively associated with any metabolite or protein nor is the level or activity of these critically dependent upon Cr concentration. Therefore, no indirect methods are being used to reliably measure Cr status. Direct methods involving determination of Cr content of biological samples have been used with limited success. The only reliable method to assess Cr status is retrospective involving Cr supplementation for 2 weeks to 3 months and evaluating individual responses to Cr supplementation. Methods to assess Cr status have been reviewed.[45,46]

SUMMARY

Aging is associated with a decrease in the Cr concentration of most tissues. Accompanying this loss of Cr are apparent signs associated with diabetes, cardiovascular disease, and related maladies. Proper Cr nutrition in man and animals often alleviates or delays the onset of many of these age-associated deleterious changes. However, further studies are needed to establish a direct link between aging and proper Cr nutrition.

REFERENCES

1. **Anderson, R. A.,** Nutritional role of chromium, *Sci. Total Environ.,* 17, 13, 1982.
2. **Mertz, W.,** Chromium occurrence and function in biological systems, *Physiol. Rev.,* 49, 163, 1969.
3. **Anderson, R. A. and Mertz, W.,** Glucose tolerance factor: an essential dietary agent, *Trends Biochem. Sci.,* 2, 277, 1977.
4. **Jeejeebhoy, K. N., Chu, R. C., Marliss, E. B., Greenberg, G. R., and Bruce-Robertson, A.,** Chromium deficiency during total parenteral nutrition, *JAMA,* 241, 496, 1978.
5. **Freund, H., Atamian, S., and Fisher, J. E.,** Chromium deficiency during total parenteral nutrition, *JAMA,* 241, 4916, 1979.
6. **Glinsmann, W. H. and Mertz, W.,** Effect of trivalent chromium on glucose tolerance, *Metabolism,* 15, 510, 1966.
7. **Levine, R. A., Streeten, D. H. P., and Doisy, R. J.,** Effects of oral chromium supplementation on the glucose tolerance of elderly human subjects, *Metabolism,* 17, 114, 1968.
8. **Hopkins, L. L., Jr. and Price, M. G.,** Effectiveness of chromium (III) in improving the glucose tolerance of middle-aged Americans, in *Proc. West. Hemisphere Nutr. Congr.,* Vol. II., 1968, 40.
9. **Railes, R. and Albrink, M. J.,** Effect of chromium chloride supplementation on glucose tolerance and serum lipids including high-density lipoprotein of adult men, *Am. J. Clin. Nutr.,* 34, 2670, 1981.
10. **Anderson, R. A., Polansky, M. M., Bryden, N. A., Roginski, E. E., Mertz, W., and Glinsmann W.,** Chromium supplementation of human subjects: effects on glucose, insulin and lipid parameters, *Metabolism,* 32, 894, 1983.
11. **Liu, V. J. K. and Morris, J. S.,** Relative chromium response as an indicator of chromium status, *Am. J. Clin. Nutr.,* 31, 972, 1978.
12. **Offenbacher, E. G. and Pi-Sunyer, F. X.,** Beneficial effect of chromium-rich yeast on glucose tolerance and blood lipids in elderly subjects, *Diabetes,* 29, 919, 1980.
13. **Sherman, L., Glennon, J. A., Brech, W. J., Klombert, G. H. and Gordon, E. S.,** Failure of trivalent chromium to improve hyperglycemia in diabetes mellitus, *Metabolism,* 17, 439, 1968.
14. **Wise, A.,** Chromium supplementation and diabetes, *JAMA,* 240, 2045, 1978.
15. Food and Nutrition Board, Recommended Dietary Allowances, 9th ed., Food and Nutrition Board, Nutrition Research Council, National Academy of Sciences, Washington, D.C., 1980.
16. **Kumpulainen, J. T., Wolf, W. R., Veillon, C. and Mertz, W.,** Determination of chromium in selected United States diets, *J. Agric. Food Chem.,* 27, 490, 1979.
17. **Koivistoinen, P., Ed.,** Mineral element composition of Finnish foods: N, K, Ca, Mg, P, S, Fe, Cu, Mn, Zn, Mo, Co, Ni, Cr, F, Se, Si, Rb, Al, B, Br, Hg, As, Cd, Pb, and Ash. *Acta Agric. Scand. Suppl.,* 22, 1980.
18. **Doisy, R. J., Streeten, D. H. P., Freiberg, J. M., and Schneider, A. J.,** Chromium metabolism in man and biochemical effects, in *Trace Elements in Human Health and Disease,* Prasad, A. S., Ed., Academic Press, N.Y., 1976, 79.
19. **Durfor, C. N. and Becker, E.,** Public Water Supplies of the 100 Largest Cities in the United States, U.S. Government Printing Office, Washington, 1962, 1812.

20. **Levine, R. A., Streeten, D. H. P., and Doisy, R. J.,** Effects of oral chromium supplementation on the glucose tolerance of elderly human subjects, *Metab. Clin. Exp.*, 17, 114, 1968.

21. **Kumpulainen, J.,** Determination of Chromium in Diets and some other Biological Materials by Graphite Furnace Atomic Absorption, Ph.D. thesis, University of Helsinki, Finland, 1980.

22. **Offenbacher, E. G. and Pi-Sunyer, F. X.,** Effects of pH and temperature on chromium concentration in water and fruit juices exposed to stainless steel, *Fed. Proc.*, 41, 391, 1982.

23. **Anderson, R. A. and Bryden, N. A.,** Concentration, insulin potentiation and absorption of chromium in beer, *J. Agric. Food Chem.*, 31, 308, 1983.

24. **Polansky, M. M. and Anderson, R. A.,** Rapid absorption of chromium, *Fed. Proc.*, 37, 895, 1978.

25. **Doisy, R. J., Streeten, D. H. P., Souma, M. L., Kalafer, M. E., Rekant, S. L., and Dalakos, T. G.,** Metabolism of ^{51}chromium in human subjects and diabetics, in *Trace Substances in Environmental Health*, Mertz, W. and Cornatzer, W. E., Eds., Marcel Dekker, N.Y., 1971 chap. 8.

26. **Anderson, R. A., Polansky, M. M., Bryden, N. A., Patterson, K. Y., Veillon, C., and Glinsmann, W. H.,** Effects of chromium supplementation on urinary chromium excretion of human subjects and correlation of Cr excretion with selected clinical parameters, *J. Nutr.*, 113, 273, 1983.

27. **Hopkins, L. L., Jr., Ransome-Kuti, O., and Majaj, A. S.,** Improvement of impaired carbohydrate metabolism by chromium III in malnourished infants, *Am. J. Clin. Nutr.*, 21, 203, 1968.

28. **Gurson, C. T. and Saner, G.,** Effect of chromium on glucose utilization in marasmic protein-calorie malnutrition, *Am. J. Clin. Nutr.*, 24, 1313, 1971.

29. **Gurson, C. T. and Saner, G.,** Effect of chromium supplementation on growth in marasmic protein-calorie malnutrition, *Am. J. Clin. Nutr.*, 26, 988, 1973.

30. **Albanese, A. A.,** Nutrition and health of the elderly, *Nutr. News*, 39, 5, 1976.

31. **Anderson, R. A., Polansky, M. M., Bryden, N. A., Roginski, E. E., Patterson, K. Y., Veillon, C., and Glinsmann, W. H.,** Urinary chromium excretion of human subjects: effects of chromium supplementation and glucose loading, *Am. J. Clin. Nutr.*, 36, 1184, 1982.

32. **Schroeder, H. A., Balassa, J. J., and Tipton, I. H.,** Abnormal trace metals in man — chromium, *J. Chron. Dis.*, 15, 941, 1962.

33. **Schroeder, H. A.,** The role of chromium in mammalian nutrition, *Am. J. Clin. Nutr.*, 21, 230, 1968.

34. **Canfield, W. K. and Doisy, R. J.,** Chromium and diabetes in the aged, in *The Biomedical Role of Trace Elements in Aging*, Hsu, J. M., Davis, R. L., and Neithamer, R. W., Eds., Eckerd College Gerontology Center, St. Petersburg, Fla., 1976, 119.

35. **Gurson, C. T., Saner, G., Mertz, W., Wolf, W. R., and Sokuci, S.,** Nutritional significance of chromium in different chronological age groups and in populations differing in nutritional background, *Nutr. Rep. Int.*, 12, 9, 1979.

36. **Hambidge, K. M.,** Chromium nutrition in man, *Am. J. Clin. Nutr.*, 27, 505, 1974.

37. **Hambidge, K. M., Rodgerson, D. O., and O'Brien, D.,** The concentration of chromium in the hair of normal children with juvenile diabetes mellitus, *Diabetes*, 17, 517, 1968.

38. **Hambidge, K. M.,** Chromium nutrition in the mother and growing child, in *Newer Trace Elements in Nutrition*, Mertz, W. and Cornatzer, W. E., Eds., Marcel Dekker, N.Y., 1971, chap. 9.

39. **Seltzer, H. S.,** Diagnosis of diabetes, in *Diabetes Mellitus: Theory and Practice*, Ellenberg, M. and Rifkin, H., Eds., McGraw-Hill, N.Y., 1970, 1070.

40. **Davidson, M. B.,** The effect of aging on carbohydrate metabolism: a review of the English literature and a practical approach to the diagnosis of diabetes mellitus in the ederly, *Metabolism*, 28, 688, 1979.

41. **Andres, R. and Tobin, J. D.,** Aging and the disposition of glucose, *Adv. Exp. Med. Biol*, 61, 239, 1975.

42. **Halter, J. B. and Chen, M.,** Nutrition and age-related changes of carbohydrate metabolism, in *Nutritional Approaches to Aging Research*, Moment, G. B., Ed., CRC Press, Boca Raton, Fla., 1982, chap. 2.

43. **Pfeifer, M. A., Halter, J. B., and Porte, D., Jr.,** Insulin secretion in diabetes mellitus, *Am. J. Med.*, 70, 579, 1981.

44. **Anderson, R. A., Polansky, M. M., Bryden, N. A., Roginski, E. E., Mertz, W., and Glinsmann, W. H.,** Effect of chromium supplementation on subjects with marginally elevated or depressed blood glucose following a glucose load, *Am. J. Clin. Nutr.*, 35, 840, 1982.

45. **Anderson, R. A. and Mertz, W.,** Assessment of chromium status of man and animals, in *The Use of Isotopes to Detect Moderate Mineral Imbalances in Farm Animals*, IAEA-TEC DOC-226, International Atomic Energy Agency, Vienna, 1982, 147.

46. **Shapcott, D.,** The detection of chromium deficiency, in *Chromium in Nutrition and Metabolism*, Shapcott, D. and Hubert, J., Eds. Elsevier/North Holand, N.Y., 1979, 113.

TRACE ELEMENT REQUIREMENTS
OF
THE ELDERLY

Shobha A. Udipi and Ronald R. Watson

INTRODUCTION

Current recommended dietary allowances[39] classify adults into two age categories — namely 23 to 50 years and 51 years and upward. The nutrient needs of the elderly, who belong to the latter group, have been essentially extrapolated from the nutrient needs of younger adults, mainly because of the lack of adequate studies of nutrient requirements among the elderly.

There is much indirect evidence that intakes and stores of trace elements are reduced in the aging. For example, studies of adult populations demonstrate that body composition changes throughout life. With increasing age, lean body mass declines progressively. Forbes and Reina[21] have demonstrated that the rate of decline tends to accelerate in later years and is greater in males. By age 65 to 70 the average male has 12 kg less and the female has 5 kg less lean body mass than at age 25. There is also a decline associated with aging in the average values for many other physiological functions.[54] Cross-sectional studies by Exton-Smith and Stanton[20] revealed a decrease in the intake of all nutrients with advancing age. Decreased taste acuity, decreased salivation, gastrointestinal function, and a number of physiological, social, and emotional factors may result in decreased food and nutrient intake by the elderly.

Both cross-sectional and longitudinal surveys have not included intakes or requirements of trace elements with the exception of iron, zinc, selenium, and chromium.

This chapter will summarize the limited information on dietary requirements for the following trace elements: manganese, silicon, vanadium, nickel, molybdenum, cobalt, iodine, and fluorine. Except for the latter two, little is known about the requirements of the elderly for these trace elements.

The National Academy of Sciences (NAS) has set up recommendations for trace elements such as silicon, vanadium, nickel, and molybdenum which have proven to be essential for life in many animal species. To date, no deficiency symptoms due to the lack of any of these elements have been reported in man.[39] The NAS[1] has suggested a range of intakes for manganese, fluoride, and molybdenum with the recommendation that the upper level of the range should not be exceeded due to risk of toxicity. Requirements for nickel, vanadium, silicon, and cobalt have been established for animal species but have not been quantified. The animal data appears insufficient to estimate adult human requirements.[39]

The general biochemical functions and intake and minimum requirements in animals for each of the trace elements dealt with in this chapter are summarized in Table 1.

MANGANESE (Mn)

Manganese is an essential element for many animal species. Mn deficiency may lead to abnormal formation of bone and cartilage and impaired glucose tolerance. Since Mn is necessary for protein and energy metabolism, it must be assumed that it is essential for man. Pleban and Pearson[44] reported concentrations of 9.03 ± 2.25 µg/ℓ in whole blood and 1.82 ± 0.64 µg/ℓ in serum in man. No correlation was found between blood Mn concentrations and age or sex. Schroeder and Nason[52] confirmed the observation that Mn concentrations did not vary with age or sex during luxus consumption of this mineral element. An

Table 1
TRACE ELEMENTS IN THE ELDERLY[38,54]

	Vanadium	Silicon	Fluorine	Molybdenum	Iodine	Cobalt	Manganese
Biological effects	Functions most likely as a oxidation reduction catalyst. Inhibits cholesterol synthesis and caries. Lowers phospholipid and cholesterol in blood and catalyzes nonenzymatic oxidation of catecholamines	Structural, but possibly also matrix for formation of organic compounds. Cross linking agent, contributing to stability of acid mucopolysaccharides and other connective tissue components. Catalytic effect on peptides	Incorporated in bone crystal structure forming fluorapatite plays a role in prevention of dental caries and possibly periodontal disease and osteoporosis	Deficiency states in humans unknown. Essential for function of enzymes involved in production of uric acid and oxidation of sulfates and aldehydes	An integral part of the thyroid hormones, thyroxin and triiodothyronine	Functions as an integral part of Vitamin B-12	An essential part of several enzyme systems necessary for protein and enery metabolism and information of mucopolysaccharides
Human intake	2 mg daily on ''well-balanced diet'' for 75 kg weight 032 μg/g of diet	Not determined: much of intake possibly not biologically available. Clayeaters (geophagy)	Intake varies with fluoride content of water. 1.73 - 3.44 mg/day in areas with fluorided water supply and 0.91 ± 0.05 in nonfluoridated area	0.1 to 0.46 mg	64-677 μg/day[3]	—	2 to 9 mg/day[3]

| Require-ment (in animals) | 0.1 ppm in the diet as sodium vanadate. Large differences in potency of various vanadium compounds and oxidation states | 50 mg % in the diet as $Na_2SiO_3 \cdot 9H_2O$ Other forms possibly more potent. Present in all living beings. Highest level in epidermis and connective tissue. Also present in bone (region of active calcification, liver, heart and muscle (2—10 ppm) and blood (approx. 5 ppm) | 1.5—4.0 mg | 0.15—5.0 mg | 150 μg | 2. 3-NAS. 1980 | 2.5—5.0 mg |

assumption appears valid that some binding sites could be unfilled on the enzyme for which the metal is a cofactor when tissue concentrations are low.[51] This suggestion needs to be demonstrated in vivo.

The World Health Organization has recommended 2 mg/day as an adequate Mn intake for adults. The NAS recommends a somewhat higher intake of 2.5 to 5 mg of Mn per day based on studies by McLeod and Robinson[35] and Schroeder.[52] McLeod and Robinson[35] conducted balance studies in adult human subjects and found the subjects to be in equilibrium or positive balance on intakes of 2.5 mg/day or greater, and in negative balance resulting from intakes of 0.7 mg daily. This may be a normal intake in adults. Guthrie and Robinson[24] found that the average daily intake for Mn in a small number of women, 19 to 50 years of age was 2.7 ± 0.1 mg/day. Foods of plant origin, particularly cereals, are the major contributors of Mg to the daily diet. Guthrie[24] has reported that a diet high in animal products, low in vegetables, and containing 78% extraction flour would probably supply adequate amounts of Mn for New Zealanders.

We found that no intermediate levels have been studied and there have been no reports concerning requirements or adequacy of Mn in the elderly population. Schroeder et al.[52] reported steady (state) tissue concentrations of Mn in the U.S. population.

In man, toxicity has been reported only in people exposed to high concentrations of Mn dust. In animals 1000 μg of Mn per gram of diet must be fed to produce signs of toxicity. Thus, the NAS has recommended that the Mn intake of adults be in the range mentioned above[39] which includes a margin of safety and allows for the possibility of increased needs. Lower intakes and very limited human studies of Mn requirements of the aged show that they require further research before any separate recommendations could be made.

SILICON (Si)

Silicon serves as an important cross-linking agent by forming bridges or linking within and between individual polysaccharide chains to proteins. Thus, Si aids in maintaining the structural integrity of connective tissues as well as bone.[5] Levels of Si in blood and intestinal tissues of male and female rats are affected by age, sex, castration, adrenalectomy, and thyroidectomy. The Si content of the aorta, skin, and thymus is found to decline significantly with age, whereas the Si content of tissues such as heart, kidney, muscle, and tendon remain relatively unchanged in rats, pigs, and chickens.[4] Loeper et al.[31] found that the Si content of normal human aorta also decreases with age. In addition, a decrease in the Si level in the arterial wall is seen with the development of atherosclerosis.[31] Experimental data in rats suggests that the decreased Si could be due to a decrease in absorption of silicon with increasing age.[4,53]

Since silicon participates in bone formation, maintenance of structural integrity of connective tissues, and possibly wound healing, it may be of importance in degenerative conditions which are associated with significant changes in mucopolysaccharides such as atherosclerosis and osteoarthritis.[6]

Schwarz[53] has stated that populations in all developed countries, including the American population, are possible target groups for the effects of low Si levels (if it is essential in humans).

In its natural state, grain is very rich in Si, but 65% extraction flour contains only about 2% Si. Much of the Si in grain is discarded with bran and germ during processing. Mastication problems and often incorrect food choices may lead to increased consumption of refined foods by the elderly. In view of the above findings, Si deficiency (especially in relation to high incidence of atherosclerosis in developed countries) may prove to be a point of increased concern if Si is proved to be an essential element in the future.

Increased consumption of fat and simple carbohydrates and a concomitant decrease in fiber consumption may lead to changes (and possibly decreases) in intakes of trace elements such as Si. Various kinds of dietary fiber have been reported to reduce cholesterol and blood-lipid levels and bind bile acids in vitro. Fiber products active in these tests were found to contain high amounts of Si (1000 to 25,000 ppm) in contrast to negligible amounts present in the inactive fiber products.[54] It has been suggested that silicate-silicon may be the active agent in dietary fiber which affects the development of atherosclerosis. The possible mechanisms via which silicate-silicon could have a beneficial effect on atherosclerosis are: (1) binding of bile acids in the digestive tract, thus enhancing the elimination of metabolic and products of cholesterol or cholesterol itself, (2) functions as an essential constituent of connective tissue and would thus contribute to the integrity and stability of the arterial wall, and (3) an activated form of silicic acid could participate directly in the intermediary metabolism of steroids and bile acids.

In animals, 50 mg of Si per 100 g of diet has been shown to improve growth rates in rats.[5] Nevertheless, Si has not been shown to be clearly essential in man; hence, no firm statement regarding the requirements of the elderly can be made.

VANADIUM (V)

Vanadium is a ubiquitous element of low atomic weight. The body exerts homeostatic control over the V content and changes in V intakes cause some physiological changes.[22] Several studies have shown that dietary V deficiency results in elevated cholesterol levels in chicks[48] and rabbits.[10,13] There are several reports of alterations in lipid metabolism by V in man. Diamond et al.[13] reported that V had no effect on serum lipids in contrast to a reduction in serum cholesterol and body cholesterol pools found by Curran and co-workers[9]. However, V has not been found to lower blood cholesterol in older humans.[26]

Most of the V ingested by man is excreted via the feces. Balfour et al.[2] and Hansen[25] have shown that vanadate may control the response of the sodium pump to potassium. Thus, V may play a role in the retention of salt and water in conditions such as nutritional edema.[22] Normal levels of V in plasma are 35 to 48 μg/dℓ.[48] A low concentration of circulating V is a measure of deficiency state.[22] The V levels in liver, kidney, muscle, and bone have been shown to be responsive to changes in dietary levels in the high or toxic range.[58] Localization of V in tooth structure and upon i.v. injection of radiovanadium suggests that this element may play a role in the prevention of dental caries.[26] Animal studies indicate that the maximum requirement that prevents deficiency may be approximately 100 ppb of V on a purified diet.[53]

Myron and co-workers[38] reported that the V content of five hospital diets ranged from 4.7 to 10.6 μg/1000 cal. Daily intakes ranged from 12.4 μg on a low calorie diet to 28 μg (general diet). In chicks, deficiency signs i.e., elevation of plasma cholesterol and hematocrit values occur on diets containing 0.030 to 0.035 μg of V per gram of diet. The average V concentration of the nine human diets was 0.032 μg/g of diet.[38] If the V requirement for humans is similar to that of rats and chicks, adequacy of V in our diets should not be taken for granted. A diet consisting of milk, meat, and vegetables could contain less than 0.10 μg V per gram of diet.[40]

Few studies have been conducted on V in older animals and/or humans. Also, difficulty arises in transposing values for individual foods into daily intakes due to (1) disagreement in values reported by different researchers, (2) incomplete data, and (3) influence of methods of preparation and cooking.

Thus, under natural dietary conditions, V requirements in man may be higher than the levels shown to be essential in animals. The NAS[39] has stated that the human requirement for V cannot be established on the basis of available knowledge.

NICKEL (Ni)

Nickel is probably necessary for the metabolism and structure of membranes, as well as to stabilize the structure of RNA and DNA. Ni is nontoxic to animals except in astringent doses.[41] Homeostatic mechanisms are implied by serum levels, excretion rates, and lack of excessive accumulation. Deficiency of Ni has been produced in two or more animal species. However, the production of Ni deficiency in experimental animals necessitates strict control of dietary and environmental contamination. This suggests that requirements are probably low and easily met by diets presently consumed.

Studies on Russian men reported an average intake of 289 ± 23 µg Ni per day with fecal excretion of 258 ± 23 µg/day.[41] This may indicate poor absorption of Ni from ordinary diets.[41] Horak and Sunderman[27] reported similar values for fecal excretion of Ni in ten healthy subjects who consumed various homemade diets. In contrast, Perry[43] and Sunderman[60] reported that approximately 20 µg is excreted daily via the feces. Myron and co-workers[38] reported that the Ni content of institutionalized diets in adults was 0.19 to 0.41 µg Ni per gram (dry weight). The mean intake of Ni was 165 ± 11 µg/day or 75 ± 10 µg/1000 cal. Ni content tended to be lower in breakfast than other meals. About 0.05 to 0.08 µg of dietary Ni as $NiCl_2$ per gram of diet is necessary to prevent deficiency signs in chicks and rats. If animal data can be extrapolated to man, the dietary requirement would probably be in the range of 50 to 80 ng/g of diet. Most diets would provide this amount.

In plants, Ni has been shown to translocate as a stable anionic amino acid complex. Nielsen[40-41] states that it remains to be determined whether these organic complexes are the usual compounds of Ni in plant tissues and whether they influence bioavailability. Grains, which are rich in Ni, are also rich in phytic acid. Ni forms a stable complex with phytic acid.[40] Thus, phytate in grains and other vegetables may decrease the availability of Ni. In contrast, foods of animal origin contain relatively little Ni which is more available. Ni nutriture may conceivably be of concern in the aged with diseases that interfere with intestinal absorption or who are under extreme physiological stress. For example, relatively large amounts of Ni are lost in sweat. Thus, conditions which result in large losses of sweat may increase the need for Ni.

MOLYBDENUM (Mo)

Deficiencies of Mo in soil are readily reflected in the food chain. Pronounced differences are found in the Mo concentration of human tissues in diverse geographic regions.[36] Deficiency states in human beings are not known. Since no disturbances in functions of important enzymes for which Mo is essential have been reported, it has been suggested that the human requirement for Mo is low enough to be adequately fulfilled by routine diets.[39] Balance studies conducted on humans have shown retention of Mo on intakes of approximately 0.1 to 0.15 mg daily. However, Engel and co-workers[18] and Robinson et al.[45] reported a negative balance on diets furnishing 0.1 mg daily. The recommended safe and adequate intake ranges from 0.15 to 0.5 mg daily in adults.[39]

Dietary intakes of 10 to 15 mg of Mo per day in the Ankavan region of Russia are associated with elevated blood xanthine oxidase activity and doubling of uric acid concentration in blood and urine.[36] The high Mo intake appears to be responsible for elevated enzymatic activity. Studies on experimental animals and humans suggest that a high intake of copper (Cu) accompanying the high levels of Mo in the diet may protect against Mo exposure.

Studies conducted in India[12,39] and Russia[36] showed that intakes exceeding 10 mg/day were detrimental to the health of adults whereas 1.5 mg/day did not affect health noticeably.

Excretion of Cu increases as the Mo content of the diet is raised. Excretion of Cu was 24 µg/day and increased to 42 µg when the Mo intake was raised to 540 µg. When the daily intake of Mo was raised to 1540 μg, an additional increase was observed in the excretion of Cu. Thus, it appears in adult and aging humans on normal Cu intakes that an intake of 0.5 to 1 mg of Mo daily can be assumed as safe. Oster[42] has shown that xanthine oxidase causes depletion of plasmalogens. Plasmalogens are widely distributed in the human body and are present in cardiac and skeletal muscle tissue and intima of the arterial wall. It has therefore been suggested that Mo may possibly be linked to coronary heart disease in man.

Tsongas and co-workers[67] estimated the average daily dietary intake of Mo in the U.S. by analyzing foods collected in grocery baskets. They reported a level of intake of 120 to 240 μg of molybdenum per day, which was found to vary with age, sex, and income. No separate recommendations have been made for the group 51 years and above. As Mo may have a link with coronary heart disease, which is found more in older people and has greater incidence in developed countries, it is perhaps possible that the Mo requirements of older people may be different from those of adults below the age of 50. Resolution of these questions awaits further research regarding both the essentiality of Mo for humans and its adequacy for the elderly.

COBALT (Co)

Cobalt is physiologically active as a part of the molecule of Vitamin B-12. A 1-μg dose of Vitamin B_{12} containing 43 μg of Co is sufficient to prevent fatal pernicious anemia. Inorganic Co has been shown to have a variety of effects in several tissues, although its importance for normal function and well-being has not been demonstrated.[64]

Methodological problems occur in the measurement of low concentrations of Co as has been seen in other trace elements, leading to conflicting reports.

Schroeder, Nason, and Tipton[50] and Tipton and co-workers[66] reported that Western diets contain approximately 150 to 450 μg of Co per day. Other studies have estimated Co intakes to range between 140 and 1770 μg daily. In a study on eight subjects, Schroeder et al.[50] found excretion of 50 to 190 μg/ℓ of Co in urine and much smaller amounts (23 to 69 μg/day) in feces. Urine appears to be the major route of excretion with loss of Co via feces, sweat, and hair being minimal. Studies by Schroeder and Nason[51] and Sorbie[59] showed that the majority of dietary Co is not absorbed.

Schroeder et al.[50] found that tissue concentrations of Co were unaltered by age. Total body content has been estimated to be 1.1 mg, with the major portion being present in muscle and soft tissues. Co produces polycythemia in man. Duckham and Lee[14] treated anephric, refractory anemic, dialysis patients with 25 to 50 mg of Co chloride daily for 12 to 32 weeks. They reported an increase in hemoglobin concentration and a decrease in the need for blood transfusions. Although Co is not greatly used in the treatment of anemia, reports of Co therapy reflect that this element may be required as an erythropoietic stimulant.

Co induces hyperlipemia in rabbits and chickens.[15,64-65] In Co-treated anephric patients, it produced a prompt elevation of serum triglycerides with slower, less marked rises in the concentration of cholesterol and free fatty acids.[63] However, following administration of Co, pancreatic alpha-cell degeneration and destruction was observed.[64] Eaton[16] reported elevated concentrations of plasma glucagon as well as increased sensitivity of rats to insulin after treatment with Co. Al-Tamer[64] found no differences in the levels of plasma glucagon in patients treated with Co and in those who were not. In anephric patients the lipemic response is not maintained beyond the period of therapy.

Co interferes with thyroid function and reduces thyroidal ^{131}I uptake. Cardiomyopathy may result in heavy beer drinkers due to a combination of high level of Co intake, high consumption of alcohol, and low intakes of protein or thiamin.[1,11,23]

Treatment with cobalt chloride has been reported to result in the development of sarcoma in rats.[64] Development of tumors has been reported in 2 out of 61 patients treated with Co

(300 mg/day for 2 days) for anemia. However, it remains to be seen whether smaller doses of Co may be used to treat patients with safety.[64]

Alloys of Co, Cr, Ni, and Mo have been successfully used in joint prostheses.[64] As expected, the incidence of metal sensitivity has been shown to be higher in mature adults with implants than in subjects without implants.[17] Winter[71] reported the presence of particulate matter containing alloy fragments around surgical implants. Elevated levels of Co and Cr have been found in the blood and urine of patients with metallic hip replacements.[8] McKee et al.[34] found a low incidence of failure and many patients with metal sensitivity had perfectly functioning joints.

The NAS[39] has stated that requirements of inorganic Co for man cannot be established since man must meet his Co requirements through Vitamin B_{12} intake.

Additional studies concerning aging would need to be performed before any conclusions could be drawn regarding requirements for the elderly. No deficiencies of Co have been reported to date; hence, it appears that dietary intakes even lower than the average are adequate to meet man's need for Co.

IODINE (I)

Iodine is an integral part of the thyroid hormones, thyroxin, and triiodothyronine. It is an essential micronutrient for man and deficiency leads to enlargement of the thyroid gland (goiter). Endemic goiter is a cause for concern in many parts of the world. In the U.S. this problem has been resolved to a large extent by the introduction of iodized table salt. However, surveys[32,33] have reported an incidence of 6.6 and 5.4% in Michigan and Texas, respectively.

Evered et al.[19] measured the serum thyroxin (T_4), triiodothyronine (T_3), and thyrotropin levels in 2779 adults of both sexes, 18 to 75 years of age. Serum T_4 and T_3 levels were higher in females and did not seem to change with age. However, in males the levels of T_3 were observed to decrease as age increased (after 65 years) in contrast to a decrease in T_4, levels with age. There is no evidence regarding the importance or the role these changes may play and how they in turn may affect I requirements.

The daily I requirement for the prevention of goiter in adults is 50 to 75 μg i.e., 1 μg/kg of body weight.[39] Thus, including the extra margin for safety, 150 μg is recommended for adults of both sexes.

An intake between 50 to 1000 μg of I can be considered safe. There have been reports of thyrotoxicosis resulting from prolonged and high intakes of I in Tasmania and Japan.

No evidence of adverse reactions due to increased intake has been reported. Therefore, the NAS states that the present intake by the majority of the population in the U.S. is adequate and safe.

Although studies have reported changes in the concentrations of thyroid hormones with age, no studies have been reported so far on I requirements of the elderly. Therefore, the allowance set by the NAS/NRC for adults may be considered applicable to the elderly population as well.

FLUORINE (F)

Although the results of studies on the essentiality of F for growth are not fully confirmed, it is considered essential for the growing organism as an anticariogenic element. The main target organs for F in man are bones and the enamel of teeth, where it is incorporated into hydroxyapatite. Incorporation of F into tissues may give some measure of protection against diseases of older age such as periodontal disease and osteoporosis.[3]

Husdan et al.[28] reported that serum ionic F ranged from 0.5 to 2.0 μmol/ℓ in men and 0.5 to 2.3 μmol/ℓ in women. After the age of 45 years, the concentration of serum ionic

F tended to be linearly, but inversely correlated with age. Husdan suggested that decreased renal function with increasing age and increased release of previously deposited bone F may be responsible for the higher serum F levels. In addition, in aging women, the rise noted is probably due to enhanced release for F from bone after menopause. Suzuki[61] reported that the F concentration of bones in Japanese subjects increased throughout their 20s and reached a plateau at about 60 years of age. Plasma F concentrations in individuals from communities with widely differing F levels are essentially constant, suggesting a homeostatic regulation of F concentration.

Ingested F is almost completely absorbed via the stomach and deposited in mineralized tissue or excreted via the urine.[7,11] Fluoride alone or in combination with vitamin D and Ca may be useful for treating osteoporosis and other metabolic diseases.[29,62] The average daily intake of F is 0.9 ± 0.05 mg in nonfluoridated areas with higher intakes being observed in communities having fluoridated water supplies,[30] (i.e., 2.63 ± 0.17 mg). The contribution of fluoridated drinking water would be approximately 1 to 1.5 mg, whereas in nonfluoridated areas the contribution would be only 0.1 to 0.6 mg. Thus, the intake could range from 1 to 4 mg daily. Spencer and co-workers[56] found that in humans the use of relatively small amounts of aluminium hydroxide (which is used in antacids) results in increased fecal fluoride excretion and a decreased net intestinal absorption of F.

Studies on the possible prophylactic and therapeutic effects of F on osteoporosis in animals and man have yielded inconsistent results so far.[57] Rao et al.[47] studied the effect of F administration from weaning age, early adulthood, and middle age on the severity of osteoporosis in aged mice. Although cortical thickness was observed to decrease with age, the provision of up to 10 ppm of F in drinking water throughout the latter half of life was found to have no significant effect on cortical thickness. However, providing 10 ppm of F from weaning in water appeared to prevent the decline in breaking load at 25 months of age. However, this effect was not statistically significant. Thus, F apparently did not influence the amount of bone tissue present at senescence. F is toxic when consumed in excessive amounts; the daily intake required to produce chronic toxicity after prolonged consumption for years is 20 to 80 mg. This level is far in excess of the average intake in the U.S.

In view of the studies presented here, the recommendations made by the NAS should be used as guidelines for all adults, including the elderly. The adequacy of F in the diet may be of concern, however, for those elderly individuals consuming antacids. Due to a lack of further evidence however, no suggestions or recommendations can be made regarding F requirements in the aged.

SUMMARY

This article has attempted to review studies concerning aging and the trace elements included herein. Few conclusions have been drawn or recommendations made regarding requirements for these various minerals in the elderly. Further research regarding the true essentiality of these minerals for man is required before any allowances can be set up.

The production of deficiencies of these elements depends on strict control of dietary and environmental conditions for experimental animals. Currently, it appears that the diets consumed by the majority of the adult human population are adequate in their trace element supply. As food consumption patterns change with increasing intakes of refined foods, the adequacy of these diets with regard to their content of trace elements may become a matter of concern.

Various studies with the elderly have demonstrated decreased intakes of other nutrients, apathy, and decreased efficiency of GI function. These factors would, therefore, have to be considered before determining the requirements of trace elements. The problem of selecting a proper nutritious diet for an elderly person is not simple because one or more of the environmental factors such as deficient dentition, low economic status, ingrained eating habits, excessive introspection, and loss of independence may influence their food selection and eating habits. Thus, consideration of the elderly as a separate group seems warranted while reviewing studies and establishing dietary standards for a population.

REFERENCES

1. **Alexander, C. S.,** Cobalt-beer cardiomyopathy. A clinical and pathologic study of 28 cases, *Am. J. Med.,* 53, 395, 1972.
2. **Balfour, W. E., Granthan, J. J., and Glynn, I. M.,** Vandate-stimulated natriuresis, *Nature (London),* 275 (5682), 768, 1978.
3. BEAP, Committee on Biologic Effects of Atmospheric Pollutants, National Research Council. Fluorides, National Academy of Sciences, Washington, D.C., 1971, 295.
4. **Carlisle, E. M.,** Essentiality and function of silicon, in *Trace Element Metabolism in Animals,* Vol 2, Hoekshra, W. G., Suttie, J. W., Ganther, H. E., and Mertz, W., Eds., U.K. Butterworths, London, 1974, 407.
5. **Carlisle, E. M.,** Silicon, *Nutr. Rev.,* 33, 257, 1975.
6. **Carlisle, E. M.,** Silicon as an essential element. *Fed. Proc.,* 33, 1758, 1974.
7. **Carlson, C. H., Armstrong, W. D., and Singer, L.,** Distribution and excretion of radiofluoride in the human, *Proc. Soc. Exp. Biol. Med.,* 104, 235, 1960.
8. **Coleman, R. F., Herrington, J., and Scales, J. T.,** Concentration of wear products in hair, blood and urine after total hip replacement, *Br. Med. J.,* 1, 527, 1973.
9. **Curran, G. L., Azarnoff, D. L., and Bolinger, R. E.,** Effect of cholesterol synthesis inhibition in normo-cholesteremic young men, *J. Clin. Invest.,* 38, 1251, 1959.
10. **Curran, G. L. and Costello, R. L.,** Reduction of excess of cholesterol in the rabbit aorta by inhibition of endogenous cholesterol synthesis, *J. Exp. Med.,* 103, 49, 1956.
11. **Cznarnecki, S. K. and Kritchevsky, D.,** Trace elements, in *Nutrition and the Adult Micronutrients,* Vol. 3B, Czarnecki, S. K. and Kritchevsky, D., Eds., Plenum Press, 1980, 319.
12. **Deosthale, Y. G. and Gopalan, C.,** The effect of molybdenum levels in sorghum (Sorghum vulgare pers) on uric acid and copper excretion in man, *Br. J. Nutr.,* 31, 351, 1974.
13. **Diamond, G., Caravaca, J., and Berchimol., A.,** Vanadium, excretion, toxicity, lipid effect in man, *Am. J. Clin. Nutr.,* 12, 49, 1963.
14. **Duckham, J. M. and Lee, H. A.,** The treatment of refractory anemia of chronic renal failure with cobalt chloride, *Q. J. Med.,* 45, 277, 1976.
15. **Eaton, R. P.,** Cobalt chloride-induced hyperlipemia in the rat: effects on intermediary metabolism, *Am. J. Physiol.,* 222, 1550, 1972.
16. **Eaton, R. P.,** Glucagon secretion and activity in the cobalt chloride-treated rat, *Am. J. Physiol.,* 225, 67, 1973.
17. **Elves, M. M., Wilson, J. M., Scales, J. T., and Kerns, M. B. S.,** Incidence of metal sensitivity in patients with total joint replacements, *Br. Med. J.,* 4, 376, 1975.
18. **Engel, R. W., Price, N. O., and Miller, R. F.,** Copper, manganese of cobalt and molybdenum balance in pre-adolescent girls, *J. Nutr.,* 92, 197, 1967.
19. **Evered, D. C., Tunbridge, W. M. G., Hall, R., Appleton, D., Boewis, M., Clark, F., Manuel, P., and Young, E.,** Thyroid hormones concentration in a large scale community survey. Effect of age, sex, illness and medication, *Clin. Chim. Acta,* 83, 223, 1978.
20. **Exton-Smith, A. N. and Stanton, B. R.,** Report on an investigation in the dietary of elderly women living alone, King Edward's Hospital Fund, London.
21. **Forbes, B. G. and Reina, C. J.,** Adult lean body mass declines with age: some longitudinal observations, *Metabolism,* 19, 653, 1970.
22. **Golden, M. H. N. and Golden, B. E.,** Trace elements. Potential importance in human nutrition with particular reference to zinc and vanadium. *Br. Med. Bull.,* 37, 31, 1981.
23. **Grice, H. C., Mungo, I. C., Wiberg, C. S., and Heggtveit, H. A.,** The pathology of experimentally induced cobalt, *Clin. Toxicol.,* 2, 273, 1969.
24. **Guthrie, B. E. and Robinson, M. F.,** Cardiomyopathy: a comparison with beer drinkers' cardiomyopathy. The nutritional status of New Zealanders with respect to manganese, copper, zinc and cadmium. A review, *N.Z. Med. J.* 87, 3, 1978.

25. **Hansen, O.**, Facilitation of ouabain binding to (Na + + Kt) − ATPase by vanadate at in vivo concentrations, *Biochem. Biophys. Acta*, 568, 265, 1979.
26. **Hopkins, L. L. and Mohr, M. E.**, Vanadium as an essential nutrient, *Fed. Proc.*, 33, 1773, 1974.
27. **Horak, E. and Sunderman, F. W.**, Distribution and excretion of nickel-63 administered intravenously to rats, *Clin. Chem.*, 19, 429, 1973.
28. **Husdan, H., Vogl, R., Oreopoulos, D., Gryfe, C., and Rapoport, A.**, Serum ionic fluoride: normal range and relationship to age and sex, *Clin. Chem.*, 22, 1884, 1976.
29. **Jowsey, J., Riggs, B. L., Kelley, P. J., and Hoffman, D. L.**, Effect of combined therapy with sodium fluoride, vitamin D and calcium in osteoporosis, *Am. J. Med.*, 53, 43, 1972.
30. **Kramer, L., Osis, D., Wiatrowski, E., and Spencer, H.**, Dietary fluoride in different areas in the United States, *Am. J. Clin. Nutr.*, 27, 590, 1974.
31. **Loeper, J., Leoper, J., and Lemaine, A.**, Etude du silicium en biologie animale et au cours de L'atherome, *Presse Med.*, 74, 865, 1966.
32. **Matovinovic, J.**, Extent of Iodine Insufficiency in the United States, in Iodine Nutriture in the United States, Food and Nutrition Board, National Research Council, National Academy of Sciences, Washington, D.C.
33. **McGanity, W.** Extent of iodine insufficiency in the United States. In: Iodine nutriture in the United States. Food and Nutriton Board, National Research Council, National Academy of Sciences, Washington, D.C.
34. **McKee, G. K.**, Metal sensitivity in patients with joint prostheses, (Letter), *Br. Med. J.*, 4, 645, 1975.
35. **McLeod, B. E. and Robinson, M. F.**, Metabolic balance of manganese in young women, *Br. J. Nutr.*, 27, 221, 1972.
36. **Mertz, W.**, Defining trace element deficiencies and toxicities in man, in *Molybdenum in the Environment*, Vol. 1, Chappell, W. R. and Peterson, K. K., Eds., Marcel Dekker, N.Y., 1976.
37. **Mountain, J. T., Stockell, F. R., and Stokinger, M. E.**, Effects of ingested vanadium on cholesterol and phospholipid metabolism in the rabbit, *Proc. Soc. Exp. Biol. Med.*, 92, 582, 1956.
38. **Myron, C. E., Zimmerman, T. J., Shuler, T. R., Klevay, L. M., Lee, D. E., and Nielsen, F. H.**, Intake of nickel and vanadium by humans. A survey of selected diets, *Am. J. Clin. Nutr.*, 31, 527, 1978.
39. National Academy of Recommended Dietary Allowances Sciences, National Research Council, Washington, D.C., 1980.
40. **Nielsen, F. H. and Sandstead, H. H.**, Are nickel vanadium, silicon, fluorine and tin essential for man? A review, *Am. J. Clin. Nutr.*, 27, 515, 1974.
41. **Nielsen, F. H.**, Essentiality and function of nickel, in *Trace Element Metabolism in Animals*. II., Hoekstra, W. G., Suttie, J. W., Ganther, H. E., and Mertz, W., Eds., U.K. Butterworths, 1974, 381.
42. **Oster, K. A.**, Plasmalogens diseases: a new concept of the etiology of the atheroscerotic process, *Am. J. Clin. Res.*, 2, 30, 1971.
43. **Perry, H. M. and Perry, E. F.**, Normal concentrations of some trace metals in human urine: changes produced by ethylene-diamine tetracetate, *J. Clin. Invest.*, 38, 1452, 1959.
44. **Pleban, P. A. and Pearson, K. H.**, Determination of manganese in whole blood and serum, *Clin. Chem.*, 25, 1915, 1979.
45. **Robinson, M. F., McKenzie, J. M., Thomson, C. D., and Van Riji, A.**, Metabolic balance of zinc, copper, cadmium, iron, molybdenum and selenium in young New Zealand women, *Br. J. Nutr.*, 30, 195, 1973.
46. **Roe, F. J. C. and Lancaster, M. C.**, Natural, metallic and other substances as carcinogens, *Br. Med. Bull.*, 20, 127, 1964.
47. **Rao, G. V. G. K., Tsao, K., and Draper, H. H.**, The effect of fluoride on some physical and chemical characteristics of the bones of aging mice, *J. Geronotol.*, 27, 183, 1972.
48. **Schroeder, H. A., Balassa, J. J., and Tipton, I. H.**, Abnormal trace elements in man: vanadium, *J. Chron. Dis.*, 16, 1047, 1963c.
49. **Schroeder, H. A., Balassa, J. J., and Tipton, I. H.**, Essential trace elements in man: manganese, *J. Chron. Dis.*, 19, 545, 1966.
50. **Schroeder, H. A., Nason, A. P., and Tipton, I. H.**, Essential trace metals in man: cobalt, *J. Chron. Dis.*, 20, 869, 1967.
51. **Schroeder, H. A. and Nason, A. P.**, Trace element analysis in clinical chemistry, *Clin. Chem.*, 17, 461, 1971.
52. **Schroeder, H. A. and Nason, A. P.**, Interactions of trace elements in rat tissues. Cadmium and nickel with Zn, Cr, Cu and Mn, *J. Nutr.*, 104, 167, 1974.
53. **Schwarz, K.**, New essential trace elements (Sn, V, F, Si); progress report and outlook, in *Trace Element Metabolism in Animals*, Vol. 2, Hoekstra, W. G., Suttie, J. W., Ganther, H. E., and Mertz, W., Eds., U.K. Butterworths, 1974, 355.
54. **Schwarz, K.**, Silicon, fiber and atherosclerosis, *Lancet*, 26, 454, 1977.
55. **Shock, N. W.**, Energy metabolism, caloric intake and physical activity of the aging, in *Nutrition Old Age*, Carlson, L., Almquist, A., and Wiksell, Uppsala.

56. **Spencer, H., Kramer, L., Norris, C., and Wiatrowski, E.,** Effect of aluminum hydroxide on fluoride metabolism, *Clin. Pharmacol. Ther.,* 28, 529, 1980.

57. **Spencer, H., Osis, D., and Lender, M.,** Fluoride metabolism and aging, in *Handbook of Geriatric Nutrition,* Hsu, J. and Davis, R., Eds., Noyes Public N.J., 1982, 250.

58. **Spivey-Fox, M. R. and Tsao, S. H.,** Mineral content of human tissues from a nutrition perspective, *Fed. Proc.,* 40, 2130, 1981.

59. **Sorbie, J., Olatiumboxun, D., Corbett, W. E. N., and Valberg, L. S.,** Cobalt excretion test for the assessment of body iron stores, *Can. Med. Assoc. J.* 104, 777, 1971.

60. **Sunderman, F. W.,** Measurements of nickel in biological materials by atomic absorption spectrometry, *Am. J. Clin. Pathol.,* 44, 182, 1965.

61. **Suzuki, Y.,** The normal levels of fluorine in the bone tissue of Japanese subjects, *Tohoku J. Exp. Med.,* 129, 327, 1979.

62. **Taves, D. R.,** New approach to the treatment of bone disease with fluoride, *Fed. Proc.,* 29, 1185, 1970.

63. **Taylor, A., Marks, V., Shabaan, A. A., Mahmood, M. A., Duckham, J. M., and Lee, H. A.,** Cobalt induced lipermia and erythroporesis, in *Clinical Chemistry and Chemical Toxicology of Metal,* Brown, S. S., Ed., Amsterdam, Elsevier/North Holland, Amsterdam, 1977, 105.

64. **Taylor, A. and Marks, C.,** Cobalt: a review. *J. Hum. Nutr.,* 32, 165, 1978.

65. **Telib, M. and Schmidt, F. H.,** Effect of cobaltous chloride in laboratory animals. II. Effect on blood sugar, plasma insulin and plasma lipids in rabbits. *Endocrinologie,* 61, 395, 1973.

66. **Tipton, I. H., Stewart, P. L., and Martin, P. G.,** Trace elements in diets and excreta, Hlth. Phys. 12, 1683, 1966.

67. **Tsongas, T. A., Meglen, R. R., Waltravens, P. A., and Chappell, W. R.,** Molybdenum in the diet: an estimate of average daily intake in the United States, *Am. J. Clin. Nutr.,* 33, 1103, 1980.

68. **Underwood, E. J.,** Molybdenum in animal nutrion, in *The Biology of Molybdenum,* Chappell, W. R. and Petersen, K. K., Ed., Marcel Dekker, N.Y., 1976.

69. **Underwood, E. J.,** *Trace Elements in Human and Animal Nutrion,* 4th ed., Academic Press, N.Y., 1977, 545.

70. **Vohra, P. G., Gray, A., and Kratzer, F. H.,** Phytic acid metal complex, *Proc. Soc. Exp. Biol. Med.,* 120, 447, 1965.

71. **Winter, G. D.,** Tissue reaction to metallic wear and corrosion products in human patients, *J. Biomed. Mater. Res.,* 8, 11, 1974.

72. World Health Organization (INHO) Trace Elements in Human Nutrition, Technical Report Series No. 532. Geneva, 1973.

VITAMIN C AND THE ELDERLY

Lorraine Cheng, Marvin Cohen, and Hemmige N. Bhagavan

INTRODUCTION

Vitamin C (ascorbic acid) inadequacy, eventually leading to scurvy, is one of the oldest maladies of mankind. The decline of this dreaded disease began with the introduction of root crops in Europe during the Middle Ages. It was also learned in the past several hundred years that fruits, fruit juices, and vegetables could be used to treat scurvy, and today this has become a relatively rare disease, especially in the developed countries.

It has been over 50 years since vitamin C, the active material in fruit juices that would prevent scurvy, was first isolated.[223] Although we have learned a great deal about the biochemical effects of vitamin C deficiency during this time, there is still a wide gap in our understanding of its biochemical and physiological functions. The presence of vitamin C in all eukaryote organisms suggests that it plays an essential role in cellular function.

Although the effects of vitamin C are involved in many physiological functions (e.g., collagen synthesis, immune function, drug metabolism, folate metabolism, cholesterol catabolism, iron metabolism, carnitine biosynthesis), clear-cut evidence for a biochemical role is available only with respect to collagen biosynthesis (hydroxylation of proline and lysine). There is also controversy concerning the requirements for vitamin C in health and disease. There is some evidence in the literature suggesting increased vitamin C requirements under conditions of stress (e.g., disease status). Data are also available indicating that the vitamin C requirements of the elderly are much higher than the currently recommended dietary allowances.[81] While we know the minimum amount of vitamin C needed to prevent scurvy, there is no consensus on the optimum level needed to maintain good health in all age groups.

VITAMIN C STATUS IN THE ELDERLY

Some surveys of the elderly have indicated that poor vitamin status is common in this segment of the population.[99,189,225] Other evidence leading to this conclusion includes decreased ascorbic acid levels in whole blood,[138,139] serum,[167] plasma,[30,133] and leukocytes.[6,30,129,133,155] Low blood ascorbic acid levels have been attributed to impaired absorption or poor utilization in some cases,[139,189] but inadequate dietary intake appears to be a major factor.[190]

A report on the elderly[222] indicated that while the mean intakes of nutrients generally satisfied the Recommended Daily Allowances (RDA), dietary intake of vitamin C was below the RDA level. A similar trend was noted in a survey of elderly people living in New York City;[130] 40% of the people surveyed were deficient in vitamin C. Lyons and Trulson[163] found that approximately 25% of low-income, elderly subjects had diets deficient in vitamin C and other nutrients. LeBovit[146] found a mean vitamin C intake of 71% RDA among 283 elderly persons who lived alone in Rochester, N.Y.

According to Loh and Wilson,[157] 35 mg/day of ascorbic acid may not satisfy the hematopoietic needs of the elderly. Adequate supplementation in a 5:1 ratio of ascorbic acid to iron is recommended to prevent anemia if no other deficiencies are present.

A syndrome termed "bachelor scurvy" is frequently seen in elderly men living alone.[50,109,149,208] Burr[37] found that men who ate alone tended to have a poorer ascorbic acid status than those whose meals were cooked or shared by another person. LeBovit[146] found a mean ascorbic acid intake of 71% RDA among 283 elderly persons in Rochester, N.Y. who lived alone; another New York State survey by Jordan et al.[130] found 40% of the subjects

Table 1
ASCORBIC ACID (AA) STATUS OF INSTITUTIONALIZED ELDERLY

Site	Results	Ref.
England	Survey of 136 subjects. Rank order of leukocyte AA levels was: small welfare home > large hospital > living at home > large welfare hospital > small hospital	7
England	Leukocyte AA levels in female inpatients lower than in outpatients; patients at large hospital had lower levels than those at smaller institution	8a
Maryland, New Jersey	4.9% of 44 institutional elderly had blood AA levels <0.5 mg%; 23.5% of 51 elderly living at home had blood AA levels <0.4 mg %	13
England	Plasma AA levels in 18 geriatric patients lower than in 10 healthy subjects living at home	20
New York State	Lowest AA levels found in men living in residential county home or patients at veteran's hospital	32
England	Low AA levels found in hospitalized geriatric patients	33
England	41% of elderly were deficient in AA at admission to hospital; 27% of subjects at home showed deficiency	100
U.S.A.	6/13 nursing homes provided <70—75 mg/day of dietary AA	105
Western U.S.A.	Serum AA levels acceptable (>0.2 mg %) in 70 women living in nursing or private homes	107
New Zealand	35 subjects; residents of a home had lower plasma AA levels than individuals living alone; no difference in leukocyte AA levels	171
U.S.A.	44 men living in a county home had serum AA levels and daily intake below values found for 232 men living in private homes	177
Finland	Survey of 48 residents of municipal institution and 87 living in rural communes or private homes showed institutionalized men had lower plasma AA levels; less marked differences among women	194
Switzerland	Mean daily AA intake of 67 institutionalized subjects was <55—60 mg	202
England	Lower plasma and leukocyte AA levels in 72 female hospital patients than in 35 elderly women living at home	206
Ireland	Survey of 86 institutionalized subjects and 36 elderly living at home showed institutionalized men had lower plasma AA levels and lower AA intake than subjects in private homes	232, 233
Australia	Survey of 17 institutionalized elderly and 12 in private homes showed women but not men in private homes had greater AA intake	246
Oregon	20 residents of a retirement village had lower AA intake than 80 individuals living in private homes	248

deficient in ascorbic acid. McClean et al.[170] found evidence of vitamin C deficiency in >34% of elderly men living alone. Although this phenomenon appears to be less common among women living in a similar situation,[37] Davidson[58] did not find a sex difference, and Walker[235] reported several cases of scurvy in elderly women.

Many surveys of the elderly in an institutional setting (e.g., old age or nursing homes) have indicated that ascorbic acid deficiencies are more prevalent under these conditions. Table 1 summarizes some representative studies that have examined this factor. Individuals living at home usually had higher ascorbic acid levels than elderly subjects in a hospital, welfare home, or other institution.[7,194,206,248] The incidence of low ascorbic acid status in these elderly populations may represent a need for nutritional intervention.

Effect of Age and Sex on Ascorbic Acid Status

Many studies have concluded that the elderly generally are in poor vitamin C status.[79,99,189,190,225] The data for this conclusion have been acquired principally from measurements of blood ascorbic acid levels. A review of several surveys[138] indicated that serum ascorbic acid levels tend to decrease with age, but statistically significant correlations were not always obtained.

Studies in which blood ascorbic acid levels of the elderly have been compared with younger subjects have generally shown lower levels in older individuals.[6,30,129,133,138,155,167] Kirk[137] reported that age differences in blood ascorbic acid tended to show up with low or moderate intakes, but not with very low or very high intakes. In a study examining whether low blood ascorbic acid levels might be a physiological characteristic of aging, Kirk and Chieffi[139] were able to raise blood ascorbic acid levels 3 to 4 times by orally administering 100 mg/day of ascorbic acid. When the supplement was withdrawn, ascorbic acid blood levels quickly dropped to baseline values. The authors questioned the desirability of raising ascorbic acid levels in the elderly since the low levels found were not associated with obvious signs of scurvy. Other investigators have suggested that symptoms in the elderly such as susceptibility to infection,[79] purpuric spots, indolent ulcers or multiple bruises,[133,175] and normocytic anemia[175] may indicate a latent ascorbic acid deficiency. However, attempts to relate blood ascorbic acid levels in the elderly to signs of scurvy have not been very successful.[6,8,10]

Impaired absorption or utilization of ascorbic acid by the elderly has been suggested as a cause of low blood ascorbic acid levels,[139,189] but inadequate dietary intake appears to be a major factor.[7,137,184,190] Other factors that have been shown to affect ascorbic acid status include seasonal variation[23,173] and economic level.[48,178,184]

Surveys of elderly men and women have generally shown that men tend to have lower blood ascorbic acid levels.[2,99,138,173,178] Roderuck et al.[193] also found no change in blood, serum, or plasma ascorbic acid levels with age in 569 women aged 20 to 99 years. Milne et al.[173] on the other hand, found a significant decrease of leukocyte ascorbic acid levels with age in women but not men. Loh and Wilson[155] found that a more marked reduction in leukocyte ascorbic acid occurred in elderly women than in men. Morgan et al.[178] suggested that the vitamin C requirements of men >50 years old may be significantly greater than those of women. Observations by Kirk and Chieffi[138] that elderly men tended to be more debilitated than elderly women led them to suggest that the degree of debilitation might be a contributing factor to a sex difference in ascorbic acid status.

Table 2 summarizes literature on relationships between age and ascorbic acid levels in plasma or leukocytes. Table 3 summarizes data relating to sex differences in ascorbic acid levels among the elderly and other populations.

Vitamin C Requirements of the Elderly

It is generally recognized that a daily intake of 10 mg of ascorbic acid will prevent scurvy in adults, but the amount needed for the maintenance of good health is much larger.[123] Daily intake of at least 60 mg ascorbic acid by a 70-kg adult has been recommended for maintaining a plasma concentration of >0.6 mg%.[106] Estimates of vitamin C requirements for the elderly have been based on studies of blood and urinary levels after varying intakes. Roderuck et al.[193] found that a daily intake of 1.1 mg/kg or more of ascorbic acid provided elderly women with sufficient levels of the vitamin. In a study of amounts required for tissue saturation, Morse et al.[180] found that 57 mg/day of ascorbic acid resulted in saturated leukocytes in both younger and older women. A few earlier studies have indicated that vitamin C requirements of the elderly may be different from younger individuals. Gander and Niederberger,[79] on the basis of urinary ascorbic levels after oral doses, concluded that the elderly required approximately 50% more vitamin C than younger subjects. Similar conclusions were reported by Kirchmann[136] and Rafsky and Newman.[190]

Elderly subjects generally have lower blood and leukocyte[7,30,38] ascorbic acid levels than younger individuals and some surveys have concluded that vitamin C deficiency is common among the elderly.[98] Burr et al.[38] reported that 40% of males aged 70 years or more were vitamin C deficient by both blood (plasma levels <0.1 mg %) and dietary assessments.

Table 2
EFFECT OF AGE ON ASCORBIC ACID (AA) LEVELS

Subjects and site	Results	Ref.
102 subjects, 25—80 years Bangladesh	No age-related differences in blood AA	1
473 elderly and 204 younger subjects Maryland, New Jersey	9.7% of elderly and 2.5% of controls had blood AA levels <0.5 mg %	13
Subjects aged 18—80 years England	No effect of age on leukocyte or skin AA levels	16
889 subjects, 18—94 years France, Switzerland	Incidence of AA deficiency correlated with age	27
138 subjects, 17—36 years England	Significant decline of plasma AA but not leukocyte AA levels with age	35
830 elderly subjects Wales	Plasma and leukocyte AA levels lower than values reported for younger subjects	38
50 elderly subjects	Mean AA levels approximately 1/3 of younger group	61
214 elderly subjects New York State	Plasma AA levels decreased slightly for each decade of life	62
Healthy, elderly 65—87 years old England	No correlation between leukocyte AA and age	66
60 elderly women living alone, England	Subjects >75 years old had lower AA intake than those <75 years old	73
722 normal subjects 20—50 years old and 89 hospital patients >60 years old Norway	Elderly subjects had 1/2—1/5 AA levels of younger subjects	101
142 hospital patients aged 40—103 years St. Louis	Significant decrease in blood AA levels with age in men but not women; AA levels in elderly approximately 1/2 those of younger subjects	138
135 subjects 18—87 years Dublin	Decreased leukocyte and plasma AA levels with age	151
210 subjects, 11—92 years Ireland	No age-related differences in leukocyte or plasma AA levels	152
156 subjects, 18—87 years Ireland	Plasma but not leukocyte AA levels decreased with age	155
138 subjects, 11—92 years Dublin	Highest leukocyte AA levels in adolescence and lowest in elderly	
487 subjects, 62—90 years Edinburgh	More elderly >75 years old had diets providing <50% RDA AA than in subjects <75 years old	156
48 denture wearers 19 to >60 years old Finland	Subjects >60 years old did not have lower serum AA levels than younger individuals	167
178 healthy men 17—68 years old New Zealand	Plasma and leukocyte AA levels declined with age	169
475 elderly subjects Edinburgh	Age-related decline in leukocyte AA for women but not men	173
525 subjects 50 to >75 years old California	No relationship of age to serum AA levels or AA intake	177
569 women aged 20—90 years North Central States	No relationship between age and AA blood levels or AA intake	193
Postmortem tissue St. Louis	Correlation between age and AA levels in pituitary and cerebrum, but not other tissues	200
31—60 geriatric hospital patients England	Plasma and leukocyte AA levels were lower than a younger control group but similar to a hospitalized younger group	203
155 geriatric hospital patients England	Plasma and leukocyte AA levels lower than in younger subjects	205
Cases of sudden death Canada	AA levels in elderly subjects lower than in younger subjects	238a
200 geriatric hospital admissions	No effect of age on leukocyte AA levels in range 60 to	242

Table 2 (continued)
EFFECT OF AGE ON ASCORBIC ACID (AA) LEVELS

Subjects and site	Results	Ref.
England Postmortem tissue Pennsylvania	90 years Age-related decrease in AA for all tissues examined	247

Table 3
SEX DIFFERENCES IN ASCORBIC ACID (AA) STATUS

Subjects and site	Results	Ref.
102 subjects 25—80 years Bangladesh	No sex-related differences in blood AA levels	1
473 elderly subjects New Jersey, Maryland	No sex differences in blood AA levels	13
Subjects aged 18—80 years England	No correlation between leukocyte AA levels and sex	16
63 residents of homes for aged Germany	Women had consistently higher blood AA levels than men	31
234 elderly subjects New York State	Women had significantly higher AA levels than men	32
138 subjects, 17—65 years	Women consistently had higher plasma and leukocyte AA levels	35
830 elderly subjects Wales	Women had consistently higher plasma and leukocyte AA levels	38
102 elderly subjects New York State	Women had higher blood AA levels than men	62
2130 men, 2865 women 4 to >20 years U.S.	Men began to show lower blood AA levels than women in adolescence; difference persisted in older subjects	63
Elderly subjects living at home England	No correlation between leukocyte AA levels and sex	66
308 elderly subjects England	English women had higher serum AA levels than English men; no difference in Asian subjects	70
270 elderly subjects New Mexico	Plasma AA levels significantly higher in women than men	81
165 residents of homes for the aged	Women had higher blood AA levels than men	83
295 elderly subjects Colorado, Oregon	No difference in mean serum AA levels of men and women	108
136 elderly subjects Missouri	No difference in serum AA levels for men and women	141
656 hospital admissions France	Men had plasma AA levels <0.2 mg % more often than women	148, 152
156 subjects 18—87 years Ireland	No consistent difference in leukocyte or plasma AA levels in men and women at any age group	156
138 subjects, 11—92 years Ireland	Women showed greater decrease in leukocyte AA level with age than men	155, 156
135 subjects, 18—87 years Ireland	No sex differences in leukocyte AA levels; female students and adults had higher plasma AA levels; male elderly had higher plasma AA levels than females	157, 158
210 subjects, 11—91 years	No sex-related differences in leukocyte or plasma AA levels	152
487 subjects, 64—90 years Edinburgh	No sex differences in mean AA intake or proportion with intake <50% RDA	159

Table 3 (continued)
SEX DIFFERENCES IN ASCORBIC ACID (AA) STATUS

Subjects and site	Results	Ref.
475 elderly subjects Edinburgh	No sex differences in leukocyte AA levels; more men than women >70 years old had AA intakes <30 mg/day	173
569 elderly subjects California	Women consistently had higher serum AA levels than men; no sex difference in AA intake	177
48 elderly institutional residents	Mean daily AA intake of men ¹/₃ less than for women	186
135 elderly subjects Finland	Women had higher plasma AA levels than men; no sex difference in AA intake	194
Postmortem samples St. Louis	Females had higher AA levels in pituitary and myocardium	200
680 elderly subjects Michigan	Women had higher dietary AA intake than men	227
122 elderly subjects Ireland	AA intake <0.3 mg% found more often in men than in women	232, 233
14 geriatric patients England	Mean leukocyte AA levels higher in women than in men	244
100 residents of old age home Germany	Fasting blood AA levels <0.2 mg % in 19/45 men and 1/15 women; 3 hr after 1000 mg AA mean blood AA levels in women 45% higher than in men	245
29 elderly subjects Australia	Plasma AA levels <0.2 mg % found in 8/15 men and 4/14 women	246
100 elderly subjects Oregon	Serum AA levels higher in women than men	248

Hughes[123a] suggested that the elderly may have a changed pattern of metabolism that is reflected in lower ascorbic acid levels and indicated that factors such as smoking or sex also affect levels.

Elderly people who are hospitalized for long periods of time tend to show an even greater depletion of ascorbic acid reserves.[7,133,203,205] Since reduced levels can also be found in younger hospitalized subjects,[203] factors such as reduced intake and depletion of vitamin C by disease processes probably play a major role. The low incidence of scurvy in the elderly despite low ascorbic acid levels may be due to seasonal variations in ascorbic acid intake[204] and the 80 to 120 days of zero intake required for clinical scurvy to become evident.[116]

In the absence of signs of clinical scurvy, low ascorbic acid levels have been associated with behavioral changes,[135,174] poor recovery from surgical procedures,[114] the presence of sublingual petechiae,[6,225] weight loss,[24a] and reduced immune function.[228] Supplementation improved some of these parameters. These data suggest that low vitamin C levels in body tissues may have a general deleterious effect on health in addition to the specific signs of scurvy. This may be of particular importance to the elderly who are at greater risk for chronic illness and complications due to decreased immune function.

In a recent trial, Garry et al.[81] have made an important observation on the vitamin C status and requirements of a healthy, free-living, elderly population. Their data show that elderly men and women need 150 and 75 mg of vitamin C daily, respectively, to maintain plasma vitamin C at 1 mg/dℓ. It should be noted that even in this population of health-conscious elderly who belonged to a higher socioeconomic group, about 2% were at high risk for vitamin C deficiency (plasma vitamin C values <0.2 mg/dℓ). More recently, there has been another significant observation.[94a] Mental function was evaluated in the same elderly population, and the data indicated that cognitive functioning was impaired in elderly subjects with low plasma values of vitamin C. These findings help reinforce the concept that the assessment of nutrient requirements should also take into account both physical and mental performance. Table 4 summarizes literature relating to the range of dietary ascorbic acid and intake found among various populations of elderly subjects.

Effect of Diet on Ascorbic Acid Status

Plasma levels of ascorbic acid are related to dietary intake; daily administration of 100 mg will result in plasma levels of approximately 1.2 mg%.[123] Plasma values of 0.1 mg% are associated with a high risk of scurvy.[128] Table 5 summarizes literature relating the amount of dietary ascorbic acid intake to plasma or leukocyte levels of the vitamin.

Indicators of Ascorbic Acid Status

Plasma ascorbic acid is generally accepted as an indicator of ascorbic acid status of the individual. However, leukocyte levels are considered to be a better index of ascorbic acid status.[6,99,123]

Bates et al.[26] explored relationships between ascorbic acid levels in parotid saliva and overall ascorbic acid status. Significant correlations were found between salivary ascorbic acid levels and leukocyte or plasma levels in subjects >65 years old. However, an interfering substance in saliva that affected ascorbic acid determinations made this approach of limited usefulness as a reliable measure of ascorbic acid status. Similar results were obtained by Freeman and Hafkesbring.[78] Subsequently, however, it has been shown, using selective and sensitive techniques, that ascorbic acid is not present in any measurable amount in saliva.[74a]

Bates[24] studied the relationship between proline/hydroxyproline excretion and vitamin C status in 23 elderly subjects aged 74 to 86 years. A strong positive correlation was found between plasma and buffy-coat vitamin C levels. No significant correlation was found between plasma or buffy-coat vitamin C levels and total urinary hydroxyproline or proline/hydroxyproline ratio. A significant negative correlation was noted between plasma or buffy-coat vitamin C and total urinary proline in hydrolyzed urine. The author concluded that poor vitamin C status could reflect a defect in collagen proline hydroxylation; measurement of urinary proline levels could be a means of detecting subclinical vitamin C deficiency. However, plasma, leukocyte, and ascorbic acid values provide the best direct assessment of ascorbic acid status in individuals.

Ascorbic Acid and Senile Purpura

An association between the presence of sublingual petechiae and leukocyte vitamin C content has been reported.[6] The mean level was 10.2 $\mu g/10^8$ cells (range 5.1 to 12.3) in 24 subjects with petechiae and 17.7 (range 6.4 to 17.9) in 64 patients without this clinical sign. Petechiae were also found in five subjects that had been receiving supplementary vitamin C (up to 100 mg/day) for up to 2 months and had leukocyte levels of 15.3 to 33.9 $\mu g/10^8$ cells. Taylor[225] reported that signs of vitamin deficiency were common in geriatric wards; they included senile purpura as well as dermal, follicular, or sublingual petechial hemorrhages.

Eddy and Taylor[69] surveyed 22 elderly vegetarians who had high ascorbic acid levels in plasma (mean 1.02 mg %) and leukocytes (mean 35.9 $\mu g/10^8$ cells). The incidence of sublingual petechiae and varicosities was much lower than in a general geriatric population. The authors suggested that long-term ascorbic acid deficiencies may lead to irreversible vascular changes not affected by subsequent vitamin supplementation. In a subsequent controlled study,[69a] the effect of supplementation with 200 mg/day ascorbic acid on capillary fragility in geriatric hospital patients was examined. Leukocyte ascorbic acid levels of 30 $\mu g/10^8$ cells were associated with high capillary resistance to trauma (Hess test). These data demonstrate the importance of ascorbic acid in maintaining the integrity of the capillaries.

Table 4
DIETARY ASCORBIC ACID (AA) INTAKE OF ELDERLY

Subjects and site	Results	Ref.
146 elderly living at home New Jersey, Maryland	23.5% had blood AA levels <0.4 mg %	13
100 hospital admissions >60 years old, England	20% had blood AA levels <0.2 mg % and 9% had levels <0.1 mg %	23
234 elderly subjects New York State	8% had plasma AA levels <0.1 mg %	32
28 elderly subjects Netherlands	64% had intake of <50% RDA and 29% had intake of 50—80% RDA	56
104 healthy individuals aged 51—97 years England	16% had intake <50 mg/day; 5% had intake <20 mg/day	58
102 elderly subjects New York State	37 had 66% or less RDA	62
Healthy elderly living at home England	Dietary AA intake ranged from 5—106 mg/day, leukocyte AA levels ranged from 0—63 mg/10^8 cells	66
479 elderly California	AA intake unrelated to lower mortality of health conscious subjects	72
60 elderly women England	Mean AA intake 37.6 mg/day	73
270 elderly New Mexico	3—10% received dietary AA intake <100% RDA; 2% of women received <50% RDA. Mean dietary intake was 137 mg/day (men) and 142 mg/day (women)	80, 81
Elderly hospital admissions or living at home England	27—41% of subjects showed AA deficiency	100
295 elderly subjects Colorado, Oregon	Mean dietary intake — 138—170% RDA (women), 178% RDA (men)	108
63 elderly hospital admissions or residents of old age home Australia	AA deficiency in 20—27.3%	134
136 elderly subjects Missouri	19% of both sexes had AA blood levels <0.2 mg %; mean dietary intake 135 mg/day (men) and 124 mg/day (women)	141, 142
466 elderly subjects Missouri	7% had serum AA levels <0.2 mg %; mean daily AA intake was 70% RDA	141, 142
487 elderly subjects Edinburgh	10.2—17.8% had AA intake <50% RDA	159
100 elderly subjects Boston	10% of men and 6% of women obtained <50% RDA for AA; mean dietary intake — 105 mg/day (men) and 95 mg/day (women)	163
264 elderly subjects living at home	5% had AA intake <10 mg/day	166
48 denture wearers 19—38 years old	1/11 of group > 60 years old had <0.2 mg % AA serum levels; none of younger group had low levels	167
35 elderly men living alone New Zealand	34% had evidence of AA deficiency; mean dietary intake 31 mg/day	170
441 elderly subjects Edinburgh	50—58% had daily intake of <30 mg. 23.6—28.1% had intakes <20 mg/day	173
Geriatric hospital patients England	4/64 had evidence of AA deficiency; 9/254 had confusional state associated with AA deficiency	175, 176
525 elderly subjects California	14% of men and 8% of women had serum AA <0.3 mg %; 5% of men and 7.5% of women had intake of <30 mg/day	177
Geriatric hospital admissions England	54/93 had leukocyte AA levels <15 mg/10^8 cells; no clinical evidence of scurvy.	178
100 elderly subjects Ireland	65—69% of low-income group and 25% of high-income group had AA intake <30 mg/day	184

Table 4 (continued)
DIETARY ASCORBIC ACID (AA) INTAKE OF ELDERLY

Subjects and site	Results	Ref.
48 elderly institutionalized subjects Finland	Daily AA intake — 12—68 mg (women), and 10—38 mg (men)	186
135 elderly subjects Finland	30% had AA intake <20 mg/day, 60% <30 mg/day	194
70 elderly outpatients	39% had diets with borderline or inadequate AA content	199
Elderly women Switzerland	2/12 had AA intake <30 mg/day; mean intake over 7—14 days — 20—149.8 mg/day	201, 202
680 elderly subjects Michigan	22% had inadequate dietary AA intake	227
36 elderly subjects living at home Ireland	14—20% had dietary AA intake <67% RDA; 47.2% had plasma AA levels <0.3 mg %	232
196 elderly subjects Ireland	29% had serum AA levels <0.3 mg %	233
Geriatric hospital admissions England	32/159 had leukocyte AA levels <12 mg/10^8 cells	242
Residents of old age homes Germany	20/60 had fasting leukocyte AA levels <0.2 mg %	245
29 elderly subjects Australia	Mean daily AA intake 49—69% RDA	246
100 healthy elderly subjects Oregon	9% of women had AA intake <87% RDA; mean dietary intake 237% RDA (men) and 227% RDA (women)	248

Other studies have indicated that "senile purpura" in geriatric patients does not appear to be related to ascorbic acid status. Andrews and Brook[6] found no association between the presence or absence of this sign and leukocyte ascorbic acid levels. Tattersall and Seville[224] found that excretion of vitamin C was similar in patients with or without senile purpura. Similar conclusions were reported by Arthur et al.[10] and Andrews et al.[8]

Dymock and Brocklehurst[67] administered 200 mg/day ascorbic acid to 12 geriatric patients for 1 year. No significant improvement in the appearance of the underside of the tongue was noted. However, increased hemoglobin levels were found in 7 out of 12 given the supplement, while only 12 out of 63 patients not receiving the supplement showed a similar change.

Banerjee and Etherington[14] studied platelet function in elderly patients (mean age 81.4 years) who were scorbutic and compared results to "normal" elderly subjects with and without purpuric lesions and to a younger (mean age 36 years) group. Ascorbic acid levels were not related to the presence or absence of purpura, but all elderly subjects had lower ascorbic acid values than the younger group. No difference in platelet counts was noted among groups. All elderly groups had similar values for platelet adhesiveness and aggregation responses to ADP or collagen, but values were less than those for the younger group. A significant difference in platelet factor 3 availability was noted when elderly scorbutic and elderly normal groups were compared.

Other Factors Affecting Ascorbic Acid Status

Reduced ascorbic acid levels have been associated with GI hemorrhage due to peptic ulcer or aspirin/alcohol ingestion.[55,197] Russell et al.[197] surveyed 39 patients >45 years old and found significantly lower leukocyte ascorbic acid levels in patients with GI hemorrhage. A similar relationship was found in patients who had a history of aspirin or alcohol ingestion. Subsequent animal studies indicated that the addition of aspirin to a scorbutogenic diet significantly increased the incidence of gastric mucosal bleeding.

Table 5
EFFECTS OF DIETARY ASCORBIC ACID (AA) INTAKE AND BLOOD AA LEVELS

Subjects and site	Results	Ref.
23 elderly subjects England	Strong correlation between AA intake and plasma or leukocyte AA levels	24
63 residents of old age homes Germany	Significant correlation between AA intake and blood levels in women but not men	31
34 women Wales	Increased AA intake of 5 mg/day increased blood AA levels by 0.12 mg %	38
Healthy elderly subjects living at home, England	Men, but not women, showed good correlation between dietary AA intake and leuckotye AA levels	66
Review article	Dietary levels of 15—25 mg/day AA led to plasma AA levels of 0.1—0.3 mg %; dietary levels of 75—100 mg/day led to 1—1.4 mg %.	94
135 subjects 18—87 years old Dublin	Decreased dietary AA intake related to decreased leukocyte AA levels in elderly	151
479 subjects 62—94 years old Edinburgh	Significant correlation between AA intake and leukocyte AA levels	173
569 healthy subjects aged 50 to >80 years California	Significant correlation between AA intake and serum AA levels	177
25 elderly subjects U.S.A.	Doses of AA up to 1000 mg/day p.o. produced increasing retention with increasing dose in 7 subjects, consistent retention in 2, and a biphasic response in 16	188
569 women aged 20—90 years North Central States	Daily AA intake of at least 1.1 mg/kg resulted in blood levels >0.8 mg % in 89%	193
135 elderly subjects Finland	Poor correlation between AA intake and plasma AA levels	194
122 elderly subjects in institutions or living at home	Reduced dietary AA intake correlated with reduced plasma and leukocyte AA levels	232, 233
29 elderly subjects Australia	Significant correlation between dietary AA intake and plasma AA levels	246
100 healthy, elderly residents Oregon	Significant correlation between dietary AA and serum AA levels	248

Cohen and Duncan[49] studied the effect of GI disorders on leukocyte ascorbic acid levels and dietary intakes. Tests show that 14 patients with gastroduodenal disorders (mean age 50 years) had significantly lower ascorbic acid levels than an age-matched group without GI disease or younger healthy adults. Dietary intake of ascorbic acid was also lower for subjects with ulcers or other alimentary disorders. The authors concluded that reduced ascorbic acid levels were due to reduced dietary intake rather to an increased utilization because of disease. In a related study, Williamson et al.[238] found that elderly patients with intestinal malabsorption or who had undergone gastric surgery had reduced leukocyte ascorbic acid levels.

Smoking may further deplete vitamin C stores in elderly people with levels reduced by an inadequate diet. Studies in geriatric patients as well as other age groups[65,71,164] have repeatedly shown an inverse relationship between duration of smoking and vitamin C levels. McClean et al.[169] found that the effect of smoking on plasma ascorbic acid levels was less in subjects 60 to 69 years old than in younger groups.

Pelletier[187] evaluated the vitamin C status of smokers or nonsmokers after ascorbic acid at 2.2 g/day for 6 days. Smokers had lower blood levels of ascorbic acid than nonsmokers prior to supplementation and excreted 40% less than nonsmokers when an excess was given. The amount of vitamin C excreted during 8 hr after a 1.1-g load correlated well with blood vitamin C levels of both smokers and nonsmokers.

Table 6
USE OF ASCORBIC ACID (AA) SUPPLEMENTS BY THE ELDERLY

Subjects and site	Results	Ref.
7 geriatric patients England	200 mg/day AA for 2 weeks markedly increased leukocyte AA levels	7
36 residents of an old old age home England	40—80 mg/day AA increased leukocyte AA levels to values found in younger subjects	8
381 elderly subjects Maryland, New Jersey	151/299 institutionalized and 36/82 noninstitutionalized subjects took vitamin supplements	13
Geriatric hospital patients England	200 mg/day AA produced increased leukocyte AA levels within 3 months	33
270 elderly subjects New Mexico	Mean daily supplementary AA intake was 355 mg for men and 500 mg for women	80
3192 elderly subjects Florida	11.5—16% of women and 8.7—10.4% of men used vitamin C or multivitamin supplements	102
170 elderly subjects Colorado and Oregon	Supplements taken by 51/145 women and 4/25 men supplied 460% RDA and 133% RDA, respectively	108
19 geriatric patients St. Louis, Missouri	100 mg/day AA produced 2.5- to 5-fold increase in blood AA levels	138
136 elderly subjects Missouri	AA tablets taken by 11% of men and 6% of women; multivitamins used by 13—32%.	142
156 subjects aged 18—87 years Dublin	Supplementation produced significant increases in leukocyte and plasma AA levels	155
12 elderly women Switzerland	2/12 used vitamin supplements	201, 202
100 healthy elderly residents Oregon	Supplements used by 22/75 women and 4/25 men	248

A significant increase in ascorbic acid utilization (as indicated by urinary ascorbic acid excretion) was noted by Grant et al.[95] even in light smokers (1/2 pack/day).

MacLennan and Hamilton[165] compared plasma and leukocyte ascorbic acid concentrations in 20 geriatric patients in a "stable" phase of their illness and 23 patients in an "unstable" phase. The latter group was characterized by reduced leukocyte ascorbic acid levels, but plasma levels remained unchanged. Recovery was associated with increased leukocyte ascorbic acid levels.

Elderly individuals with chronic, debilitating diseases such as rheumatoid arthritis may require a higher intake of ascorbic acid to maintain adequate tissue levels. Mullen and Wilson[181] compared plasma ascorbic acid levels in 42 patients with rheumatoid arthritis and in controls. Plasma and leukocyte ascorbic acid levels were significantly lower in rheumatoid subjects; 42% had leukocyte ascorbic acid levels in the deficient range (<20 mg/10^8 cells). Urinary excretion and dietary intake of ascorbic acid did not differ for the two groups. Administration of supplementary ascorbic acid (500 mg) produced similar increases in plasma and leukocyte ascorbic acid in both groups. The authors concluded that individuals with rheumatoid arthritis utilized ascorbic acid at a faster rate than disease-free subjects. Sahud and Cohen[198] found that plasma ascorbic acid levels were reduced more than platelet ascorbic acid levels in rheumatoid arthritis patients.

The use of supplemental vitamin C can produce a marked improvement in elderly subjects with low dietary intake of ascorbic acid. Table 6 summarizes some of the literature relating to this aspect of vitamin C status in the elderly.

Seasonal factors have often been implicated in vitamin C status in the elderly. The most consistent trends that have been reported suggest that plasma/leukocyte ascorbic acid levels tend to be lower in fall and winter and higher in spring and summer seasons. A summary of the literature relating to this topic is presented in Table 7. Ascorbic acid interactions with

Table 7
SEASONAL EFFECTS ON ASCORBIC ACID (AA) STATUS IN THE ELDERLY

Subjects and site	Results	Ref.
59—98-year-old and younger controls England	Reduced leukocyte AA levels in February compared to previous October when no supplement given	7
214 elderly subjects New York State	Plasma AA levels in spring slightly higher than in fall	62
165 residents of homes for aged Czechoslovakia	50% incidence of latent AA deficiency at end of winter season	88
89 patients >60 years old Norway	Differences in serum AA levels between older and younger patients independent of season	101
475 subjects 62—94 years old, Edinburgh	Leukocyte AA levels higher from July—December than during other months	173
135 elderly subjects Finland	Plasma AA levels in women higher in spring and fall than in winter; less marked difference for men	194
Groups of 6—21 geriatric patients England	Plasma AA levels lower during March—May than in July—September	204

many drugs and nutrients may affect ascorbic acid status. These are discussed in more detail in the section on Interactions of Ascorbic Acid.

VITAMIN C AND LIPID METABOLISM

In the U.S., diseases of the heart rank first among the causes of death.* Among the other leading causes of death, cerebrovascular diseases rank third and arteriosclerosis ranks eighth. These are primarily diseases of the aged, occurring with the greatest frequency in individuals older than 64 years of age. Hyperlipidemia is recognized as a risk factor in the above-mentioned diseases. There is now sufficient evidence implicating a role for vitamin C in lipid metabolism, thus suggesting a beneficial effect of vitamin C in coronary artery disease. Several comprehensive reviews of this subject have appeared in recent years.[84,86,87,144,230]

Vitamin C Status and Lipid Levels

An interrelationship between vitamin C status and blood lipid levels has been demonstrated in healthy aged individuals, especially in men. There was a significant negative correlation between plasma vitamin C and cholesterol levels, a significant positive correlation between plasma cholesterol and triglyceride levels, but no correlation between plasma vitamin C and triglyceride levels.[97,145] The above correlations were apparent in male subjects but no significant correlation between any of the parameters was observed in aged female subjects. The data indicated that the average plasma levels of vitamin C in healthy subjects aged 70 to 84 years approached the lower limit of normal values, especially in males, so that these subjects could be considered at risk for hypercholesterolemia.

In contrast to the above findings, Bates et al.[24] reported that total plasma cholesterol was not significantly correlated with either plasma or leukocyte vitamin C concentrations in healthy elderly subjects. However, there was a strong positive correlation between plasma vitamin C and high-density lipoprotein (HDL)-cholesterol in men; a slightly weaker, albeit positive correlation was also demonstrated for leukocyte vitamin C levels. Such correlation was not found in women, nor was it found between plasma vitamin C and low-density lipoprotein (LDL) + very low-density lipoprotein (VLDL)-cholesterol levels. The strong positive correlation between plasma vitamin C and HDL-cholesterol could also be observed

---

* Statistical Abstracts of the United States, 1981.

in elderly hospitalized men and in young healthy subjects, both parameters being higher in the latter group.

Similarly, in aged men, but not women with ischemic heart disease, leukocyte vitamin C levels were positively correlated with HDL-cholesterol concentrations.[119] The majority of patients with ischemic heart disease had subnormal leukocyte ascorbic acid levels. It was postulated that latent vitamin C deficiency may be a risk factor in the pathogenesis of ischemic heart disease. Another study demonstrated that in a group of 150 patients ranging in age from 19 to 78 years (mean age 51 years), leukocyte ascorbic acid levels were significantly lower in those with coronary atherosclerosis as compared to subjects with normal coronary arteries.[191] The data indicated that patients with abnormal coronary arteries were older. Elderly subjects who had sustained an acute myocardial infarction[125,231] or an acute cerebrovascular accident[124] had significantly depressed leukocyte and serum ascorbate values which persisted for several weeks. Autopsies performed on those who died from the disease revealed depleted ascorbate concentrations in the pituitary and adrenal glands. Blood and leukocyte vitamin C concentrations in a group of hypercholesterolemic diabetic patients between 50 and 60 years of age were also lower than in healthy subjects.[91] The above studies suggest that vitamin C deficiency predisposes an individual to hypercholesterolemia and may contribute to the development of coronary artery diseases.

TREATMENT OF HYPERCHOLESTEROLEMIA WITH VITAMIN C

Sokoloff et al.[218] postulated that while abnormal lipid metabolism, i.e., increased mean fasting blood levels of total cholesterol and triglycerides with a concomitant decrease in lipoprotein lipase (LPL), may show an age-related frequency in persons 35 to 40 years or older, it may be a condition which could be remedied by the administration of vitamin C. It is interesting to note that 63% of the subjects aged 61 to 80 years had abnormal blood lipid factors as compared to 55% of those aged 51 to 60 years, 27% of those aged 41 to 50 years, and only 11% of those aged 19 to 40 years. On the other hand, 8 subjects in the 75 to 80 age group had completely normal values. High doses of ascorbic acid (1.5 to 3.0 g daily) given to 122 patients with widely different levels of blood lipid factors produced varying responses. The administration of ascorbic acid (1.5 g daily) for 4 to 5 months produced no significant effect in 40 patients with normal blood lipid factors, but normalized blood lipid levels in 14 to 22 patients with moderate deviations of lipid metabolism. In 60 patients with pronounced hypercholesterolemia and/or cardiac disease, high-dose ascorbic acid (2.0 to 3.0 g daily for 12 to 30 months) produced no effect in 10, but resulted in definite improvement in the remaining 50, with an average of 100% increase in LPL activity and a 50 to 70% decrease in triglycerides. However, there was no significant change in total cholesterol. A subsequent study from the same laboratory[217] examined the variability of lipid parameters after varying periods (5 to 24 months) of vitamin C (2.0 to 3.0 g/day) administration in individuals of different age groups. Older patients with the most pronounced abnormalities in lipid metabolism appeared to have a better response.

Ginter[89] studied a group of subjects aged 42 to 65 years with low blood vitamin C levels and hypercholesterolemia (232 to 312 mg %). The administration of 300 mg vitamin C daily for 47 days significantly reduced blood cholesterol levels; the effect was most pronounced in subjects with cholesterol values >240 mg %. Ascorbic acid administered at a dose of 1.0 g/day for 3 months to subjects aged 50 to 75 years with plasma cholesterol levels <200 mg % did not alter plasma cholesterol levels.[82] However, the same dose of ascorbic acid given for 6 months to a group of subjects similar in age but with plasma cholesterol concentrations of 206 to 327 mg %, resulted in a very significant decline in cholesterolemia. Similar results were obtained when the treatment periods were extended to 9 months and 1 year in subjects with initial plasma cholesterol concentrations >230 mg %.[83] Heine and

Nordon[112] showed that in 63 patients aged 35 to 73 years with arteriosclerosis obliterans, the hypocholesterolemic effect of long-term (3 to 53 months) vitamin C administration at 1 g/day was evident only if the initial serum cholesterol value did not exceed 300 mg %. HDL-cholesterol was elevated by vitamin C administration, whereas there was no change in serum triglyceride values. Intake of ascorbic acid at 1.0 g/day for 3 months to 1 year led to a significant reduction of plasma triglycerides in subjects with an average plasma triglyceride level of 331 mg %.[83] Ascorbic acid levels in whole blood were elevated by such treatment.

In a group of hypercholesterolemic patients, predominantly between 50 and 60 years of age and with maturity-onset diabetes mellitus, the administration of ascorbic acid at 500 mg/day resulted in a significant reduction of mean serum cholesterol levels and a moderate but significant reduction of mean serum triglycerides after 6 months and 1 year.[91] Concomitantly, the significantly lower initial blood and leukocyte vitamin C concentrations as compared to healthy subjects were reversed. However, vitamin C administration for 1 year proved ineffective in lowering serum cholesterol in about one third of the patients, whereas in 60% of the patients, cholesterol concentration decreased by a minimum of 40 mg % to >100 mg %. The authors suggested that the hypocholesterolemic effect of ascorbic acid was due to increased hepatic levels that led to a greater degradation rate of cholesterol to bile acids.

The hypocholesterolemic effect of ascorbic acid persisted for at least 6 weeks after cessation of intake, whereas whole blood ascorbic acid concentration declined 3 weeks after the end of ascorbic acid administration.[88] However, in certain subjects, hypercholesterolemia was resistant to long-term administration of high doses of vitamin C.

The optimal doses of vitamin C that can prevent hypercholesterolemia remain to be determined. Hanck and Weiser[104] found that 4.0 g of ascorbic acid daily for 3 weeks significantly depressed plasma cholesterol levels in healthy volunteers aged 25 to 45 years. In a group of male subjects ranging in age from 65 to 90 years studied by Fidanza et al.,[76] the administration of vitamin C at a dose of 3 g/day for 3 weeks led to a significant reduction of cholesterol, total lipids, and triglycerides in normolipidemic as well as in hyperlipidemic individuals; concomitantly, both plasma and leukocyte vitamin C concentrations increased. Similar results were demonstrated in a more recent study by Fidanza et al.[75] in a group of male subjects aged 44 to 79 years; in addition, they noted that levels of β- and pre-β-lipoproteins decreased while α-lipoproteins were increased. Alterations in free fatty acid composition of different lipid fractions were also found. Bordia[29] also found significant decreases in serum cholesterol and β-lipoprotein levels accompanied by an increase in α-lipoprotein during a 6-month period of vitamin C intake at 2 g/day in patients (mean age 50.8 years) with a past history of myocardial infarction. Vitamin C also increased fibrinolytic activity and decreased platelet adhesiveness; these effects indicate the potential of this vitamin in the management of coronary artery disease. Ascorbic acid treatment at 1 g daily for 6 weeks increased HDL-cholesterol concentration in aged patients with ischemic heart disease.[119] In male patients, total serum cholesterol and LDL-cholesterol were reduced and triglyceride levels were not significantly altered. However, in female patients, both total serum and VLDL-triglycerides were depressed.

Studies reviewed thus far confirm earlier studies[183,212] and provide evidence of a role for vitamin C as a hypocholesterolemic agent in the prevention or treatment of coronary artery disease. The hypocholesterolemic effect of vitamin C appears to be a function of initial pretreatment plasma cholesterol concentrations. In order for vitamin C to normalize plasma cholesterol levels, initial plasma concentrations must be greater than 230 mg % and must not exceed 300 mg %. A negative correlation between vitamin C intake and mortality of coronary heart disease has been suggested.[85,104] Ginter attributed the marked decline in coronary mortality in the U.S. since 1958 to the high intake of synthetic ascorbic acid.

VITAMIN C AND CANCER

After heart disease, malignant neoplasms rank second among the leading causes of death in the U.S.* The death rate for cancer increases exponentially with age in both males and females, and is the highest in those aged 65 years and over. Death rates from cancer in the U.S. have been steadily increasing during the period 1940 to 1978, particularly in males 65 years of age and older. The interrelationship between vitamin C and cancer in the aged has been the subject of several studies within the past decade.

Depletion of Vitamin C in Cancer Patients

Hypovitaminosis C has been demonstrated in geriatric patients with cancer. Leukocyte and plasma ascorbic acid concentrations were considerably lower in a 60-year-old male patient with pulmonary carcinoma when compared to mean values obtained from 7 normal geriatric subjects.[131] A biopsy sample of a secondary tumor of the skin from this patient showed an ascorbic acid value of 46 μg/g, whereas the value for the surrounding unaffected skin was 25 μg/g. The unaffected skin, however, contained about 31% of the ascorbic acid concentration reported for skin of normal subjects. Leukocyte and plasma ascorbic acid levels were also lower than normal in patients with a variety of cancers, including cancers of the skin, lung, buccal regions, bladder, breast, and rectum.[132] Ascorbic acid concentrations in tumor tissues were found to be 33 times higher than the levels in the adjoining normal tissues in the 7 patients with skin cancer and 1 patient with cancer of the cervix studied. These observations suggest that the accumulation of ascorbic acid by growing tumor tissue in humans may contribute to hypovitaminosis C in cancer patients.

Higher ascorbic acid levels in tumor tissues as compared to adjacent normal tissues were also demonstrated in 29 of 30 patients with epithelial tumors.[179] The mean tumor tissue ascorbic acid level was 12.1 μg/100 mg, whereas the mean ascorbic acid level for the surrounding normal tissue was 4.95 μg/100 mg, demonstrating a ratio of 2.4:1. A correlation of tumor ascorbic acid levels with plasma levels was also observed. These patients ranged from 22 to 89 years of age; 19 were over 65 years of age. The older patients had a tendency towards lower tumor tissue ascorbic acid levels than did younger patients.

Similar results were obtained in a more recent study[9] examining a larger group of patients. Plasma and buffy-coat vitamin C levels were below the threshold of hypovitaminosis C (0.35 mg % and 18 μg/10^8 cells, respectively) in most of the 139 patients (aged 33 to 80 years, mean age 63 years) with bronchial carcinoma; plasma and buffy-coat values below the thresholds for incipient clinical scurvy were found in 64 and 25%, respectively; frank clinical scurvy was also found. Plasma and buffy-coat vitamin C values rose in six patients given vitamin C supplementation. Four patients received 1 g/day for 3 days, then 200 mg/day for 2 weeks, and 2 patients received vitamin C from the time of resection at a level of 1 g/day for 6 days followed by 300 mg/day for 3 months. The vitamin C content of 13 primary lung tumor surgical specimens was found to be higher than that of normal lung tissue (111.6 ± 55.1 vs. 58.5 ± 20.4 μg/g tissue). These data support the reports of other workers[131,179] that tumors tend to accumulate vitamin C.

Among a group of male and female subjects aged 48 to 85 years,[21] breast cancer patients with skeletal metastases had the lowest mean leukocyte ascorbic acid levels (12 μg/10^8 W.B.C.) when compared to patients with advanced malignant tumors at other sites without any skeletal metastases (17 μg/10^8 W.B.C.), patients with nonmalignant diseases (21 μg/10^8 W.B.C.), and healthy subjects (33 μg/10^8 W.B.C.). An inverse relationship between leukocyte ascorbic acid levels and urinary hydroxyproline concentrations was observed, with the highest level found in patients with breast tumors and bone metastases (135 mg/g creatinine). Administration of an oral 1-g loading dose of ascorbic acid lowered urinary

* Statistical Abstracts of the United States, 1982.

hydroxyproline levels in breast cancer patients with skeletal metastases, but produced no change in patients with tumors at other sites without bone metastases. It was suggested that bone metastases caused an increase in the rate of bone collagen degradation resulting in an increased requirement for ascorbic acid in order to maintain the turnover rate of synthesis of bone collagen.

Physical signs indicative of subclinical scurvy, such as sublingual petechiae, abnormal forearm or abdominal hairs, or follicular hyperkeratosis, were observed more frequently in 30 of 50 patients with malignant diseases having low leukocyte ascorbic acid levels (<12.5 $\mu g/10^8$ W.B.C.) than in the remaining 20 patients with higher leukocyte ascorbic acid levels.[143] In addition, those patients with the lowest levels also had a significant decrease in capillary fragility. It was also noted that this group of cancer patients had lower leukocyte ascorbic acid levels when compared with age related controls (16.56 $\mu g/10^8$ W.B.C.) and healthy young adults (29.5 $\mu g/10^8$ W.B.C.). The low levels were possibly due to dietary ascorbic acid deficiency resulting from anorexia known to occur in patients with neoplasms. It was suggested that patients with malignancy should receive vitamin supplements. Low leukocyte and plasma ascorbic acid concentrations in a group of cancer patients (aged 58 $\pm$ 12 years) rose significantly after radiotherapy.[132a]

In a more recent study[96] hypovitaminosis C was also demonstrated in the majority of 59 patients (aged 38 to 83 years) with lung or bladder cancer. Some of these patients were given large doses of vitamin C orally (5 g/day) for several weeks, resulting in an increase of ascorbic acid blood levels beyond the upper range of normal (1500 μg %).

Treatment of Cancer with Vitamin C

The most outspoken proponents advocating the value of vitamin C in the treatment of cancer are Cameron and Pauling.[43,44] They first suggested that the utilization of massive doses of vitamin C might have potential value in the supportive treatment of cancer, since vitamin C may be involved in various physiological processes mediating host defense mechanisms.[41] They postulated that the incidence, morbidity, and mortality of cancer could be diminished if the natural resistance of the host against neoplastic diseases could be enhanced, and claimed that vitamin C held great promise in improving the general management of cancer because of its important role in host resistance.

Basu,[18] in discussing the possible role of vitamin C in cancer therapy, postulated that high-dose vitamin C may exert its antitumor effects through (1) decreasing the availability of amino acids such as lysine and cysteine, and of trace elements, particularly zinc, which are believed to be required for tumor cell proliferation; and (2) increasing tissue concentration of cyclic AMP, which has been found to inhibit tumor growth.

Cameron and Campbell[39] proceeded to study the effects of high-dose vitamin C supplements in 50 patients (aged 40 to 93 years) with various advanced cancers. Many of these patients received initial treatment with continuous i.v. infusion of ascorbic acid at 10 g/day for up to 10 days; a few received 45 g/day. The great majority of patients received ascorbic acid orally at 10 g/day, either as their sole treatment, or following the i.v. regimen. Several patients received oral treatment for more than 2 years without any significant ill effects. The total dose amounted to more than 9994 g in a 69-year-old male patient with liver metastases after palliative resection of colon carcinoma. His liver became impalpable within 10 weeks, and liver function tests returned to normal within 6 months. He was alive and well after 2.5 years of ascorbic acid treatment, and there was no indication of malignancy. In the entire group of 50 patients, 17 had no response, 10 had minimal response, 11 had growth retardation, 3 had cytostasis ("stand still effect"), 5 had tumor regression, and 4 had tumor hemorrhage and necrosis. Subjective and objective benefits were also noted. The authors considered it important that 10% of the patients had tumor regression, and that tumor hemorrhage and necrosis represented a very strong defense reaction. Some of the patients appeared to have survived much longer than reasonable clinical expectation. It was pointed out, however, that large doses of ascorbic acid must be prescribed with caution in patients with advanced cancer, since side effects did occur in a number of cases. Indeed, potentially dangerous symptoms such as acute pyrexial illness and mediastinal compression have been observed following massive doses of ascorbic acid in cancer patients.[45]

In a later study,[41] the 50 patients who had received high-dose ascorbic acid treatment were combined with 50 other terminal cancer patients given similar supplemental ascorbate treatment. The mean survival time of these 100 patients was >210 days, or 4.2 times longer than that of a selected control group of 1000 similar patients not given vitamin C. In the latter group, the mean survival time was only 50 days. In addition, an improvement in the quality of life was also observed in the patients given vitamin C. In order to enhance the reliability of the results, Cameron and Pauling[42] replaced 10 patients with rare forms of cancer in the original ascorbate-treated group of 100 with 10 others, and selected a new group of 1000 matched controls. Comparison of these revised groups showed that the mean survival time for patients treated with high-dose vitamin C was >300 days longer than that of the control group.

Prolongation of survival times and improvement in the quality of life were also observed in a group of Japanese patients with terminal cancer given high-dose ascorbate (5 to 60 g/day) when compared to those given low-dose ascorbate (0.5 to 4 g/day). Of the 55 high-ascorbate patients, 3 were still alive with an average survival of 1550 days.[182] The results were in agreement with those reported by Campbell and Pauling.

On the other hand, Creagan et al.[54] failed to find evidence that high-dose vitamin C therapy was beneficial to patients with advanced cancer aged >45 to >65 years who had previously received irradiation or chemotherapy. In a controlled double-blind study, 60 evaluable patients received vitamin C orally at 10 g/day and 63 received lactose placebo. There was no significant difference between the two groups with respect to survival, symptom reduction, or well-being. The median survival was about 7 weeks for all patients.

In view of the controversial findings with respect to the use of megadoses of vitamin C in the management of cancer, Hodges[115] suggested that further well-planned, randomized clinical trials be carried out to assess the therapeutic value of vitamin C in malignant diseases.

VITAMIN C AND IMMUNE RESPONSE

There is evidence showing that the aging process is associated with an impairment of the cellular immune mechanisms.[168] Response of lymphocytes from aged human subjects to plant mitogens was depressed when compared to the response of lymphocytes from younger persons.[237] This depressed response was attributable to a markedly reduced number of mitogen-responsive cells in the lymphocyte population from elderly individuals.[128] These mitogen-responsive cells had a reduced capacity for repeated cell divisions when compared to mitogen-responsive lymphocytes from young donors.[111,127] Additionally, suppressor lymphocytes were diminished in aged humans.[103] A recent study demonstrated that lymphocytes from old persons were more sensitive to [3H]thymidine-induced cell cycle arrest and chromosomal damage.[220]

Vitamin C may be involved in certain aspects of host defense.[228] In vitro studies have implicated a role for vitamin C in enhancing the cellular immune function of humans.[60,92,185,211,214a] In healthy adults, the ingestion of vitamin C at dosages of 1.2 or 3 g/day for 1 week[4] or 5 g/day for 3 days[249] produced a stimulation of the lymphocyte responsiveness to mitogens. Daily doses of 2 or 3 g also enhanced neutrophil motility to a chemotactic stimulus.[4]

Vitamin C appeared to have no effect on humoral immunity. In certain breast cancer patients receiving chemotherapy, vitamin C at 5 g/day was found to increase delayed hypersensitive skin reactivity as well as lymphocyte blastogenesis in response to mitogen.[249] In a 58-year-old diabetic patient with malignant external otitis due to a *Pseudomonas* infection, both the ear lesion and impairment of neutrophil chemotaxis responded to treatment with vitamin C at 3 g/day for 1 month.[51] The above results indicate that vitamin C may have a beneficial effect on the immune function in the elderly through preventing or reversing the impairment of cell-mediated immunity associated with the aging process. Further studies in this area are certainly warranted.

OTHER USES OF ASCORBIC ACID

Ascorbic acid has been used with some success in the treatment of Paget's disease.[22,213] Doses of 3 g/day administered for 2 weeks produced a lessening of pain in 50 to 73% of subjects. Hydroxyproline excretion increased following ascorbic acid therapy, and the authors noted that the highest excretion rates were associated with the greatest relief of pain.

Ascorbic acid was reported to reduce intraocular pressure,[150,192,234] but administration of 4.5 to 5 g/day to 6 patients with open angle glaucoma did not improve the effects of standard medical therapy.[77] In other studies, no significant difference in blood levels of ascorbic acid was found in patients with glaucoma or controls,[11] and patients with glaucoma or senile cataracts did not have reduced aqueous humor levels of ascorbic acid.[147]

Andrews and Wilson[5] determined vitamin C levels in 538 patients on admission to a geriatric unit. Patients with more coronary episodes prior to admission did not have lower vitamin C levels than other patients. Administration of vitamin C to 32 patients raised vitamin C levels when compared to a placebo group, but no effect was found on the amount of time preceding a cardiovascular episode after vitamin C administration had begun.

Altman et al.[3] treated 132 psychogeriatric patients with a vitamin B complex + C preparation or placebo for 6 weeks. A striking decrease in excitement was found in a nonschizophrenic subgroup (primarily organic brain syndrome). However, the effectiveness of the different components of the multivitamin preparation was not determined.

The addition of vitamin C (200 mg/dose) to sodium fluoride treatment of osteoporosis reduced the incidence of side effects.[47] Ascorbic acid may be of value in elderly individuals at increased risk for deep vein thrombosis due to surgery. Spittle[219] conducted a double blind study of the incidence of venous thrombosis in elderly subjects receiving ascorbic acid. Administration of 1 g/day for 8 to 42 days prior to surgery resulted in an overall 33% incidence of thrombosis in 30 patients. The placebo group had a 60% incidence of thrombotic complications. However, in a more recent double-blind study,[226] no difference was found in the incidence of deep vein thrombosis for patients given ascorbic acid for 7 days prior to surgery. The authors noted that leukocyte ascorbic acid levels on post-surgical days 6 or 9 were significantly lower in patients with thrombosis. The authors suggested that any protective effect produced by ascorbic acid might be related to changes in the ability of leukocytes to adhere to venous endothelium.

INTERACTIONS OF ASCORBIC ACID

Animal studies have indicated that ascorbic acid deficiency can reduce the rate at which various pharmacologic agents are metabolized.[251] This suggests that elderly individuals with inadequate intake of vitamin C would be at increased risk for adverse drug effects.

Effects on Drug and Hormone Levels

Antacids have not been shown to affect GI absorption of vitamin C.[46,117] The excretion of ascorbic acid can be increased by aspirin,[57,239] barbiturates such as barbital,[59] and tetracycline.[210]

Wilson et al.[239] studied the pharmacokinetics of antipyrine in 14 healthy subjects who had been given ascorbic acid in doses of 300 to 4800 mg/day for up to 14 days. No significant difference was found in antipyrine elimination rate or other pharmacokinetic parameters. No effect of diphenylhydantoin pharmacokinetics was found in two patients given 1200 mg vitamin C/day for 2 weeks.

Ginter and Vejmolova[90] examined the effect of long-term ascorbic acid administration (500 mg/day × 1 year) on antipyrine pharmacokinetics in subjects aged 48 to 76 years. A significantly increased elimination constant and decreased half-life were noted. Plasma levels in individuals receiving supplements were 2 to 3 times higher than a placebo group who relied entirely on dietary intake. An inverse correlation was found between serum vitamin C levels and antipyrine half-life or elimination constant. Similar results were found by Smithard and Loveman.[215,216]

Houston[120] found that repeated doses of vitamin C (1 g/day × 7 days) significantly shortened the saliva half-life of antipyrine in 5 healthy male subjects.

Aspirin may inhibit GI absorption of vitamin C. Basu[19] found that 900 mg aspirin could block increased plasma, leukocyte, and urinary ascorbic acid levels produced by the ingestion of 500 mg ascorbic acid. In vitro studies with guinea pigs produced a similar antagonism between the two compounds. This type of interaction may be relevant to elderly populations where many individuals have suboptimal vitamin C levels (see Section II). Sahud and Cohen[198] found that rheumatoid arthritis patients who were taking 12 or more aspirin tablets daily showed significantly lower plasma and platelet ascorbic acid levels. This decline in ascorbic acid status could be reversed with supplementary doses of the vitamin given for at least 3 months. Indomethacin showed an effect similar to aspirin, so that this phenomenon may be a characteristic of anti-inflammatory therapy.

Oral administration of 3 g ascorbic acid to 5 healthy volunteers 1.5 hr after acetaminophen produced a rapid and pronounced decrease in the excretion rate and an increase in the biological half-life of the latter drug.[121] This change was associated with a decrease in the fraction of acetaminophen excreted as the sulfate metabolite; concomitant administration of sodium sulfate prevented the change in metabolic rate. Concomitant administration of salicylamide and ascorbic acid produced a decrease in the conversion of salicylamide to the sulfate metabolite.[121]

Windsor et al.[243] administered 1 g/day tetracycline to 14 men aged 73 to 94 years; a fall in leukocyte ascorbic acid levels was noted over a 5-day period. A control group treated with phenobarbital, phenylbutazone, or aloxiprin did not show a similar response. Shah et al.[210] observed that tetracycline therapy over a 4-day period led to a significant depletion of leukocyte ascorbic acid levels by day 3 and serum levels by day 4. The mechanism of this interaction appears to be increased urinary excretion of ascorbic acid.[243]

An interaction between fluphenazine and ascorbic acid was reported by Dysken et al.[68] Administration of 1000 mg/day ascorbic acid produced a gradual decrease in blood levels of fluphenazine over a 19-day period, while ascorbic acid levels increased from 0.3 to 0.7 mg %.

The ability of ascorbic acid to increase plasma levels of ethinyl estradiol in women taking contraceptive preparations[12] suggests that elderly individuals taking steroids for various conditions may experience adverse effects if daily intake of vitamin C is increased to 1 g/day or more.

Ascorbic acid deficiency has been associated with corticotrophin therapy.[118,221] Case reports by Kitabchi and Duckworth[140] indicated that pituitary-adrenal axis function was normal in 2 elderly subjects with scurvy; similar results in 19 elderly individuals with low ascorbic acid levels were reported by Dubin et al.[64]

Five patients on long-term warfarin therapy received 1 g/day ascorbic acid for 14 days; no change was found in blood coagulability.[125] In two case reports,[195,214] changes in sensitivity to warfarin were associated with ascorbic acid administration; in one case the dose was 2 g/day for several weeks. In a controlled study, Feetam et al.[74] found that high doses of ascorbic acid (3 to 10 g/day) reduced total plasma warfarin levels, but no significant antagonism of hypoprothrombinemic acitivity was noted.

Trang et al.[229] investigated potential interactions of ascorbic acid and caffeine in 10 elderly men. Reduced ascorbic acid intake or supplementation had no significant effect on several pharmacokinetic parameters when caffeine was administered i.v.

Effects on Nutrients

Individuals deficient in ascorbic acid show an increased excretion of urinary folate; this pattern is reversed when ascorbic acid is administered.[52] Dietary and plasma ascorbic acid levels appear to be inversely related to serum folic acid concentrations.[152] Ascorbic acid supplementation may also precipitate an overt folic acid deficiency.[53,250] Bates et al.[25] studied the interaction between folate status and vitamin C in 21 healthy elderly subjects. Prior to supplementation, ascorbic acid intake was significantly correlated with folate intake; plasma, but not red cell folate, was significantly correlated to plasma and buffy-coat ascorbic acid levels. During a period of vitamin C supplementation, no significant changes in folate status were detected.

Ascorbic acid facilitates the uptake of iron from the intestine, participates in various phases of iron distribution, and potentiates the incorporation of iron into protoporphyrin.[93,157,162] An optimal ratio of 5:1 between ascorbic acid and iron appears to be needed for this beneficial interaction to occur.[28]

Daily administration of iron to geriatric patients over a 14-week period produced a gradual fall in leukocyte ascorbic acid levels in men.[154] Women showed a more complex response that suggested a metabolic readjustment to ascorbic acid demands. In a related study,[155] an inverse relationship between leukocyte ascorbic acid and hemoglobin levels was found in elderly subjects.

Some studies have suggested a relationship between chronically low ascorbic acid levels and the development of osteoporosis.[126,161,236] However, Banerjee et al.[15] determined bone mass and leukocyte ascorbic acid levels in 80 geriatric subjects (aged 69 to 91 years) and did not find a significant correlation between the two parameters. A progressive decline with age was noted, but rates were not simultaneous in each individual studied.

In a study of relationships between glucose tolerance and plasma ascorbic acid, Setyaadmadja et al.[209] found that elderly subjects with low plasma ascorbic acid levels (<0.5 mg %) showed a decreased tolerance for a glucose load. Elderly subjects with higher plasma ascorbic acid levels (>0.6 mg %) showed responses similar to those in younger age groups.

CONCLUSION

Although we have acquired a good deal of knowledge on various factors affecting the vitamin C status of the elderly, the optimum vitamin C intake to maintain peak physical and mental health needs to be established. The current RDA may not be adequate as shown by the recent findings of Garry and co-workers.[80,81,94a] We should also be cognizant of the clinical, behavioral, physiological, and biochemical signs of marginal vitamin C status in the elderly, knowing that their dietary intake is generally inadequate to meet the needs for several nutrients. This is especially so in the institutionalized elderly. If the desired amount cannot be provided by diet alone, supplementation should then be considered to improve their health.

ACKNOWLEDGMENT

The skillful assistance of Mrs. O. Buchko in the preparation of this manuscript is gratefully acknowledged.

REFERENCES

1. **Ahmad, T.**, Changes in the process of ageing. II, *Bangladesh Med. Res. Counc. Bull.*, 4, 49, 1978.
2. **Allen, M. A., Andrew, J., and Brook, M.**, A sex difference in leukocyte vitamin c status in the elderly, *Nutrition*, 21, 136, 1969.
3. **Altman, H., Mehta, D., Evenson, R. C., and Sletten, I. W.**, Behavioral effects of drug therapy on psychogeriatric inpatients, *J. Am. Geriatr. Soc.*, 21, 249, 1973.
4. **Anderson, R., Oosthuizen, R., Maritz, R., Theron, A., and Van Rensburg, A. J.**, The effects of increasing weekly doses of ascorbate on certain cellular and humoral immune functions in normal volunteers, *Am. J. Clin. Nutr.*, 33, 71, 1980.
5. **Andrews, C. T. and Wilson, T. S.**, Vitamin C and thrombotic episodes, *Lancet*, 2, 39, 1973.
6. **Andrews, J. and Brook, M.**, Leucocyte-vitamin-C content and clinical signs in the elderly, *Lancet*, 1, 1350, 1966.
7. **Andrews, J., Brook M., and Allen, M. A.**, Influence of abode and season on the vitamin C status of the elderly, *Gerontol. Clin.*, 8, 257, 1966.
8. **Andrews, J., Letcher, M., and Brook, M.**, Vitamin C supplementation in the elderly: a 17-month trial in an old persons' home, *Br. Med. J.*, 2, 416, 1969.
8a. **Andrews, J., Atkinson, S. J., Ridge, B. D., and Wyn-Jones, C.**, A comparison of vitamin C status of elderly in-patients with that of elderly patients, *Proc. Nutr. Soc.*, 32, 45A, 1973.
9. **Anthony, H. M. and Schorah, C. J.**, Severe hypovitaminosis C in lung-cancer patients: the utilization of vitamin C in surgical repair and lymphocyte-related host resistance, *Br. J. Cancer*, 46, 354, 1982.
10. **Arthur, G., Monro, J. A., Poore, P., Rilwan, W. B., and Murphy, E. L.**, Trial of ascorbic acid in purpura and sublingual haemorrhages, *Br. Med. J.*, 67, 732, 1967.
11. **Asregadoo, E. R.**, Blood levels of thiamine and ascorbic acid in chronic open-angle glaucoma, *Ann. Ophthalmol.*, 11, 1095, 1979.
12. **Back, D. J., Breckenridge, A. M., MacIver, M., Orme, M. L., Purba, H., and Rowe, P. H.**, Interaction of ethinyloestradiol with ascorbic acid in man, *Br. Med. J.*, 282, 1516, 1981.
13. **Baker, H., Frank, O., Thind, I. S., Jaslow, S. P., and Louria, D. B.**, Vitamin profiles in elderly persons living at home or in nursing homes, versus profile in healthy young subjects, *J. Am. Geriatr. Soc.*, 27, 444, 1979.
14. **Banerjee, A. K. and Etherington, M.**, Platelet function in elderly scorbutics, *Age Ageing*, 3, 97, 1974.
15. **Banerjee, A. K., Lane, P. J., and Meichen, F. W.**, Vitamin C and osteoporosis in old age, *Age Ageing*, 7, 16, 1978.
16. **Barton, G. M. G., Laing, J. E., and Barisoni, D.**, The effect of burning on leucocyte ascorbic acid and the ascorbic acid content of burned skin, *Int. J. Vitam. Nutr. Res.*, 42, 524, 1972.
17. **Basu, T. K.**, Possible toxicological aspects of megadoses of ascorbic acid, *Chem. Biol. Interact.*, 16, 247, 1977.
18. **Basu, T. K.**, Possible role of vitamin C in cancer therapy, *Int. J. Vitam. Nutr. Res.*, (suppl. 19), 95, 1979.
19. **Basu, T. K.**, Vitamin C-aspirin interactions, *Int. J. Vitam. Nutr. Res.*, 23, 83, 1982.
20. **Basu, T. K., Jordan, S. J., Jenner, M., and Williams, D. C.**, Blood values of some vitamins in long-stay psycho-geriatric patients, *Int. J. Vitam. Nutr. Res.*, 46, 61, 1976.
21. **Basu, T. K., Raven, R. W., Dickerson, J. W. T., and Williams, D. C.**, Leucocyte ascorbic acid and urinary hydroxyproline levels in patients bearing breast cancer with skeletal metastases, *Eur. J. Cancer*, 10, 507, 1974.
22. **Basu, T. K., Smethurst, M., Gillett, M. B., Donaldson, D., Jordan, S. J., Williams, D. C., and Hicklin, J. A.**, Ascorbic acid therapy for the relief of bone pain in Paget's disease, *Acta Vitaminol. Enzymol.*, 32, 45, 1978.
23. **Batata, M., Spray, G. H., Bolton, F. G., Higgins, G., and Wollner, L.**, Blood and bone marrow changes in elderly patients, with special reference to folic acid, vitamin B12, iron, and ascorbic acid, *Br. Med. J.*, 2, 667, 1967.
24. **Bates, C. J.**, Proline and hydroxyproline excretion and vitamin C status in elderly human subjects, *Clin. Sci. Mol. Med.*, 52, 535, 1977.

24a. **Bates, C. J., Rutishauser, I. H., Black, A. E., Paul, A. A., Mandal, A. R., and Patnaik, B. K.,** Long-term vitamin status and dietary intake of healthy elderly subjects. II. Vitamin C, *Br. J. Nutr.*, 42, 43, 1979.

25. **Bates, C. J., Fleming, M., Paul, A. A., Black, A. E., and Mandal, A. R.,** Folate status and its relation to vitamin C in healthy elderly men and women, *Age Ageing*, 9, 241, 1980.

26. **Bates, J. F., Hughes, R. E., and Hurley, R. J.,** Ascorbic acid status in man: measurements of salivary, plasma and white blood cell concentration, *Arch. Oral Biol.*, 17, 1017, 1972.

27. **Bermond, P.,** Letter: clinical symptoms of malnutrition and plasma ascorbic acid levels, *Am. J. Clin. Nutr.*, 29, 493, 1976.

28. **Booth, J. B. and Todd, G. B.,** Subclinical scurvy — hypovitaminosis C, *Geriatrics*, 27, 130, 1972.

29. **Bordia, A. K.,** The effect of vitamin C on blood lipids, fibrinolytic activity and platelet adhesiveness in patients with coronary artery disease, *Atherosclerosis*, 35, 181, 1980.

30. **Bowers, E. F. and Kubik, M. M.,** Vitamin C levels in old people and the response to ascorbic acid and to the juice of the acerola (Malphighia punctifolia), *Br. J. Clin. Pract.*, 19, 141, 1965.

31. **Bramkamp, K. and Wirths, W.,** Sex specific vitamin C status of whole blood in elderly humans, *Int. J. Vitam. Nutr. Res.*, 43, 479, 1973.

32. **Brin, M., Dibble, M. V., Peel, A., McMullen, E., Bourquin, A., and Chen, N.,** Some preliminary findings on the nutritional status of the aged in Onondaga County, New York, *Am. J. Clin. Nutr.*, 17, 240, 1965.

33. **Brocklehurst, J. C., Griffiths, L. C., Taylor, G. F., Marks, J., Scott, D., and Blackley, J.,** The clinical features of chronic vitamin deficiency. A therapeutic trial in geriatric hospital patients, *Gerontol. Clin.*, 10, 309, 1968.

34. **Brocklehurst, J. C. and Griffiths, L. L.,** Vitamin C and the elderly, *Br. Med. J.*, 2, 824, 1969.

35. **Brook, M. and Grimshaw, J. J.,** Vitamin C concentration of plasma and leukocytes as related to smoking habit, age, and sex of humans, *Am. J. Clin. Nutr.*, 21, 1254, 1968.

36. **Burr, M. L., Elwood, P. C., Hole, D. J., Hurley, R. J., and Hughes, R. E.,** Plasma and leukocyte ascorbic acid levels in the elderly, *Am. J. Clin. Nutr.*, 27, 144, 1974.

37. **Burr, M. L., Milbank, J. E., and Gibbs, D.,** The nutritional status of the elderly, *Age Ageing*, 11, 89, 1982.

38. **Burr, M. L., Sweetnam, P. M., Hurley, R. J., and Powell, G. H.,** Effects of age and intake on plasma-ascorbic-acid levels, *Lancet*, 1, 163, 1974.

39. **Cameron, E. and Campbell, A.,** The orthomolecular treatment of cancer. II. Clinical trial of high-dose ascorbic acid supplements in advanced human cancer, *Chem. Biol. Interact.*, 9, 285, 1974.

40. **Cameron, E. and Pauling, L.,** The orthomolecular treatment of cancer. I. The role of ascorbic acid in host resistance, *Chem. Biol. Interact.*, 9, 273, 1974.

41. **Cameron, E., and Pauling, L.,** Supplemental ascorbate in the supportive treatment of cancer: prolongation of survival times in terminal human cancer, *Proc. Natl. Acad. Sci. U.S.A.*, 73, 3685, 1976.

42. **Cameron, E., and Pauling, L.,** Supplemental ascorbate in the supportive treatment of cancer: reevaluation of prolongation of survival times in terminal human cancer, *Proc. Natl. Acad. Sci. U.S.A.*, 75, 4538, 1978.

43. **Cameron, E. and Pauling, L.,** Cancer and vitamin C, The Linus Pauling Institute of Science and Medicine, Menlo Park, CA, 1979.

44. **Cameron, E., Pauling, L., and Leibovitz, B.,** Ascorbic acid and cancer: a review, *Cancer Res.*, 39, 663, 1979.

45. **Campbell, A. and Jack, T.,** Acute reactions to mega ascorbic acid therapy in malignant disease, *Scott. Med. J.*, 24, 151, 1979.

46. **Chamberlain, D. T., and Perkin, H. J.,** The level of ascorbic acid in the blood and urine of patients with peptic ulcer, *Am. J. Drig. Dis.*, 5, 493, 1938.

47. **Chlud, K.,** Treatment of osteoporosis with a delayed-action sodium fluoride preparation, *Z. Rheumatol.*, 36, 126, 1977.

48. **Chope, H. D. and Dray, S.,** The nutritional status of the aging, *Calif. Med.*, 74, 105, 1951.

49. **Cohen, M. M. and Duncan, A. M.,** Ascorbic acid nutrition in gastroduodenal disorders, *Br. Med. J.*, 4, 516, 1967.

50. **Connelly, T. J., Becker, A., and McDonald, J. W.,** Bachelor scurvy, *Int. J. Dermatol.*, 21, 209, 1982.

51. **Corberand, J., Nguyen, F., Fraysse, B., and Enjalbert, L.,** Malignant external otitis and polymorphonuclear leukocyte migration impairment. Improvement with ascorbic acid, *Arch. Otolaryngol.*, 108, 122, 1982.

52. **Cox, E. V.,** The anemia of scurvy, *Vitam. Horm.*, 26, 635, 1968.

53. **Cox, E. V., Meynell, M. J., Northam, B. E., and Cook, W. T.,** The anaemia of scurvy, *Am. J. Med.*, 42, 220, 1967.

54. **Creagan, E. T., Moertel, C. G., O'Fallon, J. R., Schutt, A. J., O'Connell, M. J., Rubin, J., and Frytak, S.,** Failure of high-dose vitamin C (ascorbic acid) therapy to benefit patients with advanced cancer. A controlled trial, *N. Engl. J. Med.*, 301, 687, 1979.

55. **Croft, D. N.,** Aspirin, vitamin-C deficiency, and gastric haemorrhage, *Lancet*, 2, 831, 1968.
56. **Dalderup, L. M., van Dam, B. E., Schiedt, K., Keller, G. H., and Schouten, F.,** Intake of vitamins and some other nutrients in aged people, adults and children, *Int. Z. Vitaminforsch.*, 40, 553, 1970.
57. **Daniels, A. L. and Everson, G. J.,** Influence of acetylsalicylic acid on urinary excretion of ascorbic acid, *Proc. Soc. Exp. Biol. Med.*, 35, 20, 1936.
58. **Davidson, C. S., Livermore, J., Anderson, P., and Kautman, S.,** The nutrition of a group of apparently healthy aging persons, *Am. J. Clin. Nutr.*, 10, 181, 1962.
59. **Dayton, P. G. and Weiner, M.,** Ascorbic acid and blood coagulation, *Ann. N.Y. Acad. Sci.*, 92, 302, 1981.
60. **DeChatelet, L. R., McCall, C. E., Cooper, M. R., and Shirley, P. S.,** Ascorbic acid levels in phagocytic cells, *Proc. Soc. Exp. Biol. Med.*, 145, 1170, 1974.
61. **Denson, K. W. and Bowers, E. F.,** A comparison of W.B.C. ascorbic acid and phenolic acid excretion in elderly patients, *Clin. Sci.*, 21, 157, 1961.
62. **Dibble, M. V., Brin, M., Thiele, V. F., Peel, A., Chen, N., and McMullen, E.,** Evaluation of the nutritional status of elderly subjects, with a comparison between fall and spring, *J. Am. Geriatr. Soc.*, 15, 1031, 1967.
63. **Dodds, M. L.,** Sex as a factor in blood levels of ascorbic acid, *J. Am. Diet. Assoc.*, 54, 32, 1969.
64. **Dubin, B., MacLennan, W. J., and Hamilton, J. C.,** Adrenal function and ascorbic acid concentrations in elderly women, *Gerontology*, 24, 473, 1978.
65. **Durand, C. H., Audinot, M., and Frajdenrajch, S.,** Latent hypovitaminosis C and tobacco, *Concours Med.*, 84, 4801, 1962.
66. **Dymock, I. W.,** Ascorbic acid status of healthy old people, *Int. Z. Vitaminforsch.*, 40, 555, 1970.
67. **Dymock, S. M. and Brocklehurst, J. C.,** Clinical effects of water soluble vitamin supplementation in geriatric patients, *Age Ageing*, 2, 172, 1973.
68. **Dysken, M. W., Cumming, R. J., Channon, R. A., and Davis, J. M.,** Drug interaction between ascorbic acid and fluphenazine (letter), *JAMA*, 241, 2008, 1979.
69. **Eddy, T. P. and Taylor, G. F.,** Sublingual varicosities and vitamin C in elderly vegetarians, *Age Ageing*, 6, 6, 1977.
69a. **Eddy, T. P.,** Difficulties in the clinical assessment of vitamin C therapy, Int. J. Vitam. Nutr. Res., Suppl. 19, 103, 1979.
70. **Elwood, P. C., Burr, M. L., Hole, D., Harrison, A., Morris, T. K., Wilson, C. I., Richardson, R. W., and Shinton, N. K.,** Nutritional state of elderly Asian and English subjects in Coventry, *Lancet*, 1, 1224, 1972.
71. **Elwood, P. C., Hughes, R. E., and Hurley, R. J.,** Ascorbic acid and serum cholesterol, *Lancet*, 2, 1197, 1970.
72. **Enstrom, J. E., and Pauling, L.,** Mortality among healthy-conscious elderly Californians, *Proc. Natl. Acad. Sci. U.S.A.*, 79, 6023, 1982.
73. **Exton-Smith, A. N. and Stanton, B. R.,** Report of an investigation into the diet of elderly women living alone, King Edward's Hospital Fund for London, 1965.
74. **Feetam, C. L., Leach, R. H., and Meynell, M. J.,** Lack of a clinically important interaction between warfarin and ascorbic acid, *Toxicol. Appl. Pharmacol.*, 31, 544, 1975.
74a. **Feller, R. P., Black, H. S., and Shannon, I. L.,** Evidence for absence of ascorbic acid in human saliva, *Arch. Oral Biol.*, 20, 563, 1975.
75. **Fidanza, A., Audisio, M., and Mastroiacovo, P.,** Vitamin C and cholesterol, *Int. J. Vitam. Nutr. Res.*, Suppl. 23, 153, 1982.
76. **Fidanza, A., Floridi, S., Martinoli, L., Mastroiacovo, P., Servi, M., DiVirgilio, D., and Ravallese, F.,** Therapeutic action of vitamin C on cholesterol metabolism, *Boll. Soc. Ital. Biol. Sper.*, 55, 553, 1979.
77. **Fishbein, S. L. and Goodstein, S.,** The pressure lowering effect of ascorbic acid, *Ann. Ophthalmol.*, 4, 487, 1972.
78. **Freeman, J. T. and Hafkesbring, R.,** Comparative study of ascorbic acid levels in gastric secretion, blood, urine and saliva, *Gastroenterology*, 18, 224, 1951.
79. **Gander, J. and Niedenberger, W.,** The vitamin C requirements of old people, *Muench. Med. Wochenschr.*, 83, 1386, 1936.
80. **Garry, P. J., Goodwin, J. S., Hunt, W. C., Hooper, E. M., and Leonard, A. G.,** Nutritional status in a healthy elderly population: dietary and supplemental intakes, *Am. J. Clin. Nutr.*, 36, 319, 1982.
81. **Garry, P. J., Goodwin, J. S., Hunt, W. C., and Gilbert, B. A.,** Nutritional status in a healthy elderly population: vitamin C, *Am. J. Clin. Nutr.*, 36, 332, 1982.
82. **Ginter, E.,** Ascorbic acid in cholesterol and bile acid metabolism, *Ann. N.Y. Acad. Sci.*, 258, 410, 1975.
83. **Ginter, E.,** Vitamin C and cholesterol, *Int. J. Vitam. Nutr. Res.*, Suppl. 16, 53, 1977.
84. **Ginter, E.,** Marginal vitamin C deficiency, lipid metabolism, and athrogenesis, *Adv. Lipid Res.*, 16, 167, 1978.

85. **Ginter, E.,** Decline of coronary mortality in United States and vitamin C, *Am. J. Clin. Nutr.,* 32, 511, 1979.

86. **Ginter, E., Bobek, P., Babala, J., Jakubovsky, J., Zaviacic, M., and Lojda, Z.,** Vitamin C in atherosclerosis, *Int. J. Vitam. Nutr. Res.,* Suppl. 19, 55, 1979.

87. **Ginter, E., Bobek, P., Kubec, F., Vozar, J., and Urbaova, D.,** Vitamin C in the control of hypercholesterolemia in man, *Int. J. Vitam. Nutr. Res.,* Suppl. 23, 137b, 1982.

88. **Ginter, E., Cern'a, O., Budlovsk'y, J., Bal'az, V., Hrub'a, F., Roch, V., and Sasko, E.,** Effect of ascorbic acid on plasma cholesterol in humans in a long-term experiment, *Int. J. Vitam. Nutr. Res.,* 47, 123, 1977.

89. **Ginter, E., Kajaba, I., and Nizner, O.,** The effect of ascorbic acid on cholesterolemia in healthy subjects with seasonal deficit of vitamin C, *Nutr. Metab.,* 2, 76, 1970.

90. **Ginter, E. and Vejmolova, J.,** Vitamin C-status and pharmacokinetic profile of antipyrine in man (letter), *Br. J. Clin. Pharmacol.,* 12, 256, 1981.

91. **Ginter, E., Zdichynec, B., Holzerova, O., Ticha, E., Kobza, R., Koziakova, M., Cerna, O., Ozdin, L., Hruba, F., Novakova, V., Sasko, E., and Gaher, M.,** Hypocholesterolemic effect of ascorbic acid in maturity-onset diabetes mellitus, *Int. J. Vitam. Nutr. Res.,* 48, 368, 1978.

92. **Goetzl, E. J., Wasserman, S. I., Gigli, I., and Austen, K. F.,** Enhancement of random migration and chemotactic response of human leukocytes by ascorbic acid, *J. Clin. Invest.,* 53, 813, 1974.

93. **Goldberg, A.,** The enzymic formation of haem by the incorporation of iron into protoporphyrin, *Br. J. Haematol.,* 5, 150, 1959.

94. **Goldsmith, G. A.,** Chemical measurements in relation to physical evidence of malnutrition, *Fed. Proc.,* 8, 553, 1949.

94a. **Goodwin, J. S., Goodwin, J. M., and Garry, P. J.,** Association between nutritional status and cognitive functioning in a healthy elderly population, *JAMA,* 249, 2917, 1983.

95. **Grant, F. W., Cowen, M. A., Ozerengin, M. F., and Bigelow, N.,** Nutritional requirements in mental illness. I. Ascorbic acid retention in schizophrenia. A reexamination, *Biol. Psychiatry,* 5, 289, 1973.

96. **Greco, A. M., Gentile, M., DiFilippo, O., and Coppola, A.,** Study of blood vitamin C in lung and bladder cancer patients before and after treatment with ascorbic acid. A preliminary report, *Acta Vitaminol. Enzymol.,* 4, 155, 1982.

97. **Greco, A. M., and LaRocca, L.,** Correlation between chronic hypovitaminosis C in old age and plasma levels of cholesterol and triglycerides, *Int. J. Vitam. Nutr. Res.,* Suppl 23, 129, 1982.

98. **Griffiths, L. L.,** *Vitamins and the Elderly,* Exton-Smith, A. N. and Scott, D. L., Eds., Briston-Wright, England, 1968, 34.

99. **Griffiths, L. L., Brocklehurst, J. C., Scott, D. L., Marks, J., and Blackley, J.,** Thiamine and ascorbic acid levels in the elderly, *Gerontol. Clin.,* 9, 1, 1967.

100. **Griffiths, L. L., Brocklehurst, J. C., MacLean, R., and Fry, J.,** Diet in old age, *Br. Med. J.,* 1, 739, 1966.

101. **Hagtvet, S.,** Serum ascorbic acid values in healthy Norwegians from 1940-1942, *Nord. Med.,* 28, 2335, 1945.

102. **Hale, W. E., Stewart, R. B., Cerda, J. J., Marks, R. G., May, F. E.,** Use of nutritional supplements in an ambulatory elderly population, *J. Am. Geriatr. Soc.,* 30, 401, 1982.

103. **Hallgren, H. M., and Yunis, E. J.,** Suppressor lymphocytes in young and aged humans, *J. Immunol.,* 118, 2004, 1977.

104. **Hanck, A., and Weiser, H.,** Vitamin C and lipid metabolism, *Int. J. Vitam. Nutr. Res.,* Suppl. 16, 67, 1977.

105. **Hankin, J. H. and Antonmatter, J. C.,** Survey of food service practices in nursing homes, *Am. J. Public Health Natl. Health,* 50, 1137, 1960.

106. **Harper, A. E.,** The recommended dietary allowances for ascorbic acid, *Ann. N.Y. Acad. Sci.,* 258, 491, 1975.

107. **Harrill, I., and Cervone, N.,** Vitamin status of older women, *Am. J. Clin. Nutr.,* 30, 431, 1977.

108. **Harrill, I., Schutz, H. G., Standal, B. R., Day, M.-L., and Yearick, E. S.,** Nutritional status studies in the western region: selected ethnic and elderly groups, *Nutr. Rep. Int.,* 25, 189, 1982.

109. **Hashimoto, K., Kitabchi, A. E., Duckworth, W. C., and Robinson, N.,** Ultrastructure of scorbutic human skin, *Acta Derm. Venereol.,* 50, 9, 1970.

110. **Hayes, K. C. and Hysted, D. M.,** Toxicity of the vitamins, in *Toxicants Occurring Naturally in Food,* 2nd ed., National Academy of Sciences, Washington, D.C., 1973, 235.

111. **Hefton, J. M., Darlington, G. J., Casazza, B. A., and Weksler, M. C.,** Immunologic studies of aging. V. Impaired proliferation of PHA responsive human lymphocytes in culture, *J. Immunol.,* 125, 1007, 1980.

112. **Heine, H. and Norden, C.,** Vitamin C therapy in hyperlipoproteinemia, *Int. J. Vitam. Nutr. Res.,* Suppl. 19, 45, 1979.

113. **Herbert, V. and Jacob, E.,** Destruction of vitamin B_{12} by ascorbic acid, *JAMA,* 230, 241, 1974.

114. **Hill, G. L., Blackett, R. L., Pickford, I., Borkinshaw, L., Young, G. A., Warren, J. V., Schorah, C. J., and Morgan, D. B.,** Malnutrition in surgical patients, *Lancet,* 1, 689, 1977.

115. **Hodges, R. E.,** Vitamin C and cancer, *Nutr. Rev.,* 40, 289, 1982.

116. **Hodges, R. E., Hood, J., Canham, J. E., Sauberlich, H. E., and Baker, E. M.,** Clinical manifestations of ascorbic acid deficiency in man, *Am. J. Clin. Nutr.,* 24, 432, 1971.

117. **Hoffmann, W. S. and Dyniewicz, H. A.,** The effect of alumina gel upon the absorption of amino acids, ascorbic acid, and neutral fat from the intestinal tract, *Gastroenterol.,* 6, 50, 1946.

118. **Holley, H. L. and McLester, J. S.,** Manifestations of ascorbic acid deficiency after prolonged corticotropin administration, *Arch. Int. Med.,* 88, 760, 1951.

119. **Horsey, J., Livesley, B., and Dickerson, J. W.,** Ischaemic heart disease and aged patients: effects of ascorbic acid on lipoproteins, *J. Hum. Nutr.,* 35, 53, 1981.

120. **Houston, J. B.,** Effect of vitamin C supplement on antipyrine disposition in man, *Br. J. Clin. Pharmacol.,* 4, 236, 1977.

121. **Houston, J. B., and Levy, G.,** Modification of drug biotransformation by vitamin C in man, *Nature (London),* 255, 78, 1975.

122. **Houston, J. B. and Levy, G.,** Drug biotransformation interactions in man. VI. Acetaminophen and ascorbic acid, *J. Pharm. Sci.,* 65, 1218, 1976.

123. **Hsu, J. M.,** The current status of ascorbic acid, vitamin B_6, folic acid and vitamin B_{12} in the elderly, in *Handbook of Geriatric Nutrition,* Hsu, J. M. and Davis, R. L., Eds., 1981, 128.

123a. **Hughes, R. E.,** Assessment of vitamin C status, *Proc. Nutr. Soc.,* 32, 243, 1973.

124. **Hume, R., Vallance, B. D., and Muir, M. M.,** Ascorbate status and fibrinogen concentrations after cerebrovascular accident, *J. Clin. Pathol.,* 35, 195, 1982.

125. **Hume, R., Weyers, E., Rowan, T., Reid. D. S., and Hillis, W. S.,** Leucocyte ascorbic acid levels after acute myocardial infarction, *Br. Heart J.,* 34, 238, 1972.

126. **Hyams, D. E. and Ross, E. J.,** Scurvy, megaloblastic anaemia and osteoporosis, *Br. J. Clin. Pract.,* 17, 332, 1963.

127. **Inkeles, B., Innes, J. B., Kuntz, M. M., Kadish, A. S., and Weskler, M. E.,** Immunological studies of aging. III. cytokinetic basis for the impaired response of lymphocytes from aged humans to plant lectins, *J. Exp. Med.,* 145, 1176, 1977.

128. **Irwin, M. A. and Hutchins, B. K.,** A conspectus of research on vitamin C requirements of man, *J. Nutr.,* 106, 821, 1976.

128a. **Irwin, M. A. and Hutchins, B. K.,** A conspectus of research on vitamin C requirements of man, *J. Nutr.,* 106, 821, 1976.

129. **Jacobs, A., Greenman, D., Owens, E., and Covilli, I.,** Ascorbic acid status in iron-deficiency anemia, *J. Clin. Pathol.,* 24, 694, 1971.

130. **Jordan, M., Kyses, M., Hayes, R., and Hammond, W.,** Dietary habits of persons living alone, *Geriatrics,* 9, 230, 1954.

131. **Karkar, S. and Wilson, C. W.,** Ascorbic acid metabolism in human cancer, *Proc. Nutr. Soc.,* 33, 110A, 1974.

132. **Karkar, S. C. and Wilson, C. W. M.,** Ascorbic acid values in malignant disease, *Proc. Nutr. Soc.,* 35, 9A, 1976.

132a. **Karkar, S. C., Wilson, C. W., and Moriarty, M. J.,** The relationship between cancer, radiotherapy and vitamin C, *Ir. J. Med. Sci.,* 146, 289, 1977.

133. **Kataria, M. S., Rao, D. B., and Curtis, R. C.,** Vitamin C levels in the elderly, *Gerontol. Clin.,* 7, 189, 1965.

134. **Kien, L. J. and McDermott, F. T.,** Screening ascorbic acid deficiency by a lingual test, *Med. J. Aust.,* 2, 420, 1972.

135. **Kinsman, R. A. and Hood, J.,** Some behavioral effects of ascorbic acid deficiency, *Am. J. Clin. Nutr.,* 24, 455, 1971.

136. **Kirchmann, L. L.,** The importance of vitamin C in clinical medicine, *Ergebn. Inn. Med. Kinderheelk.,* 56, 101, 1930.

137. **Kirk, J. E.,** Blood and urine vitamin levels in the aged, in *Symp. Problems Gerontology,* Series 9, National Vitamin Foundation Inc., N.Y., 1954, 73.

138. **Kirk, J. E. and Chieffi, M.,** Vitamin studies in middle-aged and old individuals, XI. The concentration of total ascorbic acid in whole blood, *J. Gerontol.,* 8, 301, 1953a.

139. **Kirk, J. E. and Chieffi, M.,** Vitamin studies in middle-aged and old individuals. XII. Hypovitaminemia C, effect of ascorbic acid administration on the blood ascorbic acid concentration, *J. Gerontol.,* 8, 305, 1953.

140. **Kitabchi, A. E. and Duckworth, W. C.,** Pituitary adrenal axis evaluation in human scurvy, *Am. J. Clin. Nutr.,* 23, 1012, 1970.

141. **Kohrs, M. B.,** Evaluation of nutrition programs for the elderly, *Am. J. Clin. Nutr.,* 36 (Suppl.), 812, 1982.

142. **Kohrs, M. B., O'Neal, R., Preston, A., Eklund, D., and Abrahams, O.,** Nutritional status of elderly residents in Missouri, *Am. J. Clin. Nutr.,* 31, 2186, 1978.

143. **Krasner, N. and Dymock, I. W.,** Ascorbic acid deficiency in malignant diseases: a clinical and biochemical study, *Br. J. Cancer,* 30, 142, 1974.

144. **Krumdieck, C. and Butterworth, C. E., Jr.,** Ascorbate — cholesterol — lecithin interactions: factors of potential importance in the pathogenesis of atherosclerosis, *Am. J. Clin. Nutr.,* 27, 866, 1974.

145. **La Rocca, L., Greco, A. M., D'aponte, D., and Bozza, P.,** Metabolic correlations between blood ascorbic acid, cholesterol and triglycerides in the aged, *Boll. Soc. Ital. Biol. Sper.,* 53, 2344, 1977.

146. **LeBovit, C.,** The food of older persons living alone, *J. Am. Diet. Assoc.,* 46, 285, 1965.

147. **Lee, P., Lam, K. W., and Lai, M.,** Aqueous humor ascorbate concentration and open-angle glaucoma, *Arch. Ophthalmol.,* 96, 308, 1977.

148. **Lemoine, A., LeDevehat, C., Cadaccioni, J. L., Monges, A., Bermond, P., and Salkeld, R. M.,** Vitamin B_1, B_2, B_6 and C status in hospital inpatients, *Am. J. Clin. Nutr.,* 33, 2595, 1980.

149. **Leung, F. W. and Guze, P. A.,** Adult scurvy, *Ann. Emerg. Med.,* 10, 652, 1981.

150. **Linner, E.,** Intraocular pressure regulation and ascorbic acid, *Acta. Soc. Med. UPS,* 69, 225, 1964.

151. **Loh, H. S.,** The relationship between dietary ascorbic acid intake and buffy coat and plasma ascorbic acid concentrations at different ages, *Int. J. Vitam. Nutr. Res.,* 42, 80, 1972.

152. **Loh, H. S. and Dempsey, P. M.,** Ascorbic acid, folic acid inter-relationships in geriatric subjects, *J. Ir. Med. Assoc.,* 67, 247, 1974.

153. **Loh, H. S., Odumosu, A., and Wilson, C. W. M.,** Factors influencing the metabolic availability of ascorbic acid. I. The effect of sex, *Clin. Pharmacol. Therap.,* 16, 390, 1974.

154. **Loh, H. S. and Wilson, C. W.,** Iron and vitamin C, *Lancet,* 2, 768, 1971.

155. **Loh, H. S. and Wilson, C. W.,** The relationship between leucocyte ascorbic acid and haemoglobin levels at different ages, *Int. J. Vitam. Nutr. Res.,* 41, 259, 1971.

156. **Loh, H. S. and Wilson, C. W.,** Relationship between leucocyte and plasma ascorbic acid concentrations, *Br. Med. J.,* 3, 733, 1971.

157. **Loh, H. S. and Wilson, C. W.,** Anemia in the elderly, *Br. Med. J.,* 4, 612, 1973.

158. **Loh, H. S. and Wilson, C. W.,** Vitamin C and thrombotic episodes, *Lancet,* 2, 317, 1973b.

159. **Lonergan, M. E., Milne, J. S., Maule, M. M., and Williamson, J.,** A dietary survey of older people in Edinburgh, *Br. J. Nutr.,* 34, 517, 1975.

160. **Lowry, O. H., Bessey, O. A., Brock, M. J., and Lopez, J. A.,** The interrelationship of dietary serum, white blood cell and total body ascorbic acid, *J. Biol. Chem.,* 166, 111, 1946.

161. **Lynch, S. R., Berelowitz, I., Seftel, H. C., Miller, G. B., Krawitz, P., Charlton, R. W., and Bothwell, T. H.,** Osteoporosis in Johannesburg Bantu males; its relationship to siderosis and ascorbic acid deficiency, *Am. J. Clin. Nutr.,* 20, 799, 1967.

162. **Lynch, S. R. and Cook, J. D.,** Interaction of vitamin C and iron, *Ann. N.Y. Acad. Sci.,* 355, 32, 1980.

163. **Lyons, J. C., and Trulson, M. F.,** Food practices of older people living at home, *J. Gerontol.,* 11, 66, 1956.

164. **MacCormack, W. J.,** Ascorbic acid as chemotherapeutic agent, *Arch. Pediat.,* 69, 151, 1952.

165. **MacLennan, W. J. and Hamilton, J. C.,** The effect of acute illness on leucocyte and plasma ascorbic acid levels, *Br. J. Nutr.,* 38, 217, 1977.

166. **MacLeod, C. C., Judge, T. G., and Caird, F. I.,** Nutrition of the elderly at home II. Intakes of vitamins, *Age Ageing,* 3, 209, 1974.

167. **Makila, E.,** The vitamin status of elderly denture wearers, *Int. Z. Vitaminforsch.,* 40, 81, 1970.

168. **Makinodan, T. and Kay, M. M. B.,** Age influence on the immune system, *Adv. Immunol.,* 29, 287, 1980.

169. **McClean, H. E., Dodds, P. M., Abernethy, M. H., Stewart, A. W., and Beaven, D. W.,** Vitamin C concentration in plasma and leucocytes of men related to age and smoking habit, *N.Z. Med. J.,* 83, 226, 1976.

170. **McClean, H. E., Dodds, P. M., Stewart, A. W., Beaven, D. W., and Riley, C. G.,** Nutrition of elderly men living alone. II. Vitamin C and thiamine status, *N.Z. Med. J.,* 84, 345, 1976.

171. **McClean, H. E., Stewart, A. W., Riley, C. G., and Beaven, D. W.,** Vitamin C status of elderly men in a residential home, *N.Z. Med. J.,* 86, 379, 1977.

172. **Mengel, C. E. and Greene, H. L., Jr.,** Ascorbic acid effects on erythrocytes, *Ann. Int. Med.,* 84, 490, 1976.

173. **Milne, J. S., Lonergan, M. E., Williamson, J., Moore, F. M., McMaster, R., and Percy, N.,** Leucocyte ascorbic acid levels and vitamin C intake in older people, *Br. Med. J.,* 4, 383, 1971.

174. **Milner, G.,** Ascorbic acid in chronic psychiatric patients — a controlled trial, *Br. J. Psychiat.,* 109, 294, 1963.

175. **Mitra, M. L.,** Vitamin C deficiency in the elderly and its manifestations, *J. Am. Geriatr. Soc.,* 18, 67, 1970.

176. **Mitra, M. L.,** Confusional states in relation to vitamin deficiencies in the elderly, *J. Am. Geriatr. Soc.,* 19, 536, 1971.
177. **Morgan, A. F., Gillum, H. L., and Williams, R. I.,** Nutritional status of the aging. III. Serum ascorbic acid and intake, *J. Nutr.,* 55, 431, 1955.
178. **Morgan, A. G., Kelleher, J., Walker, B. E., Losowsky, M. S., Droller, H., and Middleton, R. S.,** A nutritional survey in the elderly: blood and urine vitamin levels, *Int. J. Nutr. Res.,* 45, 448, 1975.
179. **Moriarty, M. J., Mulgrew, S., Malone, J. R., and O'Connor, M. K.,** Results and analysis of tumour levels of ascorbic acid, *Ir. J. Med. Sci.,* 146, 74, 1977.
180. **Morse, E. H., Potgieter, M., and Walkers, G. R.,** Ascorbic acid utilizaton by women, *J. Nutr.,* 58, 291, 1956.
181. **Mullen, A., and Wilson, C. W.,** The metabolism of ascorbic acid in rheumatoid arthritis, *Proc. Nutr. Soc.,* 35, 8A, 1976.
182. **Murata, A., Morishige, F., and Yamaguchi, H.,** Prolongation of survival times of terminal cancer patients by administration of large doses of ascorbate, *Int. J. Vitam. Nutr. Res.,* Suppl. 23, 103, 1982.
183. **Myasnikov, A. L.,** Vitamins in the development and prophylaxis of atherosclerotic heart disease, *6th Int. Congr. Nutr.,* Edinburgh, Aug. 9-15, 1963; 1964, 123.
184. **O'Sullivan, D. J., Callaghan, N., Ferriss, J. B., Finucane, J. F., and Hegarty, M.,** Ascorbic acid deficiency in the elderly, *Ir. J. Med. Sci.,* 7, 151, 1968.
185. **Panush, R. S. and Delafuente, J. C.,** Modulation of certain immunologic responses by vitamin C, *Int. J. Vitam. Nutr. Res.,* Suppl. 19, 179, 1979.
186. **Pekkarinen, M. and Roine, P.,** Studies on the intake of vitamin C and thiamine among aged people in Finland, *Int. Z. Vitaminforsch.,* 40, 555, 1970.
187. **Pelletier, O.,** Vitamin ''C'' status of cigarette smokers and nonsmokers, *Am. J. Clin. Nutr.,* 23, 520, 1970.
188. **Rafsky, H. A. and Newman, B.,** Vitamin C studies in the aged, *Am. J. Med. Sci.,* 201, 749, 1941.
189. **Rafsky, H. A. and Newman, B.,** Nutritional aspects of aging, *Geriatrics,* 2, 101, 1947.
190. **Rafsky, H. A. and Newman, B.,** A quantitative study of diet in the aged, *Geriatrics,* 3, 267, 1948.
191. **Ramirez, J. and Flowers, N. C.,** Leukocyte ascorbic acid and its relationship to coronary artery disease in man, *Am. J. Clin. Nutr.,* 33, 2079, 1980.
192. **Ray, S. S. and Thomas, A.,** Intravenous glycerol-sodium ascorbate combination as osmotic agent to reduce intraocular pressure, *Indian J. Ophthalmol.,* 25, 27, 1977.
193. **Roderuck, C., Burrill, L., Campbell, L. J., Brakke, B. E., Childs, M. T., Leverton, R., Chaloupka, M., Jebe, E. H., and Swanson, P. P.,** Estimated dietary intake, urinary excretion and blood vitamin C in women of different ages, *J. Nutr.,* 66, 15, 1958.
194. **Roine, P., Koivula, L., Pekkarinen, M., and Rissanen, A.,** Vitamin C intake and plasma level among aged people in Finland, *Int. J. Vitam. Nutr. Res.,* 44, 95, 1974.
195. **Rosenthal, G.,** Interaction of ascorbic acid and warfarin, *JAMA,* 215, 1671, 1971.
196. **Russell, R. I. and Goldberg, A.,** Effect of aspirin on the gastric mucosa of guinea pigs on a scorbutogenic diet, *Lancet,* 2, 606, 1968.
197. **Russell, R. I., Williamson, J. M., Goldberg, A., and Wares, E.,** Ascorbic acid levels in leucocytes of patients with gastrointestinal haemorrhage, *Lancet,* 2, 603, 1968.
198. **Sahud, M. A. and Cohen, R. J.,** Effect of aspirin ingestion on ascorbic-acid levels in rheumatoid arthritis, *Lancet,* 1, 937, 1971.
199. **Saint, E. G., Abrecht, H. F., and Turner, C. N.,** Old age: a clinical, social, and nutritional survey of 70 patients over 65 years of age seen in a hospital out-patient department in Melbourne, *Med. J. Aust.,* 1, 757, 1953.
200. **Schaus, R.,** The ascorbic acid content of human pituitary cerebral cortex, heart and skeletal muscle and its relation to age, *Am. J. Clin. Nutr.,* 5, 39, 1957.
201. **Schlettwein-Gsell, D., Brubacher, G., and Vuilleumier, J. P.,** On the study of nutrient supply of food in a selected group of elderly women, *Int. Z. Vitaminforsch.,* 37, 515, 1967.
202. **Schlettwein-Gsell, D., Vuilleumier, J. P., and Brubacher, G.,** On the provision of a selected group of 12 elderly women with some nutrients via the daily diet, *Int. Z. Vitaminforsch.,* 38, 227, 1968.
203. **Schorah, C. J.,** An assessment of the prevalence and importance of vitamin C depletion in an urban population, *Int. J. Vitam. Nutr. Res.,* Suppl. 19, 167, 1979.
204. **Schorah, C. J.,** The level of vitamin C reserves required in man: towards a solution to the controversy, *Proc. Nutr. Soc.,* 40, 147, 1981.
205. **Schorah, C. J., Newill, A., Scott, D. L., and Morgan, D. B.,** Clinical effects of vitamin C in elderly inpatients with low blood-vitamin-C levels, *Lancet,* 1, 403, 1979.
206. **Schorah, C. J., Tormey, W. P., Brooks, G. H., Robertshaw, A. M., Young, G. A., Talukder, R., and Kelly, J. F.,** The effect of vitamin C supplements on body weight, serum proteins, and general health of an elderly population, *Am. J. Clin. Nutr.,* 34, 871, 1981.

207. **Schrauzer, G. N. and Rhead, W. J.,** Ascorbic acid abuse: effects of long-term ingestion of excessive amounts on blood levels and urinary excretion, *Int. J. Vitam. Nutr. Res.,* 43, 201, 1973.
208. **Scobie, B. A.,** Scurvy in the adult, *N. Z. Med. J.,* 70, 398, 1969.
209. **Setyaadmadja, A. T., Cheraskin, E., and Ringsdorf, W. M., Jr.,** Ascorbic acid and carbohydrate metabolism. I. The cortisone glucose tolerance test, *J. Am. Geriatr. Soc.,* 13, 924, 1965.
210. **Shah, K. V., Barbhaiya, H. C., and Skrinwasan, V.,** Ascorbic acid levels in blood during tetracycline administration, *J. Indian Med. Assoc.,* 51, 127, 1968.
211. **Siegel, B. V.,** Enhancement of interferon production by poly(rI)-poly(rC) in mouse cell cultures of ascorbic acid, *Nature (London),* 254, 531, 1975.
212. **Simonson, E. and Keys, A.,** Research in Russia on vitamins and atherosclerosis, *Circulation,* 24, 1239, 1961.
213. **Smethurst, M., Basu, T. K., Gillett, M. B., Donaldson, D., Jordan, S. J., Williams, D. C., and Hicklin, J. A.,** Combined therapy with ascorbic acid and calcitonin for the relief of bone pain in Paget's disease, *Acta Vitaminol. Enzymol.,* 3, 8, 1981.
214. **Smith, E. C., Skalski, R. I., Johnson, G. C., and Rossi, G. V.,** Interaction of ascorbic acid and warfarin, *JAMA,* 221, 1166, 1972.
214a. **Smith, W. B., Shohet, S. B., Zagajeski, E., and Lubin, B. H.,** Alteration in human granulocyte function after in vitro incubation with L-ascorbic acid, *Ann. N.Y. Acad. Sci.,* 258, 329, 1975.
215. **Smithard, D. J. and Langman, M. J.,** Vitamin C and drug metabolism (letter), *Br. Med. J.,* 1, 1029, 1977.
216. **Smithard, D. J. and Langman, M. J.,** The effect of vitamin supplementation upon antipyrine metabolism in the elderly, *Br. J. Clin. Pharmacol.,* 5, 181, 1978.
217. **Sokoloff, B., Hori, M., Saelhof, C., McConnell, B., and Imai, T.,** Effect of ascorbic acid on certain blood fat metabolism factors in animals and man, *J. Nutr.,* 91, 107, 1967.
218. **Sokoloff, B., Hori, M., Saelhof, C. C., Wrzolek, T., and Imai, T.,** Aging, atherosclerosis and ascorbic acid metabolism, *J. Am. Geriatr. Soc.,* 14, 1239, 1966.
219. **Spittle, C. R.,** Vitamin C and deep-vein thrombosis, *Lancet,* 2, 199, 1973.
220. **Staiano-C., L., Darzynkiewicz, Z., Hefton, J. M., Dutkowski, R., Darlington, G. J., and Weksler, M. E.,** Increased sensitivity of lymphocytes from people over 65 to cell cycle arrest and chromosomal damage, *Science,* 219, 1335, 1983.
221. **Stefanini, M. and Rosenthal, M. C.,** Hemorrhagic diathesis with ascorbic acid deficiency during administration of anterior pituitary corticotropic hormone (ACTH), *Proc. Soc. Exp. Biol. Med.,* 75, 806, 1950.
222. **Steinkamp, R. C., Cohen, N. L., and Walsh, H. E.,** Re-survey of an aging population: a 14-year follow up, *J. Am. Diet. Assoc.,* 46, 103, 1965.
223. **Szent-Györgi, A.,** Observations on the function of peroxidase system and the chemistry of the adrenal cortex. Description of a new carbohydrate derivative, *Biochem. J.,* 22, 1387, 1928.
224. **Tattersall, R. N. and Seville, R.,** Senile purpura, *J. Med.,* 19, 151, 1950.
225. **Taylor, G.,** Diet of elderly women, *Lancet,* 1, 926, 1966.
226. **Taylor, T. V., Raftery, A. T., Elder, J. B., Loveday, C., Dymock, I. W., Gibbs, A. C., Jeacock, J., Lucas, S. B., and Pell, M. A.,** Leucocyte ascorbate levels and postoperative deep venous thrombosis, *Br. J. Surg.,* 66, 583, 1979.
227. **Templeton, C. L.,** Nutrition counseling needs in a geriatric population, *Geriatrics,* 33, 59, 1978.
228. **Thomas, W. R. and Holt, P. G.,** Vitamin C and immunity: an assessment of the evidence, *Clin. Exp. Immunol.,* 32, 370, 1978.
229. **Trang, J. M., Blanchard, J., Conrad, K. A., and Harrison, G. G.,** The effect of vitamin C on the pharmacokinetics of caffeine in elderly men, *Am. J. Clin. Nutr.,* 35, 487, 1982.
230. **Turley, S. D., West, C. E., and Horton, B. J.,** The role of ascorbic acid in the regulation of cholesterol metabolism and in the pathogenesis of atherosclerosis, *Atherosclerosis,* 24, 1, 1976.
231. **Vallance, B. D., Hume, R., and Weyers, E.,** Reassessment of changes in leucocyte and serum ascorbic acid after acute myocardial infarction, *Br. Heart J.,* 40, 64, 1978.
232. **Vir, S. C. and Love, A. H.,** Vitamin C status of institutionalized and non-institutionalized aged, *Int. J. Vitam. Nutr. Res.,* 48, 274, 1978.
233. **Vir, S. C. and Love, A. H.,** Nutritional status of institutionalized and noninstitutionalized aged in Belfast, Northern Ireland, *Am. J. Clin. Nutr.,* 32, 1934, 1979.
234. **Virno, M., Bucci, M. G., Pecori-Giraldi, J., and Cantore, G.,** Oral treatment of glaucoma with vitamin C, *EENT Monthly,* 46, 1502, 1967.
235. **Walker, A.,** Chronic scurvy, *Br. J. Dermatol.,* 80, 625, 1968.
236. **Wapnick, A. A., Lynch, S. R., Seftel, H. C., Charlton, R. W., Bothwell, T. H., and Jowsey, J.,** The effect of siderosis and ascorbic acid depletion on bone metabolism with special reference to the osteoporosis in the Bantu, *Br. J. Nutr.,* 25, 367, 1971.
237. **Wcksler, M. E. and Huetterroth, T. H.,** Impaired lymphocyte function in aged humans, *J. Clin. Invest.,* 53, 99, 1974.

238. **Williamson, J. M., Goldberg, A., and Moore, F. M.,** Leucocyte ascorbic acid levels in patients with malabsorption or previous gastric surgery, *Br. Med. J.,* 2, 23, 1967.

238a. **Willis, G. C. and Fishman, S.,** Ascorbic acid content of human arterial tissue, *Can. Med. Assoc. J.,* 72, 500, 1955.

239. **Wilson, C. W.,** Ascorbic acid metabolism and the clinical factors which affect tissue saturation with ascorbic acid, *Acta Vitaminol. Enzymol.,* 31, 35, 1977.

240. **Wilson, C. W. M. and Nolan, C.,** The diets of elderly people in Dublin, *Ir. J. Med. Sci.,* 3, 345, 1970.

241. **Wilson, J. T., VanBoxtel, C. J., Alvan, G., and Sjoquist, F.,** Failure of vitamin C to affect the pharmacokinetic profile of antipyrine in man, *J. Clin. Pharmacol.,* 16, 265, 1976.

242. **Wilson, T. S., Weeks, M. M., Mukherjee, S. K., Murrell, J. S., and Andrews, C. T.,** A study of vitamin C levels in the aged and subsequent mortality, *Gerontol. Clin.,* 14, 17, 1972.

243. **Windsor, A. C. M., Hobbs, C. B., Treby, D. A., and Astley-Cowper, R.,** Effect of tetracycline on leucocyte ascorbic acid levels, *Br. Med. J.,* 1, 214, 1972.

244. **Windsor, A. C. and Williams, C. B.,** Urinary hydroxyproline in the elderly with low leucocyte ascorbic acid levels, *Br. Med. J.,* 1, 732, 1970.

245. **Wirths, W.,** Supply of thiamine and ascorbic acid investigated in aged people and apprentices living in homes, *Int. Z. Vitaminforsch.,* 40, 556, 1970.

246. **Woodhill, J. M.,** Australian dietary surveys with special reference to vitamins, *Int. Z. Vitaminforsch.,* 40, 520, 1970.

247. **Yavorsky, M., Almaden, P., and King, C. G.,** The vitamin C content of human tissues, *J. Biol. Chem.,* 106, 525, 1934.

248. **Yearick, E. S., Wang, M. S., and Pisias, S. J.,** Nutritional status of the elderly: dietary and biochemical findings, *J. Gerontol.,* 35, 663, 1980.

249. **Yonemoto, R. H.,** Vitamin C and immune responses in normal controls and cancer patients, *Int. J. Vitam. Nutr. Res.,* Suppl. 19, 143, 1979.

250. **Zalusky, R. and Herbert, V.,** Megaloblastic anemia in scurvy with response to 50 mg folic acid daily, *N. Engl. J. Med.,* 265, 1033, 1961.

251. **Zannoni, V. G. and Sato, P. H.,** Effects of ascorbic acid in microsomal drug metabolism, *Ann. N.Y. Acad. Sci.,* 258, 119, 1975.

Macronutrient Intakes and Functions

FIBER: THE TREATMENT AND PREVENTION OF CONSTIPATION IN THE AGED

Dorothy L. Brooks

INTRODUCTION

Medical practitioners and caretakers who work with the elderly cite constipation as the most common ailment of this age group in Western industrialized countries. The magnitude of the problem promises to increase as the elderly segment of our society grows to an expected 12% of the population by the end of this century. To alleviate their constipation, Americans spent $37,870 million on various over the counter laxatives in 1981.[1] Although the beneficial use of diets high in fiber has been recognized by astute practitioners for several decades,[2-4] significant interest in this mode of treatment was not aroused until Burkitt et al.[5-9] dramatically publicized epidemiologic studies showing an inverse relationship between a group of so-called "Western-risk" diseases and the amount of fiber in the diet.

The incidence of constipation, diverticular disease, and cancer of the colon, which are among the "risk" diseases, is low among the Third World groups. Evidence for the role of fiber in these diseases was based upon the differences in fecal frequency, consistency, wet weight, and transit time when the quantity of ingested fiber varies, as it does when the refined "Western" diet is compared to that of rural Third World groups. Since the *fiber hypothesis* became popular during the late 1960s and early 1970s, numerous experimental and clinical studies have been conducted to try to define the precise role of fiber in the etiology, prevention, and treatment of constipation and other intestinal disorders.

THE PROBLEM OF CONSTIPATION

Many factors allegedly contribute to the development of constipation with aging. Among young healthy individuals, changes in normal daily activities, alterations in diet and fluid intake, and travel are frequently identified causes of constipation. According to Brockle-hurst,[10] constipation is no more common in the healthy, active, old person than in the young; however, as most individuals age, their daily physical activity declines, thereby depressing normal muscular stimulus to bowel activity. Diet and fluid intake may be altered because of (1) physiological changes, such as poor dentition, (2) economic difficulties which restrict ability to buy usual food, or (3) psychological or emotional depression which reduces regularity, variety, and quantity of food intake. The resulting diet usually consists of high carbohydrate, highly refined, easy-to-prepare food.[11] Loss of smooth muscle fibers as a result of aging also affects intestinal motility. This decreased motility, the low fiber diet, and a lack of physical activity combine to produce constipation, manifested by (1) passage of fewer bowel movements than previously, or (2) the passage of a small dry stool after considerable difficulty and straining.

The scope of the problem of constipation in young and old alike is uncertain because wide variations exist in bowel activity among individuals or even in a single individual on a day-to-day basis. Godding[12] states that a "physiological yardstick, in numerical terms, to human bowel function is badly needed". The number of bowel movements per day cannot be used as a measure since many individuals who defecate two or more times per day may still be constipated if bulk, consistency, and ease of passage are also considered. He suggests that studies are needed to quantify the frequency, consistency, wet weight, and transit time (FCWT) for normal healthy subjects who are ingesting a "good fiber" diet. Variations from these norms can then be used to diagnose and treat constipation.

Over the past few years, epidemiologic and experimental studies have provided considerable data on the FCWT of various populations and study groups when different quantities of fiber are eaten. High fiber intake increases stool weights and decreases intestinal transit times. Stool weights may range from 300 to 600 g on high fiber diets to less than 100 g on refined diets. Transit times on refined diets exhibit considerable variation, ranging from 72 hr to 2 weeks, while on a high fiber intake excretion occurs rather predictably within 30 to 35 hr.[5,13]

Assessing this data, Godding[12] has proposed that until more comprehensive studies are conducted, good bowel function should be defined as F = 1 movement per day, C = soft, W = 150 to 200 g/day, and T = 48 hr.

DEFINITION OF FIBER

An intake of dietary fiber of sufficient quality and quantity will prevent constipation and achieve or maintain good bowel function. Unfortunately, information about the exact amounts and kinds of fiber contained in foods is not readily available. Until the early 1970s, analyses were performed and composition reported as "crude fiber." Since this extraction process resulted in the loss of one fifth to one half of fiber components,[14] these values are now generally regarded as useless. The currently accepted term, "dietary fiber", coined by Trowell,[8] includes several compounds of different chemical and physical characteristics which apparently are resistant to the action of human digestive enzymes. Among the components of dietary fiber are (1) structural polysaccharides, including celluloses, hemicelluloses, and pectic substances derived from plant cell walls, (2) nonstructural polysaccharides such as gums and mucilages, and (3) lignins, a noncarbohydrate structural material. Subsequently, Trowell et al.[12,15] proposed that a broader term, edible fiber, be adopted. Added to the substances described above would be three additional fiber sources ingested by man, namely: (1) partially synthetic polysaccharides, such as methyl cellulose, (2) indigestible fiber-like material derived from connective tissue and internal organs of animals (aminopolysaccharides), and (3) pharmaceutical preparations from plants, e.g., ispaghula husk. Acceptance and use of the term "edible fiber" would probably add some measure of consistency and precision to the literature on this subject.

Various methods for determining the fiber constituents of foodstuffs are being pursued by a number of investigators[16-18] who hope to develop a standardized method of analysis which will yield complete and reliable composition data. Until this data becomes available, comparison of the intake of the different fiber components in the diets of Western industrialized and Third World populations will be unsatisfactory. Burkitt,[13-19] utilizing the fiber composition data which is presently available, has estimated that the crude fiber intake of rural Africans/Indians is approximately 20 g/day and the dietary fiber is probably 60 g. Without further analyses it is not possible to pinpoint these constituents more precisely. For purposes of comparison, however, we can refer to the analysis of food consumption data in the U.S. presented by Heller and Hackler.[20] They used this information to calculate the per capita decline in crude fiber consumption from 6.8 in 1909 to 4.9 g/day in 1975. Of the food groups consumed, cereals, dried peas and beans, and potatoes decreased most markedly. According to Burkitt's hierarchy of plant fiber sources, these food items are the ones most essential to maintaining a good fiber intake.

Of the fiber components, both cellulose and hemicellulose have water-holding properties which, by increasing bulk and decreasing transit time, alleviate constipation. The pentose fractions of the hemicellulose in particular have significant water-absorbing capacity. Of these, arabinoxylans, consisting of equal proportions of arabinose and xylose and present in large quantities in bran, exhibit especially high water-holding capabilities. The higher proportion of arabinoxylan in bran than in legumes, fruits, or vegetables accounts for the greater effectiveness of this foodstuff in alleviating constipation.

TREATMENT WITH FIBER

In recent years many clinical trials in which the dietary fiber intake of the study groups was manipulated have been reported. In the cases of constipation, the fiber decreased transit time and increased frequency and wet weight of the stool. In individuals with rapid intestinal transit times, e.g., those with irritable bowel syndrome,[21] transit was slowed to a more normal range. A decrease in the symptoms of diverticular disease has been described and the need for surgical intervention has apparently diminished.[22] The ''bulking'' effect of a fiber-rich diet apparently lowers the intracolonic pressure which contributed to the development of the diverticula.[19]

Many geriatric hospitals and nursing homes in the U.K. have adopted the fiber regimen outlined by Burkitt and Meisner[13] for their clients. This regimen consists of the addition of bran and whole grain breads and cereals sufficient to add 6 to 10 g dietary fiber to the daily diet. The usual method of administration is to serve 20 g of bran (3 heaping tablespoons) to patients at breakfast along with their whole grain cereal. This level of fiber supplementation evidently is enough to remedy constipation even in those patients with severe intractable constipation who had previously required regular laxatives.

Many reports in the literature are variations of the regimen described above. Painter[22] treats constipation with 8 teaspoons of bran (16 g) served 2 to 3 teaspoons at each meal, plus wholemeal bread and whole grain cereals. After this regimen was instituted, the bowel activity of most patients in his geriatric facility met the criteria of normality proposed by Godding, i.e., F = one movement daily, W = 175 g, and T = 41 hr. Hull et al.[11] supplemented the diets of their institutionalized geriatric patients with bran and whole grains to supply a minimum of 8 g/day of additional dietary fiber, thereby eliminating the use of most laxatives in the institution. Emphasis was placed on producing palatable bran-supplemented products for routine service to all residents.

An unexpected difference in response to bran-supplementation by sex was reported by Clark and Scott.[23] A mean dose of 15 g of bran daily caused a significant increase in F (number of movements) and a significant reduction in the number of constipated days for men but not for women. The quantity of dietary fiber supplementation needed to improve bowel function appears to be rather consistent, according to most reports. This quantity, 6 to 10 g/day, can be supplied by the addition of 15 to 20 g of bran each day. The choice of wheat bran and the use of whole grain breads and cereals is logical since their fecal bulking capacity is greater than fiber from other sources.[24] Administration of uncooked bran of large particle size increases its effectiveness. Coarse bran produced significantly shorter T and greater W than equal quantities of finely ground bran.[25] A comparison of the effects of cooked (Kellogg's All Bran) and raw bran[26] suggests that cooked bran has little effect upon F, W, or T. In this study, however, doses of raw bran in excess of 22 g were needed to significantly decrease transit time (T) while even this quantity did not influence wet weight or stool frequency. Anderson et al.[27] compared the changes in T when a bulk laxative or 20 g of wheat bran was administered. Using radioopaque pellets, they traced the digesta as it traversed the colon. The length of time required for the fecal mass to pass through the ascending, transverse, and descending colon was the same after the ingestion of either bran or laxative. The transit time through the rectosigmoid, however, was significantly less (23 hr compared to 62 hr) with the bran diet. From this study, it appears that the effect of bran on transit time is primarily in the sigmoid colon and rectum — a fact which may be significant since this ''flushing out'' may remove potential carcinogens from prolonged contact with colonic mucosa.

PROBLEMS

Flatulence, distention, and abdominal discomfort are among the problems reported by individuals whose diets are modified to include bran supplements. A gradual introduction may minimize these side effects. The symptoms may persist 2 to 4 weeks or longer[12] before gradually subsiding. Although the effect of a high fiber diet on FCWT may be viewed as beneficial by physicians and other caretakers of the elderly, the nutrient-binding effects of fiber must also be weighed. The absorption of several nutrients, among which are certain minerals, protein, bile acids, and cholesterol, appear to be affected by fiber. Loss of these absorbable nutrients may be caused by more rapid transit of the digesta, or fiber may bind the nutrients and thus render them unavailable for absorption. Each of the fiber components binds different nutrients and the exact mechanism by which they influence nutrient availability undoubtedly varies. Bran has been noted by some investigators to increase fecal losses of calcium, magnesium, zinc, iron, and phosphorus.[28] Cellulose, both naturally occurring and refined, has a similar effect upon calcium and magnesium.[29] These observations are not universal and the discrepant reports[30] make interpretation of the results difficult. One difficulty is the widely varying study periods used by the investigators.[14] Nutrient losses of considerable magnitude after short-term administration of fiber have been reported by some researchers. With prolonged ingestion of a high fiber diet, however, loss of some minerals seems to stabilize, indicating that adaptation may occur over time.[29-30] On the other hand, the occurrence of certain deficiency diseases among populations who have chronically eaten high fiber diets has also been reported.[28] The nutrient losses, while not insignificant, might be overlooked in well-nourished individuals. In the elderly, whose diets may be inadequate or at best marginally adequate, the question of fiber/nutrient interaction deserves attention.

Constipation among the elderly is not a problem that will disappear or that can be ignored. Laxatives are probably the primary alternative to the use of dietary fiber in the treatment of the disorder since a program of physical activity is not feasible for most aged individuals. Laxatives, however, produce water and electrolyte losses,[31-32] which can lead to such problems as dehydration and hypokalemia. The chronic use of purgatives also results in atrophy of the smooth muscle of the colon and loss of its intrinsic innervation.[33] Finally, laxatives focus only on the treatment of an established condition, not upon its prevention. Consequently, the advantages of the use of a fiber-rich diet for constipation, as contrasted to other modes of treatment, appear to outweigh the disadvantages.

REFERENCES

1. **Cardinale, V., Ed.,** Digestive aids, in *Drug Topics*, Medical Economics, Oradell, N.J.
2. **Cowgill, G. R. and Sullivan, A. J.,** Further studies on the use of wheat bran as a laxative, 100, 795, 1933.
3. **Williams, R. D. and Olmstead, W. H.,** The effect of cellulose, hemicellulose, and lignin on the weight of the stool, *J. Nutr.*, 11, 433, 1936.
4. **Dimock, E. M.,** The prevention of constipation, *Br. Med. J.*, 1, 906, 1937.
5. **Burkitt, D. P., Walker, A. R. P., and Painter, N. S.,** Dietary fiber and disease, *JAMA*, 229, 1068, 1974.
6. **Burkitt, D. P. and Trowell, H. C., Eds.,** *Refined Carbohydrate Foods and Disease*, Academic Press, N.Y., 1975.
7. **Burkitt, D. P.,** Relationships between diseases and their etiological significance, *Am. J. Clin. Nutr.*, 30, 262, 1977.
8. **Trowell, H.,** The development of the concept of dietary fiber in human nutrition, *Am. J. Clin. Nutr.*, 31, S3, 1978.

9. **Connell, A. M.,** Natural fiber and bowel dysfunction, *Am. J. Clin. Nutr.,* 29, 1427, 1976.

10. **Brocklehurst, J. C.,** Disorders of the lower bowel in old age, *Geriatrics,* 35, 47, 1980.

11. **Hull, C., Greco, R. S., and Brooks, D. L.,** Alleviation of constipation in the elderly by dietary fiber supplementation, *J. Am. Geriat. Soc.,* 28, 410, 1980.

12. **Godding, E. W.,** Physiologic yardsticks for bowel function and the rehabilitation of the constipated bowel, *Pharmacology,* 20 (Suppl. 1), 88, 1980.

13. **Burkitt, D. P. and Meisner, P.,** How to manage constipation with high-fiber diet, *Geriatrics,* 34, 33, 1979.

14. **Kelsay, J. L.,** A review of research on effects of fiber intake on man, *Am. J. Clin. Nutr.,* 31, 142, 1978.

15. **Trowell, H., Godding, E., Spiller, G., and Briggs, G.,** Fiber bibliographies and terminology, *Am. J. Clin. Nutr.,* 31, 1489, 1978.

16. **Southgate, D. A. T., Bailey, B., Collinson, E., and Walker, A. F.,** A guide to calculating intakes of dietary fibre, *J. Hum. Nutr.,* 30, 303, 1976.

17. **Southgate, D. A. T.,** Dietary fiber: analysis and food sources, *Am. J. Clin. Nutr.,* 31, S107, 1978.

18. **Van Soest, P. J.,** Dietary fibers: their definition and nutritional properties, *Am. J. Clin. Nutr.,* 31, S12, 1978.

19. **Burkitt, D. P.,** The protective properties of dietary fiber, *N. C. Med. J.,* 42, 467, 1981.

20. **Heller, S. N. and Hackler, R.,** Changes in the crude fiber content of the American diet, *Am. J. Clin. Nutr.,* 31, 1510, 1978.

21. **Holloway, W. D., Tasman-Jones, C., and Bell, E.,** The hemicellulose component of dietary fiber, *Am. J. Clin. Nutr.,* 33, 260, 1980.

22. **Painter, N. S.,** Constipation, *Practitioner,* 224, 387, 1980.

23. **Clark, A. N. G. and Scott, J. F.,** Wheat bran in dyschezia in the aged, *Age Ageing,* 5, 149, 1976.

24. **Cummings, J. H., Southgate, D. A. T., Branch, W., Houston, H., Jenkins, D. J. A., and James, W. P. T.,** Colonic response to dietary fiber from carrot, cabbage, apple, bran and guar gum, *Lancet,* 1, 5, 1978.

25. **Heller, S. N., Hackler, L. R., Rivers, J. M., Van Soest, P. J., Roe, D. A., Lewis, B. A., and Robertson, J.,** Dietary fiber: the effect of particle sizes of wheat bran on colonic function in young adult men, *Am. J. Clin. Nutr.,* 33, 1734, 1980.

26. **Wyman, J. B., Heaton, K. W., Manning, A. P., and Wicks, A. C. B.,** The effect on intestinal transit and the feces of raw and cooked bran of different doses, *Am. J. Clin. Nutr.,* 29, 1474, 1976.

27. **Andersson, H., Bosaeus, I., Falkheden, T., and Melkersson, M.,** Transit time in constipated geriatric patients during treatment with bulk laxative and bran: a comparison, *Scand. J. Gastroenterol.,* 14, 821, 1979.

28. **Reinhold, J. G., Faradji, B., Abadi, P., and Ismail-Beigi, F.,** Decreased absorption of calcium magnesium, zinc and phosphorus by humans due to increased fiber and phosphorus consumption as wheat bread, *J. Nutr.,* 106, 493, 1976.

29. **Slavin, J. L. and Marlett, J. A.,** Influence of refined cellulose on human bowel calcium and magnesium balance, *Am. J. Clin. Nutr.,* 33, 1932, 1980.

30. **Stasse-Wolthuis, M., Albers, H. F. F., Van Jeveren, J. G. C., deJong, J. W., Hautvast, Hermus, R. J. J., Katan, M. B., Brydon, W. G., and Eastwood, M. A.,** Influence of dietary fiber from vegetables and fruits, bran, or citrus pectin on serum lipids, fecal lipids, and colonic function, *Am. J. Clin. Nutr.,* 33, 1745, 1980.

31. **Fingl, E.,** Laxatives and cathartics, in *Goodman and Gilman's The Pharmacologic Basis of Therapeutics,* Gilman, A. G., Goodman, L. S., and Gilman, A., Eds., MacMillan, N.Y., 1980, 1002.

32. **Sladen, G. E.,** Effects of chronic purgative abuse, *Proc. R. Soc. Med.,* 65, 12, 1972.

33. **Smith, B.,** Pathology of cathartic colon, *Proc. R. Soc. Med.,* 65, 12, 1972.

WATER NEEDS IN THE ELDERLY

Maury Massler

INTRODUCTION

Water Needs

Water is more important to health than any other single nutrient. Man can survive for weeks or months when deprived of carbohydrates, proteins, fats, vitamins, or minerals before deficiencies become clinically evident. Body reserves may be slowly depleted of other nutrient reserves, but man cannot survive on a desert or after a shipwreck without water for more than 10 to 14 days.[1-10] This is particularly true in the elderly.

All animals must find watering holes to survive. This is more important to survival than grass to the herbivores or meat to the carnivores. The water in grass, vegetables, or in meats is not enough to meet all their water needs.

When mammals emerged from the sea, where water was ever-present, they had to bring with them water in their tissues and had to ingest water daily to replace the water lost from their tissues through urine, feces, perspiration, and lungs. This means that daily water loss (about 2500 mℓ/day in temperate climates) and mild activity must be balanced by water intake or else dehydration will result.[1-10]

Water is so essential to continued health, recovery from illness, and the prevention of premature aging that the body provides the thirst response[5,8] as an immediate signal that water is needed. No similar signal exists to warn the patient as quickly that deficiencies are occurring in other essential nutrients; the clinician must wait until gross tissue changes become evident before he/she can recognize nutritional deficiencies and correct them.

WATER NEEDS IN THE ELDERLY

Slow chronic dehydration occurs during aging.[2,3] While acute fluid loss over a short period of time (e.g., hemorrhage, diarrhea, fever) can be reversed by replacing blood and/or subcutaneous fluids, slow chronic dehydration (e.g., prolonged illnesses or aging) is not easily halted, much less reversed, because severe prolonged water loss ultimately removes water (and potassium) from the cells of the body. When this intracellular water loss reaches 25% loss of body water, death usually results.[3,5]

Water needs are even more critical in the elderly than in the healthy young. Negative water balance in the elderly stems primarily from degenerative changes in the kidney.[1,5,8] With age, the glomerulae lessen their ability to filter out the toxins from the body.[3] These toxins may be endogenous metabolites, wastes from disease or fevers, or exogenous toxins from food additives, medications, antibiotics, or alcohol. Even medications essential to functions of the heart, blood vessels, and liver or medications essential to combat disease must ultimately be detoxified in the liver lest they damage or even destroy essential kidney function. This filtrate (about 120 mℓ/min in the young) is then concentrated in the tubules from 5 to 100 times, to preserve body water.[1,5]

With age the ability of the renal tubules to recycle the water filtered by the glomerulae back to the blood stream is lessened. About 1 ℓ/min of blood passes through the kidneys (about $^1/_5$ of the total cardiac output). If the renal tubules fail to recycle the water from the plasma as evidenced by failure to concentrate the urine, uremia (toxic products in the blood) results. Death soon follows unless cleaning the blood by dialysis is available. For details of kidney functions in the elderly, see Brocklehurst.[3]

The colon is also important to the conservation of water in the elderly. The colon is an area in which water is reabsorbed from the feces. Vegetable fibers are hydrophilic and hold

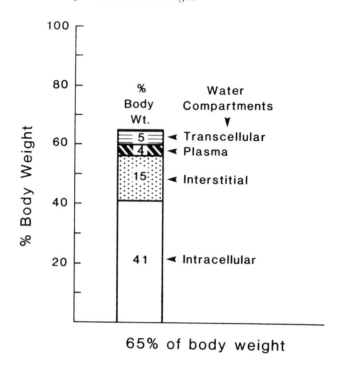

FIGURE 1.   Diagram showing the four major water compartments of the body and the percent of total body water contained in each compartment. Note that larger storehouse of water lies within the cells (the intracellular compartment, 41%) and in the extracellular tissues (15%). Although the amounts in the blood serum and in the synovial, cerebraspinal, and ocular fluids are relatively small (4 to 5%), water loss in these cells are significant in the elderly.

ingested water until it reaches the colon where it can be reabsorbed. Lack of fiber in the diet is therefore a significant factor in progressive dehydration of the elderly. Lack of fiber results in a hard dry fecal mass and constipation caused by delayed evacuation time.[11-13]

For details on the role of fiber in the diet of the aged, see "Fiber: The Treatment and Prevention of Constipation in the Aged" by Dorothy Brooks in this volume, and "Dehydration in The Elderly" by Maury Massler.[9]

BODY WATER COMPARTMENTS

Water constitutes 65 to 70% of body weight depending upon the amount of fat in the body.[7] Water is distributed in three compartments of the body (Figure 1).

The outer compartment contains only 10% of the total body water in the form of the blood serum and lymph, with some water within the blood cells.

The extracellular water compartment consists of the water bound in the interstitial connective tissues of the body. Connective tissues contain four times the amount of water in the blood. Most of this water is held in a gel between the connective tissue fibers. This is the area where sodium is dissolved in the tissue water. It is an important area for water balance via exchange between the intracellular water and the blood serum.

The intracellular water compartment comprises the most important inner compartment. The cells collectively contain twelve times the amount of water in the blood. Cellular integrity and function are more critical to survival at all ages than the water in the other two areas (Tables 1 and 2).

Table 1
THE WATER COMPARTMENTS

Water compartment	Body weight (%)
Intracellular water	41
Interstitial tissue water	15
Plasma water	4
Transcellular water[a]	5
Total	65

[a] Cerebrospinal, ocular, and synovial fluids.

Table 2
INTAKE, OUTPUT, AND RECYCLED WATER (AVERAGE PER DAY)

Water intake (mℓ)		Water output (mℓ)	
Ingested fluid	1200	Urine	1500
Water in foods	1000	Faces	150
Metabolic water	300	Lungs	350*
Total	2500	Sweat	500[a]
		Total	2500

Recycled water (mℓ)	
Gastric juice	2500
Saliva	1500
Pancreatic juice	700
Bile	500
Total	5200

[a] Water loss in hot sun and vigorous exercise, nude, may reach 3000 mℓ/day.

Fluid exchange between each water compartment is continuous. Nutrients, hormones, enzymes, and vitamins pass from the outer blood compartment to the middle tissue compartment. From the middle compartment these nutrients, hormones, enzymes, and vitamins pass into the cells. Then, from the intracellular compartment, the metabolites and wastes are discharged into the middle tissue compartment, thence into the blood stream, and excreted into the urine and the feces. Significant amounts of water are discharged via the lungs (volatile substances) and sweat (Table 2).

The membranes around the cells limit the free diffusion of substances into and from the cells. Cell membranes are highly selective to a variety of ions. Therefore, one other factor must always be considered in writing on water needs, especially in the elderly. The minerals and food nutrients must be dissolved in water. Water has been called a "universal solvent" and all substances must be dissolved in water before it can be carried from the GI tract into the blood stream and then into the cells. Therefore, it is not possible to speak of body water as a discreet and separate entity alone, or as free water. All body waters contain minerals, soluble nutrients, enzymes, etc. Together they constitute the body fluids. For a detailed discussion of body fluids, see Elkinton and Danowski,[5] Lindeman,[8] and Robinson.[9] Lindeman properly speaks of fluid balance as a whole rather than water balance alone. This chapter

Table 3
PROPORTIONS OF WATER IN VARIOUS TISSUES

	Total water (%)	Intracellular water (%)	Extracellular water (%)
Muscle	50	40	10
Skeleton	60	38	22
Skin	45	7	38
Liver	68	38	30
Whole blood	10	6	4

intends to focus more specifically upon the water needs and water balance from the point of view of its essentiality as a primary and critical nutrient need.

Tissue Water

All tissues contain water (Table 3). For example, muscle contains 50% water. Body water and therefore tissue water diminishes significantly with age.[8] This reduction reflects decreased intracellular as well as extracellular water. Since muscle is 50% water, this tissue becomes thin, stringly, and weak with age. Muscle breakdown can be measured by loss of creatinine and/or 3-methylhistidine. The intracellular water loss can be estimated by excretion of potassium.

Transcellular Water

Transcellular fluids are secreted by cells into small compartments where they are held as in containers: cerebrospinal fluid, ocular fluid, and synovial sacs. Synovial fluids act to lubricate the joints and tendons as oil does in an automobile (Table 1).

WATER LOSS

Water loss is about 2500 mℓ/day. The major loss is via the urine. In the elderly, this loss can be excessive if the kidney filtrate mechanism in the glomerulae and reabsorption in the tubules become inefficient. The average 80-year-old person has half the renal function of a normal 30 year old.[8] When the tubules fail to recycle the water back into the blood stream, then negative water balance ensues and dehydration results.[8] In the elderly there is a significant decrease in the ability to concentrate the urine. Urine volume may be decreased to 300 to 500 mℓ/day. Kidney failure becomes the most important factor in preventive medicine and proper geriatric patient care.[3,5]

Water may also be lost through the feces if water is not reabsorbed from the colon. This failure may result from inadequate water intake, especially through the lack of water-bound vegetable fiber in the diet of the elderly. Almy[11] and Mitchell and Eastwood[12] explain in detail the effects of inadequate water intake. These include constipation, diverticulosis, irritable colon syndrome, and even malignancies of the bowel in severe, long-term lack of water intake in the elderly (Table 6).[11,12] In the geriatric years, diseases of the large bowel become serious concerns of both patient and doctor.

The lungs are also implicated in water loss. This type of "insensible perspiration" via the breath can result in a great deal of water loss (about 350 mℓ/day) especially in the elderly living in hot climates (Table 2).[1]

Water loss via sweat is about 300 to 500 mℓ/day. This can increase in hot climates (Table 2). It is interesting to note that water loss in persons who sunbathe in the nude is relatively great when compared to persons sitting in the shade or covered with clothes.[5] The elderly are therefore advised to avoid too much exposure to the sun and "tanning" at the beach or pool. Nomads living in the desert learned centuries ago to wear heavy, loose-fitting clothes to insulate against the heat of the sun and to reduce water loss through sweating. The southern belles of a generation ago wore wide-brimmed hats and parasols in the sun to avoid wrinkling of the facial skin through water loss.

A camel can go for 2 weeks without water in the desert. The camel conserves body water in remarkable ways to survive. The kidney concentrates the urine to a remarkable degree so that the water can be recycled. The kidneys even use the urea (which man excretes) to change into protein. The feces contain 1/40th the amount of water compared to cows. A camel can produce milk when water is scarce and other animals die of dehydration. The camel conserves water also by *not* sweating during the heat of day (body temperature rises to 105°F) and releasing this body heat during the chill of the desert night when the body temperature falls to 93°F. Even the moisture in its breath is recycled in the nose so this water is not lost.

In this respect man's attempts at water conservation are primitive and inefficient. In the elderly, even man's feeble attempts at conserving water by concentrating his urine are defeated by degeneration of the kidney glomerulae and tubules so that water intake becomes more critical and essential to health, especially after mild illnesses, fevers, or physical stress.[14]

Water loss by sweating is also significant during exercise and vigorous games. It is common practice during vigorous exercises to drink beverages or plain water plus salt (NaCl) since loss through sweat can be considerable in both.

Sweat evaporation is the air conditioning system for the body. However, poor temperature regulation is common in the elderly (as in young babies) so that care must be taken to avoid sweating in the summer and hypothermia in the winter. Hypothermia in the elderly can be deadly.

Table 2 shows the percentage of water loss through urine, feces, lungs, and skin in the young. The water loss through saliva (except in spitters) and tears is very small since these fluids are recycled by swallowing (Table 2).

Excessive water loss in the elderly (dehydration) can be seen clinically as xerostomia (dry mouth), xeropthalmia (dry eyes), and dry, scaly, wrinkled skin (Table 6).[9] In the young, thirst warns the patient as to water needs; in the elderly, this signal becomes less demanding, probably because when 12% of body water is lost the mouth is dry and swallowing becomes difficult. Increased water intake in the elderly is therefore very desirable, especially in the form of vegetable soups.[9] When water intake is decreased in the elderly, there is an effort by the kidneys to conserve body water. The amount of urine is greatly diminished from 1500 to 500 mℓ in 24 hr.

WATER INTAKE

Water intake must be approximately 2500 mℓ/day to balance water loss as outlined above. At least 800 to 1200 μℓ must come from tap water and beverages, exclusive of moist foods (Table 2). The most common source of water in the U.S. and Europe are the daily beverages (tea, coffee, and "pop"), but a considerable amount of water is ingested through foods such as vegetables, cheeses, and fish (Table 5). Perhaps the most efficient method of water intake is by vegetable soups. Soups are full of dissolved minerals, soluble proteoses, and carbohydrates, and include the fiber so essential to water retention and recycling in the elderly.[9] Cultures which serve soups at least once a day show the effects in terms of lesser dehydration in the elderly in these countries (South America and Europe).

Water derived from the metabolism of fats, starch, and proteins contribute to water intake until these body stores are depleted (Tables 2 and 4).

Table 4
WATER PRODUCED BY
METABOLISM

100 g of	Water (g)
Fat	107.01
Starch	55.1
Protein	41.3
Alcohol	117.4

Table 5
PERCENTAGE OF WATER
IN COMMON FOODS

Vegetables	90
Whole milk	87
Fresh fruits	85
Cooked oatmeal	85
Eggs	75
Lean beef	60
Bread	36

Table 6
EFFECTS OF EXTRACELLULAR FLUID LOSS IN THE ELDERLY

Thirst
Low urine volume
Breakdown of body reserves (yields endogenous water from fats and carbohydrates)
Loss of weight
Eyes become sunken due to loss of retro-ocular fat (which contains 20% water)
Joints become stiff due to decrease in secretion of synovial fluids
Skin becomes dry, scaly, and wrinkled; looseness of skin is marked
Gland secretions are suppressed especially salivary glands leading to xerostomia (parched mouth and throat, extreme
  thirst); xerophthalmia leading to eye irritations and infections); sweat (dry skin); secretion of all GI juices decreased
Intracellular water loss with potassium depletion followed
  by apathy, depression, and confusion

Exogenous deficiency of water intake or endogenous loss of water by damaged kidneys leads to the clinical picture of drying-up, shrivelling, and wrinkling during advanced aging (Table 6). In the not so chronologically old, but aging person, loss of subcutaneous tissue water leads to wrinkling of the skin, loss of the connective tissue pad, and discomfort under artificial dentures.[9] In the Far East with its hot climate, dehydration killed more people than did starvation.

CONCLUSIONS

Slow but continual dehydration with age in the elderly is seen clinically as the loss of subcutaneous tissue with wrinkling, thinning, and weakening of muscles and a decrease in most of the secretions of the body (saliva, tears, sweat, and all GI secretions).

The loss of water from the middle extracellular water compartment and ultimately from the inner cellular compartment results from a reduction in the filtration rate in the glomerulae of the kidney and decreased resorption of water in the renal tubules. All investigators agree with the evidence that the kidneys are the main organs responsible for water balance.[1-10]

Therapy and prevention must therefore be aimed at maintaining optimal water balance in the elderly, including all electrolytes.[10]

Increasing water intake and electrolytes where possible and not contraindicated by diseases in the elderly is therefore of prime consideration. From the meager scientific evidence available, daily vegetable soups containing dissolved minerals and soluble organic substances plus hydrophilic fiber seem to be most frequently used empirically to reduce dehydration in the geriatric patient.

REFERENCES

1. **Best, C. H. and Taylor, N. B.**, Fluid distribution and exchange, in *Physiological Basis of Medical Practice, Section 4*, 9th ed., Brobeck, J. R., Ed., Williams & Wilkins, Baltimore, 1973, 4.
2. **Bosschieter, E. B.**, Fluid deficiency in the older person, *Ned. Tidjdschr. Gerontol.*, 10(2), 97, 1979.
3. **Judge, T. G.**, The milieux interieur and aging, in *Textbook of Geriatric Medicine* and Gerontology, Broklehurst, J. C., Ed., Churchill Livingstone, Edinburgh, 1973, 113.
3a. **Durnin, J. V. G. A.**, Nutrition and changes in total body mass and in body composition, in *Textbook of Geriatric Medicine and Gerontology*, Brocklehurst, J. C., Ed., Churchill Livingstone, Edinburgh, 1973, 390.
4. **Cooper, L. F., Barber, E. M., and Mitchell, H. S.**, *Nutrition in Health and Disease*, 9th ed., J. B. Lippincott, Philadelphia, 1943, 109.
5. **Elkinton, J. R. and Danowski, T. S.**, *The Body Fluids*, Williams & Wilkins, Baltimore, 1955, pp. 57, 158.
6. **LeMaire, M. and Perlier, G.**, Frequent dehydration problems in aged subjects, *SOINS*, 24, 19-21 July 1979.
7. **Lesser, G. T. and Marofsky, J.**, Body water compartments with human aging using fat-free mass as the reference standard, *Am. J. Physiol.*, 23, 215, 1979.
8. **Lindeman, R. D.**, Application of fluid and electrolyte balance principles to the older patients, in *Clinical Aspects of Aging*, Reichel, W., Ed., Williams & Wilkins, Baltimore, 1978, 213.
9. **Massler, M.**, Geriatric nutrition. II. Dehydration in the elderly, *J. Prosth. Dent.*, 42, 489, 1979.
10. **Robinson, J. R.**, *Body Fluid Dynamics in Mineral Metabolism*, Part A, Vol. 1, Comar, C. L. and Bronner, F., Eds., Academic Press, N.Y., 1960, 205.
11. **Almy, T. P.**, The role of fiber in the diet, in *Nutrition and Aging*, Winick, M., Ed., John Wiley & Sons, N.Y., 1976, 155.
12. **Mitchell, W. D. and Eastwood, M. A.**, Dietary fiber and colon function, in *Fiber in Human Nutrition*, Spiller, G. A. and Amen, R. J., Eds., Plenum Press, N.Y., 1978, 185.
13. **Seymour, D. G., Henschke, P. J., Cape, R. D., and Campbell, A. J.**, Acute confusional states and dementia in the elderly: the role of age, *Aging*, 9, 137, 1980.
14. **Anon.**, *Science 82*, 3(7), 182, 1982.

ALCOHOL AND PSYCHIATRIC PROBLEMS

John P. Wattis

INTRODUCTION

The myth that there are few elderly alcoholics dies hard. In 1851 Huss[1] stated: "It is a rare exception to meet an alcoholic who is over 60 years of age." Since then there has been an enormous increase in the number of people surviving into old age. Even in the last 30 years the number of old people in the industrialized nations has increased dramatically. In England and Wales under 5% of the population were over 65 in 1901, 11% in 1951, and 15% in 1981.[2] The last 30 years have also seen an increase in alcohol consumption and alcohol-related problems in many countries. In the U.K. for example, adult alcohol consumption per person increased by 87% between 1950 and 1976.[3]

Some alcoholics do not survive into old age[4] and Drew[5] characterized alcoholism as a "self-limiting" disease. He showed that detected alcoholism in the over 40-year-old age group was much less than predicted from cumulative incidence figures and hypothesized that this was due to high mortality, spontaneous remission, and some treatment effect. Despite this, other workers[6] have shown that not only do some chronic alcoholics survive into old age, but also that another group of people start to abuse alcohol for the first time in old age.

EPIDEMIOLOGY

Surveys of drinking habits usually distinguish between light and heavy drinkers on the basis of frequency and/or quantity of alcohol consumed. Some surveys also aim to detect those who are suffering from definite physical, social, or psychiatric harm as a result of excessive drinking. These people are usually called "alcoholics". When interview or questionnaire survey methods are applied to a population containing old people there are several possible systematic sources of error. Old people may be less accustomed to surveys than younger people and may be more reluctant to disclose details of "personal" matters such as drinking habits. When interviews are used, lonely old people may be motivated to give the interviewer the information they feel he wants or needs. There may even be fear that a "bad report" could lead to curtailment of social security benefits.[7a] In addition, a larger proportion of elderly heavy drinkers may have memory defects to render their self-report unreliable.

Surveys only give a picture of drinking habits at a given point in time. They therefore generally fail to distinguish between changing drinking habits as individuals grow older, the cohort effect as a generation of light or heavy drinkers ages, and possibly varying effects of social strains and pressures on different age groups. Knupfer and Room[8] asked people about past as well as present drinking habits. They found that abstention was slightly more common among elderly than young former drinkers, but that lifelong abstention was much more common in the older age group. One explanation for this would be that the older group had been in their teens and twenties during the era of Prohibition and had therefore never acquired the habit of taking alcohol (Table 1). Two surveys conducted in London's Camberwell area in 1965 and 1974 gave further information about changes in drinking patterns.[9] They showed an increase of nearly 50% in mean total consumption of alcohol across the age groups with a nearly 100% increase in the over 55 age group. This, again, could be explained as a cohort effect although the alternative explanation that social pressures were acting differently on different age groups cannot be discounted.

Table 1
PERCENTAGE OF FORMER DRINKERS WHO ABSTAIN AND OF LIFELONG ABSTAINERS BY AGE AND SEX

Former drinkers who abstain				Lifelong virtual abstainers			
Age 21—39		60 +		Age 21—39		60%	
Male	Female	Male	Female	Male	Female	Male	Female
3	6	11	9	6	20	20	56

From Knupfer, G. and Room, R., *Q. J. Stud. Alcoholism,* 31, 108, 1970. With permission.

Table 2
PERCENTAGE OF OLD PEOPLE WHO ARE REGULAR OR HEAVY DRINKERS

Country	Definition	Age (years)	Percentage		Ref.
			M	F	
Continental U.S.	Heavy, moderate, and light-frequent	65 +	46	25	11
Sydney, Australia	Heavy and moderately heavy	60 +	58	8	12
Scotland	Regular	66 +	43	22	13

Table 3
PERCENTAGE OF OLD PEOPLE WHO ARE LIGHT INFREQUENT DRINKERS OR ABSTAINERS

Country	Definition	Age (years)	Percentage		Ref.
			Male	Female	
Continental U.S.	Abstainers and infrequent	65 +	54	75	11
Sydney, Australia	Abstainers and light infrequent	60 +	30	67	12
Scotland	Occasional and abstainers	66 +	57	78	13

The definition of "heavy" drinking is a further source of confusion. Although this varies from study to study so that results are not always comparable, all workers use the same definition for young and old. Yet evidence is growing that age as well as sex is an important determinant of one's ability to "handle" alcohol. Vestal et al.[10] infused a dose of alcohol intravenously into volunteers. They found that the doses, individually calculated on the basis of surface area, produced higher peak ethanol concentrations in older subjects and attributed this to differences in lean body mass and volume of body water.

With these reservations, Table 2 gives the percentage of old people who were regular or heavy drinkers in three representative surveys. The excess of male over female regular or heavy drinkers is noteworthy in all three studies, especially the Australian one. It is possible that cultural biases may have prevented women from reporting heavy drinking. Table 3 gives the percentage of light, infrequent drinkers or abstainers from the same surveys.

Table 4
CONTRIBUTION OF ALCOHOLICS AND ALCOHOLIC PSYCHOSES TO ELDERLY HOSPITAL ADMISSIONS/ ATTENDANCES

Location	Age	% Alcoholic	Ref.
U.S. county psychiatric screening ward	60+	44	17
U.S.A. Harlem outpatients	65+	12	18
Suburban Rockland county mental health center	65+	17	18
Queensland, Australia, psychiatric admissions	65+	8	19
Psychiatric hospital admissions, England	65+	1.2	20
Psychiatric hospital admissions, Scotland	65+	4.9	21

Some surveys have aimed to find those alcoholics who are suffering from definite damage as a result of heavy drinking. The Cambridgeshire survey[14] found that 9% of alcoholics were elderly, giving a prevalence of less than 0.5% for alcoholism in the elderly. Edwards et al.[15] explored the relationship between the number of alcoholics known to agencies and the number detected in a careful community survey in an area of London. They found a reported prevalence for alcoholism in the over 60s of 0.55% for men and 0.25% for women, but estimated that overall "needy" cases outnumbered reported cases between four- and ninefold. They also found that whereas male age-specific prevalence rates for alcoholism peaked in the 50s, for women the highest prevalence was in the over 70s. A study in New York gave an overall prevalence for alcoholism of around 2.2% in the 65 to 74 year age group but in this area, age-specific prevalence rates for women peaked at 45 to 54 and for men at 65 to 74.[16] This may reflect differences in male:female age-specific prevalence trends between London and New York or changes with time or differences in detection rates.

Psychiatric hospital admission rates for alcoholism and alcoholic psychoses vary in how well they reflect community prevalence. Some figures are summarized in Table 4. It may be said that studies which have been specially conducted such as those in the U.S. and Australia tend to yield higher figures than routinely collected statistics like those for England and Scotland.

Routine detection of alcohol abuse in general hospital patients tends to be poor. Jarmann and Kellett[22] found that for all ages, nearly 20% of hospital patients were "problem drinkers", a prevalence six times greater than that in the surrounding community. Furthermore, nearly 90% of the patients detected did not have a diagnosis usually associated with alcoholism. Another general hospital study[23] found that the admitting doctor had recorded no details of alcohol intake for 40% of patients and only vague general comments in a further 25% of cases. Indirect and unusual presentations of alcohol abuse in old people, especially self-neglect, repeated falls, confusion, and depression[24] may lead to an even lower level of detection.

PHYSIOLOGY OF ALCOHOL IN OLD PEOPLE

Table 5 demonstrates the wide range of alcohol concentrations in different drinks. The rate of alcohol absorption depends on several factors. The higher the concentration of alcohol, the more rapidly it is absorbed; carbon dioxide accelerates and sweetness slows absorption. Food in the stomach slows absorption markedly. There are no known age differences in absorption. After absorption, alcohol is distributed throughout the body and differences in lean body mass and total body water probably result in higher concentrations per unit of alcohol consumed in older people.[10] Most alcohol is metabolized in the liver and only 2 to 5% is eliminated unchanged via the kidneys and lungs. Chronic heavy consumption of

Table 5
ALCOHOL CONTENT OF
DIFFERENT BEVERAGES FROM
ROYAL COLLEGE

Group	Examples	Alcohol(%)
Beers	Lager	3—6
	Ales	3—6
	Stout	4—8
Table wines	(Still) red, white, and rose	8—14
	(Sparkling) Champagne	12
Distilled spirits	Brandy	40
	Whisky	37—40
	Rum	40
	Gin	37—40
	Vodka	57.5
Liquers		20—55

Adapted from Special Committee of the Royal College of Psychiatrists, *Alcohol and Alcoholism*, Tavistock Publications, 1979.

alcohol initially induces liver enzymes so that metabolism is accelerated and there is probably a synergistic effect with smoking. When liver *damage* supervenes, decreased metabolic capacity may be one of the factors related to change in tolerance. Although ethanol is the main drug in alcoholic beverages, they also contain various amounts of different ''congeners'' which as well as imparting characteristic flavors, may also, themselves, produce pharmacological effects.

General Effects

High concentrations of alcohol damage human tissue. Drinks of 40% concentration or more cause inflammation of the gastric mucosa and lower concentrations may cause damage in predisposed patients by stimulating gastric secretions. Once absorbed, alcohol causes a small transient increase in heart rate and a dilatation of cutaneous blood vessels. This can lead to hypothermia in old people, especially as larger doses may inhibit the temperature-regulating mechanism in the brain. Very high concentrations of alcohol can cause respiratory or circulatory failure and death. The long-term toxic effects of alcohol on skeletal and cardiac muscle[25] and on the liver[26,27] and other organs will not be considered in detail here. There is no evidence to suggest that older people are more at risk of liver damage[28] or cardiomyopathy[29] because of age rather than because of duration of drinking.

Effects on the Central Nervous System

In low doses, alcohol produces a sense of well-being and increased alertness. This may be related to different effects on different neurotransmitter systems or possibly to biphasic or even triphasic effects on individual systems. At the psychological level, these effects may be attributed to disinhibition. Table 6 summarizes the progressive effects on brain function as blood alcohol level rises. The question of whether the aging brain reacts differently to alcohol is only beginning to be explored. Helderman et al.[30] have shown that whereas alcohol simply suppresses the secretion of antidiuretic hormone in younger subjects, in older people this hormone is initially suppressed but then rebounds to nearly twice the basal level. This

Table 6
EFFECTS OF INCREASING LEVELS OF BLOOD ALCOHOL

Blood alcohol level (mg%) (approximately)	Effects
30	Driving skill affected
50	Disinhibition and increased reaction time
80	Risks of road accidents doubled
100	Clumsiness and emotional lability
200	Clear impairment of movement and emotion
300	Gross intoxication
500—800	Death

Adapted from Special Committee of the Royal College of Psychiatrists, *Alcohol and Alcoholism,* Tavistock Publications, 1979.

change in response by the hypothalamic-pituitary-renal neurohormonal system in older people may reflect other neurohormonal differences in response.

DRUG INTERACTIONS

In the early 1970s old people represented about 12% of the U.K. population but accounted for 30% of the National Health Service expenditure on drugs.[31] In a U.K. general practice study, 87% of people aged 75 or over were having regular drug treatment and 44% were believed to be taking three or more drugs regularly.[32] The risk of interactions between prescribed drugs and between prescribed and nonprescribed drugs (including alcohol) is increased. The acute effect of alcohol is generally to inhibit drug metabolism so that action is prolonged. Sustained heavy drinking first leads to enzyme induction which may speed metabolism, then as liver damage supervenes, metabolism is impaired. Table 7 summarizes some of the main drug interactions with alcohol.

Aging itself produces changes in drug metabolism. Absorption is unaltered, but reduced body weight and body water and an increased proportion of body fat together with some lowering of plasma albumin produce complicated effects on distribution. Oxidation of some drugs (e.g., chlormethiazole) is also decreased whereas others (e.g., diazepam) appear to be oxidized as readily as in younger subjects.[33] Renal elimination of some drugs is decreased. The final result in any elderly individual taking alcohol and other drugs depends on complicated interactions, but the potential for harm is clear.

DAMAGE

General

Mellström and colleagues[35] examined a sample of 468 70-year-old men and reexamined 342 of them 5 years later together with a cohort comparison group of 489 70-year-old men. About 10% in all three examination groups met their criteria of registration twice or more with the temperance board, for previous alcohol abuse or large-scale alcohol consumption. The morbidity of diabetes and chronic bronchitis (corrected for cigarette consumption), the consumption of care, and the mortality rate were higher among the recidivists. They also had a lower ability in cognitive tests, muscle strength, "gonadal function" (measured by the plasma lecithin: lysolecithin ratio), pulmonary function, visual actuity, and walking ability. In addition, skeletal density was lower and obvious trends or significant differences

Table 7
DRUG INTERACTIONS WITH ALCOHOL

Nature of interaction	Drugs involved
Acute effects of alcohol; inhibition of metabolism and/or synergism potentiating and prolonging action of drug	Phenylbutazone Barbiturates Mepro b amate Benzodiazepenes Chloral hydrate Opiates Antidepressants (sedative) Beta blockers Antihistamines (sedative) Neuroleptics (sedative and dystonic reactions) Sulphonylureas Warfarin
Dangerous interactions with drugs frequently used in suicide attempts	Barbiturates Opiates Diazepam Chlormethiazole Paracetamol Dextropropoxyphene
Chronic effects of alcohol: enzyme induction leading to decreased action of drug (when cirrhosis supervenes this may be reversed)	Phenytoin Barbiturates Meprobamate Diazepam Warfarin Tolbutamide
Drugs which interfere with alcohol metabolism	Chlorpromazine Chloral hydrate Disulfiram Chlorpropamide Metronidazole

Adapted from **Anon.**, *Br. Med. J.*, 1, 507, 1980.

were found in blood glucose, plasma free fatty acids, and serum aspartate aminotransferase activity which were all higher in the recidivists. Unfortunately, the recidivists, while similar to the nonrecidivist group in nearly all demographic variables, did have a shorter time in education and lower incomes than the controls so that it is possible that some of the differences were due to factors other than alcohol consumption.

There is no evidence that aging in itself increases the risk of damage to the GI tract, liver, pancreas, or skeletal and cardiac muscle but neither does it confer immunity from this damage.

Neuropsychiatric Complications

Alcohol Dementia

This is probably much more common than is usually realized.[36] Although alcoholic dementia is usually described as a ''presenile'' dementia, it appears that the age of the subject is the most critical determinant of the risk of developing alcoholic dementia.[37] It is therefore logical to assume that elderly drinkers are most at risk for this condition, but because other causes of dementia are common at this age,[2] because presentation may be masked, and because of difficulties in clinical differential diagnosis, alcoholic dementia is probably underdiagnosed amongst elderly patients. Some 17% of "indigent" elderly people admitted to a county psychiatric screening ward in the U.S. were diagnosed as alcoholic while a further 27% had alcoholism with an associated chronic brain syndrome.[17] Of 35 elderly alcoholics presented to psychiatric, psychogeriatric, and alcoholism services in a Scottish study, 17 had alcoholism, 1 had Korsakoff's psychosis, and a further 6 had alcoholism with organic or senile dementia.[38] In neither study were the authors able to consistently differentiate between alcoholic dementia and dementia due to other causes.

The clinical picture is of gradual personality change with deteriorating memory. Excessive drinking is often denied, sometimes with the collusion of family members.[24] The mood may be euphoric, depressed, or labile, and there is usually little "depth" to the feeling except sometimes in depression. Judgement is impaired and sexual and social disinhibition may be more marked than in patients with Alzheimer's or multi-infarct dementia. Specific damage to the frontal lobe may account for some of these symptoms although the debate continues about localization of alcoholic brain damage.[37] Probably the best way of differentiating between alcoholic and nonalcoholic dementia is that progress of the former is arrested when alcohol consumption stops and there may even be slight improvement. Confirmation of this and further clarification of the differences between alcoholic and other dementias in old age would be extremely useful to future research. Cortical atrophy and ventricular dilation have been detected on computerized axial tomography and at postmorten without any consistent focal distribution.[37]

Wernicke's Encephalopathy/Korsakoff Psychosis

Wernicke's encephalopathy, due to a deficiency of thiamine (Vitamin B_1) is relatively rare. The classic triad of symptoms is ophthalmoplegia, confusion, and ataxia, of which ophthalmoplegia is probably most important diagnostically. The condition may develop during a long spell of heavy drinking or during early withdrawal. Some old people may have a relative lack of thiamine even without the complication of alcohol abuse. Older and Dickerson[39] showed that patients undergoing elective hip replacement operations had a normal preoperative thiamine which fell at 48 hr after the operation and returned to normal by 14 days, whereas in a group of patients with fractured neck of femur, thiamine deficiency remained at this time. The metabolic strains imposed by operation or other physical illness may be particularly important determinants of thiamine deficiency in elderly alcoholics.

If Wernicke's encephalopathy is not prevented or treated promptly and efficiently, Korsakoff's psychosis supervenes. There is severe memory impairment with anterograde and some retrograde amnesia. Confabulation is common but not diagnostic as it also occurs in senile dementia.[40] Only 14% of a group of patients with Korsakoff psychosis improved over a 5-year period.[41] It is uncertain whether Korsakoff's psychosis is more common in older people, but older alcoholics are more likely to have a temporary Korsakoff-like memory defect in the first few weeks following alcohol withdrawal than their younger counterparts.[42]

Intoxication Phenomena

Violent or sexually disinhibited behavior out of keeping with the patient's personality is sometimes referred to as "pathological drunkness". Old people, especially if they have a degree of brain damage, may be especially susceptible to this problem. The elderly male heavy drinker may become physically aggressive to his wife after a relatively small amount of alcohol which he could previously handle.

Memory blackouts when, for example, the whole of an evening's drinking is forgotten, can occur in any drinker who attains a sufficiently high blood-alcohol concentration to block consolidation of memory. Heavy drinkers are especially at risk, and decreased biochemical reserves together with increased blood concentrations per unit dose of alcohol[10] are likely to increase the vulnerability of old people.

Withdrawal States

In those who have become physically dependent on alcohol, tremor and nausea, sometimes accompanied by feelings of guilt and dread follow within 12 hr of withdrawal. Delirium tremens develop in less than half of dependent patients withdrawn from alcohol and is characterized by clouding of consciousness, visual and auditory hallucinations, and a subjective feeling of terror. The patient may misinterpret efforts to help him and may be irritable and aggressive. Cardiovascular changes are present and epileptic fits may occur. Untreated, the condition may be fatal. There is a correlation between clouding and hallucinations across all age groups, but, in the elderly, clouding becomes relatively more common and hallucinations relatively less so.[43] Delirium tremens should always be suspected in elderly patients who develop unexplained acute confusion, especially if this comes on soon after hospital admission.

Functional Psychiatric Syndromes

Affective

The unpleasant mood state occurring during alcohol withdrawal is sometimes prolonged.[37] Depression may also be reactive to the social and personal difficulties encountered by the heavy drinker. On the other hand, reactive or endogenous disturbance of mood may precede heavy drinking. The old person who is socially isolated may become reactively depressed and start drinking heavily in a futile attempt to lift the depression.[24] Heavy drinking may also occur in the elated or depressed phase of a manic-depressive psychosis. The association of affective illness and alcoholism in different members of the same family led Winokur et al.[44] to suggest the concept of "depressive spectrum disease", a hereditary disorder manifesting as early onset depression in women and as alcoholism or antisocial behavior in men. Subsequent research has, however, suggested depression in female children of alcoholics may be an environmental rather than a genetic effect, and the question of a genetic contribution to alcoholism has recently been reviewed by Murray and Gurling.[45] In parallel with other conditions, it is likely that late onset alcoholics have less of a genetic loading than the early onset group.

Schizophreniform

Alcoholic hallucinosis, characterized by auditory hallucinations, often of a sexual or derogatory nature, emerges during uninterrupted drinking. Thought disorder and passivity phenomena are absent although Schneiderian "first rank" symptoms of schizophrenia can occur during alcohol withdrawl. Paranoid delusions, often of morbid jealousy, also occur during heavy drinking. It has been suggested that decreased sexual performance due to the effects of alcohol may contribute to the delusion that the spouse (almost invariably the wife) is being unfaithful. Morbid jealousy can be a dangerous condition as the paranoid person may resort to violence against his wife or her supposed lover.

Cutting[37] believes that alcoholic hallucinosis and alcoholic paranoia are rare. Beyond the fact that most of these problems emerge after years of heavy drinking, there is no special reason to believe that they have a higher frequency or different prognosis in older alcoholics.

Other Complications

Subdural Hematoma

This condition occurs mainly in elderly people and is notoriously difficult to diagnose. It has been suggested that alcoholism also predisposes to subdural hematoma, perhaps through the mechanism of repeated falls. Noltie and Denham[46] studied 23 elderly patients with subdural hematoma, 8 acute, 5 subacute, and 10 chronic. They found a history of head injury in 7 of the acute but only 5 of the chronic cases. Hemiparesis and/or dysphasia or a steady deterioration in general physical condition were the most common symptoms. Headache and pupillary abnormality were not particularly helpful diagnostically and 8 cases had no neurological signs. Only two thirds of the cases were diagnosed before death and the authors commented on the many difficulties in making a diagnosis of subdural hematoma in old people. Their study did not, unfortunately, look for a history of alcohol abuse. In any elderly alcoholic patient with unexplained deterioration or neurological signs, subdural hematoma should be excluded. Perhaps the most useful investigation is computerized axial tomography although even this can be misleading.

Cerebellar degeneration — This condition may occur as part of the Wernicke-Korsakoff syndrome but sometimes cerebellar ataxia develops as an isolated finding over a period of several weeks or months.

Peripheral neuropathy — Chronic abuse and malnutrition can cause a painful, subacute sensorimotor neuropathy. This is symmetrical and usually most pronounced in the legs. A "burning" sensation on the soles of the feet is commonly described and distal muscle weakness and tenderness occurs. If the patient is or has been a secret drinker, the cause of the neuropathy may remain undiagnosed.

MANAGEMENT

Prevention

There is irrefutable evidence that harm from alcohol consumption is related to national drinking habits.[47] The Royal College of Psychiatrists Report on Alcohol and Alcoholism[3] makes firm recommendations concerning prevention. It proposes that national per capita consumption of alcohol should not be allowed to rise beyond the present level and that agreement should be reached to bring down alcohol consumption over the next decade with concurrent monitoring of indexes of harm (e.g., road traffic accidents involving alcohol and cirrhosis death rates). Other recommendations include coordinated efforts by different government departments to use public revenue policies to achieve these aims and a greatly enhanced commitment to education, persuasion, and relevant research, especially into the impact of advertising.

At a community and personal level, employers and individuals are asked to review the extent to which work and social activities press people to drink more than is good for them. Special efforts should be made in those trades and professions at risk. Leisure activities which do not engender pressure to drink alcohol should be encouraged. Smith[47] comments on the vested interests which make these practical suggestions politically hard to implement.

Bereavement, retirement, and loneliness may be potent contributors to late onset heavy drinking[24] and improved leisure and activity opportunities with a healthier social attitude to bereavement and counseling help for the bereaved when needed should help. Many old people obtain alcohol from relatives and "helpers" who may be unaware of other sources of supply or deny that there is any problem.[24,48] Education is needed that excessive drinking in old people is not a harmless foible but a source of damage and disability.

Detection

Specific questions should always be asked about alcohol consumption when an old person presents to the medical services. Reliable relatives should be interviewed whenever possible and information should be gathered from other sources such as home help services, social workers, and the family practitioner if there is any reason to suspect alcohol abuse. A home visit frequently reveals clues in the form of empty bottles or glasses. Suspicion should be especially high in cases of self-neglect, unexplained confusion, repeated falls, and atypical affective illness and where close relatives are known to be abusing alcohol.[24]

Treatment

Acute Phase

When alcohol abuse is established, treatment should begin with inpatient admission for alcohol withdrawal. Physical illness, especially pneumonia, peptic ulceration, and subdural hematoma may complicate alcohol withdrawal. Because they may present atypically in old people, it is especially important to look for them. In view of the risk of Wernicke-Korsakoff syndrome and other complications of poor nutrition and vitamin deficiency, early vitamin supplementation initially with high potency intravenous B and C preparations is mandatory. Oral preparations should subsequently be continued. In patients who have been drinking heavily immediately before coming into the hospital, withdrawal symptoms (and the risk of delirium tremens) should be minimized by the use of a decreasing regime of chlormethiazole, chlordiazepoxide, or other suitable sedative. Oral chlormethiazole should be given in reduced doses in older people[33] and in those with cirrhosis.[49] It can be given in an i.v. infusion but very careful supervision is essential to avoid oversedation. Hypokalaemia, dehydration, and other electrolyte imbalance should be corrected, if necessary, by suitable oral fluids or i.v. infusion.[50] Hypoglycemia may occur soon after withdrawal and may be corrected by i.v. injection of glucose. Thiamine should also be given intravenously to prevent precipitation of the Wernicke-Korsakoff syndrome.[50] Sympathetic nursing in an appropriately lighted environment is essential during this phase. Functional symptoms should be treated in their own right, but in the acute withdrawal phase drugs which might precipitate epilepsy should be avoided as far as possible. Subdural hematoma is usually treated surgically. The pain of alcoholic neuropathy which may not respond to conventional analgesics may be amenable to treatment with carbamazepine. In the author's opinion, the treatment of elderly alcoholics with drugs such as disulfiram is not usually indicated and the most realistic aim of treatment is abstention rather than controlled drinking.

Social Rehabilitation

Where drinking is of late onset in response to remediable social stress, prognosis is relatively good, especially if the patient and relatives acknowledge the problem. Detailed restructuring of the social network can remove both the need and the opportunity for further heavy drinking. Where excessive drinking has been continuing since early life or where there is collusion by alcohol-abusing family members,[24] prognosis is poor. When brain damage impairs judgment, deliberate restrictions of alcohol supply through institutional care may be the only way to prevent further damage. As long ago as 1964, Droller[48] suggested that treated elderly alcoholics should be discharged to a communal living situation such as residential care.

BENEFITS

Mishara and Kastenbaum[7b] devote a chapter of their book on alcohol and old age to possible beneficial effects of alcohol on the elderly. After reviewing a number of studies they conclude that moderate, voluntary use of alcohol by patients in an institutional setting may be beneficial. They also report on possible beneficial effects of a regular glass of wine on the functioning and sleep patterns of self-sufficient elderly people.

CONCLUSIONS

Since biblical times the beneficial effects of moderate amounts of alcohol[51] have been reported as well as the detrimental effects of excessive consumption.[52] The industrialized production and high-pressure advertising of alcohol are relatively recent phenomena which have pushed up consumption of alcohol in many modern societies. Today's medical and social services are also facing a unique challenge because of increasing numbers of very old people in the population. This chapter has been directed to refuting the idea that elderly people are immune to alcohol problems. Education and preventative measures are now vital if unnecessary disabilities induced by alcohol and consequent pressure on overworked services are to be avoided. At the same time there must be good treatment for those old people who are already victims of the present epidemic of alcohol abuse.

REFERENCES

1. **Huss, M.,** Chronische Alcoholskrankheit oder Alcoholismus Chronicus. Stockholm and Leipzig, 1851, cited by Drew, 1968, and Rix, 1980.
2. **Wells, N. E. J.,** Dementia in Old Age, Office of Health Economics, London 1979.
3. Special Committee of the Royal College of Psychiatrists, *Alcohol and Alcoholism,* Tavistock Publications, London, 1979.
4. **Peterson, B., Kristenson, H., Sternby, N. H., Trell, E., Fex, G., and Hood, B.,** Alcohol consumption and premature death in middle-aged men, *Br. Med. J.,* 1, 1403, 1980.
5. **Drew, L. R. H.,** Alcoholism as a self-limiting disease, *Q. J. Stud. Alcohol,* 29, 956, 1968.
6. **Rosin, A. J. and Glatt, M. M.,** Alcohol excess in the elderly, *Q. J. Stud. Alcohol,* 32, 53, 1971.
7a. **Mishara, B. L. and Kastenbaum, R.,** *Alcohol and Old Age,* Grune & Stratton, N.Y., 1980, 45.
7b. **Mishara, B. L. and Kastenbaum, R.,** *Alcohol and Old Age,* Grune & Stratton, N.Y., 1980, 127.
8. **Knupfer, G. and Room, R.,** Abstainers in a metropolitan community, *Q. J. Stud. Alcohol.,* 31, 108, 1970.
9. **Cartwright, A. K. J., Shaw, S. J., and Spratley, T. A.,** The relationships between per capita consumption, drinking patterns and alcohol related problems in a population sample, 1965-1974. I. Increased consumption and changes in drinking patterns, *Br. J. Addict.,* 73, 237, 1978.
10. **Vestal, R. E., McGuire, E. A., Tobin, J. D., Andres, R., Norris, A. H., and Mezey, E.,** Ageing and ethanol metabolism, *Clin. Pharmacol. Ther.,* 21, 343, 1977.
11. **Cahalan, D., Cisin, I. H., and Crossley, H. M.,** *American Drinking Practices: A National Study of Drinking Behavior and Attitudes,* Monographs of the Rutgers Center of Alcohol Studies, No. 6, New Brunswick, N.J., 1969.
12. **Encel, S., Kotowicz, K. C., and Resler, H. E.,** Drinking patterns in Sydney, Australia, *Q. J. Stud. Alcohol.,* Suppl. 6, 1, 1972.
13. **Dight, S.,** *Scottish Drinking Habits: A Survey of Scottish Drinking Habits and Attitudes Towards Alcohol,* Office of Population Censuses and Surveys, Her Majesty's Stationery Office, London, 1976.
14. **Moss, N. C. and Davies, E.,** *A Survey of Alcoholism in an English County,* Geigy Scientific, London, 1967.
15. **Edwards, G., Hawker, A., Hensman, C., Peto, J., and Williamson, V.,** Alcoholics known or unknown to agencies: epidemiological studies in a London suburb, *Br. J. Psych.,* 123, 169, 1973.
16. **Bailey, M. B., Haberman, P. W., and Allisne, H.,** The epidemiology of alcoholism in an urban residential area, *Q. J. Stud. Alcohol.,* 26, 19, 1965.
17. **Gaitz, C. M. and Baer, P. E.,** Characteristics of elderly patients with alcoholism, *Arch. Gen. Psychiatry,* 24, 372, 1971.
18. **Zimberg, S.,** The elderly alcoholic, *Gerontologist,* 14, 221, 1974.
19. **Daniel, R.,** A five-year study of 693 psychogeriatric admissions in Queensland, *Geriatrics,* 27, 132.
20. Department of Health and Social Security, Inpatient Statistics from the Mental Health Enquiry for England, 1976, Statistical and Research Report Series 22, Her Majesty's Stationery Office, London, 1979.
21. Scottish Health Services, Common Services Agency, Information Services Division, *Scottish Health Statistics 1977,* Her Majesty's Stationery Office, Edinburgh, 1979.
22. **Jarman, C. M. B. and Kellett, J. M.,** Alcoholism in the general hospital, *Br. Med. J.,* 2, 469, 1979.
23. **Barrison, I. G., Viola, L., and Murray-Lyon, I. M.,** Do housemen take an adequate drinking history?, *Br. Med. J.,* 2, 1040, 1980.
24. **Wattis, J. P.,** Alcohol problems in the elderly, *J. Am. Geriatr. Soc.,* 3, 131, 1981.
25. **Portal, R. W.,** Alcoholic heart disease, *Br. Med. J.,* 283, 1202, 1981.
26. **Anon.,** How does alcohol abuse damage the liver?, *Br. Med. J.,* 2, 1733, 1978.
27. **Heaton, K. W.,** Alcoholic liver disease, *Br. J. Hosp. Med.,* 3, 118, 1977.
28. **Ishino, H., Suwaki, H., and Itoshina, T.,** Liver function tests in addictive alcoholics, *Jpn. J. Stud. Alcohol,* 9, 65, 1974; *J. Stud. Alcohol.,* 36 (Abstr.), 1072, 1975.

29. **Portal, R. W.**, personal communication, 1981.
30. **Helderman, J. H., Vestal, R. E., Rowe, J. W., Tobin, J. D., Andres, R., and Robertson, G.**, The response of arginine vasopressin to intravenous ethanol and hypertonic saline in man: the impact of ageing, *J. Gerontol.*, 33, 39, 1978.
31. **Judge, T. G. and Caird, F. I.**, *Drug Treatment of the Elderly Patient*, Pitman Medical, London, 1978.
32. **Law, R. and Chalmers, C.**, Medicines and elderly people: a general practice survey, *Br. Med. J.*, 1, 565, 1976.
33. **Ramsay, L. E. and Tucker, G. T.**, Drugs and the elderly, *Br. Med. J.*, 1, 125, 1981.
34. **Anon.**, Drugs and alcohol, *Br. Med. J.*, 1, 507, 1980.
35. **Mellstrom, D., Rundgren, A., and Svanborg, A.**, Previous alcohol consumption and its consequences for ageing, morbidity and mortality in men aged 70-75, *Age Ageing*, 10, 277, 1981.
36. **Anon.**, Minor brain damage and alcoholism, *Br. Med. J.*, 283, 2455, 1981.
37. **Cutting, J.**, Neuropsychiatric complications of alcoholism, *Br. J. Hosp. Med.*, 27, 4, 335, 1982.
38. **Rix, K. J. B.**, Alcoholism and the elderly, M. Phil. thesis, Edinburgh University, 1980.
39. **Older, M. W. J. and Dickerson, J. W. T.**, Thiamine and the elderly orthopaedic patient, *Age Ageing*, 11, 101, 1982.
40. **Berlyne, N.**, Confabulation, *Br. J. Psychiatry*, 120, 31, 1972.
41. **Cutting, J.**, The relationship between Korsakov's syndrome and alcoholic dementia, *Br. J. Psychiatry*, 132, 240, 1978.
42. **Cermack, L. S. and Ryback, R. S.**, Recovery of short-term memory in alcoholics, *J. Stud. Alcohol.*, 37, 1, 46, 1976.
43. **Gross, M. M., Rosenblatt, S. M., Lewis, E., Malenowski, B., and Broman, M.**, Hallucinations and clouding of the sensorium in alcohol withdrawal, *Q. J. Stud. Alcohol.*, 32, 1061, 1971.
44. **Winokur, G., Reich, T., Rinmer, J., and Pitts, F. N.**, Alcoholism, *Arch. Gen. Psychiatry*, 23, 104, 1970.
45. **Murray, R. M. and Gurling, H. M. D.**, Polygenic influence on a multifactorial disorder, *Br. J. Hosp. Med.*, 27/4, 328, 1982.
46. **Noltie, K. and Denham, M. J.**, Subdural hematoma in the elderly, *Age Ageing*, 10, 241, 1981.
47. **Smith, R.**, The politics of alcohol, *Br. Med. J.*, 284, 1392, 1982.
48. **Droller, H.**, Some Aspects of Alcoholism in the Elderly, *Lancet*, 2, 137, 1964.
49. **Pentikaunen, P. J., Neuvonen, P. J., Taspila, S., and Syvalahti, E.**, Effects of cirrhosis of the liver on the pharmacokinetics of chlormethiazole, *Br. Med. J.*, 2, 861, 1978.
50. **Rix, K. J. B.**, Alcohol withdrawal states, *Hosp. Update*, 403, 1978.
51. The Bible, Psalm 104:15.
52. The Bible, Proverbs 23:30.

ALCOHOL AND ITS EFFECTS ON THE NUTRITION OF THE ELDERLY

Joseph J. Barboriak and Carol B. Rooney

INTRODUCTION

During recent years the U.S. has experienced a marked reduction in mortality among the elderly. Consequently, the number of men and women reaching this age stratum is increasing both in absolute numbers and relative terms. In 1979, the proportion of elderly in the age bracket of 65 years or over reached 11% of the total U.S. population.[1] In 1979, the estimated absolute number of elderly in this age range was 24.7 million and is expected to increase to 32 or even 37 million by the year 2000.[1] Such changes in the population will necessitate considerable modifications and adaptions in socioeconomic needs and health care delivery. This will place increasing stress on efforts to prevent rather than treat the infirmities of old age.

One area of health care of the elderly which is already receiving increasing attention concerns their nutritional assessment and changing nutritional needs.[2-5] However, examination of the data from studies dealing with nutritional needs of this age group indicate that most of them have neglected to consider alcohol intake, a dietary component which, in addition to its calorigenic effects, may impair the nutritional status in numerous ways.

DEFINITION OF TERMS

Before considering the possible alcohol-nutrition interaction in the elderly in greater detail, one needs to define the two terms, i.e., the age limits for "elderly" and the nature of alcohol consumption, since the opinions on their meanings are far from unanimous.

Age can be defined on a chronological,[6] functional,[7] or biological[8] basis. While each of the bases has some advantages and drawbacks, most investigators agree that the best overall indicator of the aging process is the chronological age. It is easy to assess, it can be used universally, and can be reliably ascertained. The boundary between "middle" and "old" age varies by social customs, laws, and regulations. For the purpose of this paper, the age of 60 years and over was selected as indicating "old" since this is the age when, according to Title VII of the Older Americans Act, individuals are eligible for the supported nutrition programs.[6] Other commonly used lower limit for the "older" individuals begins with the age of 65 years.[6]

Literature dealing with alcohol intake tends to focus on alcohol abuse or alcoholism.[9-12] These terms are not synonymous, since the definition of alcoholism includes some personality traits which may preceed alcohol abuse.[13] For the purpose of this review, alcohol consumption will be considered either in the general sense without regard to the amount taken, or in the form of moderate or excessive alcohol intake. According to several authors,[14] moderate alcohol consumption denotes intake of up to four drinks daily. Higher alcohol amounts will be included in the category of alcohol abuse. It could be expected that the nutritional status of elderly alcohol abusers will be affected more severely. However, it should be noted that elderly alcohol abusers are not a homogenous group. They can be divided into several different subgroups. The *essential* alcoholics have usually started abusive drinking early in youth and continue this habit into their older age.[15] The second subgroup are *reactive* alcoholics. They are mostly individuals who drink because of some special traumatic event such as death of a spouse, loss of a friend, or because of boredom after retirement.[16] Finally, there are former alcoholics who, for various reasons connected either with health problems, edicts of courts, or pressures of family, gave up drinking and may still suffer the effects of previous abusive drinking.

Table 1
REPORTS ON ALCOHOL INTAKE IN ELDERLY POPULATION GROUPS

Geographic area	Sample size	Age limits (years)	Imbibers (%)	Abusers (%)	Ref.
U.S.	624	≥60	53	6	17
Boston	158	≥65	68	9	18
Boston	928	≥60	47	4	19
New York City	169	>65	49	11	22
Milwaukee	122	>60	57	4	20
Southern California	185 men	>60	82	—	21
England and Wales	173 men	≥65	91	3	23
	221 women	≥60	76	1	

EXTENT OF DRINKING AMONG THE ELDERLY

Available information on the extent of drinking among the elderly indicates that a considerable proportion of this population group consumes alcoholic beverages (Table 1). Calahan et al.,[17] in their comprehensive report on drinking habits of the U.S. population, estimate that 53% of subjects 60 years of age or older consumed alcohol regularly and approximately 6% were alcohol abusers. Other reports have suggested regional and national differences. In the Boston metropolitan area, 68% of the participants in a study of subjects 65 or older were drinking alcoholic beverages and 9% were characterized as heavy drinkers.[18] Another more recent study from the same metropolitan area[19] has found that 47% of participants 60 or older consumed alcoholic beverages. In Milwaukee County, Wisc., in a proportionate stratified population sample of men 60 or older, 57% consumed alcoholic beverages.[20] In a male population sample from California, over 82% of the participants in a similar age bracket reported regular alcohol consumption.[21] In the Upper East Side of New York City, 49% of the respondents of a survey reported imbibing alcoholic beverages.[22]

In a recent examination of drinking habits in England and Wales,[23] 60% of the male population age 55 or over reported drinking during the week prior to the survey, 27% were judged to be moderate to severe drinkers. The corresponding figures for women of similar age were 41% and 13%, respectively. Therefore, it appears that at least 50% of the general population of subjects 60 or over imbibe in alcoholic beverages and approximately 5 to 10% may be considered heavy drinkers or alcohol abusers. Specialized populations, such as hospitalized patients (especially those in mental hospitals), show a considerably higher proportion of elderly individuals suffering from alcohol abuse. Of the 100 consecutive admissions in a Texas psychiatric screening ward, 44% of those 60 or older were diagnosed as alcoholics.[24] In the Psychiatric Unit of the San Francisco General Hospital, 57% of the patients 60 or over were alcohol drinkers and 23% were abusers.[25] The recent compilation of data from the Veterans Administration Hospital System[26] indicates a rising trend of hospitalized patients with problems related to alcohol abuse. The proportion of patients 65 or over diagnosed as alcoholics increased from 7% in 1970 to 12% in 1980.

In our study of patients who have had coronary arteriography because of symptoms suggestive of coronary heart disease,[20] 76% of the male patients 60 or older and 55% of female patients of the same age range reported consumption of alcoholic beverages. Approximately 12% of the male and 3% of the female patients consumed 3 to 5 drinks daily; 1.3% of the men and 0.7% of the women reported consumption of six or more drinks daily.

In view of the frequently reported malnutrition among the hospitalized patients,[27,28] the relatively high proportion of alcohol abuse in this population group deserves special attention.

Despite the relatively large number of reports dealing with the effect of alcohol on nutritional status[29-34] or with the nutritional needs of the elderly,[2-5] there are relatively few

Table 2
ALCOHOL AND KILOCALORIE CONTENT OF POPULAR ALCOHOLIC BEVERAGES

Beverage	Measure	Weight (g)	kcal	Alcohol (g)
Distilled liquors				
Brandy	1 brandy glass	30	73	10.5
Gin, rum, vodka				
Whiskey (rye/scotch)				
80-proof	1 jigger, 1½ oz	45	104	15.0
100-proof	1 jigger, 1½ oz	45	133	19.1
Liquors				
Anisette	1 cordial glass	20	75	7.0
Benedictine	1 cordial glass	20	70	6.6
Creme de menthe	1 cordial glass	20	67	7.0
Curacao	1 cordial glass	20	55	6.0
Malt liquors				
Beer	12 oz	360	145—175	12.1—13.3
Beer, Budweiser	12 oz	360	150	17.6
Beer, natural light	12 oz	360	100	14.4
Beer, near	12 oz	360	65	1.3
Wines				
Champagne, domestic	1 wine glass	120	85	11.0
Dessert (18.8% alcohol by volume)	1 wine glass	100	137	15.3
Red, California	1 wine glass	100	85	10.0
Sauterne, California	1 wine glass	100	85	10.5
Sherry, dry, domestic	1 wine glass	60	85	9.0
Table (12.2% alcohol by volume)	1 wine glass	100	85	9.9
Cocktails				
Daiquiri	1 cocktail	100	125	15.1
Highball	8 oz	240	165	24.0
Manhattan	1 cocktail	100	165	19.2
Martini	1 cocktail	100	140	18.5
Old-fashioned	4 oz	100	180	24.0
Tom Collins	10 oz	300	180	21.5
Whiskey sour	1 cocktail	75	138	15.3

From Pennington, J. A. and Church, H. N., *Bowes and Church's Food Values of Portions Commonly Used*, J. B. Lippincott, Philadelphia, 1980. With permission.

papers focusing specifically on modification of the nutrient intake by alcohol in the older groups. Therefore, this report will concentrate on the areas of nutritional needs which, in addition to being affected by alcohol, are also frequently found to require special attention in the elderly.

NUTRIENT-ALCOHOL INTERACTION IN THE ELDERLY

A considerable amount of information exists about the metabolic effects of alcohol on the modification of nutrient utilization by alcohol, as well as about the effect of progressing age on the processing of ingested nutrients. We will discuss these topics as they relate to the main theme of this chapter.

Caloric Requirements and Alcohol

The contribution of alcohol to the total dietary caloric intake in the elderly is frequently neglected when dietary interviews or surveys are undertaken (Table 2). National consumption

Table 3
NUTRIENT VALUE OF ALCOHOLIC BEVERAGES

Nutrient	Beer (40 oz)	Whiskey, 80 proof (8 oz)	Table wine, 12.2% Alcohol by volume (20 oz)
Kilocalories	505	520	497
Protein (g)	3.5	—	0.6
Fat (g)	0	0	0
Carbohydrate (g)	45.5	Trace	24.6
Calcium (mg)	60.0	—	51.4
Phosphorus (mg)	360	—	57.1
Iron (mg)	Trace	—	2.3
Sodium (mg)	85	Trace	28.6
Potassium (mg)	300	8	537
Vitamin A (IU)	—	—	—
Thiamin (mg)	0.05	—	Trace
Riboflavin	0.35	—	0.06
Niacin (mg)	7.0	—	0.6
Ascorbic acid (mg)	—	—	—

From Adams, C. F., Nutritive Value of American Foods in Common Units, USDA Agriculture Handbook No. 456, 1975.

figures indicate that alcohol accounts for 4.5% of caloric intake in the average diet. In alcoholics, approximately 30 to 60% of their caloric intake is consumed as alcohol. Some alcoholic beverages contain small amounts of carbohydrates, minerals, and vitamins. However, this constitutes only a very small proportion of daily nutrient requirements and, in terms of total demands, the calories derived from alcohol are, in a practical sense, "empty", i.e., without any significant noncaloric nutritive values (Table 3).

The caloric density of alcohol (7.1 kcal/g) is intermediate between the caloric density for fat (9 kcal/g) and for carbohydrate and protein (4 kcal/g). However, alcohol calories do not provide caloric food value equivalent to carbohydrate. Studies by Pirola and Lieber[35] have demonstrated that weight gain is significantly lower when alcohol is substituted for carbohydrate on a caloric basis. Loss of body weight was observed when dietary carbohydrate was substituted isocalorically by alcohol, accounting for 50% of the total ingested calories. Alcoholics supplemented daily with an additional 2000 kcal in the form of alcohol or chocolate failed to gain weight on ethanol calories.[35] This observation may be explained by an increased energy requirement following alcohol consumption. Animal studies have indicated that rats fed excessive quantities of alcohol increased their oxygen consumption, showed growth retardation, and demonstrated an increase in the activity of microsomal enzymes.[36] It is possible that the metabolic processing of ethanol is "energy wasteful" because it is not connected with the energy-conserving mechanisms such as synthesis of ATP.[37] Although alcohol dehydrogenase is the main pathway for the metabolism of alcohol to acetaldehyde, other pathways may metabolize as much as 20 to 25% of ingested ethanol to acetaldehyde. The energy produced in this manner is only a fraction of that produced by the oxidation of alcohol to CO_2 and H_2O via the citric acid cycle.

The influence of alcohol on body weight is frequently determined by the role of alcohol in the total dietary intake. If alcohol becomes part of the daily diet at the expense of other caloric sources, it may cause a decrease in body weight. Rats fed unrestricted amounts of food and offered diluted ethyl alcohol in place of drinking water have responded with a decrease in food intake which, on a caloric basis, was equivalent to consumed alcohol.[38] Despite the unchanged total caloric intake, the animals receiving alcohol gained less weight than the water-drinking controls.[38,39] Most alcoholics have a history of weight loss over a period of years and a weight gain following abstinence. Mezey and Faillace[40] reported grossly deficient diets in 68% of a group of 56 alcoholics with fatty liver on biopsy. The diets consisted primarily of carbohydrate with inadequate amounts of protein and vitamins. After the alcoholics were hospitalized for 3 weeks, they exhibited a mean weight gain of 3.1 kg. Morgan et al.[41] studied 55 middle-aged, middle-class alcoholics with liver disease. Clinical malnourishment was evident in 16 of the 55 patients (29%). The mean daily intake of protein, fat, and carbohydrate were similar to the nonalcoholics. However, diets of the alcoholics showed a significant reduction in the contributions to total energy from the three major nutrients because of the caloric contribution from alcohol.

Barboriak et al.[42] found that regular consumption of alcoholic beverages by the elderly reduced their intake of nonalcohol calories, even when food was in ample supply and served free of charge. When compared with a group of abstainers, the imbibers ate fewer food calories (2140 vs. 2415 kcal) daily. The lower caloric intake extended to the three primary energy sources (carbohydrate, protein, and fat). The average amount of alcohol consumed by the subjects were calculated to be equivalent to 473 kcal/day.

If alcohol is added to the diet without change in the nonalcohol calories, body weight will likely increase. Hurt et al.[43] studied a group of middle-aged, middle-class alcoholics admitted to an alcoholism treatment program. They found that 88% of the patients met or exceeded ideal weight and the nutrient intake of these patients generally met the recommended dietary allowances. Jones et al.,[21] on the other hand, did not find that drinkers were more obese than nondrinkers. They conducted a community study of caloric and nutrient intake in drinkers and nondrinkers of alcoholic beverages. Based on a 24-hr dietary recall, 51% of the 691 men and women (30 to 90 years of age) consumed an average of 30 g of alcohol during the preceding 24 hr. Generally, the alcohol-derived calories were in addition to the usual diet and did not replace calories derived from other nutrients. Similar findings were reported by Neville et al.[44] who carried out dietary studies on alcoholics free of major pathology. They found that the nutritional status of alcoholics was not markedly inferior to that of nonalcoholics, particularly when health and economic status were similar. The mean energy intake was 2710 kcal/day in the males and 2578 kcal/day in the females with alcohol providing 36% and 22% of total calories, respectively. The mean daily protein intake was 68 g in males and 72 g in females.

The caloric content of the alcoholic beverages has special implications for the imbibing elderly because of the reduced caloric requirements of this group. If consumed at the expense of regular food calories, alcohol may result in inadequate intake of vitamins and/or minerals. If consumed in addition to the usual food calories, the additional energy from alcohol will lead to an increase in body weight and obesity. This poses the usual danger of developing hypertension, possibly diabetes, a tendency to less physical activity, and more extensive skeletal load; all are conditions which may aggravate the frequently fragile health of the elderly.

Carbohydrate-Alcohol Interaction

The influence of alcohol on carbohydrate metabolism varies with the acuity and amount of intake, the nutritional state, and some co-existing health impairments.

The acute effects of alcohol may be due to a direct action of alcohol on enzymes, on structural components of the cell, or to changes induced during metabolic processing of alcohol. The chronic effects of alcohol may be due to accumulation, repetition, or continuation of the acute affects, adaptive changes in metabolic pathways, physiological processes, or structural alterations, or to other secondary factors associated with prolonged alcoholism such as an inadequate diet and fluid and electrolyte imbalance.

Metabolic Processing of Alcohol

The effects of ethanol on carbohydrate metabolism have been studied primarily in the liver as alcohol is predominantly metabolized by this organ. The first step in the metabolism of alcohol is its oxidation to acetaldehyde. This reaction can be catalyzed by at least three different enzyme systems — alcohol dehydrogenase (ADH), catalase, and the microsomal ethanol oxidizing system (MEOS). Alcohol dehydrogenase is the most important enzyme in the oxidation of alcohol to acetaldehyde and is dependent on nicotinamide adenine dinucleotide (NAD).[45] The following equilibrium is catalyzed by ADH:

$$NAD + C_2H_5OH \xrightarrow{ADH} NADH + H_2 + CH_3CHO$$

Catalase may also be a factor in alcohol metabolism.[46] Hydrogen peroxide is required for the following reaction:

$$C_2H_5OH + H_2O_2 \rightarrow CH_3CHO + 2H_2O$$

The hydrogen peroxide may be produced as a result of the action of aerobic dehydrogenase systems such as xanthine oxidase, glucose oxidase, or reduced nicotinamide adenine dinucleotide phosphate (NADPH) oxidase. The role of catalase is probably insignificant since chemical inhibition of catalase activity does not appear to have a significant effect in reducing the rate of alcohol metabolism.[47] The MEOS may account for 20% of alcohol metabolism. However, some reservations have been expressed concerning its importance.[48,49]

The second step in the metabolism of alcohol involves the conversion of acetaldehyde to acetate. Aldehyde dehydrogenase is the principal enzyme catalyzing this reaction, which requires NAD as a cofactor.[48,50]

$$NAD + Acetaldehyde \rightarrow Acetate + NADH + H$$

Ethanol metabolism, by increasing reduced nicotinamide adenine dinucleotide and adenosine monophosphate, affects glycolysis and gluconeogenesis. The ratio of reduced to oxidized nicotinamide-adenine dinucleotide (NADH/NAD) in the liver is increased when alcohol is ingested. This has been shown by direct analysis of the coenzyme concentrations[51] and through measurements of substrate pairs thought to be in equilibrium with the NAD system.[52,53]

Alcohol and Blood Glucose
Nonfasting State

The influence of alcohol on blood glucose depends on the nutritional status of the subject. In a well-nourished animal and in men with minimal impairment of liver metabolism, alcohol causes only a slight and transient increase in blood glucose.[54-57] This effect could, to some extent, be due to alcohol-induced hormone action on glycogenolysis in the liver such as increased release of epinephrine.[58] However, the effect cannot be due entirely to hormone action, since Forsander et al.[59] found increased glucose production in rat liver perfused with diluted blood. The increased glucose output may be caused by glycogen breakdown since the same investigators observed no such effect in livers from starved rats.

When well-fed rats were given moderate amounts of ethanol (1.5 g/kg body weight), the hexose monophosphates showed a significant increase, the glucose-6-phosphatase activity was somewhat increased, and the fructose disphosphate (FDP) concentration was slightly increased.[60]

In practical terms, the hyperglycemic effect of alcohol may be quite important; e.g., it has become a practice in some nursing homes to improve socialability and manageability of patients by serving alcohol in the form of beer or wine.[61] Health care providers for the elderly should be made aware that the routine administration of alcohol in a well-fed elderly person could result in a small and transient increase in blood sugar. With advancing age, the ability to metabolize carbohydrate is reduced, as shown by studies using various tolerance tests.[62-65] Andres and Tobin[65] published a monogram for oral glucose tolerance tests which showed the needed age correction. They recommended that this correction factor be applied rather than using an arbitrary number for all ages and misclassifying the elderly as diabetic on the basis of poor performance on the glucose tolerance test.

Fasting State

In the fasting or depleted state, the blood glucose concentration is maintained by gluconeogenesis. However, alcohol inhibits gluconeogenesis and may induce significant *hypoglycemia*.[66-71] Krebs et al.[72] reported that when alcohol produces hypoglycemia, the glycogen store of the liver is virtually exhausted and the quantity of amino acids utilizable for glucose production is reduced. Therefore, glucose formation is primarily due to gluconeogenesis from lactate, originating in the peripheral tissues through the Cori cycle.

Other studies indicate that there are factors other than inhibition of gluconeogenesis that contribute to the hypoglycemia seen after alcohol ingestion in fasting individuals. Searly et al.[73] studied glucose turnover prior to and following the administration of alcohol to obese human subjects and human subjects of normal weight fasting for 3 days. The rate of decrease in glucose concentration was significantly greater in subjects of normal weight than in the obese subjects. This difference has been shown to be due to an average increase in the rate of glucose utilization in subjects of normal weight vs. an average decrease in the rate of glucose utilization in obese subjects during the production of hypoglycemia.

When alcohol is administered under circumstances which more closely approximate the usual mode of alcohol and food intake, the data suggested that ethanol enhances glucose-stimulated insulin secretion.[74] In normal and mild diabetic subjects, the blood glucose levels following the glucose load were 30 to 80 mg/100 ml lower and the early insulin secretory early insulin secretory response (15 to 45 min) was 35 to 42% higher after ethanol ingestion. Ethanol intake had no effect on the glucagon response to glucose ingestion. The reduced blood glucose rise observed with ethanol may be related to the greater insulin response or to the decreased GI absorption of glucose. For the mild diabetic patient, moderate intake of ethanol was without acute harmful effects of carbohydrate homeostasis and may improve the blood glucose response to ingested carbohydrate.

O'Keefe and Marks[75] presented evidence that consumption of alcohol combined with soft drink mixers on an empty stomach provokes a greater insulinemia and more profound reactive hypoglycemic response than alcohol alone. Soft drinks are carbonated and have a low pH; this accelerates the absorption of oral glucose. The authors have suggested that alcohol stimulates the release of one or more of the several intestinal insulin-releasing polypeptides which augment glucose-stimulated insulin secretion but which do not themselves stimulate insulin secretion except in the presence of mild to moderate hyperglycemia. In addition, Straus et al.[76] demonstrated that alcohol, like low pH, can stimulate the release of secretin, a known augmentor of glucose-mediated insulin release. The hyperinsulinemia stimulated by alcohol has also been attributed to a direct priming effect of alcohol on the beta cells, making them more susceptible to the insulin stimulatory effect of glucose and other insulinotrophic agents.[77]

Nutritionists generally agree that the diets of elderly individuals are frequently inadequate due to poor dietary selection, low income, isolation, chronic illness, or disease. Therefore, severe hypoglycemia could result if a nutritionally depleted individual began consuming alcohol.

Alcohol and Carbohydrate Absorption and Utilization

When alcohol is administered intravenously it causes a significant increase in motility in the duodenum,[78] jejunum, and ileum.[79] These changes are believed to account for the observed changes in xylose absorption.[80] Alcohol in moderate doses lowers urinary excretion of orally administered D-xylose.[81,82] Perlow et al.[83] reported depressed levels of intestinal lactase, and lactose intolerance in well-nourished alcoholics. Blacks showed a more significant reduction in jejunal disaccharidase activity than whites. Lactose intolerance due to lactase deficiency exists in adulthood in the majority of the population of the world.[84] The frequency of lactase deficiency as a consequence of the aging process must be differentiated from that caused by alcohol which can be reversed by eliminating alcohol from the diet. Greene et al.[85] reported a significant reduction in the activity of jejunal glycolytic and gluconeogenic enzymes in normal volunteers given moderate doses of alcohol.

Fat and Alcohol Interrelationship

The modification of metabolic processing of fat by alcohol can occur at several levels. As already alluded to in the discussion of alcohol-carbohydrate interaction, alcohol affects the hepatic NADH/NAD ratio,[86] with subsequent preferred esterification rather than oxidation of free fatty acids.[87] This selective diversion of fatty acids from the citric acid cycle is associated with accumulation of fatty acid esters in the liver and may, to some extent, explain the consistent finding of post-ethanolic fatty livers.[88] A similar process of esterification of fatty acids after alcohol has been observed in the intestinal tract.[89] The alcohol-induced retention, rather than oxidation of fatty acids, is then reflected in higher levels of circulating plasma lipids, especially of the triglycerides or the very low density lipoproteins. Indeed, extensive hyperlipemia following alcohol abuse has been reported by Feigl[90] as early as 1918. The extent of hepatic accumulation of fat is influenced by the amount and kind of dietary fat.[91] Alcohol has also been reported to mobilize fatty acids from the adipose tissue and thus enhance the development of fatty liver.[92]

At the level of plasma lipids, alcohol has been reported to enhance dietary hyperlipidemia[93,94] and to increase the level of circulating high density lipoprotein (HDL),[95] as indicated by numerous reports indicating higher HDL cholesterol levels[96,97] in regular alcohol imbibers. However, it merits mention that in excessive alcohol abuse associated with severe liver impairment there was a *decrease* in plasma HDL cholesterol levels.[98] Since increased HDL cholesterol levels tend to be associated with lower risk of coronary artery disease, the possible role of alcohol intake in the dietary approach to the control of heart disease risk factors is receiving considerable attention.[99-101] The recent observation[101] that consumption of alcohol in a binge pattern, i.e., imbibing a number of drinks at one occasion,[102] does not lead to increased HDL cholesterol levels, indicates the need for caution in considering alcohol as a protective factor against coronary heart disease.

The relevance of these findings to alcohol effects in the elderly will require additional studies. One could expect, however, that the hepatic damage and changes in blood lipids will be at least as pronounced as in the younger subjects and, since the reparative power of organs in the elderly is usually less effective, the recuperative processes will require longer time periods.

Protein/Amino Acids and Alcohol

The ingestion of alcohol impairs intestinal absorption of amino acids,[102] and alters hepatocellular amino acid metabolism and protein synthesis. Acute and chronic alcohol administration differ in their effects.

Recent studies on the effect of acute alcohol administration have demonstrated a decrease in protein synthesis.[103-113] Alcohol appeared to specifically inhibit the production of secretory proteins as albumin and transferrin but not fibrinogen.[105] Rothschild et al.[113] have shown that the acute effects of alcohol on the ability of the liver to synthesize albumin are dependent on the nutritional state of the liver. In livers of fed donors, the acute effects are transient. However, in livers of fasted donors, the effects of the combined stresses of fasting and alcohol are more severe, and even providing excess amino acids does not reverse the impaired synthesis. Morland and Bessesen[111] have also reported that acute alcohol administration affected protein degradation in isolated hepatocytes.

Studies dealing with the effect of *chronic* alcohol feeding on protein synthesis have provided conflicting results in that some investigators have reported no changes,[49,114] some have seen decreases,[115-117] while others report increases in protein synthesis.[103,118,119] Chronic alcohol administration results in an increased accumulation of transport proteins such as albumin and transferrin in the liver[120] and could be a contributing factor to the liver enlargement observed in chronic alcoholics. The deposition of protein may be caused by the action of alcohol or its metabolite, acetaldehyde, on hepatic microtubules.[120]

Furthermore, alcohol ingestion can cause an increased loss of plasma proteins in the feces. Chowdhury et al.[121] oberved a significant loss of plasma proteins in patients with alcohol-induced gastritis.

The effect of alcohol on nitrogen balance is determined by its caloric contribution to the total diet. When alcohol is administered as supplementary calories, it has a nitrogen sparing effect. However, when alcohol was isocalorically substituted for carbohydrate, it increased urea excretion in the urine.[122,123]

Alterations in plasma amino acid concentration and urinary amino acid excretion occur with alcohol consumption and/or liver disease. Feeding alcohol to protein-deficient rats for 3 to 10 days resulted in an increased activity of enzymes related to degradation of methionine.[124] In both animals and man, the level of plasma α-amino-n-butyric acid (a product of methionine catabolism) was markedly increased following heavy alcohol consumption.[125] Studies in recently drinking alcoholics have also demonstrated that the plasma ratio of α-amino-n-butyric acid to leucine (A/L ratio) is increased.[126] The effect of alcohol on tryptophan metabolism has also been studied because this amino acid serves as a precursor of serotonin, a neurotransmitter, as well as of nicotinic acid. However, no consistent effect could be demonstrated.[127-129]

In chronic liver disease, as frequently seen in alcoholics, there is usually an increase in plasma concentration of amino acids normally removed by the liver and a reduction in the amino acids used by the extraheptic tissues. Elevated plasma levels of tyrosine, methionine, phenylalanine, tryptophan, glutamate, and asparate are seen in cirrhotics.[130-134] The levels of branched-chain amino acids, valine, leucine, and isoleucine, are reduced.[135-137]

Shaw and Lieber[138] have shown that plasma branched-chain amino acids in the alcoholic are affected by dietary protein deficiency and advanced cirrhosis which tend to decrease their levels, and by chronic alcohol consumption which tends to increase them. Plasma and liver alanine levels are depressed in alcoholics as alcohol stimulates the conversion of alanine to lactate.[139]

The effects of alcohol on protein metabolism, as previously described, could be aggravated in the elderly because of the alterations in protein metabolism, changes in dietary habits, and high frequency of illness. With increasing age, body protein mass declines progressively.[140-142] Turnover studies by Sharp et al.[143] indicate that protein synthesis is slowed by physiological aging. However, Young et al.[144] found that whole-body protein synthesis and breakdown were 46 to 68% greater in old than in young adults when related to creatinine excretion, 10 to 145% greater when related to whole body ^{40}K measurement, and 2 to 24% greater when related to basal caloric expenditure. These investigators interpret the findings as an indication of a lower contribution of muscle to whole body protein synthesis and breakdown in the elderly than in young adults. Since protein turnover is more rapid in visceral organs than in muscle and the quantity of muscle diminishes with age, visceral tissues contribute more to whole body protein metabolism in the elderly.[144] The serum albumin level declines with aging, while serum globulins show a compensatory rise.[145-147] Most investigators attribute this to a reduced hepatic protein synthesis in the elderly. Barboriak et al.[42] report a slightly lower serum albumin in elderly subjects imbibing alcohol than in subjects not consuming alcohol. The findings by Tuttle et al.[148] suggested that the minimum requirement for essential amino acids may be higher in older individuals or that the nonessential nitrogen source, glycine, may not be utilized as well by the elderly as by the young for the synthesis of other nonessential amino acids. Additional studies by Tuttle et al.[149] showed than an older persons' essential amino acid requirement is dependent on total nitrogen intake. Although nonessential as well as essential amino acid composition of the diet influence nitrogen retention, Tuttle et al.[150,151] suggested that there may be failure of certain nonessential pathways in the elderly. Studies have also demonstrated deficiencies of methionine and lysine in the diets self-selected by the elderly.[152,153] Furthermore, chronic alcohol consumption was shown to cause an increase in degradation of methionine.

The less than adequate protein intake among the elderly may be largely attributed to the economic factor as protein foods are traditionally regarded as more expensive.[154-156] Many elderly individuals consume a relatively large amount of foods high in carbohydrates because they are usually less expensive than protein foods and are available in forms that require little or no preparation.[157-159] Other factors that may contribute to protein imbalance in the aged include interference with intake due to loss of teeth or poor-fitting dentures, interference with absorption due to gastric achlorhydria, a decrease in production and delivery of digestive enzymes, diminished production and delivery of bile, and interference with protein storage and utilization.[160] Finally the high frequency of various illnesses in the elderly may increase protein losses and affect protein absorption.[161]

Vitamins and Alcohol

Increasingly, it is becoming apparent that alcohol intake can affect vitamin utilization and modify their metabolic effects. Relatively few reports are available on this interaction in the specific geriatric age group. However, neither is information available indicating that these populations will be spared from such unfavorable effects. Since, with aging, the need for vitamins may be increased by the reduced metabolic efficiency, and the supply is diminished because of lower food intake, vitamin needs of the elderly may be at an increased risk of being affected by alcohol. In the following, the alcohol-vitamin interactions as it pertains to the individual vitamins will be described.

Water-Soluble Vitamins and Alcohol
Ascorbic Acid

It has not been generally realized that alcohol may contribute to or potentiate ascorbic acid deficiency in man. Various investigators have reported lower leukocyte ascorbic acid levels,[162-164] lower plasma ascorbic acid levels,[165] and lower urinary ascorbic acid excretion[166] in alcoholics than in controls. The marginal ascorbic acid status seen with alcohol ingestion may be due to insufficient dietary intake as well as to the quantity of alcohol ingested. In a general population survey, a decrease in ascorbic acid levels was found with increased alcohol consumption.[167] O'Keane et al.[162] reported that 25 of 50 alcoholic subjects had low leukocyte ascorbic acid levels associated with inadequate dietary intake. There was a similar correlation between low dietary intake and plasma ascorbic acid levels in hospitalized patients, many of whom were alcoholics.[168-170] The quantity of alcohol consumed may also affect the ascorbic acid status as a significantly higher proportion of patients ingesting an equivalent of more than 70g of pure alcohol daily were found to have plasma ascorbic acid levels below 11 μmol/ℓ.[171]

The effect of alcohol on ascorbic acid absorption, metabolism, and utilization has not been investigated in detail. Furthermore, no typical clinical symptoms of ascorbic acid deficiency attributable to alcohol ingestion have been reported. However, Beattie and Sherlock[163] reported that the low level ascorbic acid levels that correlated with the low leukocyte ascorbic acid concentrations was associated with impaired drug metabolism in man.

Blood ascorbic acid levels tend to be lower in the elderly but can be increased with ascorbic acid supplements.[171-175] It has been reported that some elderly individuals have low ascorbic acid intake,[176,177] possibly due to the higher cost of many foods rich in vitamin C. The elderly have also been known to frequently use nonsteroid anti-inflammatory agents such as aspirin, indomethacin, and phenylbutazone, which inhibit ascorbic acid metabolism.[178,179] Furthermore, the elderly are occasionally prescribed corticosteroid which have been shown to increase urinary excretion of ascorbic acid.[180] A combination of the reported effects of alcohol and the tendency of older subjects to marginal ascorbic acid nutrition indicate a possible unfavorable effect of alcohol on this variable in the elderly.

Thiamin

Thiamin deficiency is the most common vitamin deficiency seen among alcoholics. In a study of 172 malnourished alcoholics, 30% had reduced blood thiamin levels.[181] Of 34 alcoholics admitted to a health care facility for treatment of alcoholism, 27% had low excretion of thiamin.[44] When 33 hospitalized alcoholics were given a glucose load, 55% had elevated blood pyruvate levels and 31% had abnormal activation of erythrocyte transketolase (ETK).[182] Alcoholism has also been found to be the major contributor to thiamin deficiency in general hospitalized patients, as 43% of 86 patients studied for vitamin deficiency had ETK values over 1.25 and 65% of these patients were alcoholics.[83]

The poor thiamin status in alcoholics may be due to deficient thiamin intake, poor availability of dietary thiamin, increased requirement, impaired intestinal absorption, impaired utilization, or genetic predisposition to thiamin deficiency.

McLaren et al.,[184] using a 7-day dietary recall, found that 70% of 73 alcoholics had an inadequate intake of thiamin. Of 13 alcoholics with a history of deficient dietary intake, 9 had low serum levels.[185] It has also been found that thiamin in foods is poorly utilized by alcoholics with liver disease.[186] Impaired intestinal absorption also contributes to the poor thiamin status of alcoholics. Reduction in thiamin absorption of 3,[187] 15,[188,189] and 148 μmol[190] have been reported. The administration of alcohol appears to change both the structure and function of the small intestine, thereby reducing thiamin absorption.[82,91] Furthermore, an active mechanism for low dosage absorption exists in man[192,193] and is impaired by alcohol.[188,189]

Alcohol tends to increase the thiamin requirement. In rabbits, the acute administration of alcohol resulted in decreased blood thiamin levels and reduced transketolase activity as well as a decline in the ratio of free thiamin to total thiamin in plasma.[194] Chronic administration of alcohol to rats on a thiamin-sufficient or deficient diet reduced brain and liver thiamin concentrations.[195]

The chronic consumption of alcohol also impairs the ability to retain thiamin. Chronic alcohol administration, to both thiamin sufficient and deficient rats, resulted in increased thiamin excretion in the urine as compared to the control rats,[196] possibly due to a reduced hepatic binding of the vitamin.[199]

Blass and Gibson[197,198] reported that a genetic variation in thiamin-dependent enzymes which impairs the ability of some individuals to use marginal amounts of thiamin may be an important factor in thiamin deficiency. They found that the binding of thiamin pyrophosphate is 12-fold higher in Wernicke-Korsakoff-derived fibroblasts than in fibroblasts from normal subjects.

Alcoholics who become thiamin-deficient show neurological, cardiovascular, and hepatic damage.[199-201] Between 3 and 12% of all hospitalized alcoholics develop Wernicke's syndrome.[184,201,202] Thiamin treatment prevents progression of the syndrome to Korsakoff's psychosis.[202,203] Thiamin pyrophosphate (TPP), the main biologically active form of thiamin, functions as a coenzyme of pyruvate oxidase in oxidative decarboxylation and forms a link between carbohydrates, fats, and proteins, and the citric acid cycle. Thiamin pyrophosphate also functions as a coenzyme with transketolase of pentose phosphate cycle. When thiamin deficiency occurs, there is an interference with the catabolism of pyruvic acid and pyruvic acid accumulates in the blood and tissues, especially after a glucose load. Increased concentrations of pyruvic and α-oxoglutaric acid correlate with the severity of neurological symptoms seen in alcoholics.[204,205]

A number of observations in the elderly suggest that they may have an increased susceptibility to thiamin deficiency. Although large cross-sectional studies of the noninstitutionalized population in the U.S. indicate that thiamin levels in the consumed diet are adequate,[206,207] approximately 5% of individuals over 60 years of age show impaired thiamin status.[208] Poverty, institutionalization, illness, medication, impaired absorption, intrinsic abnormality, and alcohol may have contributed to the thiamin deficiency. Brin et al.[209] observed impairment of thiamin biochemical status in the free-living poor. Income is a primary factor in determining the quality of a diet. Therefore, the elderly tend to purchase less expensive foods, higher in carbohydrates (bread, jam, jelly, or ready-to-eat cereals with a high sugar content). As a result, the thiamin requirement may be increased due to a high carbohydrate diet as the thiamin requirement is proportional to energy intake, especially when calories are derived primarily from carbohydrates.[210] For many elderly, breakfast consists of tea and toast. It has been reported that individuals who consume a large amount of tea, which contains a thiamin antagonist, may have an increased risk of developing thiamin deficiency.[211] A study by Baker et al.[212] showed that approximately 25% of the noninstitutionalized elderly are marginally deficient, as estimated on the basis of their thiamin and its phosphate esters levels. Studies of geriatric patients in a Veterans Administration Hospital[209] indicate that 7% were severely thiamin depleted; 50% were found depleted in a nursing home.[213] A study of 10 elderly men in a community home for the aged indicated an average intake of only 0.7 mg, i.e., approximately two thirds of the recommended intake of 1.2 mg.[213,214] The common use of diuretics by the elderly may also increase thiamin excretion and result in marginal thiamin status. Absorption difficulties may affect thiamin status in the elderly. Baker et al.[212] reported that 5 of 228 elderly nursing home residents showed a low blood thiamin level despite adequate dietary supplementation, but they repleted rapidly when thiamin was administered intramuscularly. Other studies also indicate that thiamin deficiency can be found despite the regular ingestion of the vitamin.[212,215] It appears that some of the elderly are functionally thiamin-deficient because of an intrinsic enzymatic abnormality. The above findings also indicate that alcohol, especially if taken in large amounts, may aggravate the already precarious marginal thiamin supply seen in the elderly population.

Riboflavin

Riboflavin deficiency is seen in many chronic alcoholics and low mean riboflavin blood levels,[168,181,216] as well as high mean values of the activation coefficient for erythrocyte glutathione reductase (α_{EGR})[182,217-219] have been reported. Deficiency appears to be primarily due to an inadequate intake. In a study of 51 elderly men residing in a home for veterans, Barboriak et al.[42] found that 25 subjects with a history of regular high alcohol consumption had a significantly lower (2.2 + 0.2 mg/day) riboflavin intake than 26 abstainers (2.8 + 0.2 mg/day). Furthermore, in a study of 560 hospitalized patients, Lemoine et al.[169] reported that riboflavin intake and status were significantly correlated as 23% of the patients ingesting riboflavin of less than 2.2 mg/day had a $\alpha_{EGR} > 1.20$, whereas, only 7% of those ingesting riboflavin of more than 3.6 mg/day had a risk of deficiency. However, no direct effect of alcohol on riboflavin absorption was seen in these patients and no correlation between alcohol intake and riboflavin status was observed.[186]

In rats, studies on the effects of chronic and acute alcohol administration indicated that alcohol did not significantly affect EGR values when rats were fed a nutritionally adequate diet. However, when given a riboflavin-deficient diet, alcohol potentiated the riboflavin deficiency.[220]

Despite the high incidence of reduced blood riboflavin levels, overt clinical symptoms of riboflavin deficiency are rarely seen in alcoholics. In one study, only the serum billirubin was significantly elevated.[219] However, in another study of 60 alcoholic patients with liver disease, low blood niacin and/or riboflavin values in 80% of the patients were accompanied with glossitis and cheilosis.[181] Dietary intake findings from the Health and Nutrition Examination Survey (HANES) indicate that individuals 65 years of age or older exceed the RDA for riboflavin intake.[221] The classical clinical symptoms of riboflavin deficiency are rarely seen in the elderly, even with very low dietary intake for prolonged periods of time.[222] Therefore, it appears that alcohol, if taken in excessive amounts and over prolonged period of time, may potentiate the deficiency symptoms produced by insufficient dietary intake of riboflavin. The contribution of age may reflect the higher propensity for the development of riboflavin deficiency rather than increased specific vitamin needs.

Vitamin B_6

The collective term, vitamin B_6, includes pyridoxine, pyridoxal, and pyridoxamine. Pyridoxal-5-phosphate (PLP) is the coenzyme form of vitamin B_6. It is derived primarily from hepatic synthesis and serves as a small circulating storage pool of vitamin B_6 in the body.[223] Pyridoxal-5-phosphate is bound to albumin and its concentration correlates highly with the intake of vitamin B_6 in man.[224]

Chronic alcohol consumption results in decreased levels of circulating vitamin B_6.[216,225-228] The incidence of lowered plasma PLP in alcoholic subjects without liver disease may be as high as 50%; 25 of 50[225] and 35 of 66[226] alcoholics had circulating PLP levels below the lower limit of normal range. The incidence of low plasma PLP in alcoholics with liver disease may be as high as 80 to 100%.[225]

Prolonged alcohol ingestion causes inadequate intake and interferes with vitamin B_6 absorption, storage, conversion to biologically active forms, and degradation. Although alcoholic beverages contribute a significant amount of calories, they only contain negligible quantities of the vitamin B_6[229] and may cause deficiency. Lumeng[230] presented evidence that acetaldehyde, the oxydation product of alcohol, rather than alcohol *per se,* interferes with the metabolism of vitamin B_6 by promoting the degradation of PLP. Acetaldehyde, especially at high concentrations, displaces or facilitates the dissociation of PLP from protein binding and promotes PLP degradation. Furthermore, Li et al.[231] reported that with excessive alcohol ingestion, the body storage capacity for vitamin B_6 is reduced. Liver disease can further impair vitamin B_6 status.

Baker et al.[186] reported that the intestinal tract of chronic alcoholics cannot release the bound vitamin B_6 from food. When liver disease is present, PLP degradation is accelerated.[227,228,232]

Ring sideroblasts may be one of the clinical signs of vitamin B_6 deficiency as alcoholics with the marrow change frequently had low PLP levels.[233-235] However, low circulating PLP levels do not always result in sideroblastosis.[235] It has been estimated that sideroblastic anemia is present in 20 to 30% of chronic alcoholics and is caused by multiple factors.[235-237]

Although the diagnosis of specific vitamin B_6 deficiency in the elderly is rare, vitamin B_6 status may be compromised by alcohol, as mentioned previously, by malnutrition, disease, and medications. Anticonvulsant drugs[238] and corticosteroids[239] taken by many of the elderly have been shown to increase the need for vitamin B_6. In addition, supplemental vitamin B_6 has been used to treat the polyneuritis that develops in some individuals receiving the hypotensive drug, hydralazine.[240] As seen with riboflavin, excessive alcohol intake frequently aggravates the symptoms of preexisting pyridoxine deficiency, whether it is produced by an insufficient diet or by the treatment with drugs which thereby interfere with normal pyridoxine metabolism.

Folacin and Vitamin B₁₂

Since a deficiency of either folate or vitamin B_{12} causes a megaloblastic anemia and reduced DNA synthesis of proliferating cells in the body, they will be discussed together. Alcohol has a toxic effect on folate metabolism.[82,241-246] Low folate levels are related to dietary intake, type of alcoholic beverage, quantity of alcohol ingested, and folate status. Studies have shown a correlation between dietary intake and circulating folate levels;[237,247-252] 50% of the alcoholics consuming a nutritionally marginal diet and 52% of the alcoholics consuming a poor diet had serum folate levels below 6.8 nmol/ℓ as compared to 22% of the alcoholics consuming an adequate diet.[253] Alcoholics ingesting an inadequate diet had a mean serum folate of 6.1,[246] 6.6,[248] or 8.8 nmol/ℓ[247] compared to 9.3, 15.2, or 11.1 nmol/ℓ in alcoholics ingesting an adequate diet. The type of alcoholic beverage consumed may influence the folate status as beer contains a considerable quantity of folate, while wine and whiskey contain very little.[181,253] In 21 wine and spirit drinkers, a mean folate level of 7.7 nmol/ℓ was found compared to a mean folate of 15.4 nmol/ℓ in 14 beer drinkers.[248] In another study, 45 alcoholics who drank wine or spirits had a mean circulating folate of 6.3 nmol/ℓ, whereas 29 alcoholics who drank beer had a mean circulating folate level of 11.3 nmol/ℓ.[246]

Furthermore, there is a significant correlation between the quantity of alcohol ingested and the number of patients with a depressed folate level.[249] In normal subjects, ingestion of a small amount of alcohol depressed folate levels within a few hours and the depression correlated with the quantity of alcohol ingested. Subjects who drank 6 oz of 95% alcohol had a 50% fall in serum folate; with 4 oz there was a 30 to 35% fall and, finally, subjects who drank 3 oz of alcohol had a 15 to 20% fall in serum folate.[254]

Dietary folate is composed of folic acid and polyglutamates (PGA). Folate conjugase cleaves off the terminal glutamic acid residue and forms PGA. Halsted et al.,[243,255,256] using radioisotopes, found that PGA absorption is reduced in alcoholics with folate deficiency and after recent prolonged alcohol ingestion. In addition, alcoholics have an impaired ability to convert pteroylpolyglutamic acid from food to plasma folate.[186] The ingestion of alcohol also reduces the tissue uptake of folic acid.[257]

Vitamin B_{12} is absorbed by two mechanisms. Simple diffusion accounts for a small amount and only when large quantities are ingested. Most of vitamin B_{12} is absorbed utilizing intrinsic factor and calcium ion. Studies by Lindenbaum et al.[258] suggest that alcohol impairs the intestinal uptake of vitamin B_{12}-intrinsic factor complex. In addition, reduced folate status causes a malabsorption of vitamin B_{12}. Interestingly, higher than normal blood levels of vitamin B_{12} are found in alcoholics.[225,251,252,254] Folate and vitamin B_{12} are essential for growth and proliferation of all human cells. Because folate is necessary in DNA synthesis, many tissues may be affected by a deficiency. The organs affected and the clinical lesions resulting from folate deficiency are: bone marrow (megaloblastic anemia), gonads (sterility), epithelial cell (glossitis, cheilosis, and intestinal malabsorption),[260] and nervous system (peripheral neuropathy, organic dementia, and posterior lateral column lesions).[261] A vitamin B_{12} deficieny is relatively rare and is caused by failure of absorption due to lack of intrinsic factor or malabsorption syndrome. A deficiency of vitamin B_{12} results in neurologic disorders and megaloblastic anemia.[262,263]

Folate-dependent megaloblastic anemia develops in folate-depleted individuals on a folate-deficient diet and its development is accelerated by alcohol.[237,250] In studies of alcoholics, folate-dependent megaloblastic anemia is seen more frequently in malnourished than in well-nourished alcoholics.[249,264,265]

Folate deficiency is not widespread among free-living elderly in the U.S. In the HANES study, less than 6% of 369 elderly subjects (65 to 74 years old) had serum folate below 3.0

ng/ml.[266] In the U.S. Ten-State Nutrition Survey, folate status was generally adequate as the average serum folates were above 6.0 ng/ml and red cell folates were 160 to 605 ng/ml.[267] Furthermore, results of 560 autopsies conducted in Canadian hospitals indicated that liver folate concentrations were deficient in only 2 of 164 patients over 70 years of age and the proportion of patients over 60 with liver folate less than 5 ng/ml was not different from patients 20 to 60 years of age.[268] However, elderly individuals with a low income are at a much higher risk for folate deficiency. A study of 132 elderly women in Florida showed that 60% in a lower income bracket and only 6% of those with a higher socioeconomic status had erythrocyte folate values below 140 ng/ml.[269] Another study conducted in Dade County, Florida, reported that 60% of the low income subjects had erythrocyte folate concentrations below 140 ng/ml.[270] In Britain, low folic acid levels were found in 80% of the elderly awaiting admission to welfare homes,[271] and similar findings were reported in the U.S.[272] The use of certain drugs such as diuretics, by the elderly, can interfere with folate status and cause inadequate utilization, while diphenylhydantoin (Dilantin®) can decrease folate absorption.[273] Alcohol may further potentiate acid deficiency in the elderly with poor folic acid status.

A vitamin B_{12} deficiency tends to increase with age, but it is rarely caused by dietary inadequacy.[160] The absorption of vitamin B_{12} appears to decrease with age, possibly due to a reduction in the amount of intrinsic factor which is related to gastric atrophy and decreased gastric acidity. A dietary deficiency of vitamin B_{12} can occur, however, if an individual follows a total vegetarian diet. Medications frequently affect the vitamin B_{12} status by either increasing the requirement or by interfering with absorption of the vitamin. Hypotensive drugs, dopa, and possibly methyldopa, require vitamin B_{12} metabolism.[274] The antituberculosis drug, paraaminosalicylic acid, reduces vitamin B_{12} absorption.[275] Finally, the elderly individual who consumes alcohol could become Vitamin B_{12}-deficient as alcohol reduces vitamin B_{12} absorption.

Niacin, Pantothenic Acid, and Biotin

To our knowledge, there are relatively few human studies showing a clear effect of alcohol on the absorption and utilization of niacin,[44] pantothenic acid,[276] and biotin.

Sporadic reports indicate that niacin concentration in whole blood may be reduced in alcoholics;[168,216,276,277] however, the prevalence of alcoholics with low circulating niacin levels is similar to that seen in randomly selected hospitalized patients.[272] Niacin intake does not differ greatly between imbibers and abstainers,[42,129,272] especially when the intake of its precursor, tryptophan, is included. Some clinical studies indicate that alcohol reduces whole blood pantothenic acid levels,[216,272,276] whereas other studies of alcoholic patients report a two- to threefold increase in circulating pantothenic acid levels[278] or no significant difference from the controls.[276] Studies on the effect of alcohol on circulating biotin levels indicate that they are reduced in alcoholism.[168,216,277] Liver biotin levels are also found to be reduced in alcoholics.[279]

Despite the possible changes in blood levels of these vitamins, there are no well-described signs of clinical deficiencies. Due to the possible combined effects of alcohol abuse and nutritional deficiencies, it is difficult to exclusively ascribe the changes in blood levels to either variable. Malnutrition, as such, may be the reason for low circulating levels. To our knowledge, few reports are available on the age-induced modifications in the requirements of these vitamins. A more definitive clarification of a possible interrelationship between alcohol intake and the need for these vitamins in the elderly will have to wait for findings of future studies.

Fat-Soluble Vitamins

While the sequelae of alcohol-induced deficiencies of fat-soluble vitamins may not be as common and dramatic as seen with the water-soluble vitamins, they do occur and may be associated with serious clinical impairment of patients.

Vitamin A

The interaction between alcohol and vitamin A metabolism may occur at several levels. Chronic administration of ethanol has been associated with increased release of the vitamin from the hepatic stores[280,281] when the diet contains *normal* amounts of vitamin A. With *increased* dietary vitamin A intake there was an increased deposition of the vitamin A metabolites in the liver. In addition, severe hepatic lesions, characterized by giant mitochondria, reduced activity of several hepatic enzymes, as well as reduction of fatty acid oxidation,[282] have been described. Alcohol consumption may also contribute to a vitamin A deficiency in an indirect manner by reducing the amount of available zinc, the metal functioning as an integral part of alcohol dehydrogenase, which is also needed for proper processing of vitamin A. Furthermore, the frequently seen nutrient absorption abnormalities associated with chronic alcohol abuse may interfere with proper utilization of dietary vitamin A or its precursor, beta carotene.

The main clinical symptom of vitamin A deficiency seen in alcoholics is the development of night blindness.[283] In most instances, the administration of vitamin A corrected the condition.[284] In some patients, however, treatment with both vitamin A and zinc was necessary to regain the night vision.[285] Relatively little information is available about the changes in vitamin A requirement with aging. However, the reduced fat tolerance and insufficient dietary intake make it probable that the vitamin A status in many of the elderly may be marginal; any additional insult, such as prolonged consumption of large amounts of alcohol, may contribute to the development of a more pronounced vitamin A deficiency.

Vitamin D (Cholecalciferol)

Reports in the literature indicate that the impairment of calcium metabolism and increased bone fragility seen in alcoholics are results of the combined effects of alcohol on both calcium and vitamin D, the vitamin involved in calcium utilization. The mechanism for the observed osteopenia is not clear, since several studies have shown normal hydroxylation of cholecalciferol in alcoholics.[286] However, alcoholics as a group tend to have lower blood levels of vitamin D, possibly due to less exposure to sunlight which would result in lower levels of cholecalciferol. Alcohol has also been reported to activate some hepatic drug-metabolizing enzyme system components.[287] This may account for a more rapid removal of vitamin D from the circulation.

Some of the same factors seen in alcoholics in general may contribute to the vitamin D-deficiency in the aged. The inadequate dietary intake, especially of foods which are fortified with vitamin D,[288,289] and reduced exposure to sunlight result in lower circulating levels of vitamin D in the elderly. In addition, several investigators have reported an age-dependent decrease in circulating 25-hydroxycalciferol levels, especially in women.[290] Therefore, one could expect that alcohol intake by the elderly, especially by those with greater risk of vitamin D deficiency, will enhance and aggravate the sequelae of deficiency and contribute to a speedier development of bone changes.

Minerals

Alcohol intake, especially if excessive, is usually associated with the development of deficiencies of minerals, particularly of magnesium (Mg), zinc (Zn), calcium (Ca), phosphorus (P), and potassium (K). Even under normal conditions, several of these minerals are either in greater need or not efficiently utilized by the elderly; therefore, the combination of alcohol abuse and aging may be more detrimental than either factor alone.

Magnesium

Magnesium as a component of enzymes involved in metabolism of carbohydrates, lipids, and amino acids, in energy transfer, as well as in synthesis and degradation of nucleic acids,[291] plays an important role in the maintenance of health. A Mg deficiency in alcoholics has been repeatedly observed and has been ascribed to decreased dietary intake, increased excretion, malabsorption, and deficient renal reabsorption.[292] Alcohol-induced reduction of blood and tissue levels of Mg have been reported.[292,293] The myocardium appeared to be especially affected.[294] The initial symptoms of Mg deficiency are relatively nonspecific and include apathy, nausea, and reduced appetite. In more severe deficiency, personality changes,[295] spontaneous myospasm, tremor, convulsion, and coma may develop. Since advanced Mg deficiency is usually combined with hypocalcemia and hypokalemia,[292] it is sometimes difficult to decide whether the deficiency symptoms are due solely to insufficient Mg or represent a reaction to the combined deficiencies. However, in most cases, the symptoms tend to disappear on adequate supplementation with Mg only.[295]

While the interrelationship between alcohol intake and Mg metabolism has been explored quite extensively, little or no definitive information is available on the modification of Mg metabolism by age. However, it is known that the elderly frequently use medications which are associated with Mg loss, such as diuretics for treatment of hypertension. Furthermore, the recommended increased intake of dietary fiber may also contribute to Mg loss.[296] This, combined with the usual nutritional deficiency of Mg,[296] indicates that the alcohol-induced aggravation of Mg deficiency may be more pronounced in the elderly.

Calcium

Unlike Mg, the requirements for Ca in the elderly have been rather extensively investigated, mostly due to the high prevalence of osteoporosis in the more advanced stage of deficiency[297] or to the general loss of bone mass in elderly adults.[298] Reduction in the bone mass may be seen with low dietary intake of Ca,[299] with reduced physical activity,[300] and with hormonal changes after menopause.[301] The possible role of lactose in Ca absorption is strongly indicated by the higher prevalence of osteoporosis in lactase-deficient patients.[302,303]

The consumption of alcohol, especially in large amounts, has also been reported to affect Ca utilization, possibly by affecting duodenal Ca transport. This effect is not mediated by an alcohol-induced modification of vitamin D activity, since the impairment of Ca transport by alcohol could only partially be prevented by supplementation with vitamin D.[304] A number of studies have found reduced bone mass in alcoholics,[305,306] and alcohol consumption appeared to potentiate the bone mass loss seen with progressing ages.[307]

It seems, therefore, that the elderly, already at greater risk for the development of osteoporosis because of age, hormonal changes, and reduced or modified dietary intake of Ca, may be further affected by the consumption of alcohol since the effect of this agent seems to be independent of the above factors. Additional impairment of calcium utilization due to the effect of alcohol on vitamin D metabolism could further aggravate bone health of the elderly.

Zinc

Next to Mg, Zn is probably the mineral most commonly affected by alcohol consumption. Low plasma and tissue levels of this metal have been found in chronic alcoholics.[308,309] The liver content of Zn has been found to be substantially reduced, both in man and experimental animals. Increased urinary elimination of Zn by alcoholics has also been repeatedly reported.[310,311]

Marginal or deficient dietary intake of Zn has been found in many of the elderly,[312,313] possibly due to a higher cost of the usual Zn sources, i.e., animal protein.[314,315] Similar to the conditions seen with Ca, the higher proportion of high-fiber food items consumed by the elderly may contribute to Zn deficiency since both the fiber and phytic acid commonly found in those products reduce the availability of dietary Zn.[316,317] Zn is needed for proper wound healing,[318,319] maintenance of proper immune responses,[320,321] and retention of taste acuity.[322]

One could expect that a marginal deficiency of Zn encountered especially in the institutionalized elderly, will be substantially aggravated by alcohol consumption. Since many of the conditions associated with Zn deficiency, i.e., impaired immune response and taste acuity, are quite common in the elderly, an inquiry into the drinking habits of such individuals may be helpful in identifying factors contributing to these conditions.

FACTORS AFFECTING THE NUTRIENT-ALCOHOL INTERACTION IN THE ELDERLY

The interaction between alcohol intake and nutritional needs of the elderly can occur at several levels. Socioeconomically, the elderly, being at a lower scale of economic means, may not be able to obtain a nutritionally adequate diet if their dietary needs conflict with the desire for alcoholic beverages. Furthermore, alcohol consumption, especially if excessive, may cause functional and structural impairment of organs needed for efficient utilization of nutrients provided by food. At the metabolic level, alcohol may interfere with the transformation of individual nutrients into forms which are needed for their efficient utilization. In addition, the elderly frequently suffer from diseases which have a nutritional background or components. Again, alcohol consumption by such individuals may aggravate the nutritional needs of the integrity of the organs affected by the disease process.

Socioeconomic Factors

Due to the regulations requiring compulsory retirement, many of the elderly derive most of their income from Social Security benefits and pension funds.[323] When compared with their status while gainfully employed, this usually represents a substantial reduction in their living standard. Consequently, they frequently omit higher cost foods such as meats and/or milk which are important for an adequate supply of high quality protein, Ca, and other nutrients.[324,325] If competition arises between their desire to purchase alcoholic beverages or food, they frequently opt for alcohol, thereby further reducing their nutrient intake. Furthermore, alcohol abuse may affect their intellectual and reasoning capacity,[326] and they are unable to comprehend the ultimate health-impairing effects of their dietary preferences. More recent data suggest that alcohol abuse may speed up the physiological deterioration usually associated with aging,[327] and thus further complicate provision of adequate nutrition.

Organ and Tissue Damage

Alcohol, especially if taken chronically and at high concentrations, has been reported to impair the utilization of nutients as well as the physiological function of a number of organs directly involved in the processing of food. Again, these effects can be expected to be more serious in elderly individuals who may have some impairment of these organs due to the normal age-induced structural changes or functional impairment (Table 4).

Mouth — The elderly frequently demonstrate signs of poor oral hygiene due to the loss of sensitivity to irritations of the mouth, less frequent visits to the dentist, or physical disabilities.[327] Concomitant use of alcohol may aggravate the already inadequate dental care, since alcoholics frequently lose interest in practicing proper oral hygiene and do not show concern about their appearance.[328] Chronic alcohol abuse has been reported to predispose to periodontal disease,[329] cause swelling of the salivary glands,[30] and increase the tendency to bleed profusely even from minor dental procedures.[331]

Table 4

ALCOHOL AND AGE-INDUCED ORGAN IMPAIRMENTS WHICH MAY AFFECT THE NUTRITIONAL STATUS OF THE ELDERLY

Organ	By alcohol	By age[328a]
Mouth	Periodontal disease,[329] swelling of salivary glands[330] Bleeding tendency[331]	Poor dental hygiene Loss of taste, loss of teeth
Esophagus	Motor dysfunction[332] Varices[333]	Dysphagia Hiatus hernia, esophagitis
Stomach	Delayed emptying, increased acid production[334] Back diffusion of H^+ ions,[335] gastritis, ulcers[336] Achlorhydria, atrophy[337]	Atrophy, gastritis Ulcers
Small intestine	Changes in motility,[339] steatorrhea[339] Malabsorption of nutrients[341]	Diverticula, duodenal ulcers Intestinal obstruction
Pancreas	Reduced exocrine secretion[338] Direct duct occulsion or rupture[338] Closure of sphincter of Oddi,[338] pancreatitis[338]	Reduced secretion of proteolytic activity Obstruction of pancreatic duct Pancreatitis
Liver	Changed metabolism of carbohydrate, fat, and protein[59,86,113] Fatty liver, hepatitis,[342] cirrhosis[344]	Hepatitis

Esophagus — Esophageal malfunction consisting of motor dysfunction and development of varices has been reported in a number of alcohol abusers.[333]

Stomach — Gastric reactions to alcohol range from increased acidity, possibly due to enhanced secretion of the hormone gastrin,[334] back diffusion of the hydrogen ions across the gastric mucosa,[335] gastric erosions,[336] and reduced production of the intrinsic factor[337] which may then aggravate the effect of vitamin B_{12} deficiency.

Pancreas — Alcohol abuse is frequently associated with a malfunction of the pancreas due to a constriction of the sphincter of Oddi at the pancreatic duct.[338] Furthermore, the pancreatic secretion may be affected by the changes in gastric acid production and secretin. In the final outcome, the secretion of pancreatic products is inhibited, with subsequent reduction of nutrient utilization.

Small Intestine — The integrity of the function of the small intestine is also affected by alcohol. Acutely, the presence of alcohol in the small intestine can reduce absorption of some carbohydrates, amino acids, and inhibit thiamin utilization.[339] If imbibed chronically, alcohol causes hemorrhagic lesions and reduced intestinal absorptive surface.[340] These changes are associated with reduced vitamin B_{12} absorption and reduced lactase activity.[341] The latter factor aggravates the usual lactose intolerance which is frequently seen in the elderly.

Liver — The liver, because of its central role in the metabolism of nutrients, is especially at risk in chronic or even moderate alcohol consumption. The early toxic effect leads to fatty infiltration of the organ.[342] This is normally a benign reaction reversible on cessation of alcohol consumption. The next escalation of alcohol toxicity — alcohol hepatitis — represents a more serious disease with mortality as high as 30% of the afflicted.[343] Finally, alcoholic cirrhosis is the rather common and serious form of hepatic insult directly correlated with the length and extent of alcohol consumption.[344] Due to the relatively high mortality associated with alcoholic liver disease, its prevalence among the elderly is rather limited. However, it could be expected that some of the reactive alcoholics, i.e., elderly individuals who began abusing alcohol following some traumatic event(s) or retirement, may be a high or higher risk than comparable younger individuals.

DRUG-NUTRIENT INTERACTION

Health professionals need to be aware that many drugs commonly used by the elderly can compromise nutritional status by affecting dietary intake, nutrient metabolism, and/or nutrient excretion. Some drugs cause anorexia, nausea, or gastric distress and thereby interfere with dietary intake. Some medications, when taken with food, decrease absorption of nutrients. Other drugs cause increased destruction or excretion of a nutrient. Fisher et al.[345] studied medication use by elderly noninstitutionalized Utahans. The study revealed that 28% took hypotensive drugs, 23% were prescribed heart and circulation medicines, and 21% took prescription pain medications, primarily for arthritis. Digestive and laxative aids were used by 17% of those surveyed. Sedatives and relaxants for tension and anxiety were taken by 15%. Table 5 outlines some drugs commonly taken by the elderly and how they may adversely affect nutritional status.[346-393]

CONCLUSION

While the nutritional problems of the elderly and the impairment of nutritional status by drinking were explored to considerable extent, there are inexplicable lacunae of knowledge about the effect of alcohol consumption on nutrition in the older population. The data described in this chapter indicate that a considerable proportion of the elderly *do* consume alcohol, some imbibing considerable amounts. The elderly as a group tend to have more nutritional problems, but with few exceptions,[21,42] no reliable reports are available dealing

Table 5
SOME DRUGS COMMONLY TAKEN BY THE ELDERLY THAT MAY AFFECT NUTRITIONAL STATUS AND POTENTIATE THE EFFECT OF ALCOHOL

Drug	Nutrient	Manifestation	Ref.
Antacids	Phosphorous	Phosphorous depletion symdrome	
		Bone demineralization	
		Osteomalacia	346—348
		Hypophosphatemia	
		Hypercalciura	
	Thiamine	Alkaline destruction	349
Anticonvulsants	Folacin	Low serum folate levels	350
	Vitamin D	Impaired absorption	351
		Vitamin D deficiency	351
		Osteomalacia	351
	Vitamin K	Increased requirement	352
Antibiotics	Fat	Increased excretion	353, 354
Neomycin	Nitrogen	Increased excretion	353
	Sodium	Increased excretion	353
	Potassium	Increased excretion	353
	Calcium	Increased excretion	353
	Vitamin B_{12}	Impaired absorption	354, 355
	Iron	Impaired absorption	354
	Lactose	Impaired absorption	354
	Sucrose	Impaired absorption	354
	Vitamin A	Impaired absorption	357
	Vitamin D	Impaired absorption	357
	Vitamin K	Decreased absorption (vitamin K)	356
		Potentiate coumarin anticoagulant function	356
Penicillamine	Vitamin B_6	Peripheral neuropathy	358, 359
Tetracycline	Vitamin C	Tissue desaturation	360
	Calcium	Reduced absorption	361
	Iron	Reduced absorption	362, 363
Cardiovascular drugs			
Hydralazine	Vitamin B_6	Neuropathy	358
		Polyneuritis	364
Cholestyramine	Fat	Steatorrhea	365, 366
	Vitamin A	Decreased absorption	367
	Vitamin D	Osteomalacia	368
	Vitamin K	Vitamin K deficiency	369
	Vitamin B_{12}	Decreased absorption	370, 371
Irritant cathartics	Calcium	Hypocalcemia	372, 373
		Osteomalacia	372, 373
	Vitamin D	Hypocalcemia	373
		Osteomalacia	373
	Potassium	Potassium depletion	374
Isoniazid	Vitamin B_6	Peripheral neuropathy	375, 376
	Niacin	Pellagra	377
Mineral oil	Vitamin A	Impaired absorption	378—380
	Vitamin D	Impaired absorption	378, 379, 381
		Osteomalacia	378, 379, 381
	Vitamin K	Impaired absorption	378, 379, 382
Potassium chloride	Vitamin B_{12}	Abnormal Schilling test	383
		Impaired absorption	383

Table 5 (continued)
SOME DRUGS COMMONLY TAKEN BY THE ELDERLY THAT MAY AFFECT NUTRITIONAL STATUS AND POTENTIATE THE EFFECT OF ALCOHOL

Drug	Nutrient	Manifestation	Ref.
Prednisone (other glucocorticoids)	Calcium	Impaired absorption	384, 385
	Vitamin D	Rapid turnover	384, 385
Salicylates	Vitamin C	Decreased uptake in thrombocytes and leukocytes	386—388
	Vitamin K	Hypoprothrombinemia	356
	Folacin	Reduced protein binding of folate	389
	Iron	Iron deficiency anemia	390
Triamterene	Folacin	Megaloblastic anemia	391, 392
Trimethoprim	Folacin	Minimal folate deficiency	393

with the nutrient-alcohol interaction in age groups over 60. In view of the rising alcohol consumption, especially in younger groups, and the tendency to continue this habit with aging, one can expect that the nutritional problems associated with drinking alcoholic beverages in the elderly will become more frequent in the future. It is hoped that identification of areas of overlap, where alcohol and old age seem to affect the health of the older population, will stimulate additional research and provide motivation for preventive measures.

ACKNOWLEDGMENTS

The authors wish to thank Catherine A. Walther, Patricia Arsiniega, and Lori Pakalns for their typing and editorial assistance.

REFERENCES

1. **Fingerhut, L. A. and Rosenberg, H. M.,** Mortality among the elderly, in *Health-United States,* DHHS, Publication No. (PHS) 82-1232, U.S. Department of Health and Human Services, 1980, 15.
2. **Winick, M., Ed.,** *Nutrition and Aging,* John Wiley & Sons, N.Y., 1976.
3. **Borgstrom, G., Norden, A., Akesson, B., Abdulla, M., and Jagerstad, M.,** Nutrition and old age, *Scand. J. Gastroenterol.,* 14 (Suppl. 52), 1979.
4. **Munro, H. N.,** Nutritional requirements in the elderly, *Hosp. Pract.,* 143, 1982.
5. **Shock, N. W.,** The role of nutrition in aging, *J. Am. Coll. Nutr.,* 1, 3, 1982.
6. **Butler, R. N.,** Current Definitions of Aging, in *Epidemiology of Aging,* Haynes, S. G., Feinleib, M., Ross, J. A., and Stallones, L., Eds., DHHS, Publication No. (NIH) 80-969, U.S. Government Printing Office, Washington, D.C., 1980, 7.
7. **Costa, P. T. and McCrae, R. R.,** Functional Age: A Conceptual and Empirical Critique, in *Epidemiology of Aging,* Haynes, S. G., Feinleib, M., Ross, J. A., and Stallones, L., Eds., DHHS, Publication No. (NIH) 80-969, U.S. Printing Office, Washington, D.C., 1980, 23.
8. **Adelman, R. D.,** Definition of Biological Aging, in *Epidemiology of Aging,* Haynes, S. G., Feinleib, M., Ross, F. A., and Stallones, L., Eds., DHHS, Publication No. (NIH) 80-969, U.S. Government Printing Office, Washington, D.C., 1980, 9.
9. **Mendelson, J. H. and Mello, N. K.,** Biologic concomitants of alcoholism, *N. Engl. J. Med.,* 301, 912, 1979.
10. **Kendell, R. E.,** Alcoholism: a medical or a political problem?, *Br. Med. J.,* 1, 367, 1979.
11. **Eckardt, M. J., Harford, T. C., Kaelber, C. T., Parker, E. S., Rosenthal, L. S., Ryback, R. S., Salmoiraghi, G. C., Vanderveen E., and Warren, K. R.,** Health hazards associated with alcohol consumption, *JAMA,* 246, 649, 1981.

12. **Edwards, G., Gross, M. M., Keller, M., Morer, J, and Room, R., Eds.,** Alcohol-related disabilities, WHO Offset Publication No. 32, World Health Organization, Geneva, 1977.

13. **Morey, L. C. and Blashfield, R. K.,** Empirical classiciations of alcoholism (a review), *J. Stud. Alcohol.,* 42, 925, 1981.

14. **Thompson, A. D. and Mujumdar, S. K.,** The hazard to health from moderate drinking, in *Nutrition and Health,* Turner, M. R., Ed., Alan R. Liss, N.Y., 1981, 87.

15. **Knight, R. P.,** The dynamics and treatment of chronic alcohol addiction, *Bull. Menninger Clin.,* 1, 233, 1937.

16. **Wood, W. G.,** The elderly alcoholic: some diagnostic problems and considerations, in *The Clinical Psychology of Aging,* Storand, M., Siegler, I., and Elias, M., Eds., Plenum Press, N.Y., 1978.

17. **Cahalan, D., Cisin, I. H., and Crossley, H. M.,** American drinking practices, Monograph No. 6, Rutgers Center of Alcohol Studies, *J. Stud. Alcohol,* New Brunswick, N.J., 1969.

18. **Wechsler, H., Demone, H., and Gottlieb, N.,** Drinking patterns of greater Boston adults, subgroup differences on the QFV index, *J. Stud. Alcohol,* 39, 1158, 1978.

19. **Hardford, T. C. and Gerstel, E. K.,** Age-related patterns of daily alcohol consumption in metropolitan Boston, *J. Stud. Alcohol,* 42, 1062, 1981.

20. **Barboriak, J. J., Barboriak, D. P., Anderson, A. J., and Hoffmann, R. G.,** Drinking patterns and preferences among heart patients, *Curr. Alcohol.,* 8, 293, 1981.

21. **Jones, B. R., Barrett-Connor, E., Criqui, M. H., and Noldbrook, M. J.,** A community study of calorie and nutrient intake in drinkers and nondrinkers of alcohol, *Am. J. Clin. Nutr.,* 35, 135, 1982.

22. **Johnson, L. A. and Goodrich, C. H.,** Use of alcohol by persons 65 years and over, Upper East Side of Manhattan, Report of the National Institute on Alcobol Abuse and Alcoholism, Mount Sinai School of Medicine, City University of New York, N.Y., 1974.

23. **Wilson, P.,** Drinking in England and Wales, Her Majesty's Stationery Office, London, 1980.

24. **Gaitz, C. M. and Baer, P. E.,** Characteristics of elderly patients with alcoholism, *Arch. Gen. Psychiatry,* 24, 372, 1971.

25. **Simon, A., Epstein, L. J., and Reynolds, L.,** Alcoholism in the geriatric mentally ill, *Geriatrics,* 23, 125, 1968.

26. **Stockford, M. A.,** Alcoholism and Problem Drinking, 1970-1975. A Statistical Analysis of VA Hospital Patients, 1980 Supplement, Office of Reports and Statistics, Veterans Administration, Washington, D.C., 1982.

27. **Bistrian, B. R., Blackburn, G. L., Vitale, J., Cochran, D., and Naylor, J.,** Prevalence of malnutrition in general medical patients, *JAMA,* 235, 1567, 1976.

28. **Weinsier, R. L., Hunker, E. M., Krumdieck, C. L., and Butterworth, C. E.,** A prospective evaluation of general medical patients during the course of hospitalization, *Am. J. Clin. Nutr.,* 32, 418, 1979.

29. **Thomson, A. D.,** Alcohol and nutrition, *Clin. Endocrinol. Metab.,* 7, 405, 1978.

30. **Li, T. K., Schenker, S., and Lumeng, L., Eds.,** Alcohol and Nutrition, Research Monograph No. 2, DHEW Publication No. (ADM) 79-780, U.S. Government Printing Office, Washington, D.C., 1979.

31. **Tomaiolo, P. P.,** Nutritional problems in the alcoholic, *Compr. Ther.,* 7(7), 24, 1981.

32. **Eisenstein, A. B.,** Nutritional and metabolic effects of alcohol, *J. Am. Diet. Assoc.,* 81, 247, 1982.

33. **Lieber, C. S.,** Alcohol and malnutrition in the pathogenesis of liver disease, *JAMA,* 233, 1077, 1975.

34. **Morgan, M. Y.,** Alcohol and nutrition, *Br. Med. Bull.,* 38, 21, 1982.

35. **Pirola, R. C. and Lieber, C. S.,** The energy cost of the metabolism of drugs, including ethanol, *Pharmacology,* 7, 185, 1972.

36. **Pirola, R. C. and Lieber, C. S.,** Hypothesis: energy wastage in alcoholism and drug abuse; possible role of hepatic microsomal enzymes, *Am. J. Clin. Nutr.,* 29, 90, 1976.

37. **Lieber, C. S.,** Liver adaptation and injury in alcoholism, *N. Engl. J. Med.,* 288, 356, 1973.

38. **Aebi, H. and von Wartburg, J. P.,** Comparative-biological aspects of experimental research on the effects of chronic alcohol administration, *Bull. Schweiz. Akad. Med. Wiss.,* 16, 25, 1960.

39. **Kiessling, U. H. and Filander, K.,** Biochemical changes in rat tissues after prolonged alcohol consumption, *Q. J. Stud. Alcohol.,* 22, 535, 1961.

40. **Mezey, E. and Faillace, L. A.,** Metabolic impairment and recovery time in acute ethanol intoxication, *J. Nerv. Ment. Dis.,* 153, 445, 1971.

41. **Morgan, M. Y., Camilo, M. E., Luck, W., Sherlock, S., and Hoffrand, A. V.,** Macrocytosis in alcohol-related liver disease: its value for screening, *Clin. Lab. Hematol.,* 3, 35, 1981.

42. **Barboriak, J. J., Rooney, C. B., Leitschuh, T. H., and Anderson, A. J.,** Alcohol and nutrient intake of elderly man, *J. Am. Diet. Assoc.,* 72, 493, 1978.

43. **Hurt, R. D., Higgens, J. A., Nelson, R. A., Morse, R. M., and Dickson, E. R.,** Nutritional status of a group of alcoholics before and after admission to an alcoholism treatment unit, *Am. J. Clin. Nutr.,* 34, 388, 1981.

44. **Neville, J. N., Eagles, J. A., Samson, G., and Olson, R. E.,** Nutritional status of alcoholics, *Am. J. Clin. Nutr.,* 21, 1329, 1968.

45. **Lutwak-Mann, C.,** Alcohol dehydrogenase of animal tissues, *Biochem. J.*, 32, 1364, 1938.

46. **Laser, H.,** Peroxidase activity of catalase, *Biochem. J.*, 61, 122, 1955.

47. **Nelson, G. H., Kinard, F. W., and Aull, J. C., Jr.,** Effect of amino-triazole on alcohol metabolism and hepatic enzyme activated in several species, *Q. J. Stud. Alcohol.*, 18, 343, 1957.

48. **Roach, M. K., Reese, W. N., Jr., and Creaven, P. J.,** Ethanol oxidation in the microsomal fraction of rat liver, *Biochem. Biophys. Res. Commun.*, 36, 596, 1969.

49. **Tobon, F. and Mezey, E.,** Effect of ethanol administration on hepatic ethanol and drug metabolizing enzymes and rates of ethanol degradation, *J. Lab. Clin. Med.*, 77, 110, 1971.

50. **Tephly, T. R., Tinelli, F., and Walkins, W. D.,** Alcohol metabolism: role of microsomal oxidation in vivo, *Science*, 66, 627, 1969.

51. **Smith, E. and Newman, H. W.,** The rate of ethanol metabolism in fed and fasting animals, *J. Biol. Chem.*, 234, 1544, 1959.

52. **Tygstrup, N., Winkler, K., and Lundquist, F.,** The mechanism of the fructose effect on the ethanol metabolism of the human liver, *J. Clin. Invest.*, 44, 817, 1965.

53. **Forsander, O. A., Maenpoa, P. H., and Salaspuro, M. P.,** Influence of ethanol on the lactate/pyruvate and β-hydroxybutyrate/acetoacetate ratios in rat liver experiments, *Acta Chem. Scand.*, 19, 1770, 1965.

54. **Thieden, H. I. D.,** The effect of ethanol on the concentrations of adenine nucleotides in rat liver, *FEBS Lett.*, 2, 121, 1968.

55. **Perman, E. S.,** Effect of ethanol on oxygen uptake and on blood glucose concentration in anesthetized rabbits, *Acta Physiol. Scand.*, 55, 189, 1962.

56. **Lundquist, F., Tygstrup, N., Winkler, K., Mellemgaard, K., and Munck-Petersen, S.,** Ethanol metabolism and production of free acetate in the human liver, *J. Clin. Invest.*, 41, 955, 1962.

57. **Dornhorst, A. and Ouyang, A.,** Effect of alcohol on glucose tolerance, *Lancet*, 2, 957, 1971.

58. **Perman, E. S.,** Effect of ethanol and hydration on the urinary excretion of adrenaline and nonadrenaline and on the blood sugar of rats, *Acta Physiol. Scand.*, 51, 68, 1961.

59. **Forsander, O. A., Raiha, N., Salaspuro, M., and Maenpaa, P.,** Influence of ethanol on the liver metabolism to fed and starved rats, *Biochem. J.*, 94, 259, 1965.

60. **Rawat, A. K.,** Effects of ethanol infusion on the redox state and metabolite levels in rat liver in vivo, *Eur. J. Biochem.*, 6, 585, 1968.

61. **Chien, C., Stotsky, B. A., and Cole, J. O.,** Psychiatric treatment for nursing home patients: drug, alcohol, and milieu, *Am. J. Psychiatry*, 130, 543, 1973.

62. **Smith, L. E. and Shock, N. W.,** Intravenous glucose tolerance tests in aged males, *J. Gerontol.*, 4, 37, 1949.

63. **Silverstone, F. A., Brandfonbrener, M., Shock, N. W., and Yiengst, M. J.,** Age differences in the intravenous glucose tolerance tests and the response to insulin, *J. Clin. Invest.*, 3, 504, 1957.

64. **Swerdloff, R. S., Pozefsky, T., Tobin, J. D., and Andres, R.,** Influence of age on the intravenous tolbutamide response test, *Diabetes*, 16, 161, 1967.

65. **Andres, R. and Tobin, J. D.,** *Proc. 9th Int. Congr. Gerontol.*, Vol. 1, Kiev, 1972, 276.

66. **Andres, R.,** Aging and diabetes, *Med. Clin. North Am.*, 55, 835, 1971.

67. **Arky, R. A. and Freinkel, N.,** Hypoglycemic action of alcohol, in *Biochemical and Clinical Aspects of Alcohol Metabolism*, Sardesai, V. M., Ed., Charles C Thomas, Springfield, 1969, 67.

68. **Bagade, J. D., Bierman, E. L., and Porte, D.,** Counter-regulation of basal insulin secretion during alcohol hypoglycemia in diabetic and normal subjects, *Diabetes*, 21, 65, 1972.

69. **Frienkel, N., Singer, D. L., Arky, R. A., Bleicher, S. J., Anderson, J. B., and Silbert, C. K.,** Alcohol hypoglycemia. I. Carbohydrate metabolism of patients with clinical alcohol hypoglycemia and experimental reproduction of the syndrome with pure ethanol, *J. Clin. Invest.*, 42, 1112, 1963.

70. **Madison, L. L., Lochner, A., and Wulff, J.,** Ethanol-induced hypoglycemia. II. Mechanism of suppression of hepatic gluconeogenesis, *Diabetes*, 16, 252, 1967.

71. **Madison, L. L.,** Alcohol-induced hypoglycemia, *Adv. Metab. Disord.*, 3, 85, 1968.

72. **Krebs, H. A., Freeland, R. A., Hems, R., and Stubbs, M.,** Inhibition of hepatic gluconeogenesis by ethanol, *Biochem. J.*, 112, 117, 1969.

73. **Searle, G. L., Shames, D., Cavalieri, R. R., Bagdade, J. D., and Porte, D., Jr.,** Evaluation of ethanol hypoglycemia in man: turnover studies with ^{14}C-6 glucose, *Metabolism*, 23, 1023, 1974.

74. **McMonagle, J. and Felig, P.,** Effects of ethanol ingestion on glucose tolerance and insulin secretion in normal and diabetic subjects, *Metabolism*, 24, 625, 1975.

75. **O'Keefe, S. J. D. and Marks, V.,** Lunchtime gin and tonic a cause of reactive hypoglycemia, *Lancet*, 1, 1286, 1977.

76. **Straus, E., Urbach, H. J., and Yalow, R. S.,** Alcohol-stimulated secretion of immunoreactive insulin, *N. Engl. J. Med.*, 293, 1031, 1975.

77. **Friedenberg, R., Metz, R., Mako, M., and Surmaczynski, B.,** Differential plasma insulin response to glucose and glucagon stimulation following ethanol priming, *Diabetes*, 20, 397, 1971.

78. **Pirola, R. C. and Davis, A. E.,** Effects of intravenous alcohol on motility of the duodenum and of the sphincter of Oddi, *Austr. Ann. Med.*, 19, 24, 1970.

79. **Robles, E. A., Mezey, E., Halsted, C. H., and Schuster, M. M.,** Effect of ethanol on motility of the small intestine, *Johns Hopkins Med. J.*, 135, 17, 1974.

80. **Lindenbaum, J. and Lieber, C. S.,** Effects on chronic ethanol administration on intestinal absorption of man in the absence of nutritional deficiency, *Ann. N.Y. Acad. Sci.*, 252, 228, 1975.

81. **Small, M., Longarini, A., and Zamcheck, N.,** Disturbances of digestive physiology following acute drinking episodes in "skid-row" alcoholics, *Am. J. Med.*, 27, 575, 1959.

82. **Krasner, N., Cochran, K. M., Russell, R. J., Carmichael, H. G., and Thompson, G. G.,** Alcohol and absorption from the small intestine. I. Impairment of absorption from the small intestine in alcoholics, *Gut*, 17, 245, 1976.

83. **Perlow, W., Baraona, E., and Lieber, C. S.,** Symptomatic intestinal disaccharidase deficiency in alcoholics, *Gastroenterology*, 68, 935, 1975.

84. **Bayless, T. M., Paige, D. M., and Ferry, G. D.,** Lactose intolerance and milk drinking habits, *Gastroenterology*, 60, 605, 1971.

85. **Greene, H. L., Stifel, F. B., Herman, R. H., Herman, Y. F., and Rosenweig, N. S.,** Ethanol-induced inhibition of human intestine enzyme activers reversal by folic acid, *Gastroenterology*, 67, 434, 1974.

86. **Lieber, C. S.,** Effects of ethanol upon lipid metabolism, *Lipids*, 9, 103, 1974.

87. **Ontko, J. A.,** Effects of ethanol on the metabolism of free fatty acids in isolated liver cells, *J. Lipid Res.*, 14, 78, 1973.

88. **Brodie, B. B., Butler, W. M., Jr., Horning, M. G., Maickel, R. P., and Maling, J. M.,** Alcohol-induced triglyceride deposition in liver through derangement of fat transport, *Am. J. Clin. Nutr.*, 9, 432, 1961.

89. **Carter, E. A., Drummey, G. D., and Isselbacher, K. J.,** Ethanol stimulates triglyceride synthesis by the intestine, *Science*, 174, 1245, 1971.

90. **Feigl, J.,** Neue Untersuchungen zur Chemie des Blutes bei akuter Alkoholintoxikation und bei chronischem Alkoholismus mit besonderer Berucksichtigung der Fette und Lipoide, *Biochemistry*, 92, 282, 1918.

91. **Lieber, C. S., Spritz, N., and Decarli, L. M.,** Role of dietary, adipose and endogenously synthetized fatty acids in the pathogenesis of the alcoholic fatty liver, *J. Clin. Invest.*, 45, 51, 1966.

92. **Johnson, O.,** Effect of ethanol on liver triglyceride concentration in fed and fasted rats, *Lipids*, 9, 57, 1974.

93. **Brewster, A. C., Lankford, H. G., Schwartz, M. G., and Sullivan, J. F.,** Ethanol and alimentary lipemia, *Am. J. Clin. Nutr.*, 19, 255, 1966.

94. **Ginsberg, H., Olefsky, J., Farquhar, J. W., and Reaven, G. M.,** Moderate ethanol ingestion and plasma triglyceride levels. A study in normal and hypertriglyceridemic persons, *Ann. Intern. Med.*, 80, 143, 1974.

95. **Castelli, W. P., Gordon, T., Hjortland, M. C., Kagan, A., Doyle, J. T., Hames, C. G., Hulley, S. T., and Zukel, W. J.,** Alcohol and blood lipids — the cooperative lipoprotein phenotyping study, *Lancet*, 2, 153, 1977.

96. **Barboriak, J. J., Anderson, A. J., and Hoffmann, R. G.,** Interrelationship between coronary artery occlusion high-density lipoprotein cholesterol, and alcohol intake, *J. Lab. Clin. Med.*, 94, 348, 1979.

97. **Hartung, G. H., Foreyt, J. P., Mitchell, R. E., Mitchell, J. G., Reeves, R. S., and Gotto, A. M., Jr.,** Effect of alcohol intake on high-density lipoprotein cholesterol levels in runners and inactive men, *JAMA*, 249, 747, 1983.

98. **Sabesin, S. M., Hawkins, H. L., Kuiken, L., and Ragland, J. B.,** Abnormal plasma lipoproteins and lecithin-cholesterol actyltransferase deficiency in alcoholic liver disease, *Gastroenterology*, 72, 510, 1977.

99. **Yano, K., Rhoads, G. G., and Kagan, A.,** Coffee, alcohol and risk of coronary heart disease among Japanese men living in Hawaii, *N. Engl. J. Med.*, 297, 405, 1977.

100. **Barboriak, J. J., Anderson, A. J., and Hoffmann, R. G.,** Smoking, alcohol and coronary artery occlusion, *Atherosclerosis*, 43, 277, 1982.

101. **Gruchow, H. W., Hoffmann, R. G., Anderson, A. J., and Barboriak, J. J.,** Effects of drinking patterns on the relationship between alcohol and coronary occlusion, *Atherosclerosis*, 43, 393, 1982.

102. **Israel, Y., Valenzuela, J. E., Salazar, I., and Ugarte, G.,** Alcohol and amino acid transport in the human small intestine, *J. Nutr.*, 98, 222, 1969.

103. **Kuriyama, K., Sze, P. Y., and Sauscher, G. E.,** Effects of acute and chronic ethanol administration on ribosomal protein synthesis in mouse brain and liver, *Life Sci.*, 10, 181, 1971.

104. **Rothschild, M. A., Oratz, M., Mongelli, J., and Schreiber, S. S.,** Alcohol-induced depression of albumin synthesis: reversal by tryptophan, *J. Clin. Invest.*, 50, 1812, 1971.

105. **Jeejeebhoy, K. N., Phillips, M. J., Bruce-Robertson, A., Ho, J., and Sodtke, U.,** The acute effect of ethanol on albumin, fibrinogen and transferrin synthesis in the rat, *Biochem. J.*, 126, 1111, 1972.

106. **Kirsch, R.E., Frith, L. O., Stead, R. H., and Saunders, S. J.,** Effect of alcohol on albumin synthesis by the isolated perfused rat liver, *Am. J. Clin. Nut.*, 26, 1191, 1973.

107. **Perin, A., Scalabrino, G., Sessa, A., and Arnaboldi, A.,** In vitro inhibition of protein synthesis in rat liver as a consequence of ethanol metabolism, *Biochim. Biophys. Acta,* 366, 101, 1974.

108. **Rothschild, M. A., Oratz, M., and Schreiber, S. S.,** Alcohol, amino acids and albumin synthesis, *Gastroenterology,* 67, 1200, 1974.

109. **Morland, J.,** Incorporation of labelled amino acids into liver protein after acute ethanol administration, *Biochem. Pharmacol.,* 24, 439, 1975.

110. **Morland, J. and Sjetnan, A. E.,** Effect of ethanol intake on the incorporation of labelled amino acids into liver protein, *Biochem. Pharmacol.,* 25, 2125, 1976.

111. **Morland, J. and Bessesen, A.,** Inhibition of protein synthesis by ethanol in isolated rat liver parenchymal cells, *Biochim. Biophys. Acta,* 474, 312, 1977.

112. **Sorrel, M. F., Tuma, D. J., Schafer, E. C., and Barak, A. J.,** Role of acetaldehyde in the ethanol-induced impairment of glycoprotein metabolism in rat liver slices, *Gastroenterology,* 73, 137, 1977.

113. **Rothschild, M. A., Oratz, M., and Schreiber, S. S.,** Alcohol inhibition of protein synthesis, *Clin. Toxicol.,* 8, 349, 1975.

114. **Baraona, E. and Lieber, C. S.,** Effects of chronic ethanol feeding on serum lipoprotein metabolism in the rat, *J. Clin. Invest.,* 49, 769, 1970.

115. **Banks, W. L., Kline, E. S., and Higgins, E. S.,** Hepatic composition and metabolism after ethanol consumption of rats fed liquid purified diets, *J. Nutr.,* 100, 581, 1970.

116. **Morland, J.,** Hepatic tryptophan oxygenase activity as a marker of changes in protein metabolism during chronic ethanol treatment, *Acta Pharmacol. Toxicol.,* 35, 155, 1974.

117. **Morland, J.,** Effects of chronic ethanol treatment on tryptophan oxygenase, tyrosine aminotransferase and general protein metabolism in the intact and perfused liver, *Biochem. Pharmacol.,* 23, 21, 1974.

118. **Renis, M., Giovine, A., and Bertolina, A.,** Protein synthesis in mitochondria and microsomal fractions from rat brain and liver after acute or chronic ethanol administration, *Life Sci.,* 16, 1447, 1975.

119. **Baraona, E., Leo, M. A., Borowsky, S. A., and Lieber, C. S.,** Pathogenesis of alcohol-induced accumulation of protein in the liver, *J. Clin. Invest.,* 60, 546, 1977.

120. **Baraona, E., Leo, M. A., Borowsky, S. A., and Lieber, C. S.,** Alcoholic hepatomegaly: accumulation of protein in the liver, *Science,* 190, 794, 1975.

121. **Chowdhury, A. R., Malmud, L. S., and Dinoso, V. P.,** Gastrointestinal plasma protein loss during ethanol ingestion, *Gastroenterology,* 72, 37, 1977.

122. **Klatskin, G.,** The effect of ethyl alcohol on nitrogen excretion in the rat, *Yale J. Biol. Med.,* 34, 124, 1961.

123. **Rodrigo, C., Antezana, C., and Baraona, E.,** Fat and nitrogen balances in rats with alchol-induced fatty liver, *J. Nutr.,* 101, 1307, 1971.

124. **Finkelstein, J. D., Cello, J. P., and Kyle, W. I.,** Ethanol-induced changes in methionine metabolism in rat liver, *Biochem. Biophys. Res. Commun.,* 61, 525, 1974.

125. **Shaw, S. and Lieber, C. S.,** Plasma amino acids in the alcoholic: nutritional aspects, *Alcohol. Clin. Exp. Res.,* 7, 22, 1983.

126. **Shaw, S., Stimmel, B., and Lieber, C. S.,** Plasma alpha amino-*n*-butyric acid to leucine ratio: an empirical biochemical marker of alcoholism, *Science,* 194, 1057, 1976.

127. **Orten, J. M. and Sardesai, V. M.,** Protein, nucleotide and porphyrin metabolism, in *Biology of Alcoholism,* Kissen, B. and Begleiter, H., Eds., Plenum Press, N.Y. 1974, 229.

128. **Pasquariello, G., Quadri, A., and Tenconi, L. T.,** Tryptophan-nicotinic acid metabolism in chronic alcoholics, *Acta Vitaminol.,* 18, 3, 1964.

129. **Payne, I. R., Lu, G. H., and Meyer, K.,** Relationship of dietary tryptophan and niacin to tryptophan metabolism in alcoholics and non-alcoholics, *Am. J. Clin. Nutr.,* 27, 572, 1974.

130. **Rosen, H. M., Yoshimura, N., Hodgman, J. A., and Fischer, J. E.,** Plasma amino acid patterns in hepatic encephalopathy of differing etiology, *Gastroenterology,* 72, 483, 1977.

131. **Iber, F. L., Rosen, H., Levenson, S. M., and Chalmers, T. C.,** The plasma amino acids in patients with liver failure, *J. Lab. Clin. Med.,* 50, 417, 1957.

132. **Fischer, J. E., Funovics, J. M., and Aguirre, A.,** The role of plasma amino acids in hepatic encephalopathy, *Surgery,* 78, 276, 1975.

133. **Faraj, B. A., Bowen, P. A., and Isaacs, J. W.,** Hypertyraminemia in cirrhotic patients, *N. Engl. J. Med.,* 294, 1360, 1976.

134. **Fischer, J. E., Ebeid, A. M., Rosen, H. M., James, J. H., Keane, J. M., and Soeters, P. B.,** Improvement in hepatic encephalopathy by "normalization" of plasma amino acid patterns, *Gastroenterology,* 70, 981, 1976.

135. **Siegel, F. L., Roach, M. K., and Pomeroy, L. R.,** Plasma amino acid patterns in alcoholism: the effects of ethanol loading, *Proc. Natl. Acad. Sci.,* 51, 605, 1964.

136. **Iob, V., Coon, W. W., and Sloan, M.,** Free amino acids in liver, plasma and muscle of patients with cirrhosis of the liver, *J. Surg. Res.,* 7, 41, 1967.

137. **Zinneman, H. H., Seal, U.S., and Doe, R. P.,** Plasma and urinary amino acids in Laennec's cirrhosis, *Am. J. Dig. Dis.,* 14, 118, 1969.

138. **Shaw, S. and Lieber, C. S.,** Plasma amino acid abnormalities in the alcoholic: respective role of alcohol nutrition and liver injury, *Gastroenterology,* 74, 677, 1978.

139. **Stanko, R. T., Morse, E. L., and Adibi, S. A.,** Prevention of effects of ethanol on amino acid concentrations in plasma and tissues by hepatic lipotropic factors in rats, *Gastroenterology,* 76, 132, 1979.

140. **Andrew, W., Shock, N. W., Barrows, C. H., and Yiengst, H. L.,** Correlation of age changes in histological and chemical characteristics in some tissues of the rat, *J. Gerontol.,* 14, 405, 1959.

141. **Yiengst, M. J., Barrows, C. H., and Shock, N. W.,** Age changes in the chemical composition of muscle and liver in the rat, *J. Gerontol.,* 14, 400, 1959.

142. **Neumaster, T. D. and Ring, G. C.,** Creatinine excretion and its relation to whole body potassium and muscle mass in inbred rats, *J. Gerontol.,* 20, 379, 1965.

143. **Sharp, G. S., Lassen, S., Shankman, S., Hazlet, J. W., and Kendes, M. S.,** Studies of protein retention and turnover using nitrogen-15 as a tag, *J. Nutr.,* 63, 155, 1957.

144. **Young, V. R., Perera, W. D., Winterer, J. C., and Scrimshaw, N. S.,** Protein and amino acid requirements of the elderly, in *Nutrition and Aging,* Winick, M., Ed., John Wiley & Sons, N.Y., 1976, 77.

145. **Karel, J. L., Wilder, V. M., and Beber, M.,** Electrophoretic serum protein patterns in aged, *Am. Geriatr. Soc.,* 4, 667, 1956.

146. **Eastman, G.,** The plasma proteins — recent biochemical advances, *J. Am. Geriatr. Soc.,* 10, 633, 1962.

147. **Acheson, R. M. and Jessop, W. J.,** Serum proteins in a population sample of males 65—85 years. A study by electrophoresis, *Gerontologia,* 6, 193, 1962.

148. **Tuttle, S. G., Swendseid, M. E., Mulcare, D., Griffith, W. H., and Bassett, S. H.,** Study of the essential amino acid requirements of men over fifty, *Metabolism,* 6, 564, 1957.

149. **Tuttle, S. G., Swendseid, M. E., Mulcare, D., Griffith, W. H., and Bassett, S. H.,** Essential amino acid requirements of older men in relation to total nitrogen intake, *Metabolism,* 8, 61, 1959.

150. **Tuttle, S. G., Bassett, S. H., Griffith, W. H., Mulcare, D., and Swenseid, M. E.,** Further observations on the amino acid requirements of older men. II. Methionine and lysine, *Am. J. Clin. Nutr.,* 16, 229, 1965.

151. **Tuttle, S. G., Bassett, S. H., Griffth, W. H., Mulcare, D., and Swendseid, M. E.,** Further observations of the amino acid requirements of older men. I. Effects of non-essential nitrogen supplements fed with different amounts of essential amino acids, *Am. J. Clin. Nutr.,* 16, 225, 1965.

152. **Mertz, E. T.,** Essential amino acids in self-selected diets of older women, *J. Nutr.,* 46, 313, 1952.

153. **Albanese, A. A., Higgons, R. A., Orto, L. A., and Zavattaro, D. N.,** Protein and amino acid needs of the aged in health and convalescence, *Geriatrics,* 12, 465, 1957.

154. **Kohrs, M. B., Preston, A., O'Neal, R., Eklund, D., and Abrahams, O.,** Nutritional status of elderly residents in Missouri, *Am. J. Clin. Nutr.,* 31, 2186, 1978.

155. **O'Hanlon, P. and Kohrs, M. B.,** Dietary studies of older Americans, *Am. J. Clin. Nutr.,* 31, 1257, 1978.

156. **Kohrs, M. B., O'Hanlon, P., Krause, G., and Nordstom, J.,** Title VII-nutrition program for the elderly. II. Relationship of socioeconomic factors to one day's nutrient intake, *J. Am. Diet. Assoc.,* 75, 537, 1979.

157. **Ohlson, M. A., Jackson, L., Boek, J., Cederquist, D. C., Brewer, W. D., and Brown, E. G.,** Nutrition and dietary habits of aging women, *Am. J. Public Health,* 40, 1101, 1950.

158. **Vinther-Paulsen, N.,** Investigations of actual food intake of elderly, chronically hospitalized patients, *J. Gerontol.,* 5, 331, 1950.

159. **Todhunter, E. N.,** Life-style and nutrient intake in the elderly, in *Nutrition and Aging,* Winick, M., Ed., John Wiley & Sons, N.Y., 1976, 119.

160. **Kart, C. S., Metress, E. S., and Metress, J. F.,** *Aging and Health: Biologic and Social Perspectives,* Addison-Wesley, Menlo Park, California, 1978.

161. **Young, V. R. and Scrimshaw, N. S.,** Protein needs of the elderly, *Urban Health,* 4, 37, 1975.

162. **O'Keane, M., Russell, R. I., and Goldberg, A.,** Ascorbic acid status of alcoholics, *J. Alcohol.,* 7, 6, 1972.

163. **Beattie, A. D. and Sherlock, S.,** Ascorbic acid deficiency in liver disease, *Gut,* 17, 571, 1976.

164. **Dow, J., Krasner, N., and Goldberg, A.,** Relation between hepatic alcohol dehydrogenase activity and the ascorbic acid in leukocytes of patients with liver disease, *Clin. Sci. Mol. Med.,* 49, 603, 1975.

165. **Rossouw, J. E., Lobadarios, D., Davis, M., and Williams, R.,** Watersoluble vitamins in severe liver disease, *S. Afr. Med. J.,* 54, 183, 1978.

166. **Lester, D., Buccino, R., and Bizzocco, D.,** The vitamin C status of alcoholics, *J. Nutr.,* 70, 278, 1960.

167. **Brubacher, G. and Ritzel, G., Eds.,** in *Zur Ernhrungssituation der Schweizerischen Bevölkerung,* Verlag Hans Huber Bern Stuttgart, Wien, 1975.

168. **Leevy, C. M., Thompson, A., and Baker, H.,** Vitamins and liver injury, *Am. J. Clin. Nutr.,* 23, 493, 1970.

169. **Lemoine, A., Monges, A., CoDaccioni, J. L., Bermond, P., and Korner, W. F.,** Alcoholisme chronique et vitamines, *Rev. Alcohol.,* 18, 199, 1972.

170. **Russell, R. I., Williamson, J. M., Goldberg, A., and Wares, E.,** Ascorbic acid levels in leucocytes of patients with gastrointestinal hemorrhage, *Lancet,* 2, 603, 1968.

171. **Brook, M. and Grimshaw, J. J.,** Vitamin C concentration of plasma and leukocytes as related to smoking habit, age and sex of humans, *Am. J. Clin. Nutr.,* 21, 1254, 1968.

172. **Salvatore, J. E., Vinton, P. W., and Rapuano, J. A.,** Nutrition in the aged; review of the literature, *J. Am. Geriatr. Soc.,* 17, 790, 1969.

173. **Burr, M. L., Hurley, R. J., and Sweetnam, P. M.,** Vitamin C supplementation of old people with low blood levels, *Gerontol. Clin.,* 17, 236, 1975.

174. **Roine, P., Koivula, L., and Pekkarinen, M.,** 1975, Plasma vitamin C level and erythrocyte tranketolase activity compared with vitamin intakes among old people in Finland, in *Proc. 9th Int. Congr. Nutr. Mexico,* Vol. 4, S. Karger, Basel, 1972, 116.

175. **Irwin, M. I. and Hutchins, B. K.,** A conspectus of research on vitamin C requirements of man, *J. Nutr.,* 106, 823, 1976.

176. **Theuer, R. C.,** Nutrition and old age: a review, *J. Dairy Sci.,* 54, 627, 1971.

177. **Dibble, M. V., Bin, M., Thicle, V. F., Peel, A., Chen, N., and McMullen, E.,** Evaluation of the nutrition status of elderly subjects, with comparison between fall and spring, *J. Am. Geriatr. Soc.,* 15, 1031, 1967.

178. **Hodges, R. E. and Baker, E. M.,** *Ascorbic Acid in Modern Nutrition and Disease,* Goodhart, R. S. and Shils, M. E., Eds., Lea & Febiger, Philadelphia, 1973.

179. **Edelson, J. and Douglas, J. F.,** Measurement of gastrointestinal blood loss in the rat: the effect of aspirin, phenylbutazone and seclazone, *J. Pharmacol. Exp. Ther.,* 184, 449, 1973.

180. **Mason, M., Matyk, P. M., and Doolan, S. A.,** Urinary ascorbic acid excretion in postoperative patients, *Am. J. Surg.,* 122, 808, 1971.

181. **Leevey, C. M., Baker, H., TenHove, W., Franke, O., and Cherrick, G. R.,** B-complex vitamins in liver disease of the alcoholic, *Am. J. Clin. Nutr.,* 16, 339, 1965.

182. **Baines, M.,** Detection and incidence of B and C vitamin deficiency in alcohol-related illness, *Ann. Clin. Biochem.,* 15, 307, 1978.

183. **Truswell, A. S., Konno, T., and Hansen, J. D. L.,** Thiamine deficiency in adult hospital patients, *S. Afr. Med. J.,* 46, 2079, 1972.

184. **McLaren, D. S., Docherty, M. A., and Boyd, D. H. A.,** Assessment of thiamine status of patients with an alcohol problem, *Proc. Nutr. Soc.,* 39, 24A, 1980.

185. **Kershaw, P. W.,** Blood thiamin and nicotinic acid levels in alcoholism and confusional states, *Br. J. Psychiatry.,* 113, 387, 1967.

186. **Baker, H., Frank, O., Zetterman, R. K., Rajan, K. S., TenHove, W., and Leevey, C. M.,** Inability of chronic alcoholics with liver disease to use food as a source of folates, thiamin and vitamin B_6, *Am. J. Clin. Nutr.,* 28, 1377, 1975.

187. **Tomasulo, P. A., Kater, R. M. H., and Iber, F. L.,** Impairment of thiamin absorption in alcoholism, *Am. J. Clin. Nutr.,* 21, 1341, 1968.

188. **Thomson, A. D. and Leevey, C. M.,** Observations on the mechanisms of thiamin hydrochloride absorption in man, *Clin. Sci.,* 43, 153, 1972.

189. **Thomson, A. D., Baker, H., and Leevey, C. M.,** Patterns of 35S-thiamin hydrochloride absorption in the malnourished alcoholic patient, *J. Lab. Clin. Med.,* 76, 34, 1970.

190. **Baker, H. and Frank, O.,** Absorption, utilization and chemical effectiveness of allithiamines compared to water-soluble thiamine, *J. Nutr. Sci. Vitaminol. (Suppl.),* 22, 63, 1976.

191. **Lindenbaum, J., Rybak, B., Gerson, C. D., Rubin, E., and Lieber, C.,** Effects of ethanol on the small intestine of man, *Clin. Res.,* 18, 385, 1970.

192. **Rindi, G. and Ferrari, G.,** Thiamine transport by human intestine in vitro, *Experientia,* 33, 211, 1977.

193. **Hoyumpa, A. M.,** Characterization of normal intestinal thiamin transport in animals and man, *Ann. N.Y. Acad. Sci.,* 378, 337, 1982.

194. **Takabe, M. and Itokawa, Y.,** An experimental study of thiamine metabolism in acute ethanol intoxication, *Experientia,* 36, 327, 1980.

195. **Abe, T. and Itokawa, Y.,** Effect of ethanol administration on thiamine metabolism and transketolase activity in rats, *Int. J. Vitam. Nutr. Res.,* 47, 307, 1977.

196. **Sorrell, M. F., Baker, H., Barak, A. J., and Frank, O.,** Release by ethanol of vitamins into rat liver perfusates, *Am. J. Clin. Nutr.,* 27, 743, 1974.

197. **Blass, J. P. and Gibson, G. E.,** Genetic factors in Wernicke-Korsakoff syndrome, *Acoholism,* 3, 126, 1979.

198. **Blass, J. P. and Gibson, G. E.,** Abnormality of thiamine requiring enzyme in patients with Wernicke-Korsakoff syndrome, *N. Engl. J. Med.,* 297, 1367, 1977.

199. **Hell, D., Six, P., and Salkeld, R.,** Vitamin B_1 deficiency in chronic alcoholics and its clinical correlation, *Schweiz. Med. Wochenschr.,* 106, 1466, 1976.

200. **Tomasi, A., Pasqualicchio, A., Battocchia, A., and Ghindi, O.,** Su di un caso di cardiopatia da beri-beri, *Acta Vitamol. Enzymol.,* 22, 157, 1968.

201. **Wood, B. and Breen, K. J.,** Vitamin deficiency and alcoholism with particular reference to thiamine deficiency, *Clin. Exp. Pharmacol. Physiol.,* 6, 457, 1979.

202. **Hell, D. and Six, P.,** Zur Klinik des Wernicke-Korsakoff Syndroms, *Praxis,* 65, 707, 1976.

203. **Victor, M. and Adams, R. D.,** The etiology of alcoholic neurologic disease with special reference to the role of nutrition, *Am. J. Clin. Nutr.,* 9, 379, 1961.

204. **Buckle, R. M.,** Blood pyruvic and alpha ketoglutaric acids in thiamine deficiency, *Metabolism,* 14, 141, 1965.

205. **Buckle, R. M.,** Studies on blood keto acids in vitam deficiency, *Proc. Roy. Soc. Med.,* 60, 48, 1967.

206. National Center for Health Statistics: Caloric and Selected Nutrient Values for Persons 1-74 Years of Age, First Health and Nutrition Examination survey, United States 1971-1974. Vital and Health Statistics, Ser. 11, No. 209, DHEW publication No. (DHS) 79-1657, U.S. Government Printing Office, Washington, D.C., 1979.

207. National Center for Health Statistics: Dietary Intake Source Data, United States 1976-1980, Vital and Health Statistics, Ser. II, Hyattsville, Maryland, in press.

208. **Iber, F. L., Blass, J. P., Brin, M., and Leevy, C. M.,** Thiamin in the elderly-relation to alcoholism and to neurological degenerative disease, *Am. J. Clin. Nutr.,* 36, 1067, 1982.

209. **Brin, M., Dibble, M. V., Peel, A., McMullen, E., Bourquin, A., and Chen, N.,** Some preliminary findings on the nutritional status of the aged in Onandaga County, New York, *Am. J. Clin. Nutr.,* 17, 240, 1965.

210. **Sauberlich, H. E., Herman, Y. F., and Stevens, C. O.,** Thiamin requirement of the adult human, *Am. J. Clin. Nutr.,* 23, 671, 1970.

211. **Vimokesant, S. L., Nakornchi, S., Dhanamitta, S., and Hilker, D. M.,** Effect of tea consumption on thiamin status in man, *Nutr. Rep. Int.,* 9, 371, 1974.

212. **Baker, H., Frank, O., and Jaslow, S. P.,** Oral versus intramuscular vitamin supplementation for hypovitaminosis in the elderly, *J. Am. Geriatr. Soc.,* 28, 42, 1980.

213. **Brin, M., Schwartzberg, S. H., and Arthur-Davis, D.,** A vitamin evaluation program as applied to 10 elderly residents in a community home for the aged, *J. Am. Geriatr. Soc.,* 12, 493, 1964.

214. Recommended Dietary Allowance, Revised 1980, Food and Nutrition Board, National Academy of Sciences, National Research Council, Washington, D.C.

215. **Hoorn, R. K. J., Frilkweert, J. P., and Westerink, D.,** Vitamin B-1, B-2 and B-6 deficiencies in geriatric patients, measured by coenzyme stimulation of enzyme activities, *Clin. Chem. Acta ,* 61, 151, 1975.

216. **Dastur, D. K., Santhadevi, N., Quadros, E. V., Avari, F. C. R., Wadia, N. H., Desai, M. M., and Bharucha, E. P.,** the B-vitamin in malnutrition with alcoholism. A model of intervitamin relationships, *Br. J. Nutr.,* 36, 143, 1976.

217. **Bayoumi, R. A. and Roaslki, S. B.,** Evaluation of methods of coenzyme activation of erythrocyte enzymes for detection of deficiency of vitamins B_1, B_2 and B_6, *Clin. Chem.,* 22, 327, 1976.

218. **Hell, D., and Six, P.,** Thiamin, Riboflavin und Pyridoxin — Versorgung bei Chronischem Alkoholismus, *Dtsch. Med. Wochenschr.,* 102, 962, 1977.

219. **Rosenthal, W. S., Adham, N. F., Lopez, R., and Cooperman, J. M.,** Riboflavin deficiency in complicated chronic alcoholism, *Am. J. Clin. Nutr.,* 26, 858, 1973.

220. **Pekkanen, L. and Rusi, M.,** The effects of dietary niacin and riboflavin on voluntary intake and metabolism of ethanol in rats, *Pharmacol. Biochem. Behav.,* 11, 575, 1979.

221. U.S. Department of Health, Education and Welfare, Health and Nutrition Examination Survey, Dietary Intake Findings United States, 1971-1974, DHEW Publication No. (HRA) 77-1647, Series 11, 202, 52, 1977.

222. **Whanger, A.D.,** Vitamins and vigor at 65 plus, *Postgrad. Med.,* 53, 167, 1973.

223. **Lumeng, L., Brashear, R. E., and Li, T. K.,** Pyridoxal-5-phosphate in plasma: source, protein-binding and cellular transport, *J. Lab. Clin. Med.,* 84, 334, 1974.

224. **Brown, R. R., Rose, D. P., Leklem, J. E., Linkswiler, H., and Anand, R.,** Urinary 4-pyridoxic acid, plasma pyridoxal phosphate, and erythrocytic aminotransferase levels in oral contraceptive users receiving controlled intakes of vitamin B_6, *Am. J. Clin. Nutr.,* 28, 10, 1975.

225. **Davis, R. E. and Smith, K. K.,** Pyridoxal and folate deficiency in alcoholics, *Med. J. Aust.,* 2, 357, 1974.

226. **Lumeng, L. and Li, T. K.,** Vitamin B_6 metabolism in chronic alcohol abuse: pyridoxal 5'-phosphate levels in plasma and the effects of acetaldehyde on pyridoxal phosphate synthesis and degradation in human erythrocytes, *J. Clin. Invest.,* 53, 693, 1974.

227. **Labadarios, D., Rossouw, J. E., McConnell, J. B., Davis, M., and Williams, R.,** Vitamin B_6 deficiency in chronic liver disease — evidence for increased degradation of pyridoxal 5'-phosphate, *Gut,* 18, 23, 1977.

228. **Mitchell, D., Wagner, C., Stone, W. J., Wilkinson, G. R., and Schenker, S.,** Abnormal regulation of plasma pyridoxal 5'-phosphate in patients with liver disease, *Gastroenterology,* 71, 1043, 1976.

229. **Leake, C. D. and Silverman, M.,** The chemistry of alcoholic beverages, in *The Biology of Alcoholism,* Vol. 1, Kissen, B. and Begleiter, H., Eds., Plenum Press, N.Y., 1971, 575.

230. **Lumeng, L.,** Effect of Ethanol on Vitamin B_6 Metabolism, in Research Monograph No. 2 Alcohol and Nutrition, Li, T. K., Schenker, S., and Lumeng, L., Eds., DHEW Publication No. (ADM) 79-780, U.S. Department of Health, Education, and Welfare, Washington, D.C., 1977, 251.

231. **Li, T. K., Lumeng, L., and Vietch, R. L.,** Regulation of pyridoxal 5-phosphate metabolism in liver, *Biochem. Biophys. Res. Commun.,* 61, 677, 1974.

232. **Rossouw, J. E., Labadarios, D., McConnell, J. B., Davis, M., and Williams, R.,** Plasma pyridoxal phosphate levels in fulminant hepatic failure and the effects of parenteral supplementation, *Scand. J. Gastroenterol.,* 12, 123, 1977.

233. **Hines, J. D.,** Hematologic abnormalities involving vitamin B_6 and folate metabolism in alcoholic subjects, *Ann. N.Y. Acad. Sci.,* 252, 316, 1975.

234. **Chillar, R. K., Johnson, C. S., and Beutler, E.,** Erythrocyte pyridoxine kinase levels in patients with sideroblastic anemia, *N. Engl. J. Med.,* 295, 881, 1976.

235. **Pierce, H. I., McGuffin, G., and Hillman, R. S.,** Clinical studies in alcoholic sideroblastosis, *Arch. Intern. Med.,* 136, 283, 1976.

236. **Hines, J. D. and Cowan, D. H.,** Studies on the pathogenesis of alcohol induced sideroblastic bone marrow abnormalities, *N. Engl. J. Med.,* 283, 441, 1970.

237. **Eichner, E. R. and Hillman, R. S.,** The evolution of anemia in alcoholic patients, *Am. J. Med.,* 50, 218, 1971.

238. **Reinken, L., Hohenauer, L., and Ziegler, E. E.,** Activity of red cell glutamic oxalacetic transaminase in epileptic children under antiepileptic treatment, *Clin. Chim. Acta,* 36, 270, 1972.

239. **Theuer, R. C.,** Drug and nutrition interaction, *Wis Diet.,* 38, 4, 1974.

240. **Kirkendall, W. M. and Page, E. B.,** Polyneuritis occurring during hydralazine therapy, *JAMA,* 167, 427, 1958.

241. **Davis, R. E. and Leake, E.,** Histidine metabolite excretion and serum folate levels, *Aust. J. Exp. Biol. Med. Sci.,* 43, 185, 1965.

242. **Gardner, A. J.,** Folate status of alcoholic in-patient, *Br. J. Addict.,* 66, 183, 1971.

243. **Halstead, C. H., Robles, E. A., and Mezey, E.,** Decreased jejunal uptake of labeled folic acid (^{3}H-PGA) in alcoholic patients: roles of alcohol and nutrition, *N. Engl. J. Med.,* 285, 701, 1971.

244. **Heilmann, E. and Bonninghoff, E.,** Folsauerbestimmungen in Serum und in den Erythrozyten mittels einer kimpetitiven Proteinbindungs-methode mit 125-Jod-Markierung, *Schweiz. Med. Wochenschr.,* 106, 1821, 1976.

245. **Heilmann, E. and Koschatzki, J.,** Folsaure und Vitamin B_{12} bei chronischen Alkoholikern, *Schweiz. Med. Wochenschr.,* 108, 1920, 1978.

246. **Wu, A., Chanarin, I., Slavin, G., and Levi, A. J.,** Folate deficiency in the alcoholic — its relationship to clinical and hematological abnormalities, liver disease and folate stores, *Br. J. Haematol.,* 29, 469, 1975.

247. **Carney, M. W.,** Serum folate and cyanocobalamin in alcoholics, *Q. J. Stud. Alcohol.,* 31, 816, 1970.

248. **Deller, D. J., Kimber, C. L., and Ibbotson, R. N.,** Folic acid deficiency in cirrhosis of the liver, *Am. J. Dig. Dis.,* 10, 35, 1965.

249. **Herbert, V., Zalusky, R., and Davidson, C. S.,** Correlation of folate deficiency with alcoholism and associated macrocytosis, anemia, and liver disease, *Ann. Intern. Med.,* 58, 977, 1963.

250. **Jarrold, T., Will, J. J., Davies, A. R., Duffy, P. H., and Bramscheiber, J. L.,** Bone marrow-erythroid morphology in alcoholic patients, *Am. J. Clin. Nutr.,* 20, 716, 1967.

251. **Kimber, C., Deller, D. J., Ibbotson, R. N., and Lander, H.,** The mechanism of anemia in chronic liver disease, *Q. J. Med.,* 34, 33, 1965.

252. **Klipstein, F. A. and Lindenbaum, J.,** Folate deficiency in chronic liver disease, *Blood,* 25, 443, 1965.

253. **Herbert, V.,** A palatable diet for producing experimental folate deficiency in man, *Am. J. Clin. Nutr.,* 12, 17, 1963.

254. **Paine, C. J., Eichner, E. R., and Dickson, V.,** Concordance of radioassay and microbiological assay in the study of ethanol-induced fall in serum folate level, *Am. J. Med. Sci.,* 266, 135, 1973.

255. **Halsted, C. H., Griggs, R. C., and Harris, J. W.,** The effect of alcohlism on the absorption of folic acid (H 3-PGA) evaluated by plasma levels and urine excretion, *J. Lab. Clin. Chem.,* 69, 116, 1967.

256. **Halsted, C. H., Robles, E. A., and Mezey, E.,** Intestinal malabsorption in folate-deficient alcoholics, *Gastroenterology,* 64, 526, 1973.

257. **Lindenbaum, J. and Pezzimenti, J. F.,** Effects of B_{12} and folate deficiency on small intestine function, *Clin. Res.,* 21, 518, 1973.

258. **Lindenbaum, J., Saha, J. R., Shea, N., and Lieber, C. S.,** Mechanism of alcohol-induced malabsorption of vitamin B_{12}, *Gastroenterology,* 64, 762, 1973.

259. **Lane, F., Goff, P., McGuffin, R., Eichner, E. R., and Hillman, R. S.,** Folic acid metabolism in normal, folate-deficient and alcoholic man, *Br. J. Haematol.,* 34, 489, 1976.

260. **Hoffbrand, A. V.,** Pathology of folate deficiency, *Proc. R. Soc. Med.,* 70, 82, 1977.
261. **Reynolds, E. H.,** Neurological aspects of folate and vitamin B_{12} metabolism, *Clin. Haematol.,* 5, 661, 1976.
262. **Hach, B.,** Neuropsychiatric clinical picture in vitamin B_{12} deficiency, *Med. Welt.,* 27, 2433, 1976.
263. **Spatz, R., Thimm, R., Heinze, H. H., Ross, A., and Konig, M.,** Zum Klinischen Gestatlswandel der Vitamin B_{12} mangelerkrankungen, *Nervenarzt.,* 47, 169, 1976.
264. **Kimber, C. L., Deller, C. J., and Lander, H.,** Megaloblastic and transitional megaloblastic anemia associated with chronic liver disease, *Am. J. Med.,* 38, 767, 1965.
265. **Hourihane, D. O. and Weir, D. G.,** Suppression of erythropoiesis by alcohol, *Br. Med. J.,* 1, 86, 1970.
266. **Lowenstein, F. W.,** Nutritional status of elderly in the United States of America, 1971-1974, *J. Am. Coll. Nutr.,* 1, 165, 1982.
267. Ten-State Nutrition Survey, 1968-1970, IV, Biochemical, DHEW Publication No. (HSM) 72-8132, United States Department of Health, Education and Welfare. 1972.
268. **Hoppner, K. and Lampi, B.,** Folate levels in human liver from autopsies in Canada, *Am. J. Clin. Nutr.,* 33, 862, 1980.
269. **Wagner, P. A., Bailey, L. B., Krista, M. L., Jernigan, J. A., Robinson, J. D., and Cerda, J. J.,** Comparison of zinc and folacin status in elderly women from different socio-economic backgrounds, *Nutr. Res.,* 1, 565, 1981.
270. **Bailey, L. B., Wagner, P. A., and Christakis, G. J.,** Folacin and iron status and hematological findings in predominantly black persons from urban low-income households, *Am. J. Clin. Nutr.,* 32, 2346, 1979.
271. **Read, A. E., Gaugh, K. R., Pardoe, J. L., and Nicholas, A.,** Nutritional studies of the entrants to an old people's home with particular reference to folic acid deficiency, *Br. Med. J.,* 2, 843, 1965.
272. **Leevy, C. M., Cardi, L., Frank, O., Gellene, R., and Baker, H.,** Incidence and significance of hypovitaminemia in a randomly selected municipal hospital population, *Am. J. Clin. Nutr.,* 17, 259, 1965.
273. **Herbert, V., Colman, N., and Jacob, E.,** Folic acid and vitamin B_{12}, in *Modern Nutrition in Health and Disease,* Goodhart, R. S. and Shils, M., Eds., Lea & Febieger, Philadelphia, 1980.
274. **Cotzias, G. C.,** Metabolic modifications of some neurologic disorders, *JAMA,* 210, 1255, 1969.
275. **Coltart, D. J.,** Malabsorption induced by para-aminosalicylate, *Br. Med. J.,* 1, 825, 1969.
276. **Tao, H. G. and Fox, H. M.,** Measurements of urinary pantothenic acid excretions of alcoholic patients, *J. Nutr. Sci. Vitaminol.,* 22, 333, 1976.
277. **Fennelly, J., Frank, O., Baker, H., and Leevy, C. M.,** Peripheral neuropathy of the alcoholic. I. Aetiological role of aneurin and other B-complex vitamins, *Br. Med. J.,* 2, 1290, 1964.
278. **Leevy, C. M., George, W. S., Ziffer, H., and Baker, H.,** Pantothenic acid, fatty liver and alcoholism, *J. Clin. Invest.,* 39, 1005, 1960.
279. **Frank, O., Luisada-Opper, A., Sorrell, M. F., Thomson, A. D., and Baker, H.,** Vitamin deficits in severe alcoholic fatty liver of man calculated from multiple reference units, *Exp. Mol. Pathol.,* 15, 191, 1971.
280. **Nadkarni, G. D., Deshpande, U. R., and Pahuja, D. N.,** Liver vitamin A stores in chronic alcoholism in rats: effect of prophylthiouracil treatment, *Experientia,* 35, 1059, 1979.
281. **Sato, M. and Lieber, C. S.,** Hepatic vitamin C depletion after chronic ethanol consumption, *Gastroenterology,* 79, 1123, 1980.
282. **Leo, M. A., Arai, M., Sato, M., and Lieber, C. S.,** Hepatotoxicity of vitamin A and ethanol in the rat, *Gastroenterology,* 82, 194, 1982.
283. **Carney, E. A. and Russell, R. M.,** Correlation of dark adaption test results with serum vitamin A levels in diseased adults, *J. Nutr.,* 110, 552, 1980.
284. **Dutta, S. K., Costa, B. S., Russell, R. M., and Connor, T. B.,** Fat soluble deficiency in treated patients with pancreatic insufficiency, *Gastroenterology,* 76, 1126, 1979.
285. **Halsted, J. A. and Smith, J. C.,** Night blindness and chronic liver disease, *Gastroenterology,* 67, 193, 1974.
286. **Lund, B., Sorenson, O. H., Hilden, M., and Lund, B.,** The hepatic conversion of vitamin D in alcoholics with varying degrees of liver infection, *Acta Med. Scand.,* 202, 221, 1977.
287. **Singlevich, T. E. and Barboriak, J. J.,** Ethanol and induction of microsomal drug-metabolizing enzymes in the rat, *Toxicol. Appl. Pharmacol.,* 20, 284, 1971.
288. **Corless, D., Gupta, S. P., Sattar, D. A., Switala, S., and Bouches, B. J.,** Vitamin D status of residents of an old people's home and long-stay patients, *Gerontology,* 25, 350, 1979.
289. **Lawson, D. E. M., Paul, A. A., Black, A. E., Cole, T. J., Mandell, A. R., and Davie, M.,** Relative contributions of diet and sunlight to vitamin D state in the elderly, *Br. Med. J.,* 2, 303, 1979.
290. **Hodkinson, H. M., Bryson, E., Klenerman, L., Clarke, M. B., and Wooton, R.,** Sex, sunlight, season, diet and the vitamin D status of elderly patients, *J. Clin. Exp. Gerontol.,* 1, 13, 1979.
291. **Thomson, A. D.,** Alcohol and nutrition, *Clin. Endocrinol. Metab.,* 7, 405, 1978.
292. **Shils, M. E.,** Magnesium, calcium and parathyroid hormone interaction, *Ann. N.Y. Acad. Sci.,* 355, 165, 1980.

293. **Flink, L. B.,** Magnesium deficiency syndrome in man, *JAMA*, 160, 1406, 1956.

294. **Wilkins, W. L. and Cullen, G. L.,** Electrolytes in human tissue. III. A comparison of normal hearts with hearts showing congestive heart failure, *J. Clin. Invest.*, 12, 1063, 1933.

295. **Shils, M. E.,** Experimental human magnesium depletion, *Medicine*, 48, 61, 1969.

296. **Seelig, M. S.,** The requirement of magnesium of the normal adult, *Am. J. Clin. Nutr.*, 14, 342, 1964.

297. **Albanese, A. A.,** Bone loss: causes, detection and therapy, in *Current Topics in Nutrition and Diseases*, Vol. 1, Albanese, A. A. and Kritchevsky, D., Eds., Alan R. Liss, N.Y., 1977.

298. **Garn, S. M., Rohmann, C. G., Wagner, B. et al.,** Population similarities in the onset and rate of adult endosteal bone loss, *Clin. Orthop.*, 65, 51, 1969.

299. **Garn, S. M., Solomon, M. A., and Friedl, J.,** Calcium intake and bone quality in the elderly, *Ecol. Food. Nutr.*, 10, 131, 1981.

300. **Heaney, R. P.,** Radiocalcium metabolism in disuse osteoporosis in man, *Am. J. Med.*, 33, 188, 1962.

301. **Heaney, R. P., Recker, R. R., and Saville, P. D.,** Menopausal changes in calcium balance performance, *J. Lab. Clin. Med.*, 92, 953, 1978.

302. **Birge, S. J., Keutmann, H. T., Cuatrecasas, P., and Whedon, G. D.,** Osteoporosis, intestinal lactase deficiency and low dietary calcium intake, *N. Engl. J. Med.*, 76, 445, 1967.

303. **Newcomer, A. D., Hodgson, S. F., McGill, D. B., and Thomas, P. J.,** Lactase deficiency: prevalence in osteoporosis, *Ann. Intern. Med.*, 89, 218, 1978.

304. **Krawitt E. L.,** Ethanol inhibits intestinal calcium transport in rats, *Nature (London)*, 243, 88, 1973.

305. **Krawitt E. L.,** Effect of ethanol ingestion of duodenal calcium transport, *J. Lab. Clin. Med.*, 85, 665, 1975.

306. **Nilsson, B. E. and Westlin, N. E.,** Changes in bone mass in alcoholics, *Clin. Orthop.*, 90, 229, 1973.

307. **Dalen, N. and Lamke, B.,** Bone mineral losses in alcoholics, *Acta Orthop. Scand.*, 47, 469, 1976.

308. **Sullivan, J. F.,** Affective alcohol and urinary zinc excretion, *Q. J. Stud. Alcohol.*, 23, 216, 1962.

309. **Henry, R. W. and Elmes, M. E.,** Plasma zinc in acute starvation, *Br. Med. J.*, 4, 625, 1975.

310. **Valee, B. L.,** Recent Advance in Zinc Biochemistry, in *Alcohol and Nutrition*, Li, T. K., Shenker, S., and Lumeng, L., Eds., United States Department of Health, Education, and Welfare, Publication No. 79-780, 1979, p. 1979.

311. **McClain, C. J., Van Thiel, D. H., Parker, S., and Holtzman, J. L.,** Alterations in zinc, vitamin A and retinal-binding protein in chronic alcoholics, *Alcohol. Clin. Exp. Res.*, 3, 135, 1979.

312. **Greger, J. L. and Sciscoe, B. S.,** Zinc nutriture of aged participants in an urban title VII feeding program, *J. Am. Diet. Assoc.*, 70, 37, 1977.

313. **Greger, J. L.,** Dietary intake and nutritional status in regard to zinc of institutionalized aged, *J. Gerontol.*, 32, 549, 1977.

314. National Center for Health Statistics, Preliminary Findings of the First Health and Nutrition Examination Survey, United States, 1971-1972; Dietary intake and biochemical findings, HRA Publication No. 74-1219-1, Washington, D.C., 1974.

315. **Nordstrom, J. W.,** Trace mineral nutrition in the elderly, in Symp. Nutr. Aging, Kohrs, M. S. and Kamath, S. D., Eds., *Clin. Nutr.*, 36, 788, 1982.

316. **Reinhold, J. G., Parsa, A., Karimian, N., Hamick, J. W., and Ismail-Beigi, F.,** Availability of zinc in leavened and unleavened whole-meal wheaten breads as measured by solubility and uptake by rat intestine in vitro, *J. Nutr.*, 104, 976, 1974.

317. **Reinhold, J. G., Ismail-Beigi, F., and Faradji, B.,** Fibre vs. phytate as determinant of the availability of calcium, zinc and iron of breadstuffs, *Nutr. Rep. Int.*, 12, 75, 1975.

318. **Wacker, W. E. C.,** Role of zinc in wound healing: a critical review, in *Trace Elements in Human Health and Disease*. I, Prasad, A. S., Ed., Academic Press, N.Y., 1976, 107.

319. **Pories, W. J., Mansour, E. G., Plecha, F. R., Flynn, A., and Strain, W. H.,** Metabolic factors affecting zinc metabolism in the surgical patient, in *Trace Elements in Human Health and Disease*, I, Prasad, A. S., Ed., Academic Press, N.Y., 1976, 115.

320. **Beisel, W. R.,** Single nutrients and immunity, *Am. J. Clin. Nutr.*, 35, 417, 1982.

321. **Gross, R. L. and Newburne, P. M.,** Role of nutrition in immunologic function, *Physiol. Rev.*, 60, 188, 1980.

322. **Henkin, R. I., Patten, B. M., Peter, K., and Bronzert, D. A.,** A syndrome of acute zinc loss: cerebellar dysfunction, mental changes, anorexia, and taste and smell dysfunction, *Arch. Neurol.*, 32, 745, 1975.

323. **Brun, J. K. and Clancy, K. L.,** Low income and the elderly population, *J. Nutr. Educ.*, 12, 128, 1980.

324. **Hamilton, E. M. and Whitney, E.,** *Nutrition Concepts and Controversies*, West Publishing, St. Paul, Minnesota, 1979, 453.

325. **Gambert, S. R. and Guansing, A. R.,** Protein-calorie malnutrition in the elderly, *J. Am. Geriat. Soc.*, 28, 272, 1980.

326. **McLeod, J. G.,** Alcohol, nutrition and the nervous system, *Med. J. Aust.*, 2, 273, 1982.

327. **Parsons, O. A. and Leber, W. R.,** Premature aging, alcoholism, and recovery, in *Alcoholism and Aging: Advances in Research*, Wood, N. G. and Ellias, M. F., Eds., CRC Press, Boca Raton, 1982, 79.

328. **Larato, D. C.**, Oral tissue changes in the chronic alcoholic, *J. Periodontol.*, 43, 772, 1972.
328a. **Kert, R. C. S., Meteress, E. S., and Meteress, J. S.**, *Aging and Health*, Addison-Wesley, Menlo Park, Calif., 1978, chap. 6.
329. **Stahl, S. S., Wisan, J. M., and Miller, S. C.**, The influence of systematic diseases on alveolar bone, *J. Am. Diet. Assoc.*, 45, 277, 1952.
330. **Borsanyi, D. D.**, Chronic asymptomatic enlargement of the parotid glands, *Ann. Otol.*, 71, 857, 1962.
331. **Williamson, R. and Davis, C. L.**, Drug-dependent, alcohol-dependent, and mental patients: clinical study of oral surgery procedures, *J. Am. Diet. Assoc.*, 86, 416, 1973.
332. **Hogan, W. J., Viegas de Andrade, S. R., and Winship, D. H.**, Ethanol-induced acute esophageal motor dysfunction, *J. Appl. Physiol.*, 32, 755, 1972.
333. **Chaput, J. C., Petite, J. P., Gueroult, N. et al.**, La fibroscopie d'urgence dans les hemorragies digestives des cirrhosis: a propos de 100 cas, *Nouv. Presse Med.*, 3, 1227, 1974.
334. **Dragstedt, C. A., Gray, J. S., Lawton, A. H. et al.**, Does alcohol stimulate gastric secretion by liberating histamines?, *Proc. Soc. Exp. Biol. Med.*, 43, 26, 1940.
335. **Fromm, D. and Robertson, R.**, Effects of alcohol on active and passive ion transport by gastric mucosa, *Surg. Forum*, 26, 373, 1975.
336. **Davenport, H. W.**, Gastric mucosal hemorrhage in dogs: effects of acid, aspirin, and alcohol, *Gastroenterology*, 56, 439, 1969.
337. **Palmer, E. D.**, Gastritis re-evaluation, *Medicine*, 33, 199, 1954.
338. **Pirola, R. C. and Davis, A. E.**, Effects of intravenous alcohol on motility of the duodenum and of the sphincter of Oddi, *Austr. Ann. Med.*, 19, 24, 1970.
339. **Gazzard, B. G. and Clark, M. L.**, Alcohol and the alimentary system, *Clin. Endocrinol. Metab.*, 7, 429, 1978.
340. **Gottfried, E. B., Korsten, M. A., and Lieber, C. S.**, Gastritis and duodenitis induced by alcohol: an endoscopic and histologic assessment, *Gastroenterology*, 70, 890, 1976.
341. **Lindenbaum, J. and Lieber, C. S.**, Alcohol-induced malabsorption of vitamin B_{12} in man, *Nature (London)*, 224, 806, 1969.
342. **Rubin, E. and Lieber, C. S.**, Early fine structural changes in the human liver induced by alcohol, *Gastroenterology*, 52, 1, 1967.
343. **Green, J., Mistilis, S., and Schiff, L.**, Acute alcoholic hepatitis, *Arch. Intern. Med.*, 112, 67, 1963.
344. **Lelbach, W. K.**, Epidemiology of alcoholic liver disease, in *Progress in Liver Disease*, Vol. 5, Popper, H. and Schaffner, F., Eds., Grune & Stratton, N.Y., 1976, 494.
345. **Fisher, S., Hendricks, D. G., and Mahoney, A. W.**, Nutritional assessment of senior rural Utahans by biochemical and physical measurements, *Am. J. Clin. Nutr.*, 31, 667, 1978.
346. **Harvey, S. C.**, Gastric antacids and digestants, in *The Pharmacological Basis of Therapeutics*, 3rd ed., Goodman, L. S. and Gilman, A., Eds., Macmillan, N.Y., 1965.
347. **Bloom, W. L. and Flinchum, D.**, Osteomalacia with pseudofractures caused by the ingestion of aluminum hydroxide, *JAMA*, 174, 1327, 1960.
348. **Lotz, M., Zisman, E., and Bertter, F. C.**, Evidence for a phosphorus depletion syndrome in man, *N. Engl. J. Med.*, 278, 409, 1968.
349. **Hethcox, J. M. and Stanaszek, W. F.**, Interactions of drugs and diet, *Hosp. Pharm.*, 9(10), 373, 1974.
350. **Klipstein, F. A.**, Subnormal serum folate and macrocytosis associated with anticonvulsant drug therapy, *Blood*, 23, 68, 1964.
351. **Hahn, T. J., Hendin, B. A., Scharp, C. R., and Haddad, J. C.**, Effect of chronic anticonvulsant therapy on serum 25-hydroxycalciferol levels in adults, *N. Engl. J. Med.*, 287, 900, 1972.
352. **Solomon, G. E., Hilgartner, M. W., and Kutt, H.**, Coagulation defects caused by diphenylhydantoin, *Neurology*, 22, 1165, 1972.
353. **Faloon, W. W., Fisher, C. J., and Duggan, K. C.**, Occurrence of a sprue-like syndrome during neomycin therapy, *J. Clin. Invest.*, 37 (Abstr.), 893, 1958.
354. **Faloon, W. W.**, Drug production of intestinal malabsorption, *N.Y. State J. Med.*, 70, 2189, 1970.
355. **Corcino, J. J., Wasman, S., and Herber, V.**, Absorption and malabsorption of vitamin B_{12}, *Am. J. Med.*, 48, 562, 1970.
356. **Koch-Weser, J. and Sellers, E. M.**, Drug interactions with coumarin anti-coagulants, *N. Engl. J. Med.*, 285, 487, 1971.
357. **Roe, D. A.**, Factors affecting nutritional requirements, in *Drug-Induced Nutritional Deficiencies*, AVI, Westport, Conn., 1976, chap. 2.
358. **Rosen, F., Michich, E., and Nichol, C. A.**, Selective metabolic and chemotherapeutic effects of vitamin B_6 antimetabolites, in *Vitamins and Hormones*, Harris, R. S., Wool, G., and Loraine, J. A., Eds., Academic Press, N.Y., 1964, chap. 22.
359. **Jaffe, I. A.**, Antivitamin B_6 effect of D-penicillamine, *Ann. N.Y. Acad. Sci.*, 166, 57, 1969.
360. **Coffey, G. and Wilson, C. W. M.**, Ascorbic acid deficiency and aspirin-induced haematemesis, *Br. Med. J.*, 1, 208, 1975.

361. **Ory, E. M.,** The tetracyclines, *Med. Clin. North Am.,* 54, 1173, 1970.

362. **Shils, M. W.,** Some metabolic aspects of tetracyclines, *Clin. Pharmacol. Ther.,* 3, 321, 1962.

363. **Gothoni, G., Neuvonen, P. J., Mattila, M., and Hackman, R.,** Iron-tetracycline interaction: effect of the time interval between the drugs, *Acta Med. Scand.,* 191, 409, 1972.

364. **Kirkendall, W. M. and Page, E. B.,** Polyneuritis occurring during hydralazine therapy: report of two cases and discussion of adverse reactions to hydralazine, *JAMA,* 167, 427, 1958.

365. **Hashim, S. A., Bergen, S. S., and VanItallie, T. B.,** Experimental steatorrhea induced in may by bile acid sequestrant, *Proc. Soc. Exp. Biol. Med.,* 106, 173, 1961.

366. **Roe, D. A.,** Essential hyperlipemia with zanthomatosis. Effects of cholestyramine and clofibrate, *Arch. Dermatol.,* 97, 436, 1968.

367. **Longenecker, J. B. and Basu, S. G.,** Effect of cholestyramine on absorption of amino acids and vitamin A in man, *Fed. Proc.,* 24, 375, 1965.

368. **Heaton, K. W., Lever, J. V., and Barnard, D.,** Osteomalacia associated with cholestyramine therapy for postilectomy diarrhea, *Gastroenterology,* 62, 642, 1972.

369. **Visintine, R. E., Michaels, G. D., Fukayama, G., Conklin, J., and Kinsell, L. W.,** Xanthomatous biliary cirrhosis treated with cholestyramine, a bile-acid-absorbing resin, *Lancet,* 2, 341, 1961.

370. **Coronato, A. and Glass, G. B.,** Depression of the intestinal uptake of radio-vitamin B_{12} by cholestyramine, *Proc. Soc. Exp. Biol. Med.,* 142, 1341, 1973.

371. **Coronato, A. and Glass, G. B. J.,** The effect of deconjugated and conjugated bile salts on the intestinal uptake of radio-vitamin B_{12} in vitro and in vivo, *Proc. Soc. Exp. Biol. Med.,* 142, 1345, 1973.

372. **Staffurth, J. S. and Allott, E. N.,** Paralysis and tetany due to simultaneous hypokalemia and hypocalcemia with other metabolic changes, *Am. J. Med.,* 33, 800, 1962.

373. **Frame, B., Guiang, H. L., Frost, J. M., and Reynolds, W. A.,** Osteomalacia induced by laxative (phenolphthalein) ingestion, *Arch. Intern. Med.,* 128, 794, 1971.

374. **Heizer, W. D., Warshaw, A. L., Waldmann, T. A., and Laster, L.,** Protein — losing gastroenterology and malabsorption associated with factitious diarrhea, *Ann. Intern. Med.,* 68, 839, 1968.

375. **Vilter, R. W.,** The vitamin B_6-hydrazide relationship, in *Vitamins and Hormones,* Harris, R. S., Loraine, J. A., and Wool, I. G., Eds., Academic Press, N.Y., 1965.

376. **Standall, B. R., Koo-Chen, S. M., Yang, G. Y., and Char, D. F. B.,** Early changes in pyridoxine status of patients receiving isoniazid therapy, *Am. J. Clin. Nutr.,* 27, 479, 1974.

377. **McConnell, R. B. and Chetham, H. D.,** Acute pellagra during isoniazid therapy, *Lancet,* 2, 959, 1952.

378. **Burrows, M. T. and Farr, W. K.,** The action of mineral oil per os on the organism, *Proc. Soc. Exp. Biol. Med.,* 24, 719, 1927.

379. **Morgan, J. W.,** The harmful effects of mineral oil (liquid petrolatum) purgatives, *JAMA,* 117, 1335, 1941.

380. **Curtis, A. C. and Ballmer, R. S.,** The prevention of carotene absorption by liquid petrolatum, *JAMA,* 113, 1785, 1939.

381. **Meulengracht, E.,** Osteomalacia of the spinal column from deficient diet or from disease of the digestive tract. III. Osteomalacia e abuse laxantium, *Acta Med. Scand.,* 101, 187, 1939.

382. **Javert, C. T. and Macri, C.,** Prothrombin concentration and mineral oil, *Am. J. Obstet. Gynecol.,* 42, 409, 1941.

383. **Palva, I. P., Salokannel, S. J., Timonen, T., and Palva, H. L. A.,** Drug-induced malabsorption of vitamin B_{12}. IV. Malabsorption and deficiency of B_{12} during treatment with slow-release potassium chloride, *Acta Med. Scand.,* 191, 355, 1972.

384. **Avioli, L. V., Birge, S. J., and Lee, S. W.,** Effects of prednisone on vitamin D metabolism in man, *J. Clin. Endocrinol. Metab.,* 28, 1341, 1968.

385. **Kimberg, D. V.,** Effects of vitamin D and steroid hormones on the active transport of calcium by the intestine, *N. Engl. J. Med.,* 280, 1396, 1969.

386. **Sahud, M. A.,** Uptake and reduction of dehydroascorbic acid in human platelets, *Clin. Res.,* 18, 133, 1970.

387. **Sahud, M. A. and Cohen, R. J.,** Effect of aspirin ingestion on ascorbic acid levels in rheumatoid arthritis, *Lancet,* 1, 937, 1971.

388. **Loh, H. S., Watters, K., and Wilson, C. W. M.,** The effects of aspirin on the metabolic activity of ascorbic acid in human beings, *J. Clin. Pharmacol.,* 13, 480, 1973.

389. **Alter, H. J., Zvaifler, N. J., and Rath, C. E.,** Interrelationship of rheumatoid arthritis, folic acid and aspirin, *Blood,* 38, 405, 1971.

390. **Callender, S. T.,** Fortification of food with iron — is it necessary or effective?, in *Nutritional Problems in a Changing World,* Hollingsworth, D. and Russell, M., Eds., Applied Science, London, 1973.

391. **Lieberman, F. L. and Bateman, J. R.,** Megaloblastic anemia possibly induced by triamterene in patients with alcoholic cirrhosis, *Ann. Intern. Med.,* 68, 168, 1968.

392. **Corcino, J., Waxman, S., and Herbert, V.,** Mechanism of triamterene-induced megaloblastosis, *Ann. Intern. Med.,* 73, 419, 1970.

393. **Kahn, S. B., Fein, S. A., and Brodsky, I.,** Effects of trimethoprim on folate metabolism in man, *Clin. Pharmacol. Ther.,* 9, 550, 1968.

DRUG-NUTRIENT INTERACTIONS IN THE AGED

Peter P. Lamy

INTRODUCTION

There is no question that improvement in human health in the 19th and 20th century has been accomplished as a result of improved social conditions, including nutrition, rather than by specific medical measures. Nutritional manipulation is now an important adjunct in managing chronic diseases, as in the dietary management of diabetes mellitus, hypertension, hyperlipedemia, and osteoporosis, for example.[1] There are now specified diets for cancer patients and those with renal impairment. Nutritional disorders and diseases of the elderly have been identified (Table 1). Clearly, nutrition plays an important, but still often overlooked, role in disease management, particularly management of chronic diseases.

On the other hand, nutritional deficiencies have been linked to increased susceptibility to disease and behavioral changes,[2] and this seems to be especially true for the elderly population. Prolonged undernutrition might result in anemia and vitamin deficiencies. Among the negative effects that might be cited are laryngitis, otitis media, bronchitis,[3] the carpal tunnel syndrome,[4] post-operative confusion,[5] and a host of symptoms ranging from bone pain to confusion, all of which might mimic what is often thought of as symptomatology of old age.[6] The elderly appear to be most at risk from under- and malnutrition. Subclinical malnutrition due to deficiencies in calcium, iron, magnesium, vitamins A, C, B complex, and others have been documented. While it is beginning to be recognized that nutrition is of primary importance in health, health care, health maintenance, and delay of dependency in older adults, little is known about nutrition and the elderly, and even less is being done about it.

Yet, 75% of the elderly have one or more chronic disease(s) and gastric surgery[7] can lead to nutritional deficiencies as can infections, by affecting protein, carbohydrate, lipid, and vitamin metabolism. Chronic diseases can further change food intake for a variety of reasons (Table 2) and certain chronic diseases and their management with drugs can heighten the risk of drug-food interactions (Table 3).

The Elderly

The baby boom is over, the senior boom has begun. It is here. But it is not enough to realize that the number of elderly is increasing. It is more important to recognize those who are most at risk to chronic diseases, nutritional depletion, and adverse drug reactions. It is the old-old, those 85 years and older, who constitute the fastest growing segment of our population, and it has been estimated that their number will increase in the next 20 years by 100 or even 120%. People are living longer, or, more correctly, more people are living longer to ages 85 and 90 than ever before. People are becoming more ancient, more fragile, more eviscerated, and take more toxic drugs. They still live, even though they may have serious liver, kidney, or gut failure.

Census Bureau statistics further document that among the population which is 85 years and older, women outnumber men by 2:1 (in community) and 3:1 (in nursing homes). Women, more so than men, live alone, and those living alone make more medication errors than those who do not. More recent data have shown that among the 6 million elderly women living alone, at least 3 million live at or below the poverty line and presumably, they must make choices between purchasing sufficient supplies to provide heat and food, or drugs. Physiologic, economic, and other changes that might change their nutritional status or dietary habits which may, in turn, bring about changed drug action (Table 4).

Table 1
DISEASES OF THE ELDERLY AND NUTRITIONAL DISORDERS

Disorders resulting from malnutrition
   Iron deficiency anemia
   Macrocytic anemias
     Folic acid deficiency
     Vitamin B_{12} deficiency
   Specific nutrient deficiency
     Avitaminoses
     Protein deficiency
   Underweight (caloric deficiency)
   Overweight (caloric excess)
Diseases influenced by nutrition
   Cardiovascular diseases
   Diabetes
   Hypertension
   Osteoporosis

Table 2
CHRONIC DISEASE AND FOOD INTAKE

Modified diet (i.e., sodium restriction)	Food may be unpalatable, elderly eat less
Condition may cause physical limitations	Food preparation and intake more difficult
Increased drug use	Drugs can interfere with nutritional status

Table 3
SOME DISEASES WHICH HEIGHTEN THE RISK OF DRUG-FOOD INTERACTIONS

Disease	Risk
Cardiovascular	Electrolyte disturbance: decreased K^+ and Mg^{++} and increased Ca^{++}
Depression	Anticholinergic (antidepressant) drugs can cause dry mouth, sour or metallic taste, constipation, and nausea and vomiting; possible reduced food intake, electrolyte disturbance
Gastrointestinal	Heavy use of antacids and laxatives can change nutritional status
Infections	Possible high sodium and potassium intake (via antibiotics)
	Disturbance of natural flora via antibiotics
	Nausea, vomiting, diarrhea (from treatment) can lead to electrolyte disturbances
Respiratory	High carbohydrate, low protein diet can change theophylline kinetics
Rheumatoid	High dose of aspirin could deplete iron stores
	Phenylbutazone can induce sodium retention

Table 4
CHANGES THAT MANDATE CAREFUL EVALUATION OF NUTRITIONAL NEEDS OF ELDERS

Change	Possible effect	Implication
Less physical activity, a more sedentary life	Change in body composition; more lipid tissue	Check caloric intake, nutrient density. More at risk to drug interactions and altered drug action
Changes in sight, odor perception, and taste perception	Less food intake, changes in salt and sweet perception may lead to increased intake	More difficult control of hypertension, diabetes
Changes in gastric milieu	Anacidity, hypochlorhydria, coupled with low intrinsic factor, may lead to anemia	Could be heightened by cimetidine use. Suggest foods rich in iron and B_{12}, altered drug action
Change in dietary pattern	Change in tubular secretion and/or reabsorption of some drugs	Drug toxicity and/or altered drug action

It is the old, unfortunately, who are also most likely to be influenced by advertisements and other means to buy nutritional supplements. They are convinced those supplements work, might keep one young, heighten one's ability to withstand disease, and give one more vitality and vigor. On the other hand, physicians in general do not feel that this is a field that they should be concerned with. It might be interesting just to look at some of the information on nutrition that the elderly might see, or in some cases, that they should see. For example, just recently, the FDA issued a warning about bone meal and Dolomite, (both frequently used and obtained in health food stores or by mail order) as sources of calcium, magnesium, or phosphorous. The FDA cautioned that there are unkown but often substantial quantities of lead in those substances.[8] The elderly are particularly at risk, especially if they take two or three times the recommended dose. There is the controversy about cholesterol. Recent reports state that low serum cholesterol increases the risk of cancer and this was a cause for great concern among the elderly. Yet, newer studies showed that low cholesterol is a symptom, not a cause, of cancer. Still more confusing must be even more recent reports that intake of dietary cholesterol may not effect body cholesterol levels at all. There is now the controversy about selenium. A distinguished geriatrician/gerontologist once mentioned on television that he is taking it as part of a regimen to extend life — and elderly rushed to buy it. Yet, the controversy about the effect of selenium deficiency in cardiomypathy and advisability of selenium oral supplementation has not yet been settled.[9,10]

Elderly read about the fact that a diet high in beta carotene may reduce the risk of lung cancer[11] and rush to buy that supplement; now they read about essential fatty acids. Marine fatty acids have made news as possible preventers of some cardiovascular problems.[12] There is also a resurgence in the use of herbal teas which are quite often thought to have remedial powers for certain disorders. Yet, sassafras, which was banned in 1960 when its major ingredient, safrole, was found to be carcinogenic, is still available in health food stores. Aside from its carcinogenicity, it may inhibit liver microsomal enzymes which could lead to secondary toxicities of drugs which are metabolized by these enzymes. Similarly, Comfrey tea, also widely used, is thought to be hepatotoxic if taken chronically. And now there is vitamin U, "the poor man's vitamin" which is supposed to benefit "stomach ailments, gas pains, ulcerous conditions, poor appetite, bladder, prostate, and other ills". The range of benefits that are supplied by it is astonishing and all of it is found in alfalfa tablets, which currently enjoy brisk sales. Clearly, the elderly are at risk not only to toxic effects from all this, but to economic loss, and those caring for elderly patients ought to be concerned with all aspects of nutrition.

There is another aspect to which more attention should be given. Federal statistics from 1974, addressing per capita drug use, showed that persons over 70 years of age received 13.7 new and refill prescriptions a year. Newer data indicate that this had risen to 17.9 prescriptions each year and that not only the number, but the size of each prescription had increased. The increase was ascribed to the increasing incidence of chronic disease.

No questions, then, that elderly are prescribed more drugs than any other segment of the U.S. population. Multiple pathology and polymedicine, though, can be dangerous. Drugs, even if used correctly, can be hazardous to the elderly. Among the reasons for heightened risk to adverse drug reactions among the elderly, and particularly the old-old, are the interacting effects of primary aging (normal age-related physiological changes), secondary aging (accumulation of trauma, injury, and pathophysiologic changes over time), and sociogenic aging (accumulation of losses which may lead to depression and other factors affecting drug action). Perhaps more importantly, changes in body composition, kidney and liver function, immune status, and others make prediction of drug action in the elderly more difficult. Older people respond to drug therapy in a much more individualized fashion than do younger ones, and there is a much greater variability.[13] Thus, anything that can reduce that variation in drug response should be pursued by those who prescribe, and among those factors is food. Food can and does change drug action, sometimes in an unanticipated manner.

THE NUTRITIONAL STATUS OF ELDERLY

The elderly are particularly at risk to some degree of malnutrition.[14-18] It is generally agreed that some senior citizens have poor food habits and consume diets inadequate in essential nutrients (the ''tea and toast syndrome''), and that elderly men, specifically, are at risk to malnourishment.[19] There is ample evidence that many elderly are, indeed, undernourished. Undernutrition, overnutrition, specific deficiencies, and imbalances have all been documented.

Studies on dietary practices of elderly, including USDA Food consumption,[20] the Ten State study,[21] and the National Health and Nutritional Examination Survey[22] have all shown that apparently healthy elderly are deficient in minerals, vitamins, proteins, and energy. Recently, it was reaffirmed that as many as 50% of the elderly population consume less than two thirds of current RDAs for many nutrients,[23] and, repeatedly, subclinical malnutrition due to deficiencies in calcium, iron, magnesium, vitamins A and C, thiamin, niacin, and folic acid have been shown.[18] Community surveys have suggested that anemia occurs in 5 to 20% of the elderly. In institutions, the incidence may be as high as 40%.[24] Clearly, subclinical malnutrition appears to be widespread among the elderly, including those living in the community and not only those in institutions.

Some Effects of Malnutrition

Specific deficiencies in vitamin B_{12}, vitamin C, folic acid, magnesium, potassium, and proteins could contribute to deficient neurotransmission and confusion, difficulty with memory storage and retrieval, and reduced learning speed.[25] Confusional states in the elderly have been corrected with vitamin supplements. Low body weight itself apparently heightens the susceptibility of elderly, particularly females, to adverse effects of drugs, such as barbiturates. Malnutrition is also the underlying and often unrecognized cause of many diseases, among them anemia, cancer, chronic infections, clotting disorders, dementia, dental disease, depression, GI disorders, goiter, hypertension, osteoporosis, sepsis, poor wound healing, and endocrine disorders (Table 5). Most importantly, the presenting symptoms of even mild nutritional deficiencies may easily be confused with age-related effects and might then not

Table 5
POSSIBLE DELETERIOUS EFFECTS OF
MALNUTRITION IN THE ELDERLY

Anemia	Increased morbidity and
CNS changes	mortality
Impaired visceral	Increased recuperation
protein enzymatic	time
function	Increased wound-healing time
Increased clotting time	Slowed bowel transit time

Table 6
PRESENTING SYMPTOMS OF
MILD MALNUTRITION

Confusion	Irritability
Depression	Lack of well-being
Disorientation	Listlessness
General malaise	Loss of appetite
Headache	Loss of body weight
Insomnia	Lowered resistance to illness

Table 7
SOME FACTORS WHICH MAY
DEPLETE NUTRITIONAL STATUS
OF ELDERS

Disease/factor	Nutrient depleted
Malabsorption of fat	Vitamin E
High carbohydrate diet	Thiamin
Hypo- and hyperthroidism	Riboflavin

be addressed vigorously and effectively, exposing the elderly to unnecessary problems (Table 6).[26,27]

FACTORS THAT MAY CONTRIBUTE TO MALNUTRITION

Many factors can and do influence the nutritional status of the aged. Among them are socioeconomic factors and racial differences. For example, an elevated rate of anemia was found among persons in sociodisadvantaged strata, and a difference in hemoglobin levels was also observed between American whites and blacks.[28]

Chronic Diseases

About three fourths of the elderly suffer from at least one chronic disease, and 40% must take at least one drug daily to manage a chronic disease. Chronic disease in the elderly can interfere with nutritional status either directly, by disruption of the nutrient chain leading to target tissues, or indirectly by a self-limited choice of foods (Table 7).

Among psychogeriatric patients, energy and/or protein undernutrition was found in 30%, but lack of teeth contributed substantially.[29] Protein undernutrition, indicated by low plasma albumin and low serum transferrin levels, was somewhat more prevalent among patients with dementia than with psychosis. Emotional stress, observed frequently in the elderly, can lead to a negative nitrogen and calcium balance[30] and varying levels of serum cholesterol.

Table 8
**SELECTED FACTORS IN PRIMARY AND
SECONDARY MALNUTRITION: POSSIBLE
ADVERSE EFFECTS OF DISEASES AND DRUGS**

Factor	Disease/drug effect
Reduced odor and taste perception	Can be further reduced
Reduced physical and mental health	Can be further reduced
Reduced food intake	Can be further reduced
Reduced digestive capability	Can be further reduced
Reduced absorption, distribution, storage, metabolism, excretion	Can be further reduced

Chronic Diseases and the Drugs Used to Treat Them

Elderly are especially at risk to malabsorption of nutrients (Table 8). They will have with a complex of symptoms, indicating involvement of many systems aside from involvement of the GI tract. Complaints and disorders may include anemia, purpura, osteopenia, symptomatic weakness, peripheral neuropathy, endocrine dysfunction, dermatitis, edema, and diarrhea. Elderly are particularly at risk since certain diseases, such as pancreatic insufficiency, are more common in the elderly than in younger adults, and pancreatic enzymes are needed in the digestive processes of proteins, carbohydrates, and triglycerides. Processing of dietary nutrients prior to their absorption by the gut wall may be negatively affected by diverticulosis of the small intestine. Whipple's disease can produce mucosal lesions, and other factors may interfere with nutrient transport after mucosal absorption. Patients who do not secrete adequate gastric acid may develop bacterial overgrowth in the proximal intestine, which is normally inhibited by gastric acid that empties into the small intestine from the stomach. Drugs can interfere with many of the absorptive and digestive processes and may, therefore, potentiate malabsorption. Alcohol alters mucosal fluid, electrolyte transport, and folate metabolism. Orally administered neomycin affects bile salt micelles and may directly injure enterocytes; cholestyramine binds bile salts to reduce the effective bile pool.[31,32]

There is a direct relationship between infection and malnutrition, and undernutrition is often associated with an increased incidence and/or severity of many bacterial, viral, and parasitic diseases. Malnutrition changes host resistance to infections, and infection exaggerates malnutrition. Infectious organisms may appropriate nutrients from the host's food or from the host's tissues, may impair the host's appetite (and thus, food intake), may disturb the host's intestinal function and impair digestion and absorption of nutrients. They may also impair nutrient metabolism and utilization. In persons with latent vitamin deficiencies, infections can precipitate avitaminoses.[33] Zinc and iron levels may be lowered, and iron deficiency itself can lead to increased susceptibility to infections, as it is an essential nutrient for the normal function of lymphocytic and phagocytic cells. Both are necessary in the maintenance of normal immunocompetence.[34]

It is clear that there are infection-induced alterations in amino acid and protein metabolism. During infections, there will be protein wasting, increased protein degradation, and decreased protein synthesis. Thus, protein intake should be increased during infections and during convalescence.[35] There are also alterations in the metabolism of free fatty acids and in electrolyte, vitamin, and mineral metabolism.[36] The devastating effect that infections can have on malnourished elderly was recently demonstrated, when it was shown that mortality in malnourished elderly with infections was 28% but only 4% in those not malnourished.[37] It should be emphasized that this study showed that 61% of elderly hospitalized patients exhibit malnutrition. Infections, of course, are freely treated with antimicrobial agents, which

Table 9
POSSIBLE INDIRECT EFFECTS OF DRUGS ON
FOOD INTAKE OF THE AGED

Effect	Drugs
Hyperphagia	Antidepressants, benzodiazepines, corticosteroids, cyproheptadine, hypoglycemics (oral), lithium
Hypophagia	Antineoplastic agents, digitalis, ferrous sulfate, hydrochlorothiazide, nonsteroidal antiinflammatory agents, penicillamine, potassium salts, reserpine, theophylline, triamterene
Hypogeusia	Allopurinol, antihistamines, clindamycin, clofibrate, griseofulvin, penicillamine
Metallic or bitter taste	Potassium iodide, streptomycin

cause the most rapid and most radical changes in the normal GI flora. The normal flora usually includes vitamin K, folate, vitamin B_{12}, pyridoxine, biotin, pantothenate, and riboflavin.[3] Disturbance of that flora, induced by antibiotics, could therefore easily lead to nutrient deficiencies. Certain infections, such as urinary tract infections, pneumonia, infectious endocarditis, decubitus ulcers, and vascular ulcers occur more frequently in the elderly than in young adults[39-41] and, since they pose an increased risk to elderly, they are often treated aggressively with antibiotics. Nutritional deficiencies should, therefore, be anticipated and prevented, if possible. Ulcer disease is now frequently treated with H_2 antagonists. Cimetidine, which should be given with food to obtain optimal acid inhibition,[42] interferes with iron absorption,[43] and patients on prolonged cimetidine therapy or those who have deficient cobalamin stores may be at risk to the effects of dietary cobalamin inhibition of absorption by cimetidine.[44,45] Reversible malabsorption of protein-bound cobalamin cannot be detected by the usual tests,[46] which is also true of vitamin B_{12} deficiency developed in the presence of hypochlorhydria.[47]

Alterations in Taste, Appetite, and Drug Effects

Although alterations in taste thresholds and reduction in the number of taste buds with age may be important in contributing to suboptimal nutrition in the elderly, gustatory, olfactory, tactile, and visual factors also play essential roles in flavor perceptions. Underlying diseases and medications are also important factors that should not be overlooked in altered taste perceptions of the elderly.[48-51] Alterations in taste response have been associated with many chronic diseases prevalent among the aged such as hypertension, diabetes, cancer, adrenal cortical insufficiency, and hypothyroidism.[52] An impaired zinc status and vitamin A deficiency can cause abnormal taste sensations, as can poor oral hygiene and denture wear. Importantly, a recent study issued by the National Institute on Aging reported that as many as 30% of elderly, mostly women, complain that drugs change taste acuity. Drugs, of course, also can induce changes in dietary intake by either increasing or decreasing a patient's appetite (Table 9).

Decreased Caloric and Insufficient Nutrient Intake

It is difficult to establish the nutritional needs of the elderly. Apparently, vitamin and mineral needs and presumably needs for other nutrients, do not decrease as an age-cohort per se.[53] However, the RDAs do not cover therapeutic nutritional needs, but meet the needs of healthy people only.[54] Yet, the increased incidence of chronic diseases among the elderly must be recognized. It must also be realized that recommended daily allowances should not

Table 10
RECOMMENDED ENERGY
INTAKE

Age (years)	Male (kcal)	Female (kcal)
23—50	2300—3100	1600—2400
51—75	2000—2800	1400—2200
76 +	1650—2450	1200—2000

From Recommended Dietary Allowances, 9th ed,
Food and Nutrition Board, National Academy of
Sciences, National Research Council, Washing-
ton, D.C., 1980.

be confused with daily need. For example, the 1980 edition of *Recommended Daily Allow-ances* estimates adult urinary and fecal loss of sodium at 23 mg daily and losses in perspiration at 46 to 92 mg/day. The total loss would, therefore, range between 69 and 115 mg/day, yet the estimated safe and adequate daily dietary intake of sodium for an adult is listed as 1100 to 3300 mg (48 to 143 mEq). There is then, quite obviously a large difference between daily need and estimated safe and adequate daily dietary allowance.

Recommended levels of daily caloric intake have decreased as the population has become more sedentary. Over a lifespan of 50 years, there is a decreased intake of approximately 600 kcal/day, one third due to reduced basal metabolism with age (or, more correctly stated, a reduction in the metabolically active tissue) and two thirds due to reduced physical activity. The recommended daily energy intake has been adjusted for that (Table 10). It is now estimated that as many as 75% of women over the age of 50 years have an intake of less than 1800 calories per day. Aside from the fact that it has been suggested that the four food groups do not ensure adequate intake of all nutrients, particularly vitamin E, vitamin B_6, iron, zinc, magnesium, and folacin, it is also clear that the proposed reduction in caloric intake makes it difficult to maintain a sufficiently high intake of protein, regulatory nutrients, and dietary fiber.[55] As a consequence, it is likely that there is an undersupply of vitamin B_6, magnesium, folacin, zinc, calcium, iron, and perhaps vitamin A and copper. Drugs given chronically in disease management can then be more hazardous to the nutritional status of elderly patients. In short, caloric deprivation (or severe illness) can produce a shift towards subclinical malnutrition, which in turn could be aggravated by drug therapy.[56-59]

DRUG-NUTRIENT INTERACTIONS

The possibilities of drug-nutrient interactions, with resulting depletion of nutrients, are increasingly being recognized.[60-65] The elderly are especially vulnerable to the adverse effects of these interactions, because they suffer from multiple pathology (which in itself can induce nutritional deficiencies), need multiple drugs in the management of these diseases (which can also induce deficiencies), and often have marginal food intake. These interactions can occur at many different physiologic sites (Table 11). Many drugs used in chronic care (Table 12) are involved in these interactions, which can cause serious clinical consequences, such as megaloblastic anemia, asymptomatic hypocalcemia or hypercalcemia, osteomalacia, peripheral neuropathy, symptomatic hypomagnesemia, muscular weakness, tetany, Wernicke-Korsakoff's psychosis, mental confusion, "pseudosenility", hypotention, and others. Tables 13 and 14 list some of these interactions.[54,64,66-70]

Specific Interactions: Some Mechanisms

Some mechanisms of nutrient interactions and interdependency have just recently been elucidated. Unquestionably, these are complex interactions in many cases. For example, it

Table 11
POSSIBLE PHYSIOLOGIC SITES FOR DRUG-FOOD INTERACTIONS

Gastrointestinal tract	Fat metabolism, thyroid hormones
Complexation	Carbohydrate metabolism, steroids, thiazides
Chelation	Vitamin metabolism, folic acid, isoniazid
Delay in gastric emptying	Kidneys
Metabolic pathways	Protein overload
Protein metabolism, glucocorticoids	Change in urinary pH, antacids, ascorbic acid

Table 12
CHRONIC CARE DRUGS THAT
MAY AFFECT THE NUTRITIONAL
STATUS OF ELDERS

Antibiotics (as used in UTIs)
Antacids
Anticholinergics
Anticonvulsants
Antihypertensives
Antituberculars
Diuretics
Hypoglycemics, oral
Laxatives
Mood-altering drugs

has been proposed that vitamin B_{12} is required for methionine synthesis and methionine is a key source of single carbon units for formate synthesis. Formate, in turn, is necessary for the formation of formyltetrahydrofolate and the folic coenzyme, folate polyglutamate. Interruption of this chain would then compromise folate levels.[71] It has also been suggested that there is a link between serum zinc levels, cytochrome P 450, and antipyrine clearance,[72] that antipyrine half-life is decreased following increased clearance due to vitamin C supplementation,[73] and that antipyrine clearance also proceeds faster in people with low hemoglobin.[74] There appears to be interrelationship between vitamin B_6 and magnesium. It is possible that the vitamin is necessary to facilitate magnesium cellular uptake and, indeed, patients with low red blood cell magnesium levels responded positively to vitamin B_6 administration.[75] The anticonvulsants disturb the metabolism of vitamin D and its metabolites, probably by inducing liver enzymes that metabolize that vitamin.[76] Osteomalacia may occur within months of the start of anticonvulsant therapy. However, folate supplementation during anticonvulsant therapy may decrease the serum and cerebrospinal fluid concentrations of anticonvulsants.[77] Malabsorption of nutrients due to drug effects plays a major role. Malabsorption can be induced by pathologic states,[78] to begin with, and can be heightened by drugs. For example, malabsorption of calcium can be induced by corticosteroids, cholestyramine, antacids, phenobarbital, phenytoin, primidone, and methotrexate. Large doses of aluminum hydroxide antacids and carbonate gels may deplete phosphate stores. That absorption of vitamins and trace elements can be adversely affected by antacid therapy is of particular importance to the elderly, who generally use antacids very frequently. Vitamin A, calcium, iron salts, and others may be complexed and made unavailable for absorption.[79] As a matter of fact, antacids can alter the GI absorption of drugs by forming nonabsorbable complexes, by alteration of the gastric pH, and by changing the gastric emptying time. For example, riboflavin is absorbed by a specialized transport system in the proximal portion of the small intestine, and alterations in gastric emptying can produce significant changes in the rate and extent of riboflavin absorption.[80]

Table 13
SOME DRUG-VITAMIN INTERACTIONS

Vitamin	Drug	Possible mechanism
A	Alcohol, cholestyramine, mineral oil, neomycin	Interference with vitamin absorption and metabolism
B_1	Alcohol	Interference with vitamin absorption and metabolism
C	Alcohol, anticonvulsants, aspirin, prednisone, tetracycline	Tissue deletion in part due to increased excretion
D	Cortisone, diphosphonates, glutethimide, glucocorticoids, phenobarbital, phenytoin, primidone	Interference with transport and metabolism
E	Cholestyramine, clofibrate, triiodothyronine	Alteration of fat absorption and lipid levels
Folate	Aminopyrine, methotrexate, pyrimethamine, triamterene, trimethoprim	Inhibition of dihydrofolate reductase
	Barbiturates, phenytoin, primidone	Interference with folate absorption
	Salicylates	Inhibition of folate plasma, protein binding
K	Anticonvulsants, cholestyramine, coumarin, mineral oil, salicylates	Antagonism, decreased absorption
Niacin	Alcohol, isoniazid, leucine (large doses)	Interference with absorption
Pyridoxine	Cycloserine, hydralazine, isoniazid, L-dopa, penicillamine	Antagonism
Riboflavin	Alcohol, chlorpromazine, thyroxine	Inhibition of absorption
B_{12}	Biguanides, cholestyramine, colchicine, neomycin, PAS, potassium chloride	Interference with absorption

SPECIFIC INTERACTIONS AND DEFICIENCY EFFECTS

Vitamins

Vitamin A deficiency can be induced by chronic diseases such as cancer, pneumonia, tuberculosis, nephritis, urinary tract infections, and prostatic disease, and can by induced by alcohol, cholestyramine, mineral oil, and neomycin. Night blindness, xerosis of the conjuctiva, nerve lesions, and increased pressure in the cerebrospinal fluid may occur.[2]

Thiamin deficiency, caused by alcohol but also by foods high in tannins, may result in "pseudosenility".[81] Vitamin C plays a role in neurotransmitter synthesis (formation of norepinephrine from dopamine and conversion of tryptophan to 5-hydroxy-tryptophan). A potent reducing agent, it enhances the absorption of iron and inhibits absorption of copper from the digestive tract.[54] Its depletion, induced by several drugs, can lead to poor wound healing, and it has been proposed that much higher doses of this vitamin are needed in certain disease states.[82] The absorption of vitamin D is decreased by agents that bind bile acids.[54] Gluocorticoids, given in high doses for a prolonged duration, can interfere with the hepatic metabolism of vitamin D. This reduces the plasma concentrations of 25-OH,D3, the precursor of the active form. All drugs that induce microsomal enzymes cause accelerated degradation of some vitamin D metabolites, preventing formation of the active form, while

Table 14
POSSIBLE DRUG INTERFERENCES WITH NUTRIENTS

Drug	Mechanism	Effect
Antihistamines		
Cyproheptadine	Appetite stimulation	Weight gain
Antihypertensives		
Ganglionic blockers	Impaired or enhanced GI mobility	Generalized malabsorption
Anti-infectives		
Aminoglycosides	Enzyme inactivaton	Decreased carbohydrate absorption
Chloramphenicol	Impaired nutrient metabolism and utilization	Inhibition of protein binding
Griseofulvin, Lincomycin (Lincocin)	Appetite suppression	Growth retardation, weight loss
Isoniazid	Altered nutrient excretion	Pyridoxine deficiency
Neomycin	Interference with bile acid activity	Decreased absorption of Vitamin B_{12}, carotene, iron, sugar, and triglycerides
Neomycin	Enzyme inactivation	Decreased carbohydrate absorption
Tetracycline	Impaired nutrient metabolism and utilization	Decreased bone growth
Trimethoprim (prolonged)	Impaired nutrient metabolism and utilization	Folate deficiency, accumulation of phenylalanine
Antineoplastics		
Aminopterin	Damage to intestinal mucosa	Decreased absorption of vitamin B_{12}, carotene, cholesterol, lactose, o-oxylose, megaloblastic anemia, steatorrhea
Methotrexate	Damage to intestinal mucosa	Decreased absorption of vitamin B_{12}, carotene, cholesterol, lactose, o-oxylose, megaloblastic anemia, steatorrhea
	Impaired nutrient absorption	Decreased folic acid synthesis
Antirheumatoid drugs		
Colchicine	Damage to intestinal mucosa	Decreased absorption of vitamin B_{12}, carotene, cholesterol, lactose, o-oxylose, megaloblastic anemia, steatorrhea
Penicillamine (Cuprimine, Depen)	Appetite suppression	Growth retardation, weight loss
	Altered nutrient excretion	Na^+ depletion in adrenally suppressed patients
CNS drugs		
Amphetamines	Appetite suppression	Growth retardation, weight loss
Benzodiazepines	Appetite stimulation	Weight gain
Levodopa	Competition for uptake at blood-brain barrier	Decreased phenylalanine and tyrosine absorption
Phenothiazines	Appetite stimulation	Weight gain
Phenytoin	Impaired nutrient metabolism and utilization	Osteomalacia
Primidone	Impaired nutrient metabolism and utilization	Folate deficiency, neurologic complications (possibly peripheral neuropathy, psychological effects ranging from irritability to paranoid psychosis)
Tricyclic anti-depressants	Appetite stimulation	Weight gain

Table 14 (continued)
POSSIBLE DRUG INTERFERENCES WITH NUTRIENTS

Drug	Mechanism	Effect
GI drugs		
Aluminum hydroxide gel	Complexation of nutrient by drug	Decreased phosphate absorption
Antacids	Altered GI pH	Thiamin deficiency
Anticholinergics	Impaired or enhanced GI motility	Generalized malabsorption
Cathartics	Impaired or enhanced GI motility	Calcium and potassium loss
Cholestyramine (Questran)	Interference with bile acid activity	Vitamin A, D, E, and K deficiencies
Clofibrate (Atromid-S)	Enzyme inactivation	Decreased carbohydrate absorption
	Appetite suppression	Growth retardation, weight loss
	Interference with bile acid activity	Decreased absorption of vitamin B_{12}, carotene, iron, sugar, triglycerides
Immunosuppressants		
Cytotoxic agents	Appetite suppression	Growth retardation, weight loss
Steroids	Altered nutrient excretion	Na^+ depletion in adrenally suppressed patients

From Lamy, P. P., *Drug Therapy*, 10(8), 82, 1980. With permission.

diphosphonates block formation of the active form in the kidney. When used improperly, this steroidal hormone can leave calcium deposits in arteries, joints, and kidneys and cause hypertension and renal failure. Large doses of vitamin E (higher than 400 mg/day) appear to act as vitamin K antagonist and potentiate the effect of warfarin. Overdoses of this vitamin have also been associated with hyptertension, severe fatigue, thrombophlebitis, and pulmonary embolism.

Low serum folate levels are frequently found in the elderly, and it has been estimated that up to 40% of elderly may suffer from folate deficiency.[2] Depletion of total body folate stores may take 4 to 5 months. Megaloblastic anemia would follow, but there may also be psychiatric effects.[2] Many drugs can adversely affect folate absorption, antagonize folate action, increase folate turnover, and increase its excretion. Folacin status can also be altered by congestive heart failure, chronic alcoholism, infections, skin diseases, certain hematologic disorders, liver malfunction, neoplasia, rheumatoid arthritis, and still others. Folate deficiency should be considered in the diagnosis of dementia, peripheral neuropathy, myelopathy, and a spectrum of neurologic disorders.[83] Among the agents that can alter the membrane transport of folate compounds, those that exist in the anionic form are active in micromolar concentrations. Among them are ethacrynic acid, sulfinpyrazone, phenylbutazone, and sulfasalazine.[84] Compounds capable of forming cations at physiologic pH must be present in at least millimolar concentrations. Among those that should be mentioned are chlorpromazine, procaine, tetracaine, and papaverine. Inadequate folate absorption may result when the polyglutamate cannot be split into the absorbable monoglutamate. This can happen when the GI pH is lowered and by increased intake of acidic foods or liquids. One might speculate that this could easily happen with elderly patients on methenamine given for urinary tract infection. Ascorbic acid is frequently given to optimize the action of methenamine. Also, elderly people concerned with potassium depletion often drink increased amounts of orange juice. In either case, folate absorption may be decreased. The clinical implications of drug-induced folate deficiency can be serious. Trimethoprim is a folic acid antagonist, now recommended for the treatment of acute, symptomatic, uncomplicated urinary tract infections. A 70-year-old woman, given 200 mg p.o. daily, presented with a generalized macular rash within 7 days which resolved upon withdrawal of the drug. On resumption of trimeth-

Table 15
VITAMIN K CONTENT OF ENTERAL SUPPLEMENTS

Product (1000 mℓ)	Vitamin K (mcg)	Product (1000 mℓ)	Vitamin K (mcg)
Citrotein	0	Precision LR	56
Compleat B	Trace	Precision HN	35
Ensure	148	Renu	150
Ensure Plus	211	Sustacal	235
Isocal	132	Sustacal HC	210
Isocal HCN	167	Travasorb std.	75[a]
Magnacal	300	Travasorb MCT	77[a]
Meritene	Trace	Vital	187
Osmolite	148	Vivonex	37
Precision isotonic	64	Vivonex HN	22

[a]   Product information is subject to change

oprim treatment, generalized edema occurred within 48 hr, which quickly forced hospitalization. Pancytopenia yielded to treatment with folinic acid, 6 mg/day, given intramuscularly.[85] Suppression of the normal intestinal floral by broad-spectrum antibiotics can decrease vitamin K synthesis, which becomes clinically important if vitamin K intake is low. Deficiency will manifest as hypoprothrombinemia and significant bleeding in some cases.[86]

Coumarin anticoagulants cause vitamin K deficiency. Some unanticipated vitamin K-coumarin interactions have recently been reported. Elderly, who may be given enteral supplements to increase nutrient density, may be affected and care must be taken with patients on warfarin therapy. Vitamin K in enteral supplements (Table 15), although it may be low, might interact with anticoagulants since it is probably present in a highly absorbable form.

Isoniazid and pyrazinamide interfere with pyridoxine. Since that vitamin is needed to convert tryptophan to niacin, these drugs therefore interfere with niacin, also. Pyridoxine, itself, is involved with neurotransmitters and deficiencies may result in disorders of the CNS. This is also of particular concern with elderly patients, where "senility" is often a waste-basket diagnosis. Drugs can decrease the function of pyridoxine in several ways, including direct competition for apoenzyme. Isoniazid, L-dopa (both frequently used in disease management of the elderly), cycloserine, and penicillamine can interfere with pyridoxine. Characteristic B_{12} neuropathy can follow gastrectomy, but not drug-induced malabsorption.[54] Drugs interfere with B_{12} by competitive inhibition or inactivation, and still other modes. Vitamin B_{12} deficiency should be considered in the differential diagnosis of peripheral neuropathies, myopathies, and anemia, but B_{12} deficiency may also be responsible for nonspecific complaints such as unexplained fatigue and confusional states.[2]

Minerals

Osteoporosis is a condition resulting from bone loss. It occurs when bone resorption is accelerated and bone formation fails to keep pace.[87] Low calcium intake, high protein intake, alcoholism, and coffee-drinking may play a part in its development. Efforts to benefit from positive drug interactions in its treatment, by combining calcium with fluoride, estrogen, and vitamin D[88] have not yet provided definitive answers since patients on those regimens experience significant adverse interactions and reactions, including rheumatic complaints, GI symptoms, and hypercalcemia.

Rheumatoid arthritis and its treatment may be associated with osteoporosis. Patients treated with corticosteroids and some postmenopausal women with rheumatoid arthritis are at risk from complications of osteoporosis.[89] Incidentally, dietary calcium intake without some physical activity, even walking, is suboptimal. Malabsorption of calcium in elderly osteoporotic women is not corrected by increasing plasma 250HD and 1:25(OH)$_2$D to levels which improve absorption in nonosteoporotic women, indicating some resistance to the action of vitamin D metabolites in the bowel.[90] The original report that lack of dietary calcium may be associated with essential hypertension[91] set off a record number of inquiries by the elderly. If lowered concentrations of serum-ionized calcium levels are, indeed, associated with hypertension,[92] perhaps that explains the action of thiazide diuretics which increase serum-ionized calcium[93] and induce a positive calcium balance.[94] Indeed diuretics, vitamin D supplementation, and lithium carbonate may all contribute to hypercalcemia.[95] This often manifests with such vague symptoms as weakness, confusion, lethargy, bone pain, and constipation, which can easily be ascribed to "aging symptoms" and could, therefore, be overlooked.

Electrolytes

Unnecessary potassium supplementation may be dangerous and may also increase the complexity of a patient's drug regimen. Routine prophylaxis with potassium supplementation should only be used for patients with known poor dietary intake, advanced liver disease, secondary hyperaldosteronism with renovascular hypertension, GI losses, or nondiuretic medication known to affect potassium status adversely, such as the steroids. Elderly often suffer from diabetes mellitus and may, therefore, have problems handling potassium,[96] as insulin deficiency interferes with potassium passage into cells. Patients are unable to benefit from the hypokalemic effect of insulin and hyperkalemia may result, which will resolve with administration of insulin. The combined use of potassium supplements with potassium-sparing diuretics may result in hyperkalemia,[97] which is promoted by significant renal impairment. Beta-adrenergic blockers may enhance the hyperkalemic effect of potassium supplements,[98] as can succinylcholine. Heparin apparently reduces urinary potassium excretion, as can any drug that reduces renal function. Among drugs that can also contribute to the development of hyperkalemia are the antibiotics which contain a heavy potassium load, such as potassium penicillin.

The concern with potassium supplements and hyperkalemia stems from the fact that excess potassium loss is often caused by diseases or drugs, such as the thiazide diuretics, the penicillins, the antinoeplastic agents, L-dopa, insulin in diabetic ketoacidosis, licorice derivatives, prolonged high-dose administration of corticosteroids, and loop diuretics. It is important to realize that it is not only the actual potassium level that may indicate clinical problems, but any change in that level.

There seems to be increasing evidence that serious complications such as ventricular ectopic activity may develop in asymptomatic hypokalemic patients, leading possibly to sudden death. Potassium also plays a fundamental role in the regulation of the action potential in myocardial and specialized cardiac cells. In turn, magnesium regulates potassium, and deficiencies in both ions can be dangerous. A magnesium deficiency results in hypokalemia and hypocalcemia, which will only resolve when the magnesium deficit has been replenished. Hypomagnesemia, like hypokalemia, appears to be a significant risk factor for the development of serious ventricular arrhythmias in acute myocardial infarction. Diuretics, particularly the loop diuretics, cause magnesium depletion. Older patients are particularly at risk, a risk increased by low magnesium intake and soft water, which is low in magnesium.[99] Thiazide diuretics also can induce magnesium losses, while potassium-sparing diuretics may produce magnesium retention. Secondary aldosteronism and protein-calorie malnutrition may also contribute to a negative magnesium balance.[100] There is a higher incidence of digitalis-induced arrhythmias in patients with hypomagnesemia, not only because of the magnesium and potassium deficiencies which adversely affect the heart, but also because magnesium

Table 16
POTASSIUM IN
MILK

Type	mg/ℓ
Whole milk,	
3.7% Fat	635
Skim	658
Partially skim,	
2% nonfat milk	
solids added	794

depletion increases myocardial uptake of digoxin. These arrhythmias can be treated with parenteral magnesium. Common presenting symptoms of hypomagnesemia are depression, muscle weakness, refractory hypokalemia, and atrial fibrillation refractory to digoxin. To a degree, this again points out the difficulty in geriatric medicine, where aging effects, disease effects, and adverse drug effects or deficiency syndromes often present in very similar fashion.

The controversy about potassium supplementation continues with a recent report that diabetic or elderly patients are at a significant risk of developing potentially lethal hyperkalemia. In this context, it is important to remember that elderly people often buy supplements at health food stores (some reports state that as many as 25% of the elderly may buy their own potassium supplements), may drink milk particularly high in potassium[101] (Table 16), and may use, in addition, salt substitutes. One of the new ones, NoSalt, by Norcliff Thayer, contains 35 mEq per $^1/_2$ level teaspoon. It is easy to see that excess potassium intake can quickly occur.

Proteins

Aging itself reduces the immunocompetence of elderly, which can be further reduced by protein-calorie malnutrition.[102-105] There is a progressive decline in cell-mediated immunity, depression of both T and B cell functions, and a profound reduction of thymic factor activity. As has been previously established, elderly often receive antibiotics for a variety of infections. Considering the reduced immunocompetence, consideration should be given to the use of bactericidal rather than bacteriostatic agents. The latter, of course, depend for positive action on a well-functioning immune system. While it has generally been questioned whether protein supplementation can counteract the effects of protein-calorie malnutrition on the immune system, a recent report indicates that 8-week supplementation with a formula containing proteins, fats, carbohydrates, minerals, and vitamins improved T-lymphocyte subpopulations and cell-mediated immunity.[106] On the other hand, increased protein intake can lead to kidney overload and, if a fixed calcium intake is maintained, to a negative calcium balance.[107] In any case, enteral supplementation is often recommended to improve nutritional value and caloric density of food intake of the elderly.[108] In that case, the sodium content and the osmolality should be carefully watched. Also of concern is the increasingly recognized lactose intolerance among the elderly. Dietary disaccharides have to be hydrolized into their component monosaccharides before absorption. Five enzymes (disaccharidases) are responsible for that hydrolysis, including lactase (beta galactosidase). Lactose, the only carbohydrate present in milk, is hydrolyzed principally by lactase. When lactase activity is low, lactose intolerance results. Hypolactasis may be caused by many factors (Table 17), many occurring frequently in the elderly. Perhaps they are responsible for the increasing lactose-intolerance in the elderly. The possibility of lactase deficiency should always be considered and investigated in elderly complaining of colicky or cramp-like abdominal pains, flatulence,

Table 17
CAUSES OF ACQUIRED (SECONDARY)
HYPOLACTASIA

Age (?)	Small intestinal resection
Gastroenteritis[a]	Tropical malabsorption
Celiac disease	Ulcerative colitis[a]
Protein-calorie malnutrition[a]	Crohn's disease
Giardiasis[a]	Whipple's disease
Post-surgical[a]	Cystic fibrosis
Partial gastrectomy	Lymphoid hyperplasia
Ileostomy	

[a]   More frequent in elderly than younger patients.

abdominal distension, and diarrhea. Stools are generally watery and acidic. These symptoms are usually associated with the ingestion of milk and resolve when milk is withdrawn. A preparation is now commercially available (Lactaid) which can be added to milk, making its consumption possible for patients with lactase deficiency. Apparently, *L.acidophilus* is not effective.

POSSIBLE ALTERATIONS OF PHARMACOKINETICS IN MALNUTRITION

A large body of literature in geriatrics/gerontology speaks to the age-related pharmacokinetic effects[109] which make prediction of drug action in the older population so difficult. It has also been reasoned that perhaps some of the observed pharmacokinetic changes are secondary to malnutrition. There are still very few data on the possible effect of malnutrition on pharmacokinetics. There are, however, many indicators which would make it seem likely that pharmacokinetics and therefore drug action can change with malnutrition.[110] While the effects of nutritional depletion on drug absorption, distribution, metabolism, and excretion are complex and still little understood, there is some evidence to support the suggestion that the dietary status of an elderly patient may well affect and change the pharmacokinetic parameters of several drugs.[111-125] It is further reasonable to speculate that if pharmacokinetics are affected by malnutrition, the elderly would be most at risk. Many of the factors of primary and secondary aging, which are thought to change pharmacokinetics as an age-related function, are also altered by malnutrition (Table 18), possibly enhancing the age-related effects on pharmacokinetics.

The functional status of the GI tract changes with malnutrition, as does gastric emptying time and intestinal transit time, theoretically leading to changes in rate and extent of absorption of drugs. Drug metabolism by bacteria in the stomach and drug metabolism in the stomach wall may be altered, changing the predicted or expected amount of a drug that would be absorbed.

In malnutrition, the heart is smaller than usual and there may be circulatory insufficiency, exaggerating the lessened distribution of drugs to the liver (the main metabolic organ) and the kidneys (the main excretory organ) that occurs with age. Body composition changes, as does total body water, plasma volume, extracellular fluid volume, and intracellular hydration, as does the electrolyte status. These changes can undoubtedly produce changes in the volume of distribution of drugs. Changes in plasma albumin may alter the action of strongly protein-bound drugs, and changes in lipoprotein and globulins may affect the distribution of still other drugs.

Dietary protein deficiency can impair liver microsomal mixed-function oxidases, while deficiencies of ascorbic acid, calcium, magnesium, riboflavin, zinc, and essential fatty acids also affect various hepatic enzyme systems. Also changed may be the hepatic first-pass effect.

Table 18
AGE AND NUTRITIONALLY RELATED CHANGES IN PHARMACOKINETICS

	Primary aging[a]	Secondary aging[b]	Malnutrition
Absorption			
Anorexia, diarrhea nausea, vomiting	+	+	+
Mucosal and vollous atrophy or damage	+	+	+
Anacidity, hypochlorhydria	+	+	+
Changed gastric emptying	+	+	+
Altered intestinal motility	+	+	+
Malabsorption and maldigestion	+	+	+
Distribution			
Change in body composition	+	+	+
Change in total body water	+	+	+
Change in plasma proteins	+	+	+
Cardiovascular changes	+	+	+
Metabolism			
Changes in liver enzymes and enzyme activity	+	+	+
Change in first pass effect	+	+	+
Excretion			
Change in renal status	+	+	+
Change in urinary pH	+	+	+

[a] Primary aging: normal age-associated physiologic changes.
[b] Secondary aging: pathophysiologic changes.

Table 19
DRUGS WITH APPARENTLY CHANGED PHARMACOKINETIC PARAMETERS IN MALNUTRITION

Acetaminophen	Phenobarbital
Cefotixin	Phenylbutazone
Chloramphenicol	Phenytoin
Digoxin	Salicylates
Estradiol	Sulfadiazine
Ferrous sulfate	Sulfisoxazole
Gentamicin	Tetracycline
Isoniazid	Thiopental
PAS	Tobramycin
Penicillin G	Warfarin

It is also reasonable to expect that both glomerular filtration and tubular reabsorption change. The latter may also change due to changes in urinary pH. Some pharmacokinetic studies have been performed in malnourished people. In general, protein binding is perhaps significantly lower, excretion more rapid, and drug half-lives shorter in malnourished than in well-nourished persons. The oral absorption of tetracycline is decreased, clearance of penicillin is decreased, clearance and half-life of gentamicin is decreased, digoxin binding is increased, isoniazid is acetylated faster, the metabolism of sex hormones is changed, the half-life of chloramphenicol is increased, but that of tetracycline decreased, as are those of antipyrine and sulfadiazine (Table 19).

VITAMIN OR MINERAL SUPPLEMENTATION: HELP OR DANGER

Vitamins

Older adults use vitamin supplements for nutritional insurance, to compensate for stress, to prevent illness, and to live longer.[126] As many as 60% or more of the elderly may be taking supplements, and many take several preparations at a time. Women tend to take more than men.[127-129] Vitamin C and E are most popular, followed by the B complex vitamins, potassium, and calcium. A recent study has documented that among elderly men and women who deliberately tried to remain healthy by taking large amounts of vitamins, no clear prolongation of life or reduction in disease could be attributed to their vitamin regimen.[130] Interestingly, either very low or very high vitamin E intake, among other factors, was associated with shortened life expectancy.

The most important problem with vitamin supplementation by the elderly is probably the suspected high use of megadoses. A regimen of a vitamin at least 10 times the recommended dietary allowance (which would be 500 mg of vitamin C daily, a dose frequently exceeded by large amounts) is megavitamin therapy. Megadoses of vitamins cannot act as vitamins; once human requirements are met, vitamins can then only act as chemicals, with the attendant possible adverse effects. Megadoses of vitamin C can induce serious renal toxicity in a small number of susceptible individuals, decrease absorption of vitamin B_{12}, increase urinary oxalate, cystein, and uric acid. They can be responsible for diarrhea and false negative reactions with glucose oxidase tests.

Vitamin D intoxication may lead to hypercalcemia, impaired renal function, deposition of calcium salt in soft tissues, and increase in plasma cholesterol, mental retardation, and hypotonia. Vitamin A megadoses can lead to toxic symptomatology involving the GI tract, the CNS, the skin, muscle, joints, and bones.[131]

Minerals

The effect of supplementation with any essential nutrient depends on the nutritional status of the recipient.[132] However, micronutrient supplementation is potentially toxic for the elderly, since metabolic turnover and organ function decrease with age. Toxicity of metals is cumulative, and toxic effect on the heart and kidney could be irreversible. In toxic levels, metals can depress cellular respiratory function.[133] In the elderly, known to take supplements freely, other potential problems should be anticipated. Some data have been published which show that unbalanced intakes of minerals and trace elements may lead to interactions which may be antagonistic or synergistic.[134] Calcium, copper, chromium, and iodine all lower cholesterol in humans. Copper deficiency, on the other hand, which can be induced by reduced copper uptake from a high zinc diet, is associated with increased cholesterol levels and copper deficiency anemia.[135,136] Phytate-mediated calcium/zinc interactions have been reported which can be responsible for decreased zinc absorption. On the other hand, calcium absorption can be inhibited by zinc, but only when calcium intake is low. Iron and zinc interact on a competitive basis, when their ratio exceeds 2:1. It is noteworthy that many commercially available vitamin/mineral preparations exceed this ratio. It is also possible that increased iron intake may lead to reduced copper levels. High dietary ratios of Ca/Mg, PO_4/Mg, and vitamin D excess cause loss of magnesium.[137] In view of the sometimes indiscriminate intake of vitamin and minerals by the elderly, these potentially serious interactions should be kept in mind, and elderly should be counseled accordingly.

PHARMACOLOGIC INTERACTIONS OF DRUGS AND DIETARY COMPONENTS

Foods may contain pharmacologically active components which may respond to the agonistic action of certain drugs. A number of foods contain pressor amines, including histamine, tyramine, tryptamine, norepinephrine, octopamine, and serotonin. Normally, these amines are inactivated by monoamine-oxidase-mediated metabolism in the liver and gut. Monoamine oxidase inhibitors, once widely used and again recommended for the treatment of depression in elderly when other drug treatment has failed, block the metabolic breakdown of these pharmacologically active substances.[138] The pressor amine then reaches the systemic circulation where it acts indirectly by displacing norepinephrine from the presynaptic storage granules. If sufficient noradrenaline is released, myocardial infarction or a cerebrovascular accident can result.[139] Among the foods implicated are avocado pears, certain cheeses, pickled herring, chicken livers, yeast-containing products, some red wines, and some beers.

Isoniazid is a drug increasingly being used to treat elderly tuberculosis patients. These patients are at risk when they eat sardinella, skipjack, or tuna, especially if the latter is slightly spoiled. The high histamine content of these foods interacts with the drug, causing severe headaches, redness and itching of the eyes and face, chills, palpitations, variations in pulse rate, and loose stools.[140-142] Similar reactions have been reported for isoniazid-treated patients who ingest very strong Cheshire or Swiss cheese.[143-144] Dopa-containing foods can also be potentially harmful. Pheochromacytoma-like symptoms can occur quickly after ingestion of only 8 oz of broadbeans in patients receiving monoamine oxidase inhibitors.[145] One final interaction of this type needs to be mentioned. Licorice contains an isomer of glycyrhetinic acid, which has mineralocorticoid, antidiuretic, and anti-inflammatory activity.[146] Ingestion of large amounts, particularly by patients on diuretic therapy that leads to potassium loss, can induce symptoms of overdoses of glucocorticoids and aldosterone.[147]

FOOD-DRUG INTERACTIONS

Various nutritional factors alter the absorption, distribution, metabolism, and excretion of drugs (Table 20). Macronutrients, such as carbohydrates, proteins, and fats, as well as trace substances such as vitamins and minerals, are often responsible. Nutrient components can change GI motility, secretions involved in the digestive processes, blood flow rates, and enzyme activity responsible for drug metabolism. Evidence is strong that foods and nutrition can be a major determinant of drug action.[148]

Drug Intake: The Use (Misuse) of Fluids

Elderly patients, particularly those with incontinence of urine or those on diuretics, often attempt to take their medications with as little fluid as possible. In that case, capsules, more so than tablets, can be delayed in the esophagus for up to 5 and even 15 min, possibly causing irritation, ulceration, stricture, or even more serious consequences. Patients particularly at risk are those with hiatus hernia, stricture, and an enlarged left atrium from mitral valve disease, even though quite often the delay occurs with no abnormal esophageal characteristics.

Recently, attention has been drawn to this potential problem.[149] As many as 50 to 60% of the elderly are thought to be a risk. It has been suggested that capsules should be swallowed only following a lubricating water bolus, then taken with water, followed by a water chaser. J. A. Brown, a physician from Augusta, Georgia, goes one step further. In the Oct. 15 issue of *JAMA*, he suggests that patients, in order to swallow capsules efficiently, should tilt their head forward before swallowing.

The problem of drug intake (or food intake) can be exacerbated by the fact that elderly suffer from a decreased flow of saliva, which can be further decreased by the use of anticholinergic drugs (psychotropic drugs, antihistamines, anti-Parkinson drugs).

Altered Absorption

The clinical effect of foods on drug absorption, particularly in chronic care, has in general not yet been studied. Most drugs, to be effective, must attain and then maintain a certain

Table 20
POSSIBLE DIETARY INTERFERENCES WITH DRUGS

Drug	Food	Mechanism	Effect
Anticoagulants			
Warfarin sodium (Coumadin, Panwarfin)	Vitamin K-containing foods (cabbage, green pea, turnip greens, broccoli)	Antagonism of anticoagulant effects of vitamin K	Anticoagulant failure
	Cooking oils with silicone additives	Complexation of drug by nutrient	Decreased drug absorption
Anti-infectives			
Aminoglycosides	Milk and dairy products, vegetables, almonds, chestnuts, coconuts, citrus fruits	Alkalinization of urine	Decreased antibacterial activity
	Bread, bacon, corn, lentils, meat, fish, fowl	Acidification of urine	Increased antibacterial activity
Cephalosporins (oral)	Coadministration with food	Altered GI motility and transit times	Lowered and delayed peak antibiotic serum levels
Clindamycin (Cleocin)	Pectin	Complexation of drug by nutrient	Decreased drug absorption
Erythromycin	Coadministration with food	Altered GI motility and transit times	Decreased drug absorption rate
Griseofulvin	High-lipid-content foods	Greater solubilization due to increased bile secretions	Increased drug absorption rate
Lincomycin (Lincocin)	Coadministration for food	Altered GI motility and transit times	Decreased drug absorption rate
	Pectin	Complexation of drug by nutrient	Decreased antibiotic absorption
Methenamine	Bread, bacon, corn, lentils, meat, fish, fowl	Acidification of urine	Optimal anti-infective activity

Drug	Dietary factor	Mechanism	Effect
Nitrofurantoin	Coadministration with food	Reduced gastric emptying rate; increased drug bioavailability	Increased drug absorption
	Low-protein diets; milk and dairy products, vegetables, almonds, chestnuts, coconuts, citrus fruit	Alkalinization of urine	Increased drug excretion
Penicillin	Coadministration with food	Altered GI motility and transit times	Decreased drug absorption rate
Sulfadiazine	Coadministration with food	Altered GI motility and transit times	Decreased drug absorption rate
Tetracycline	Milk and dairy products, iron supplements	Complexes with di-1 or trivalent ions such as calcium, magnesium, or iron	Decreased drug absorption
Bronchodilators			
Theophylline	High-protein, low-carbohydrate diets	Increased cytochrome P-450 activity	Decreased drug half-life
Cardiovascular drugs and diuretics			
Diuretics (long-term therapy)	Monosodium glutamate (MSG)	Adverse effects of MSG (tightening of chest, flushing of face) intensified by diuretics	Angina-like syndrome
Hydralazine	Coadministration with food	Reduced first-pass metabolism of drug in gut wall?	Increased drug absorption
Methyldopa (Aldomet)	Neutral amino acids	Competition for uptake at blood-brain barrier	Decreased drug availability to brain
Propranolol (Inderal)	Coadministration with food	Reduced first-pass hepatic metabolism of drug?	Increased drug absorption
Quinidine	Excessive alkaline diets	Altered drug excretion	Quinidine toxicity
Spironolactone (Aldactone)	Coadministration with food	Increased drug bioavailability	Increased drug absorption

Table 21
REDUCTION OF ABSORPTION OF SOME ANTIMICROBIALS

Amoxycillin (?)	Penicillin G
Ampicillin (?)	Penicillin V (?)
Cephalexin	Phenethicillin
Isoniazid	Pivampicillin
Methacillin	Rifampin
Nafcillin	Tetracycline
Oxytetracycline	

Adapted from Welling, P. G. and Tse, F. L. S., *J. Antimicrob. Chemother.*, 9, 7, 1982.

plasma level, the steady-state level. Whether foods affect that level negatively has not yet been elucidated. However, there are many indicators which would argue that, indeed, foods can interfere with drug absorption.

Foods and food components may interact with drugs in various ways. The overall effect, generally, is a reduction in the drug's bioavailability.[150] Thus, ingestion of food shortly before a drug is given to a patient may have a marked effect on the rate or extent of drug absorption and, presumably, on the resulting concentration in the circulation (Table 21). Foods can simply act as a mechanical barrier which prevents drug access to the mucosal surface of the GI tract or food components can complex drugs (proteins) or chelate drugs (polyvalent metal ions). Drug dissolution, decomposition of the drug, changes in GI transit time, and interference which mucosal transport may also be involved. The pharmaceutical dosage form may also be important. It might be speculated that enteric coated tablets may be most affected, while drugs in solution may well be least affected, but this is not yet clear.[151]

Acetaminophen absorption is five times faster in fasting persons than in persons who have consumed a high-carbohydrate meal.[2] Carbohydrate nutrients, particularly those containing large amounts of pectin, delay acetaminophen absorption, while proteins and lipids do not. Lipids, on the other hand, double the absorption of lipophilic agents, such as griseofulvin in contrast to proteins or carbohydrates, which do not exert this effect. Perhaps the lipophilic drugs are solubilized by bile secreted in response to the lipid intake.[152,153] Fasting causes a sevenfold increase in tetracycline absorption; food intake, on the other hand, reduces the absorption of tetracycline, ampicillin, and hydrocortisone.[154]

Correct (or incorrect) intake of fluids may again play an important part in the chain of drug absorption and disposition. Poorly water-soluble drugs will be absorbed to varying degrees depending on the volume of fluid used to swallow them.[155] Fluids may still play a more important role. It has been speculated that ice water, prevalent at the bedside in nursing homes, would delay the dissolution of capsules, which could become an important factor in the case of hypnotics, for example, where rapid onset to overcome sleep latency is desirable. The effect of food and food components on the bioavailability of drugs has been shown clearly in other cases. Most recently, it has been suggested that the RDA for zinc is not sufficient since dietary fibers (whole grain, legumes), phytates, milk, cheese, eggs, celery, and even lemon juice reduce its bioavailability. Oxalic acid, too, interferes with zinc uptake, as it forms insoluble salts with zinc, particularly in the presence of a high fiber diet. Decreased levels of zinc, the second most common element in the body, are associated with some forms of cancer, chronic liver disease, and leg ulcers, as well as with sensory disorders, such as taste and odor perception dysfunction. Foods can also interfere with the mucosal transport of drugs absorbed by active transport. Levodopa (as well as any drug that has a structure similar to amino acids) is absorbed by a transport mechanism for amino acids. A high protein diet diminishes this uptake by competition between the drug and the amino acids from the protein diet.[156] Riboflavin is absorbed by a specialized transport system in the proximal portion of the small intestine, and changes in the intestinal transit rate can produce a significant change in the rate and extent of riboflavin absorption.[80] A decrease in intestinal transit rate in the presence of food exposes the drug longer to the absorption sites, the absorption mechanism will not be saturated, and there will be increased uptake of riboflavin.[157]

In elderly, there is an age-related decrease in the acidity of the gastric fluids. Anacidity or hypochlorhydria occur ten times more frequently in older than in younger persons.[2] In elderly patients, particularly those who have difficulty with mastication or who have had gastric surgery, long-term cimetidine administration coupled with high fiber intake may lead to the formation of phytobezoars.[158] The absorption of weakly basic drugs, such as amitriptyline, diazepam, or pentazocin, takes place from the less acidic intestine. The gastric emptying rate can then become a critical factor; it is under the control of the CNS, which may lose efficiency with advancing age. Emotional stress, frequently seen in the elderly, also can inhibit gastric emptying.

It has been shown that the gastric emptying time is approximately 50 min in young, healthy volunteers, but about 123 min in elderly of age 77 years and over.[159] Food can further change the gastric emptying time. This could have several implications. Decreased gastric clearance of such drugs as penicillin, digoxin, and levodopa would cause more of these drugs to be metabolized in the stomach and less of the unchanged drug would be available for absorption, possibly causing an erratic therapeutic response. Changes in gastric emptying will mainly affect drugs which are rapidly absorbed and those which may have a short biological half-life.

It is interesting to point out that diabetic gastroparesis is now being treated with metoclopramide, which increases gastric emptying significantly. Accelerated gastric emptying may be responsible for decreased absorption of slow-release, large-particle size formulation of digoxin, for example, and may also change the time for food to reach the small intestine.[160]

Finally, it should be noted that not only the presence, but also the absence of food can bring about changes in drug action. Nutritional deficiencies of proteins, calories, vitamins, and minerals all produce structural changes in the GI tract[161] and in people with various nutritional disorders, such as protein-calorie malnutrition, tetracycline absorption is significantly reduced.[120]

DRUG DISPOSITION

Foods and food constituents can not only interfere with the absorption of drugs, but importantly, with the disposition of drugs once they have been absorbed. Distribution, metabolism, and excretion may be affected.

Altered Distribution

Distribution of drugs depends on many factors, all of which can be affected by the aging process. Among them are body composition, total body water, and plasma protein binding.[2] Chronic diseases and foods can further change drug pharmacokinetics. The interaction between drugs and protein molecules greatly influences drug activity. Albumin, the major plasma protein, is susceptible to changes in response to poor food intake and to diseases. Elderly suffer from diseases that predispose them to undernourishment and even malnutrition, in which case one can expect albumin levels to fall. Since protein-calorie malnutrition is not infrequent among the elderly[37] and infections, which occur frequently among elderly also negatively affect the protein status of a patient,[162] one can assume that in many instances albumin levels will be low. Other, nonnutritional causes of hypoalbuminemia include recent surgery, trauma, nephrotic syndrome, protein-losing enteropathy, and liver disease. One word of caution needs to be added; possibly in response to a diminished thirst mechanism, decreased fluid intake, or increased fluid excretion, elderly often present with some phase of dehydration. Dehydration leads to a diminished plasma volume and plasma albumin concentrations may, therefore, appear to be elevated (but not total albumin). Conversely, when patients suffer from congestive heart failure or renal impairment, one may see an expanded plasma volume and decreased albumin concentrations.[55,163] Furthermore, as previously mentioned, lean body mass, the metabolically active tissue, decreases with age. Muscular tissue decreases by 40%, the kidney by 9%, the liver by 18%, and the lung by 11%.[164] This decrease is accompanied by an increase in body fat, a decrease in total body water and, of course, a decrease in total protein synthesis.[165]

With highly bound drugs, a reduction in serum albumin can lead to a significant increase of the free (active) fraction of the drug. The result will depend on several factors, such as the volume of distribution or the speed of elimination, and potentiation of an effect or a decrease in the duration of action can be expected.

Decreased protein binding in the presence of lowered albumin levels has been demonstrated for phenytoin and warfarin.[166,167] Adverse reactions to phenytoin were recorded in more than 11% of patients with a serum albumin concentration of less than 3 g/100 mℓ but in only 3.8% of patients with normal albumin levels.[2] Serum albumin concentrations have also been correlated with the clearance of antipyrine, diazepam, and propranolol.[168,169] Aside from these changes, no clinically significant effect on drug distribution by foods or food components has as yet been demonstrated. However, about 5 years ago, a new concept was proposed.[170] Large neutral amino acids share a common transport system located at the blood brain barrier with methyldopa, an antihypertensive drug. The drug and the amino acid compete for brain uptake. Nutrients can affect serum amino acid patterns and can, thus, significantly influence the availability of drugs related to amino acids to the brain. Similarly, it has been tried to increase brain serotonin levels, but high protein meals were not effective, probably because they do not increase the relative concentration of tryptophan in the blood. High carbohydrate meals, on the other hand, do.

Altered Metabolism

Nutritional factors are important in the regulation of drug metabolism in humans. Weight, *per se*, does not seem to affect metabolism, as the half-lives of antipyrine and tolbutamide, drugs metabolized by the microsomal oxidation pathways, do not differ in healthy, normal, or obese persons. The same results were obtained for drugs metabolized by acetylation (sulfisoxazole and isoniazid) and by the pseudocholinesterase hydrolysis pathway (procaine).[171] Fasting enhances hepatic enzyme induction, but a long-term, low-protein diet has the opposite effect.[172] Dietary protein deficiency can decrease cytochrome P-450 activity and increase the duration of barbiturate action by decreasing rates of metabolism and tissue clearance.[173] High-protein, low-carbohydrate diets reduce the average plasma half-life of antipyrine by 41% and of theophylline by 36%, an effect counteracted by a low-protein, high-carbohydrate diet.[174] Diets that incude animal fat and protein shorten antipyrine half-life by 50%, compared to the half-life observed in vegetarians.[175]

Altered Excretion

Amino acids, on infuson, create a rise in creatinine clearance, causing interglomerular hypertension, hyperfusion, and erosion of renal function in patients with early renal disease,[176,177] which then can adversely affect the excretion of drugs eliminated unchanged by the kidneys.

Sodium levels in the body are important to the regulation of lithium excretion. Dietary restrictions in psychiatric patients receiving lithium therapy precipitate lithium toxicity if there is sodium depletion.[178]

Table 22
FOODS POTENTIALLY CAUSING CHANGING IN URINARY pH

Foods potentially contributing to the acidification of urine	Foods potentially contributing to the alkalinization of urine
Meats	Milk products
Meat, fish, fowl, shellfish	Milk, cream, buttermilk
Eggs	Fats
Cheese	Almonds, chestnuts, coconut
Peanut butter	Vegetables
Vegetables	All types
Corn and lentils	Fruit
Fat	Citrus fruits
Bacon	
Nuts: Brazil, filberts, peanuts, walnuts	
Fruit	
Cranberries	
Plums	
Prunes	
Breads	
Breads (all types), crackers	
Macaroni, spaghetti, noodles	
Dessert	
Cakes, cookies	

Alterations in urinary excretion most often results from changes in urinary pH. Tubular excretion/reabsorption of some drugs follows pH-dependent kinetics. If the urine has an acidic pH, weakly basic drugs, such as amitriptyline and chloroquine, would be excreted, since they will form water-soluble salts in the urine. Conversely, if the urine is alkaline, they would remain largely water-insoluble and would be reabsorbed. It is evident, then, that continued dosing at a predetermined level would ultimately lead to toxic plasma levels of the drug. The inverse, of course, is true for acidic drugs, whose elimination half-life would be decreased in an acidic urine.

Balanced protein diets will produce an acid urinary pH (pH 5.9) while low protein diets usually result in an alkaline urinary pH (pH 7.5). It is interesting to note that citrus fruit juices, contrary to "expectations", produce an alkaline urine.[2] Table 22 lists foods potentially causing changes in urinary pH. It should be noted that many elderly switch to low protein diets with advancing age, and it is thus reasonable to assume that drug elimination may well change simply due to that factor.

CONCLUSION

The elderly take more toxic drugs than any population segment. They respond to drugs with greater variability and have less capability to handle drugs efficiently. Any factors that can contribute to that lessened capability and the heightened hazard of drugs should, if possible, be eliminated. Nutrition is one of them.

The nutritional vulnerability of elderly individuals is wisely recognized. It could be worsened by drugs. While the clinical significance of many reported interactions is still unclear, there is a greater urgency to prevent drug-induced nutritional deficiencies in the elderly. In turn, to prevent greater variation in drug response than already exists among the elderly, the effects of diets and food components on drug action needs to be ascertained. Thus, the primary care physician faces a variety of problems relating to a patient's nutritional status and food intake[179] and, when prescribing a drug, the physician must take into account the patient's diet and nutritional status.[180]

REFERENCES

1. **Todhunter, E. N. and Darby, W. J.,** Guidelines for maintaining adequate nutrition in old age, *Geriatrics,* 33 (6), 49, 1978.
2. **Lamy, P. P.,** *Prescribing for the Elderly,* John Wright-PSG, Littleton, Mass., 1980.
3. **Nauss, K. M.,** Vitamin A and human response, *Nutr. M.D.,* 8 (10), 1, 1982.
4. **Ellis, J., Folgers, K., Levy, M. et al.,** Therapy with vitamin B_6 with and without surgery for treatment of patients having the idiopathic carpal tunnel syndrome, *Res. Comm. Chem. Pathol. Pharmacol.,* 33, 331, 1981.
5. **Older, M. W. J. and Dickerson, J. W. T.,** Thiamine and the elderly orthopaedic patient, *Age Ageing,* 11, 101, 1982.
6. **Lamy, P. P.,** Effects of diet and nutrition on drug therapy, *J. Am. Geriatr. Soc.,* 30, 11 (Suppl), S99, 1982.
7. **Young, R. S. and Blass, J. P.,** Iatrogenic nutritional deficiencies, *Ann. Rev. Nutr.,* 2, 201, 1982.
8. Advice on limiting intake of bonemeal, *FDA Drug Bull.,* 12 (1), 5, 1982.
9. **Sartiano, G. P., Lynch, W. E., Hopkins, C. B. et al.,** (Letter), Erythrocyte and plasma selenium measurements in congestve cardiomyopathy, *N. Engl. J. Med.,* 307, 558, 1982.
10. **King, W., Michel, L., Wood, W. C. et al.,** Reversal of selenium deficiency with oral selenium, *N. Engl. J. Med.,* 304, 1305, 1981.
11. **Shekelle, R. B., Lepper, M., Liu, S. et al.,** Dietary vitamin A and risk of cancer in the Western Electric study, *Lancet,* 2, 1185, 1981.
12. **Hirsch, J.,** Nutrition, *Drug Ther.,* 10 (2), 119, 1980.
13. **Lamy, P. P.,** Drug prescribing for the elderly, *Bull. N.Y. Acad. Med.,* 57, 718, 1981.
14. **Rivlin, R. S.,** Drugs, nutrition and aging, *U.S. Pharm.,* 6, 62, 1981.
15. **Brink, M. F., Speckman, E. W., and Blasley, M.,** Current concepts in geriatric nutrition, *Geriatrics,* 23 (3), 113, 1968.
16. **Corless, D.,** Diet in the elderly, *Br. Med. J.,* 4, 158, 1973.
17. **Mayer, J.,** Aging and nutrition, *Geriatrics,* 29 (5), 57, 1974.
18. **Rao, D. B.,** Problems of nutrition in the aged, *J. Am. Geriatr. Soc.,* 21, 362, 1973.
19. **Burr, M. L., Milbank, J. E., and Gibbs, D.,** The nutritional status of the elderly, *Age Ageing,* 11, 89, 1982.
20. Food and Nutrient Intake of Individuals in the United States, Spring, 1965, Household Food Consumption Survey, 1965—66, Report No. 11, Agricultural Research Service, U.S. Department of Agriculture, 1972.
21. Ten State Nutrition Survey, 1968—1970, Highlights, DHEW Pub. No. (HSM) U.S. Department of Health, Education and Welfare, 72-8134, 1972.
22. Preliminary Findings of the First Health and Nutritional Examination Survey, U.S., 1971—72: Dietary Intake and Biochemical Findings, DHEW Pub. No. (HRA) 72-1219-1, U.S. Department of Health, Education and Welfare, 1974.
23. **Beauchene, R. E. and Davis, T. A.,** The nutritional status of the aged, *Age Ageing,* 2, 23, 1979.
24. **Tsai, A. E., Cooper, J. W., and McCall, C. Y.,** Pharmacist impact on hematopoietic and vitamin therapy in a geriatric long-term care facility, *Hosp. Formul. Manage.,* 17 (2), 225, 1982.
25. **Weg, R. B.,** Prolonged mild nutritional deficiencies: significance for health maintenance, *J. Nutr. Elderly,* 1 (1), 3, 1980.
26. **Anderson, W. F.,** Unanswered questions in the nutrition of the elderly people, *Proc. Nutr. Soc.,* 27 (2), 185, 1968.
27. **Clements, F. W.,** Nutrition 7: vitamin and mineral supplementation, *Med. J. Austr.,* 1 (19), 595, 1975.
28. **Gran, S. M., Smith, N. J., and Clark, D. C.,** Race differences in hemoglobin levels, *Ecol. Food Nutr.,* 3, 299, 1974.
29. **Asplund, K., Normakr, M., and Petterson, V.,** Nutritional assessment of psychogeriatric patients, *Age Ageing,* 10, 87, 1981.
30. **Todhunter, E. N.,** The evolution of nutrition concepts — perspectives and new horizons, *J. Am. Diet., Assoc.,* 46, 123, 1965.
31. **Trier, J. S.,** Differentiating malabsorption causes, *Patient Care,* 17 (2), 96, 1983.
32. **Gebhard, R. L.,** Malabsorption — a cause of geriatric nutritional failure, *Geriatrics,* 38 (1), 97, 1983.
33. **Scrimshaw, N. S.,** Nutrition and infection, *Prog. Food Nutr. Sci.,* 1 (6), 393, 1975.
34. **Oski, F. A.,** Unusual manifestations of iron deficiency, *Nutr. M.D.,* 8 (6), 1, 1982.
35. **Beisel, W. R.,** Impact of infection on nutritional status: concluding remarks, *Am. J. Clin. Nutr.,* 30, 1564, 1977.
36. **Blackburn, G. L.,** Lipid metabolism in infection, *Am. J. Clin. Nutr.,* 30, 1321, 1977.
37. **Bienia, R., Ratcliff, S., Barbour, G. L. et al.,** Malnutrition in the hospitalized geriatric patient, *J. Am. Geriatr. Soc.,* 30, 433, 1982.

38. **Mackowiak, P. A.,** The normal microbial flora, *N. Engl. J. Med.,* 307, 83, 1982.
39. **Phair, J. P.,** Aging and infection, a review, *J. Chron. Dis.,* 32, 535, 1979.
40. **Gladstone, J. L. and Recco, R.,** Host factors and infectious disease in the elderly, *Med. Clin. North Am.,* 60, 1225, 1976.
41. **Smith, I. M. and Habte-Garb, E.,** Life-threatening infections: how to choose the right antibiotics, *Geriatrics,* 32 (3), 83, 1977.
42. **Spence, R. W., Creak, D. R., and Clestin, L. R.,** Influence of a meal on the absorption of cimetidine, *Digestion,* 14, 127, 1976.
43. **Esposito, R.,** Cimetidine and iron deficiency anemia, *Lancet,* 2, 1132, 1977.
44. **Streeter, A. M., Poulston, K. J., Bathur, F. A. et al.,** Cimetidine and malabsorption of cobalamin, *Dig. Dis. Sci.,* 27, 13, 1982.
45. **Salom, I. L., Selvis, S. E., and Doscherholmen, A.,** Effect of cimetidine on the absorption of vitamin B_{12}, *Scand. J. Gastroenterol.,* 17, 129, 1982.
46. **Steinberg, W. M., King, C. E., and Toskes, P. P.,** Malabsorption of protein-bound cobalamin but not unbound cobalamin during cimetidine administration, *Dig. Dis. Sci.,* 25, 188, 1980.
47. **King, C. E., Leibach, J., and Toskes, P. P.,** Clinically significant vitamin B_{12} deficiency secondary to malabsorption of protein-bound vitamin B_{12}, *Dig. Dis. Sci.,* 24, 397, 1979.
48. **Blackburn, G. L. and Harvey, K. B.,** Nutritional assessment as a routine in clinical medicine, *Postgrad. Med.,* 71, 46, 1982.
49. **Cashman, M. D.,** Geriatric malnutrition, *Postgrad. Med.,* 71, 185, 1982.
50. **Roe, D. A.,** *Drug-Induced Nutritional Deficiencies,* AVI, Westport, Conn., 1976.
51. **Roe, D. A.,** Interactions between drugs and nutrients, *Med. Clin. North Am.,* 63, 985, 1979.
52. **Booth, P., Kohrs, M. B., and Kamath, S.,** Taste acuity and aging: a review, *Nutr. Res.,* 2, 95, 1982.
53. **Justice, C. L., Howeve, J. M., and Clark, H. E.,** Dietary intakes and nutritional status of elderly patients, *J. Am. Diet. Assoc.,* 65, 639, 1974.
54. **Caldwell, M. D. and Kennedy-Caldwell, C.,** Normal nutritional requirements, *Surg. Clin. North Am.,* 61, 489, 1981.
55. **Munro, H. N. and Young, V. R.,** Protein metabolism in the elderly, *Postgrad. Med.,* 63 (3), 143, 1978.
56. Dietary Intake Findings, United States, 1971-1974, DHEW Pub. No. (HRA) 77-1647, U.S. Department of Health, Education and Welfare, 1977.
57. **Brown, P. T., Bergan, J. G., Parsons, E. P. et al.,** Dietary status of elderly people, *J. Am. Diet. Assoc.,* 71, 41, 1977.
58. U.S. Senate Select Committee on Nutrition and Human Needs, Dietary goals for the United States, *Nutr. Today,* 12, 20, 1977.
59. **Thiele, V. F.,** *Clinical Nutrition,* C. V. Mosby, St. Louis, 1976.
60. **Powell, M. F. and Lamy, P. P.,** Drug-dietary incompatibilities. I. Effects on nutritional status, *Hosp. Formul. Manage.,* 12, 774, 1977.
61. **Powell, M. F. and Lamy, P. P.,** Drug-dietary incompatibilities. II. Effects on drug therapy, *Hosp. Formul. Manage.,* 12, 870, 1977.
62. **Lamy, P. P.,** The food/drug connection in elderly patients, *Am. Pharm.,* NS18, 30, 1978.
63. **Lamy, P. P.,** Drug interactions and the elderly — a new perspective, *Drug Intell. Clin. Pharm.,* 14, 513, 1980.
64. **Lamy, P. P.,** How your patient's diet can affect drug response, *Drug Ther.,* 10 (8), 81, 1980.
65. **Lamy, P. P.,** Nutrition and the elderly, *Drug Intell. Clin. Pharm.,* 15, 887, 1981.
66. **Hodges, R. E.,** *Nutrition in Medical Practice,* W. B. Saunders, Philadelphia, 1980.
67. **Clark, F.,** Drugs and vitamin deficiency, *J. Hum. Nutr.,* 30, 333, 1976.
68. **Hatchcock, J. N. and Coon J.,** *Nutrition and Drug Interrelations,* Academic Press, N.Y., 1978.
69. **Ovesen, L.,** Drugs and vitamin deficiency, *Drugs,* 18, 278, 1979.
70. **Young, R. C. and Blass, J. P.** Iatrogenic nutritional deficiencies, *Ann. Rev. Nutr.,* 2, 201, 1981.
71. **Chanarin, I., Deacon, R., Lumis, M. et al.,** Vitamin B_{12} regulates folate metabolism by the supply of formate, *Lancet,* 2, 505, 1980.
72. **Hartoma, T. R., Sotaniemi, E. A., Pelkonene, O. et al.,** Serum zinc and serum copper as indices of drug metabolism in alcoholics, *Eur. J. Clin. Pharmacol.,* 12, 147, 1971.
73. **Smithard, D. J. and Langman, M. J. S.,** The effect of vitmin supplementation upon antipyrine metabolism in the elderly, *Br. J. Clin. Pharmacol.,* 5, 181, 1978.
74. **Langman, M. J. S. and Smithard, D. J.,** Antipyrine metabolism in iron deficiency, *Proc. Br. Pharmacol. Soc., C,* abstract, 16, 1977.
75. **Abraham, G. E., Schwartz, U. D., and Lubran, M. M.,** Effect of vitamin B_6 on plasma and red blood cell magnesium levels in premenopausal women, *Ann. Clin. Lab. Sci.,* 11, 333, 1981.
76. **Hahn, T. J.,** Bone complications of anticonvulsants, *Drugs,* 12, 201, 1976.
77. **Mattson, R. H., Gallagher, B. B., Reynolds, E. N. et al.,** Folate therapy in epilepsy: a controlled study, *Arch. Neurol.,* 29, 78, 1973.

78. **Exton-Smith, A. N. and Caird, F. I., Eds.,** *Metabolic and Nutritional Disorders in the Elderly,* John Wright & Sons, Bristol, 1980.

79. **Hurwitz, A.,** Antacid therapy and drug kinetics, *Clin. Pharmacokin.,* 2, 269, 1977.

80. **Feldman, S. and Hedrick, W.,** Antacid effects on the gastrointestinal absorption of riboflavin, *J. Pharm. Sci.,* 72, 121, 1983.

81. **Libow, L. S.,** Pseudo-senility: acute and reversible organic brain syndromes, *J. Am. Geriatr. Soc.,* 21, 112, 1973.

82. **Cathcart, R. F.,** Vitamin C, titrating to bowel tolerance, anascorbemia and acute induced scurvy, *Med. Hypothesis,* 7, 1359, 1981.

83. **Reynolds, E. H., Rothfield, P., and Pincus, J. H.,** Neurological disease associated with folate deficiency, *Br. Med. J.,* 2, 398, 1973.

84. **Branda, R. F. and Nelson, N. L.,** Inhibition of 5-methyltetrahydrofolic acid transport by amphipathic drugs, *Drug-Nutrient Interact.,* 1, 45, 1981.

85. **Sheehan, J.,** Trimethoprim-associated marrow toxicity, *Lancet,* 2, 692, 1981.

86. **Polk, R.,** Moxalactam (Moxam), *Drug Intell. Clin. Pharm.,* 16, 104, 1982.

87. **Raisz, L. G.,** Osteoporosis, *J. Am. Geriatr. Soc.,* 30, 127, 1982.

88. **Riggs, B. L., Seeman, E., Hodgson, S. F. et al.,** Effect of the fluoride/calcium regimen on vertebral fracture occurrence in postmenopausal osteoporosis, *N. Engl. J. Med.,* 306, 446, 1982.

89. **Raid, D. M., Kennedy, N. S. J., Smith, M. A. et al.,** Total body calcium in rheumatoid arthritis: effects of disease activity and corticosteroid treatment, *Br. Med. J.,* 285, 330, 1982.

90. **Francis, R. M., Peacock, M., Taylore, G. A. et al.,** Evidence for resistance to 25 hydroxy vitamin D in the malabsorption of elderly osteoporotic women, *Clin. Sci.,* 62, 40P, 1981.

91. **McCarron, D. A., Morris, C. D., and Cole, C.,** Dietary calcium in human hypertension, *Science,* 217, 267, 1982.

92. **McCarron, D. A.,** Low serum concentrations of ionized calcium in patients with hypertension, *N. Engl. J. Med.,* 307, 226, 1982.

93. **Duarte, C. G., Winnaker, J. L., Becker, K. L. et al.,** Thiazide-induced hypercalcemia, *N. Engl. J. Med.,* 284, 828, 1971.

94. **Brickman, A. S., Massry, S. G., and Coburn, J. W.,** Changes in serum and urinary calcium during treatment with hydrochlorothiazide: studies on mechanisms, *J. Clin. Invest.,* 51, 945, 1972.

95. **Brown, E. M.,** When you suspect hypercalcemia, *Patient Care,* 16 (2), 14, 1982.

96. **Strom, J. A.,** When diuretics affect electrolytes, *Patient Care,* 16 (8), 62, 1982.

97. **Hansten, P. D.,** Potassium supplement interactions, *Drug Interact. Newsl.,* 2 (1), 1, 1982.

98. **Skeha, J. D., Barnes, J. N., Drew, P. J. et al.,** Hypokalemia induced by a combination of a beta blocker and a thiazide, *Br. Med. J.,* 284, 83, 1982.

99. **Sheehan, J. and White, A.,** Diuretic-associated hypomagnesaemia, *Br. Med. J.,* 285, 1157, 1982.

100. **Swales, J. D.,** Magnesium deficiency and diuretics *Br. Med. J.,* 285, 1377, 1982.

101. **Watt, B. K. and Merrill, A. L.,** Composition of Foods, Agriculture Handbook No. 8, Consumer and Food Economics Institute, U.S. Department of Agriculture, Washington, D.C., 1976.

102. **Hallgren, H. M. and Buckley, C. E.,** Lymphocyte phytohemagglutinin responsiveness, immunoglobulins, and autoantibodies in aging, *J. Immunol.,* 111, 1101, 1973.

103. **Weksler, M. E. and Hutton, T. H.,** Impaired lymphocyte function in aged humans, *J. Clin. Invest.,* 53, 99, 1973.

104. **Carosella, E. D., Mochanko, K., and Brown M.,** Rosette-forming T cells in human peripheral blood at different ages, *Cell. Immunol.,* 12, 323, 1974.

105. **Kay, M. B.,** Immunodeficiency in old age, in *Immunodeficiency Disorders,* Chandra, R. K., Ed., Churchill Livingstone, Edinburgh, 1982.

106. **Chandra, R. K., Joshi, P., Au, B. et al.,** Nutrition and immunocompetence of the elderly: effect of short-term nutritional supplementation on cell-mediated immunity and lymphocyte subsets, *Nutr. Res.,* 2, 223, 1982.

107. **Munro, H. N.,** Nutritional requirements in the elderly, *Hosp. Pract.,* 17 (8), 143, 1982.

108. **Tomaiolo, P. P., Enman, S., and Kraus, V.,** Preventing and treating malnutrition in the elderly, *J. Parent. Ent. Nutr.,* 5 (1), 46, 1981.

109. **Lamy, P. P.,** Comparative pharmacokinetic changes and drug therapy in an older population, *J. Am. Geriatr. Soc.,* 30 (Suppl II.), 1982.

110. **Krishnaswamy, K.,** Drug metabolism and pharmacokinetucs in malnutrition, *Clin. Pharmacokin.,* 3, 216, 1978.

111. **Campbell, T. C. and Hayes, J. R.,** Role of nutrition in the drug-metabolizing enzyme system, *Pharmacol. Rev.,* 26, 171, 1974.

112. **Krishnaswamy, K. and Naidu, A. N.,** Microsomal enzymes in malnutrition as determined by plasma half-life of antipyrine, *Br. Med. J.,* 1, 538, 1977.

113. **Parke, D. V.,** The effects of nutrition and enzyme induction in toxicology, *World Rev. Nutr. Diet.,* 29, 96, 1978.

114. **Prasad, J. S. and Krishnaswamy, K.,** Streptomycin pharmacokinetics in malnutrition, *Chemotherapy,* 24, 333, 1978.

115. **Raghuran, T. C. and Krishnaswamy, K.,** Influence of nutritional status on plasma levels and relative bioavailability of tetracycline, *Eur. J. Clin. Pharmacol.,* 12, 28, 1977.

116. **Shastri, R. A. and Krishnaswamy, K.,** Undernutrition and tetracycline half-life, *Clin. Chim. Acta,* 66, 157, 1976.

117. **Fishman, J. and Bradlow, H. L.,** Effect of malnutrition on the metabolism of sex hormones in man, *Clin. Pharmacol. Ther.,* 22, 721, 1977.

118. **Buchanan, N., Eyberg, C., and Davis, M. D.,** Antipyrine pharmacokinetics and D-glucarix excretion in kwashiorkor, *Am. J. Clin. Nutr.,* 32, 2439, 1979.

119. **Shastri R. A. and Krishanswamy, K.,** Metabolism of sulphadiazine in malnutrition, *Br. J. Clin. Pharmacol.,* 7, 69, 1979.

120. **Raghuram, T. C. and Krishnaswamy, K.,** Tetracycline absorption in malnutrition, *Drug-Nutrition Interact.,* 1, 23, 1981.

121. **Buchanan, N., Davis, M. D., and Eyberg, C.,** Gentamicin pharmacokinetics in kwashiorkor, *Br. J. Clin. Pharmacol.,* 8, 451, 1979.

122. **Buchanan, N., Van Der Walt, L. A., and Strickwold, B.,** Pharmacology of malnutrition. III. Binding of digoxin to normal and kwashiorkor serum, *J. Pharm. Sci.,* 65, 914, 1976.

123. **Eyberg, C., Moodley, G. P., and Buchanan, N.,** Salicylate binding studies using normal serum/plasma and kwashiorkor serum, *So. African Med. J.,* 48, 2564, 1974.

124. **Buchanan, N. and van der Walt, L. A.,** Chloramphenicol binding to normal and kwashiorkor sera, *Am. J. Clin. Nutr.,* 30, 847, 1977.

125. **Buchanan, N. and van der Walt, L. A.,** Thiopentone binding to normal and kwashiorkor serum, *Br. J. Anesth.,* 49, 247, 1977.

126. **English, E. C. and Carl, J. W.,** Use of nutritional supplements by family practice patients, *JAMA,* 246, 2719, 1981.

127. **Hale, W. E., Stewart, R. B., Cerda, J. J. et al.,** Use of nutritional supplements in an ambulatory elderly population, *J. Am. Geriatr. Soc.,* 30, 401, 1982.

128. **Garry, P. J., Goodwin, J. S., Hunt, W. C. et al.,** Nutritional status in a healthy elderly population: dietary and supplemental intakes, *Am. J. Clin. Nutr.,* 36, 319, 1982.

129. **Garry, P. J., Goodwin, J. S., Hunt, W. C. et al.,** Nutritional status in a healthy elderly population: vitamin C, *Am. J. Clin. Nutr.,* 36, 332, 1982.

130. **Enstrom, J. E. and Pauling, L.,** Mortality among health conscious elderly Californians, *Proc. Natl. Acad. Sci. U.S.A.,* 79, 6023, 1982.

131. **DiPalma, J. R. and Ritchie, D. M.,** Vitamin toxicity, *Ann. Rev. Pharmacol. Toxicol.,* 17, 133, 1977.

132. **Mertz, W.,** Chromium: an essential micronutrient, *Contemp. Nutr.,* 7 (3), 1, 1982.

133. **Maines, M. D. and Kappas, A.,** Regulation of cytochrome P-450-dependent microsomal drug metabolizing enzymes by nickel, cobalt, and iron, *Clin. Pharmacol. Ther.,* 22, 780, 1977.

134. **Sandstead, H. H.,** Trace element interactions, *J. Lab. Clin. Med.,* 98, 457, 1981.

135. **Mertz, W.,** Trace minerals and heart disease, *Fed. Proc.,* 41, 2857, 1982.

136. **Solomons, N. W.,** Mineral interactions in the diet, *Contemp. Nutr.,* 7 (7), 1, 1982.

137. **Seelig, M. S.,** Magnesium requirements in human nutrition, *Contemp. Nutr.,* 7 (1), 1, 1982.

138. **Frolich, E. D.,** Inhibition of adrenergic function in the treatment of hypertension, *Arch. Intern. Med.,* 133, 1033, 1973.

139. **Blackwell, B.,** Hypertensive crisis due to monoamine oxidase inhibitors, *Lancet,* 2, 849, 1963.

140. **Uragoda, C. G.,** Histamine poisoning in tuberculosis patients after ingestion of tuna fish, *Ann. Rev. Resp. Dis.,* 121, 157, 1980.

141. **Uragoda, C. G. and Kottegoda, S. R.,** Adverse reactions to isoniazid on ingestion of fish with too high histamine content, *Tubercle,* 58, 83, 1977.

142. **Uragoda, C. G.,** Histamine poisoning in tuberculosis patients on ingestion of tropical fish, *J. Trop. Med. Hyg.,* 81, 243, 1978.

143. **Uragoda, C. G. and Lodha, S. C.,** Histamine intoxication in a tuberculosis patient after ingestion of cheese, *Tubercle,* 60, 59, 1979.

144. **Smith, C. K. and Durack, D. T.,** Isoniazid and reaction to cheese, *Ann. Intern. Med.,* 88, 520, 1978.

145. **Hodge, J. V. Nye, E. R., and Emerson, G. W.,** Monoamine oxidase inhibitors, broad beans, and hypertension, *Lancet,* 1, 1108, 1964.

146. **Gross, E. G., Dexter, J. D., and Roth, R. G.,** Hypokalemic myopathy with myoglobinuria associated with licorice ingestion, *N. Engl. J. Med.,* 274, 602, 1966.

147. **Koster, M. and David, G. K.,** Reversible severe hypertension due to licorice ingestion, *N. Engl. J. Med.,* 278, 1381, 1968.

148. **Campbell, T. C. and Hayes, J. R.,** Role of nutrition in the drug-metabolizing enzyme system, *Pharmacol. Rev.,* 26, 171, 1974.

149. **Channer, K. S. and Virjee, J.,** Effect of posture and drink volume on the swallowing of capsules, *Br. Med. J.,* 285, 1702, 1982.

150. **Welling, R. G. and Tse, F. L. S.,** The influence of food on the absorption of antimicrobial agents, *J. Antimicrob. Chemother.,* 9, 7, 1982.

151. **Rosenberg, H. A. and Bates, T. R.,** The influence of food on nitrofurantoin bioavailability, *Clin. Pharmacol. Ther.,* 20, 227, 1976.

152. **Kaumeier, S.,** The effect of the composition of food on the absorption of sulfameter, *Int. J. Clin. Pharmacol. Biopharm.,* 17, 260, 1979.

153. **Crounse, R. G.,** Human pharmacology of griseofulvin: the effect of fat intake on gastrointestinal absorption, *J. Invest. Dermatol.,* 37, 529, 1961.

154. **Barbheiya, R. H. and Welling, P. G.,** Influence of food on the absorption of hydrocortisone from the gastrointestinal tract, *Drug-Nutrient Interact.,* 1, 103, 1982.

155. **Welling, P. G.,** Influence of food and diet on gastrointestinal drug absorption: a review, *J. Pharmacokin. Biopharm.,* 5, 291, 1977.

156. **Gillespie, N. G., Mena, I., Cotzias, G. S., et al.,** Diets affecting treatment of parkinsonism with levadopa, *J. Am. Diet. Assoc.,* 62, 525, 1973.

157. **Jusko, W. J. and Levy, G.,** Absorption, metabolism, and excretion of riboflavin 5-phosphate in man, *J. Pharm. Sci.,* 56, 58, 1967.

158. **Nichols, T. W., Jr.,** Phytobezoar formation: a new complication of cimetidine therapy, *Ann. Intern. Med.,* 95, 70, 1981.

159. **Evans, M. A., Triggs, E. J., Cheung, M. et al.,** Gastric emptying in the elderly: Implications for drug therapy, *J. Am. Geriatr. Soc.,* 29, 201, 1981.

160. **Anon.,** Metoclopramide, *Med. Lett.,* 24, 67, 1982.

161. **Viteri, F. E. and Schneider, R. E.,** Gastrointestinal alterations in protein-calorie malnutrition, *Med. Clin. North Am.,* 58, 1487, 1974.

162. **Wannemacher, R. W., Jr.,** Effect of infecton on nutrient metabolism, *Nutr. M.D.,* 9 (1), 1, 193.

163. **Mitchell, C. O. and Lipschitz, D. A.,** Detection of protein-calorie malnutrition in the elderly, *Am. J. Clin. Nutr.,* 35, 398, 1982.

164. **Korenchevsky, V.,** *Physiological and Pathological Aging,* Karger, Basel, 1961.

165. **Munro, H. N.,** Nutrition and aging, *Br. Med. J.,* 37, 83, 1981.

166. **Hayes, M. J., Langman, M. J. S., and Short, A. H.,** Changes in drug metabolism with increasing age. I. Warfarin binding and plasma proteins, *Br. J. Clin. Pharmacol.,* 2, 73, 1975.

167. **Hayes, M. J., Langman, M. J. S., and Short, A. H.,** Changes in drug metabolism with increasing age. II. Phenytoin clearance and protein binding, *Br. J. Clin. Pharmacol.,* 2, 69, 1975.

168. **Levi, A. J., Sherlock, S., and Walker, D.,** Phenylbutazone and isoniazid metabolism in patients with liver disease in relation to previous drug therapy, *Lancet,* 1, 1275, 1968.

169. **Mawer, G. E., Miller, N. E., and Turnberg, L. A.,** Metabolism of amylobarbitone in patients with chronic liver disease, *Br. J. Clin. Pharmacol.,* 44, 459, 1972.

170. **Marovitz, D. C. and Fernstrom, J. D.,** Diet and uptake of aldomet by the brain: competition with natural large neutral amino acids, *Science,* 197, 1014, 1977.

171. **Reidenberg, M. M.,** Obesity and fasting effects on drug metabolism and drug action in man, *Clin. Pharmacol. Ther.,* 22, 279, 1977.

172. **Cooksley, W. G. E. and Powell, L. W.,** Drug metabolism and interaction with particular reference to the liver, *Drugs,* 2, 177, 1971.

173. **Campbell, T. C.,** Nutrition and drug-metabolizing enzymes, *Clin. Pharmacol. Ther.,* 22, 699, 1977.

174. **Conney, A. H., Pantuck, E. J., Kuntzman, R. et al.,** Nutrition and chemical biotransformation in man, *Clin. Pharmacol. Ther.,* 22, 707, 1977.

175. **Fraser, J. S., Mucklow, J. C., Bulpitt, C. J. et al.,** Environmental effects on antipyrine half-life in man, *Clin. Pharmacol. Ther.,* 22, 799, 1977.

176. **Graf, H., Stummvoll, H. K., Luger, A. et al.,** (Letter), Effect of amino acid infusion on glomerular filtration rate, *N. Engl. J. Med.,* 308, 159, 1983.

177. **Brunner, B. M., Meyer, T. W., and Hostetter, T. H.,** Dietary protein intake and the progressive nature of kidney disease: the role of hemodynamically mediated glomerular injury in the pathogenesis of progressive glomerular sclerosis in aging, renal ablation, and intrinsic renal disease, *N. Engl. J. Med.,* 307, 652, 1982.

178. **Fann, W. E.,** Some clinically important interactions of psychotropics, *South. Med. J.,* 66, 661, 1973.

179. **Schaffer, J. B.,** Getting elderly patients to eat properly, *Geriatrics,* 36 (10), 76, 1981.

180. **Sriwatanakui, K. and Weintraub, M.,** Food-drug interactions: an over-looked problem, *Drug Ther.,* 12 (2), 157, 1982.

RELATIONSHIP OF CANCER INITIATION AND GROWTH TO EFFECTS OF NUTRITIONAL STRESSES ON IMMUNE DEFENSES IN THE AGING ADULT

JC. Jackson and Ronald R. Watson

INTRODUCTION

There is an intimate and complex interplay between cancer, immune response, and nutritional state in the adult. The common desires for increased quality of life and a more full comprehension of physiological change with the natural aging process sets the stage for the following paragraphs. Moderate dietary restriction, opposed to severe malnutrition, enhances many of the body's life-threatening mechanisms. Cancer, one of life's greatest deterrents, may be minimized through dietary variables and consequent altered immune function.

ENHANCEMENT OF LONGEVITY IN THE YOUNG BY DIETARY RESTRICTION

Food restriction in rodents has proven to increase the mean and maximal lifespan and slow down the biological aging process. McCay[1] first observed food-restricted rats having increased longevity in 1935 and since then, numerous studies have helped elucidate the mechanistic activities involved. Varied regimens and combinations of protein, calorie, and vitamins have been fed to laboratory animals beginning at different developmental states. The results are quite consistent. Food restriction (but not severe malnutrition) applied to animals at any age, if chronic, usually produces a distinct increase in lifespan.[2] The most successful restriction diets limit caloric intake (25 to 50% less) while providing adequate quantities of other required nutrients such as vitamins, salts, and occasionally, protein.[3,4] To avoid confusion, a distinction in terms is made. Malnutrition and undernutrition are to be considered synonymous and in accordance to the food-restricted diet mentioned above, but opposed by severe malnutrition and severe undernutrition, which negatively affect life prolongation. Very little, however, is known about the effects of high vitamin or mineral intakes, nutritional stress, on longevity. Clearly, excessive caloric intake reduces longevity by nonimmune mechanisms with much less known about the effects of high protein intakes on immune functions or longevity.

One mechanism by which food restriction increases the lifespan of rats relates to the delay in time of onset and a change in chronological course of physiological decline.[5] Walford's immunologic theory of aging suggests that aging is not primarily a process of passive wearing out of systems, but of active self destruction mediated by the immune system. Thymus involvement and T cell function appear to be much more affected by dietary restriction than the B cell function and the humoral response.[6]

The chief organ to respond to nutritional variables is the thymus. The thymus translates nutritional causes into immunological effects. It is a lymphoid organ actively engaged in lymphopoiesis but independent of antigenic stimulation. Two major functions have been attributed to the thymus: (1) it acts by the elaboration of hormones (thymosin) to expand peripheral lymphocyte populations, and (2) it acts by direct seeding of peripheral lymphoid tissues with maturing T-lymphocytes.[7] It is feasible to describe the thymus as the master biological element by which specific nutrients influence development of the diseases of aging as they affect other organs and ultimately the length of life.[8] The thymus may be the aging clock[9] which in humans reaches its maximal size at about 15 years of age, then gradually decreases (involutes) in size until about 50, after which it remains fairly stable in size until death.[10] The involution of the thymus, an event temporally linked to the onset of puberty, is followed by a decline in serum thymosin levels. This presumably leads to an age-dependent and function-specific loss of T cell differentiative capacities. Underfeeding early in life drastically dampens thymic growth and alters the timing of involution. In normal aging, immunological changes include decreased responsiveness to exogenous antigens and increased autoimmune responses. Cellular immune mechanisms decrease, while levels of serum immunoglobulin increase.[9] Old mice after lifelong restriction, or moderately aged mice after 4.5 months of restriction, display "younger" immune systems than do age-matched, normally fed controls. Underfeeding from weaning nutritionally tampers with the timing and rates of prepubescent maturational events.[6] Food-restricted rodents show slower thymic growth rates and smaller thymuses at puberty, but less histological change at thymic involution than normally fed controls. Underfeeding results in slowed maturation of the immune system, with age-specific peak capacity being reached at a later age than in normally fed mice. Furthermore, the immune system in the underfed mice stayed younger longer, and this was associated with increased longevity.[4] Protein restriction at 4 to 5 months or calorie restriction at weaning afforded significant protection from the development of immune nephritis, and prolongation of lifespan. Autoantibody formation of calorie-restricted animals was significantly decreased compared to mice fed a normal diet. This would suggest the possibility of using moderate dietary restriction as a prophylactic or effective therapeutic approach to ongoing autoimmune disease.[11]

The effects on longevity of dietary restriction in mild undernutriton and severe malnutrition are diametrically opposed. It is well documented that severe nutritional deficiencies increase morbidity and mortality due to microbial pathogens and cancer.[12] Most observers of severe protein-calorie malnutrition have found this state of deprivation linked with increased vulnerability to infection and attendant disease. Severe nutritional deficiencies impair immune responsiveness, thus reducing antimicrobial protection from invasion.[13] Severe undernutrition in young animals and children has shown reduced number and functional capability of T lymphocytes and delayed hypersensitivity response to antigens and sensitizing chemicals.[13-15] Significant decreases in total hemolytic complement are shown in studies of severely malnourished children or animals. Severely malnourished Thai, Indian, and Colombian children presented reduced levels (35 to 50%) of S-IgA in secretions while IgG levels were unaffected.[13] Studies of severely malnourished adults or aged humans are usually complicated by surgery, cancer, and/or disease. The immune response in severely malnourished elderly patients is profoundly impaired, as is the immune response in the severely malnourished of any age.[12] Protein energy malnourished surgical patients have poor wound healing responses.[16] Appropriate food restriction in the sense of undernutrition is quite different immunologically from severe malnutrition in that many T-dependent immunologic responses are more vigorous in the underfed. This enhanced or preserved vigor is associated with inhibition of late-life diseases and with increases in longevity.[17]

EFFECTS OF NUTRITIONAL EXCESSES ON ENHANCEMENT OF LONGEVITY

Dietary excess may be associated with vigorous thymic activity early in life leading to a rapid drop in thymic vigor in the middle period of life in certain strains of genetically short-lived, autoimmune-prone mice. This accelerated thymic development has been strongly implicated in the early involution of immunity functions in these animals and in the development of their characteristic disease.[8] Nutritional excesses have shown varying effects on life prolongation depending on the dietary variable in surplus. High calorie and/or fat intakes lead to obesity which has been shown to increase one's susceptibility to many life-threatening maladies. Overfeeding in dogs leads to obesity and also to a decreased resistance to infection.[18] However, obesity *per se* does not seem to be routinely immunosuppressive, while high fat diets are.[19]

It is becoming clear that optimum host defenses in adults can not only be influenced by deficiencies of protein or energy, but also by very high intakes of vitamins. For example, high vitamin E intake inhibited carcinogenesis,[20,21] protected patients from adverse side effects of radiation therapy,[22] stimulated humoral immune responses to antigen or resistance to bacterial infection,[23,24] enhanced helper T cell activity,[25] and accelerated maturation of cellular immune functions[26] in young mice. Absence of vitamin E in the diet may eventually cause significant suppression of resistance to infection.[27-29] Why do these limited studies with young and mature animals suggest that high intakes of vitamin E improve disease resistance? Based upon our work, we believe that enhanced disease resistance and tumor resistance in mice[26,30,31] is due to altered cellular immune functions and secretory immune functions. For example, we found that there were increased amounts of S-IgA in the intestinal secretions of young mice fed a high vitamin E diet.[31] We also found upon supplementation of high dietary vitamin E in young mice an elevated number of cytotoxic lymphocytes which could be important in anticancer defenses, perhaps due to more rapid maturation of T cell functions.[26,30] Vitamin E did not function as a mitogen in athymic mice nor enhance the LPS response of spleen cells as it does in normal mice. These data suggest that vitamin E may act *with* thymic factors to produce a mature T helper cell. However, research is critically needed to define optimum intakes of vitamins and exercise to produce the most active immune functions and anticancer defenses, especially in people suffering protein or calorie malnutrition. It is necessary to understand the effect of intakes above that routinely found in what is now considered a ''reasonable and prudent'' diet, because as the amount of fat and polyunsaturated fatty acids increase in the diet, the dietary requirement for some vitamins like vitamin E should increase.

A major aim of clinical and experimental cancer research is to develop agents and ways of immunostimulation, which increase the tumor-directed immunological response of the adult.[32] Interactions between vitamins and minerals and cell-mediated immune responses have been studied and are included in an extensive review by Beisel.[33] In the case of retinoids, very low intakes are imunosuppressive, while extremely high dietary intakes may further enhance potential anticancer mechanisms.

Briefly, vitamin A deficiency in rats has been associated with decreased transformation responses to T and B cell mitogens which were reversible upon supplementation with vitamin A. In man, vitamin A deficiency may be associated with depressed delayed hypersensitivity responses. It should also be noted that vitamin A can serve as an adjuvant when injected with antigen.

Clearly, high dietary intakes of vitamin A have a stimulating effect on some aspects of cellular and humoral immunity.[34] Retinoic acid stimulation of cell-mediated immune reactions is also indicated by the enhancement of skin graft rejection in mice and induction of cytotoxic T cells in mice fed high retinoic acid intakes. Routinely, vitamin A has been shown to increase the mitogenic response in lymphocytes from animals and man. However, stimulation of mitogenesis is not universal for all retinoids and at all concentrations. For example, retinol at high concentrations inhibited human T cell mitogenesis. In mice, effector activity of spleen cells against a syngeneic tumor was specifically augmented by low doses of retinoic acid, whereas high doses had a suppressive effect.[35]

Some synthetic analogues of retinoic acid such as 13-*cis*-retinoic acid may be more effective in prevention of experimentally induced cancers and less toxic than vitamin A. Recent animal experiments showed a 10-fold increase in cell-mediated cytoxicity by 13-*cis*-retinoic acid, after challenge with suboptimal immunogen inoculum.[36] The role of altered immune defenses in the action of 13-*cis*-retinoic acid needs much additional study due to limited experimentation, particularly in humans.[37,38] In humans with unrespectable bronchogenic cancer there are immune potentiating effects of 13-*cis*-retinoic acid. These effects are as important as any direct ones on the tumor itself.[39] At high dosages, cell-mediated cytoxicity was inhibited by some other retinoids while 13-*cis*-retinoic acid increased cell-mediated cytoxicity. This suggests that correlation of cellular immune function, cancer resistance, and dose of each retinoid needs to be studied carefully at several doses in humans. The lower toxicity of 13-*cis*-retinoic acid than some other retinoids and higher potency in cell-mediated cytoxicity suggests that 13-*cis*-retinoic acid may be very suitable for cancer therapy or prevention.

Unfortunately, the effects of high retinoid intakes on immunosuppressed host defenses has been studied only to a limited extent. Vitamin A therapy increased cellular immune functions in lung cancer patients with suppressed responses.[34] In addition, vitamin A given in combination with cyclophosphamide or irradiation inhibited or decreased the immuno-suppressive effects of these regimens.[37] Vitamin A therapy also helped reverse postoperative immunosuppression. Vitamin A acted to block the depression of cellular immune function associated with operations and acted as an immunostimulant in man with high dosages. Although encouraging, use of high intakes of retinoids to stimulate suppressed immune defenses in cancer patients is still experimental and should be considered only with caution and review of possible toxic side effects of retinoids.

EFFECTS OF NUTRITONAL STRESSES ON CANCER INCIDENCE IN THE AGED

Research has identified numerous cancer effects associated with dietary variables in animals. Investigations by Tucker[40] in England have shown that mice and rats with food intake restricted by as little as 20% have significantly reduced incidence and delayed onset of tumor growth in several systems. This also resulted in increasing longevity accordingly. Calorie-restricted mice from midway through their normal lifespan showed increased mean and maximal longevities of 10 to 20%. Also, spontaneous lymphoma was inhibited by under-feeding.[3] Tannenbaum[41] illustrated that breast cancer in C3H mice can be prevented by calorie and fat restriction. Good et al.[42] confirmed this finding and also demonstrated that the fat component is the crucial dietary variable. The effect of dietary fat seems to be in the promotional rather than the initiation phase of breast cancer development. Diets high in polyunsaturated fat, relative to diets high in saturated fat, are more immunosuppressive and are better promoters of tumorigenesis.[15,43] Rats fed a high fat diet marginally deficient in lipotropes show enhanced induction of colon and hepatic tumors.[44] Only minimal information has been produced concerning human cancer initiation as a direct result of a specific dietary etiology. Similar to the work done with animals, a positive correlation in human colorectal tumor induction and high dietary fat intakes has been shown.[45] Most observations suggest a direct relationship with animal fat and beef intake and colon cancer in the adult.[46] The highest incidence of cancer of the large bowel reported by tumor registries is in the U.S. and Canada. It is quite significant that these populations consume diets high in animal fat and refined carbohydrates and low in unabsorbable fiber.[45]

DIETARY STRESSES: EFFECTS ON CANCER RESISTANCE MECHANISMS IN THE AGING ADULT

The most prominent factor determining susceptibility to cancer is age. In one study, 85% of the fatal carcinomas found at autopsy were in aged patients although they composed only 38% of the population studied.[10] One of our strongest natural defenses against cancer initiation and promotion is the immune system. The immune system consists of a complicated inter-action between T cells of thymic origen, B cells of bone marrow derivation, and phagocytic cells. The T-lymphocytes comprise about 80% of the lymphocytic mass and are considered to be largely responsible for thymic lymphocyte functions. B-lymphocytes comprise 15 to 20% of the lymphocyte mass and include the immunoglobulin-producing cells and complement receptor cells. Other T-lymphocytes probably active in tumor killing include K cells (killer cells) and null cells.[47] Cell-mediated immunity generally constitutes the functions of T cells which can act as helper cells for antibody production by B cells, as mediators of delayed hypersensitivity and mixed lymphocyte culture responses, or they act as cytotoxic cells for appropriate target cells.[15] The cellular response appears most responsible for tumor cell inhibition and is directly affected by nutritional status.[47] Cellular response vigor is measured by absolute lymphocyte multiplication in response to stimulation with phytohemmagglutinin or concanavalin A and delayed hypersensitivity skin testing to antigens. Immunologic function declines with aging, leading to increased susceptibility to disease and cancer.[42] Investigations of the interrelationship between nutritional status and the immune system have made rapid progress in the past 10 years. Protein-energy malnutriton is associated with a consistent decrease in the number of circulating thymus-dependent T-lymphocytes and impairment of cutaneous delayed hypersensitivity.[48]

Patients with squamous cancer of the head and neck have marked depression of cellular immunity even in the early stages of disease.[47] Studies of head and neck cancer have shown a general suppression of T cell levels and a depression to test measurements reflecting T cell function. However, the progression of the disease exhibited no significant depression of B cell levels.[47]

There is an interesting interplay between prostaglandin E1 and thymus activity. Recent evidence suggests prostaglandin E1 to be centrally important in the regulation of thymus development and T-lymphocyte function.[49] The production of prostaglandin E1 is dependent on nutritional factors, with linoleic acid, gamma linolenic acid, pyridoxine, zinc, and vitamins C and E playing key roles. A great majority of malignant and virally transformed cells are unable to make prostaglandin E1 because they cannot convert linolenic acid to gamma-linolenic acid.[49] Large amounts of prostaglandin E1 have been identified in the thymus and, in vitro, the metabolic behavior of prostaglandin E1 and thymosin are very similar. Considering the regulatory effect of prostaglandin E1 on T cell and cell-mediated immune function, and its nutritional modulation, prostaglandin E1 may be an effective means of cancer inhibition.

NUTRITIONAL THERAPY IN THE ADULT CANCER PATIENT: IMMUNE ENHANCEMENT

Considerable amounts of literature exist on the effects of ascorbic acid on cancer initiation, promotion, and treatment. Cancer patients have increased requirements for vitamin C and rapid metabolic depletion of vitamin C stores. Vitamin C plays a critical role in immunocompetence and is usually diminished in cancer patients.[50] It also prevents nitrosamine formation from the procarcinogenic nitrates and nitrites found in food.[26] Neoplastic cell invasiveness has been associated with hyaluronidase production, which vitamin C inhibits. Vitamin C is nontoxic to humans at any dose but cannot be synthesized by the body.[50] Scurvy, which has many semblances of cancer, has been controlled through vitamin C supplementation. The risk/benefit ratio relative to the severity of the disease as well as to other available treatments in cancer is heavily weighted in favor of vitamin C. However, at this point it is impossible to hypothesize a true therapeutic or preventative measure through diet supplements of ascorbic acid.[51]

Protein-calorie malnutrition is frequently seen in hospitalized cancer patients and should be considered as a contributing factor to immunoparesis.[52] Malnutrition also induces altered phagocytosis, macrophage activity, and depressed complement levels.[53] Of 161 cancer patients nutritionally assessed prior to receiving oncological therapy, 84% of whom were initially anergic, became immunocompetent with nutritional therapy. The detection and treatment of protein-calorie malnutrition prior to or in conjunction with oncological therapy has been associated with a decrease in mortality rate.[54] Correction of protein-calorie malnutrition and the achievement of a positive nitrogen balance in aged black esophageal cancer patients was associated with a significant increase in lymphocyte response to phytohemagglutinin and a significant increase in T-lymphocyte numbers.[55] Malignancy is a common cause of lymphocytopenia and a positive correlation has been reported between peripheral lymphocyte counts and prognosis in carcinoma. Nutritional repletion (parenteral, enteral) in patients undergoing antitumor therapy can help restore cell-mediated immunity.[56] Like malnutrition, trauma, sepsis, malignant disease, or anasthesia affect numerous aspects of immunity.[57] These complications, in addition to the intimate and complex relationship between immune response and nutritional state, mask the true effect of hyperalimentation.[58]

Experimental and clinical observations support the concept that undernutrition within certain limits is associated with increased host resistance to tumorigenesis and intracellular infection.[69] It depends for its effect upon the release from host cells of growth-inhibiting factors. This is an effort to conserve dwindling energy resources and synchronously suppress growth and replication of malignant cells.[59]

Corticosteroids have a suppressive effect on the number of circulating lymphocytes and monocytes and block the primary sensitization of lymphocytes in cell-mediated immunity.[60] However, they also suppress leukemia cell and other tumor cell proliferation.[61] Mild undernutrition, which elevates serum glucocorticoid levels, produces a distinct dicotomy relative to cancer inhibition. Thymosin, a hormone extract isolated from the thymus gland, has demonstrated effectiveness in the functional maturation and differentiation of lymphocytes in immunodeficient animals.[60] Combination of moderate dietary stress and exogenous thymosin treatment enhanced certain cell-mediated immune responses and resistance to leukemia cell growth.[60]

CONCLUSION

In summary, numerous animal studies have opened gateways to an increased length of life, while little confirmation has yet to be made in humans. Appropriate food restriction in rodents at any stage of life has shown increased longevity. Thymus activity and cell-mediated immune functions may have integral roles in disease prevention, cancer control, and aging process mediation. Nutritonal stresses, excesses and deficiencies can directly affect the functional prowess of immune activities.

ACKNOWLEDGMENTS

Support for the research which stimulated this review came from Phi Beta Psi Sorority and PHS Grant CA 27502.

REFERENCES

1. **McCay, C. M., Crowell, M. F., and Maynard, L. A.,** The effect of retarded growth upon the length of life span and upon the ultimate body size, *J. Nutr.,* 10, 63, 1935.
2. **Stuchlikova, E., Juricova-Horakova, M., and Deyl, Z.,** New aspects of the dietary effect of life prolongation in rodents. What is the role of obesity in aging?, *Exp. Gerontol.,* 10, 141, 1975.
3. **Weindruch, R. H. and Walford, R. L.,** Dietary restriction in mice beginning at 1 year of age: effect on life-span and spontaneous cancer incidence, *Science,* 215, 1415, 1982.
4. **Weindruch, R. H. and Walford, R. L.,** Aging and functions of the RES, *Reticuloendothel. Syst.,* 3, 713, 1982.
5. **Masoro, E. J., Bertrand, H., Liepa, G., and Yu, B. P.,** Analysis and exploration of age-related changes in mammalian structure and function, *Fed. Proc.,* 38, 1956, 1979.
6. **Weindruch, R. H., Kristie, J. A., Cheney, K. E., and Walford, R. L.,** Influence of controlled dietary restriction on immunologic function and aging, *Fed. Proc.,* 38, 2007, 1979.
7. **Schulof, R. S., Low, T. L. K., Thurman, G. B., and Goldstein, A. L.,** Thymosins and other hormones of the thymus gland, *Progr. Clin. Biol. Res.,* 58, 191, 1981.
8. **Good, R. A., West, A., Day, N. K., Dong, Z. W., and Fernandes, G.,** Effects of undernutrition on host cell and organ function, *Cancer Res.,* (Suppl.) 42, 737, 1982.
9. **Rabin, B. S.,** The effect of diet on immune responsiveness and aging, *Med. Hypotheses,* 8, 495, 1982.
10. **Katz, A. E.,** Immunity and aging, *Otolaryngol. Clin. North Am.,* 15(2), 287, 1982.
11. **Friend, P. S., Fernandes, G., Good, R. A., Michael, A. F., and Yunis, E. J.,** Dietary restrictions early and late. Effects on the nephropathy of the NZBxNZW mouse, *Lab. Invest.,* 38(6), 629, 1978.
12. **Good, R. A., Fernandes, G., Yunis, E. J., Cooper, W. C., Jose, D. C., Kramer, T. R., and Hansen, M. A.,** Nutritional deficiency immunologic function, and disease, *Am. J. Pathol.,* 84, 559, 1976.
13. **Watson, R. and Safranski, D.,** Dietary restrictions and immune responses in the aged, in *CRC Handbook of Immunology in Aging,* 1980, 125.
14. **Dominioni, L., Dionigi, R., Dionigi, P., Nazazi, S., Fossati, G. S., Prati, U., Tibaldeschi, C., and Panesi, F.,** Evaluation of possible causes of delayed hypersensitivity and impairment in cancer patients, *J. Parent. Enter. Nutr.,* 5(4), 300, 1981.
15. **Wood, C. and Watson, R. R.,** Interrelationships among nutritional status, cellular immunity, and cancer, unpublished data.
16. **Dionigi, P., Dionigi, R., Nazari, S., Bonoldi, A. P., Griziotti, A., Pavesi, F., Tibaldeschi, C., Cividini, F., and Gration, I.,** Nutritional and immunological evaluations in cancer patients: relationship to surgical infections, *J. Parent. Enter. Nutr.,* 4(4), 351, 1980.
17. **Weindruch, R. H., Kristie, J. A., Naeim, F., Mullen, B. G., and Walford, R. L.,** Influence of weaning-initiated dietary restriction on responses to T-cell mitogens and on splenic T-cell levels in a long-lived F_1 hybrid mouse strain, *Exp. Gerontol.,* 17, 49, 1982.
18. **Young, V. R.,** Diet as a modulator of aging and longevity, *Fed. Proc.,* 38, 1994, 1979.
19. **McMurray, D. N., Beskitt, P. A., and Newmark, S. R.,** Immunological status in severe obesity, *Int. J. Obesity,* 6, 61, 1982.
20. **Haber, S. L. and Wissler, R. W.,** Effect of vitamin E on carcinogenicity of methylchloranthrene, *Proc. Soc. Exp. Biol. Med. III.,* 774, 1962.
21. **Wattenberg, L. W.,** *Fundamentals in Cancer Prevention,* Magee, P. N. et al., Eds., University of Tokyo Press, Tokyo/University Park Press, Baltimore, 1976, 153.
22. **Black, H. S. and Chan, J. T.,** Suppression of ultraviolet light-induced tumor formation by dietary antioxidants, *J. Invest. Dermatol.,* 65, 412, 1975.
23. **Tengerdy, R. P., Heinzerling, R. H., Brown, G. L., and Mathias, M. M.,** Enhancement of humoral immune response by vitamin E, *Int. Arch. Allergy Appl. Immunol.,* 44, 221, 1973.
24. **Heinzerling, R. P., Nockels, C. F., Quarles, C. L., and Tengerdy, R. P.,** Protection of chicks against *E. coli* infection by dietary supplementation with vitamin E, *Proc. Soc. Exp. Biol. Med.,* 146, 279, 1974.
25. **Tanaka, J., Fugiwara, H., and Torisu, M.,** Vitamin E and immune response. I. Enhancement of helper T-cell activity by dietary supplementation with Vitamin E in mice, *Immunology,* 38, 727, 1979.
26. **Lim, T. S., Putt, N., Safranski, D., Chung, C., and Watson, R. R.,** Effect of vitamin E on cell mediated immune responses and serum corticosterone in young and maturing mice, *Immunology,* 44, 289, 1981.
27. **Van Vleet, J. F.,** Current knowledge of selenium-vitamin E deficiency in domestic animals, *J. Am. Vet. Med. Assoc.,* 176, 321, 1980.
28. **Teige, J., Jr., Mordstoga, K., and Aursjo, J.,** Influence of diet on experimental swine dysentery. I. Effects of a vitamin E and selenium deficient diet supplemented with 6.8% cod liver oil, *Acta Vet. Scand.,* 18, 384, 1977.
29. **Teige, J., Jr., Saxegaard, F., and Froslie, A.,** Influence of diet on experimental swine dysentery. II. Effects of a vitamin E and selenium deficient diet supplemented with 3% cod liver oil, vitamin E or selenium, *Acta Vet. Scand.,* 19, 133, 1978.

30. **Watson, R. R., Chung, C., and Petro, T. M.,** Proceeding of conference on vitamin E: biochemical, hematological and clinical aspects, *N.Y. Acad. Sci.,* 393, 205.

31. **Watson, R. R. and Messiha, N.,** submitted.

32. **Watson, R. R. and McMurray, D. N.,** Effects of malnutrition on secretory and cellular immunity, *CRC Crit. Rev. Food Sci. Nutr.,* 12, 113, 1979.

33. **Beisel, W. R.,** Single nutrients and immunity, *Am. J. Clin. Nutr.,* (Supp.) 35, 1982.

34. **Micksche, M., Cerni, C., Kokron, O., Titscher, R., and Wrba, H.,** Stimulation of immune response in lung cancer patients by vitamin A therapy, *Oncology,* 34, 234, 1977.

35. **Glaser, M. and Lotan, R.,** Augmentation of specific tumor immunity against a syngenic SV40-induced sarcoma in mice by retinoic acid, *Cell. Immunol.,* 45, 175, 1979.

36. **Lotan, R. and Dennert, G.,** Stimulatory effects of vitamin A analogs on induction of cell-mediated cytotoxicity in vivo, *Cancer Res.,* 39, 55, 1979.

37. **Cerni, C., Kokran, O., and Miscksche, M.,** Aspects of chemotherapy in a controlled clinical study for treatment of bronchogenic cancer, *Chemotherapy,* 8, 287, 1976.

38. **Zachariae, H., Grunnet, E., Thestrup-Pedersen, K., Molin, L., Schmidt, H., Stafelt, F., and Thomsen, K.,** Oral retinoid in combination with bleomycin, cyclophosphamide, prednisone and transfer factor in Mycosis Fungoides, *Acta Dermatol.,* 62, 162, 1981.

39. **Daly, J. M., Copeland, E. M., and Dudrick, S. J.,** Effects of intravenous nutrition on tumor growth and host immunocompetence in malnourished animals, *Surgery,* 84, 655, 1978.

40. **Tucker, M. J.,** The effect of long-term food restriction on tumors in rodents, *Int. J. Cancer,* 23, 803, 1979.

41. **Tannenbaum, A.,** The initiation and growth of tumors. Introduction. I. Effects of underfeeding, *Am. J. Cancer,* 38, 335, 1940.

42. **Fernandes, G., West, A., and Good, R. A.,** Nutrition, immunity, and cancer. A review. III. Effects of diet on the diseases of aging, *Clin. Bull.,* 9(3), 91, 1979.

43. **Vitale, J. J. and Briotman, S. A.,** Lipids and immune function, *Cancer Res.,* 41, 3706, 1981.

44. **Rogers, A. E. and Newberne, P. M.,** Dietary effects on chemical carcinogenesis in animal models for colon and liver tumors, *Cancer Res.,* 35, 3427, 1975.

45. **Bansal, B. R., Rhoads, J. E., Jr., and Bansal, S. C.,** Effects of diet on colon carcinogenesis and the immune system in rats treated with 1, 2-Dimethyl hydrazine, *Cancer Res.,* 38, 3293, 1978.

46. **Silver, R. T. and Williams, H. E., Eds.,** The prevention of colon cancer, *Am. J. Med.,* 68, 917, 1980.

47. **Wanebo, H. J.,** Immunobiology of head and neck cancer: basic concepts, *Head Neck Surg.,* 2, 42, 1979.

48. **Chandra, R. K.,** Cell-mediated immunity in nutritional imbalance, *Fed. Proc.,* 39, 3088, 1980.

49. **Horrobin, D. F. and Manku, M. S.,** The nutritional regulation of T lymphocyte function, *Medical Hypotheses,* 5, 969, 1979.

50. **Cameron, E., Pauling, L., and Leibovitz, B.,** Ascorbic acid and cancer: a review. *Cancer Res.,* 39, 663, 1979.

51. **Anthony, H. M. and Schorah, C. J.,** Severe hypovitaminosis C in lung cancer patients: the utilization of vitamin C in surgical repair and lymphocyte-related host resistance, *Br. J. Cancer,* 46, 354, 1982.

52. **Lambert, J. J., III,** Nutritional assessment of the head and neck cancer patient, *Otolaryngol. Head Neck Surg.,* 88, 695, 1980.

53. **Shils, M. E.,** Diet and nutrition as modifying factors in tumor development, *Med. Clin. North Am.,* 63(5), 1027, 1979.

54. **Harvey, K. B. and Bothe, A., Jr.,** Nutritional assessment and patient outcome during oncological therapy, *Cancer,* 43, 2065, 1979.

55. **Haffejee, A. A. and Agorn, I. B.,** Nutritional status and the nonspecific cellular and humoral immune response in esophageal carcinoma, *Ann. Surg.,* 475, 1979.

56. **Haffejee, A. A. and Angorn, I. B.,** Diminished cellular immunity due to impaired nutrition in oesophageal carcinoma, *Br. J. Surg.,* 65, 480, 1978.

57. **Mullin, T. J. and Kirkpatrick, J. R.,** The effect of nutritional support on immune competency in patients suffering from trauma, sepsis, or malignant disease, *Surgery,* 90, 610, 1981.

58. **Serrou, B. and Cupissol, D.,** Nutritional support and the immune system in cancer management. A critical review, *Cancer Treat. Rep.,* 65 (suppl. 5), 115, 1981.

59. **Murray, J. and Murray, A.,** Toward a nutritional concept of host resistance to malignancy and intracellular infection, *Persp. Biol. Med.,* 290, 1981.

60. **Watson, R. R., Chein, G., and Chung, C.,** Thymosin treatment: serum corticosterone and lymphocyte mitogenesis in moderately and severely protein-malnourished mice, *Am. Inst. Nutr.,* 483, 1982.

61. **Petro, T. M. and Watson, R. R.,** Resistance to L1210 mouse leukemia cells in moderately protein-malnourished BALB/c mice treated in vivo with thymosin fraction V, *Cancer Res.,* 42, 2139, 1982.

Specialized Nutrition Programs and Therapies for the Aged

EFFECTS OF DIABETIC TREATMENT AND GENDER ON SENSORY FUNCTIONS IN THE ELDERLY

Lawrence C. Perlmuter, David M. Nathan, and Malekeh K. Hakami

INTRODUCTION

There is little doubt that dietary habits are a risk factor in the onset and control of Type II, non-insulin-dependent diabetes. Indeed, 75% of the maturity onset diabetics were obese at the time of diagnosis.[1] Apparently, this relationship is partially reversible since there are marked improvements in plasma glucose levels and insulin resistance following weight reduction.[2] Furthermore, by limiting food intake in specially bred prediabetic hamsters, it was possible not only to delay the onset of diabetes, but also to produce a milder form of the disease.[3] An epidemiological study with humans living under circumstances in which food was available only in limited quantities appeared to support the results of the research with hamsters, namely, that restricted caloric intake lessened the risk of diabetes.[4]

The consequences of Type II diabetes are widely known and include lessened longevity, increased risk of micro- and macrovascular complications, neuropathies (both autonomic and peripheral), and limited cognitive impairment.[5-6] There is also some scant evidence indicating that Type II diabetes is associated with a decreased sensitivity in certain sensory systems, namely olfaction and gustation. It is these latter changes that the present chapter will address.

It is important to determine the degree to which gustation and olfaction are changed in diabetes since these senses have a major impact on dietary regulation, which in turn has an important effect on the regulation and control of diabetes.

There are at least two reasons to expect that taste and smell sensitivity may be weakened in Type II diabetes. First, it has been found that these sensory systems tend to become less responsive with age. Hinchcliffe[7] and Richter and Campbell[8] found an increase in gustatory thresholds, while Bartalena and Bocci[9] and Venstrom and Amoore[10] reported decreased olfactory sensitivity with age. On the assumption that in many respects diabetes represents an accelerated form of aging,[11] such sensory alterations might be expected to appear. Second, since neuropathies — both peripheral and autonomic — are often observed in diabetes,[12-13] it seems reasonable to expect that either or both forms of these might impair olfactory and gustatory sensitivity. That is, peripheral neuropathies would be expected to impair sensory functioning. The influence of autonomic neuropathies on sensory functioning has yet to be determined.

It seems unnecessary to make the argument that changes in taste and smell could affect dietary regulation, but it is not unimportant to speculate how this effect may be mediated. Furthermore, from a patient education perspective, understanding how changes in these sensory systems affect dietary habits should enhance the effectiveness of dietary training programs. Indeed, there is evidence that training programs which are geared to the needs of the participants are relatively more effective.[14]

The reciprocal relationship between food-related behaviors and one's psychological well-being cannot be too strongly emphasized. The excitement provided by food, the motivational quality of food, as well as the need for the varied selection of foodstuffs are crucial influences in dietary regulation. All of these reactions may be adversely affected when taste and smell sensitivity are blunted. Thus, for the Type II aging diabetic, the age-related loss in sensory sensitivity, compounded with the putative increments to this deteriorative process by maturity onset diabetes, can be expected to impair sensitivity to taste and smell stimuli.

To date, the evidence seems quite incontrovertible that olfactory sensitivity does grow weaker across the age span.[15-16] Whether this effect is attributable to changes in the smell receptors and/or changes in the neural projection areas is generally not agreed upon. Since the smell receptor cells must be renewed approximately every 30 days, it has been suggested that aging may retard or even prevent the renewal process.[17-18] From a more neural perspective, declines in sensory sensitivity could be the result of demyelination or basic metabolic disorders. For instance, Smith[19] reported that the number of fibers reaching the olfactory bulb declines by approximately 1% per year shortly after birth. Liss[20] later confirmed this degenerative process. Apparently, these net losses occur despite a continual turnover of olfactory sensory neurons.[21]

The sensation of taste appears to be mediated at the periphery via taste buds which are stimulated by food substances which have been dissolved in saliva. These taste receptors are short-lived and in young to middle-aged individuals are replaced approximately every 10 days. Renewal is slowed in the elderly and is further retarded in women suffering from severe estrogen depletion.[17] In addition to neural and physiological changes, the role of conditioned (cephalic) preparatory effects in anticipation of the ingestion of foodstuffs should not be ignored as an adjunct to the quality and intensity of taste sensations.

With aging, the loss in sensitivity appears to be relatively greater for smell than for taste.[22] With respect to taste, the loss for salt sensitivity appears earliest, followed soon after by the lessened sensitivity to sweet, leaving the bitter sensation for the aged.[17]

Although the experimental literature reflects a reasonably consistent picture with respect to the adverse effects of age on smell, this relationship is less consistent with taste. Moreover, studies of Type II diabetes and its effect on taste and smell are even less consistent and in the latter case almost nonexistent. The accelerated aging hypothesis of diabetes accordingly remains to be evaluated with respect to possible alterations in smell and taste. For example, in testing for threshold sensitivity to sucrose, Bisht et al.[23] found no differences among diabetics, close relatives of diabetics, or controls. Subjects of both sexes in their study ranged in age from 16 to 34 years. No mention was made either of the assignment to groups by gender, age, or diabetic treatment.

In a more recent and carefully controlled study, Dye et al.[24] examined threshold, sweetness ratings, and hedonics in males ranging in age from 40 to 88 years. Although the aged subjects (70 to 88 years) had higher thresholds to sucrose than the 40- to 60-year-old subjects, there were no effects of diabetes on threshold sensitivity. Likewise, disease had no effect on ratings of sweetness or pleasantness. Age had no measurable effects on rated sweetness; however, for the pleasantness measure, there was a three way interaction of sucrose concentration, age, and patient groups. Generally, young diabetics and old nondiabetics preferred the sweeter substances. The performance of young diabetics offered some support for the accelerated aging hypothesis since their preference curve resembled that of the older nondiabetics. On the other hand, the old diabetics preferred the less sweet concentrations. Dye et al. suggested that the old diabetics may have grown more sensitive to subtle flavors as a result of long-term dietary restrictions with the result that they preferred the less sweet substances. Overall, the interpretation of these results is very complex. Despite the explanation by Dye et al., no attempt was made by the authors to unconfound a number of potentially important variables in their study. For one, it is likely that disease duration was confounded with the age of the patient. Second, it is possible that the young diabetics may have had a more severe metabolic disorder (relative to the old diabetics), and thus their resemblance to the old nondiabetics does not unambiguously support the accelerated aging hypothesis. The presentation of information about neuropathies in these two age groups would have helped to bolster the accelerated aging hypothesis. These results are limited in generalizability since only males were studied. Finally, although Dye et al. made mention of the various diabetic treatments given to these patients, their possible differential effects as well as those of other medications in the study were not reported.

In addition to the study by Dye et al., there are other studies which have examined the effects of diabetes on taste. Abbasi[25] reported that diabetics were less sensitive to sucrose, urea, hydrochloric acid, and NaCL. Detection (ability to name the tastant) and recognition thresholds (ability to signify presence of the stimulus) were elevated in diabetics relative to controls. In addition, disease duration and autonomic vs. nonautonomic neuropathies each contributed to additional elevations in thresholds. However, the author failed to report any statistical treatment of these data. Thus, the reliability of these results remain in doubt.

Finally, Porte et al.[26] reported that the detection threshold for glucose and fructose was significantly higher in recently diagnosed diabetics than in controls. These effects were much greater for glucose than for fructose. Although Porte et al. failed to describe his subjects with respect to age, sex, treatment, etc. diabetics had fasting hyperglycemic levels (plasma glucose greater than 115 mg/100 mℓ), and these subjects were matched for age and gender with a control group. Porte et al. interpreted these findings as support for the theory that diabetes involves a glucoreceptor deficit in which ingested glucose is not recognized by the pancreatic beta cell resulting in impaired glucose metabolism.

We will now turn to the procedures used in the present study of maturity onset diabetes in males and females aged 55 to 74 years, and treated either with insulin, oral agents, or diet. An age-matched control group was also employed. Magnitude estimations and hedonics were measured in response to a series of suprathreshold stimuli, one primarily gustatory and one olfactory.

METHOD

Participants — The participants, examined at the Massachusetts General Hospital, were Type II diabetics treated with insulin, oral agents, or diet and were tested following an overnight fast (approximately 12 hr). Patients ranged in age from 55 to 74 years while disease duration ranged from 1 to 20 years. The age-matched controls also ranged in age from 55 to 74 years. Of the 153 diabetic participants, 78 were females, while among the 34 controls, 24 were females. Participants were paid for their participation.

Gustatory Stimuli — A measure of 5 mℓ of each six sucrose solutions was placed into 24 medicine cups (wax coated, 30 mℓ). The range of concentrations were 0.031, 0.062, 0.125, 0.25, 0.5, and 1.0 (mol/ℓ). One additional medicine cup with 5 mℓ of the 0.125 sucrose solution and a cup of rinse water was also set before the subjects. The middle concentration (0.125) was presented at the beginning of the series as an example. The subjects were informed that the sample had a magnitude of 10 based on a scale ranging from 0 to 100.

Olfactory Stimuli — The stimuli were six solutions of cold press orange oil and light paraffin oil. The range of concentrations were 0.4, 1.2, 3.7, 11.0, 33.0, and 100% of orange oil. The experimenter presented each stimulus by uncorking the flask as it reached the patient's nose in order to prevent the escape of the odor.

Procedure — Gustatory Evaluation

Magnitude estimations of the taste stimuli as well as hedonic ratings of the tastants were evaluated. The psychophysical evaluation was part of a larger test battery and was presented in the first portion of the day for each subject. For magnitude estimations, the subjects' task was to judge the strength or intensity of each of the taste solutions. The subjects were instructed to sip the solution, hold it in their mouth for a few seconds, and then spit it out. Immediately thereafter, the intensity judgment and the hedonic judgment were made in that order. Subjects were permitted to use a score of 0 for no sweetness at all, and 100 for the highest strength of the sweetness. The subjects were instructed that the number assigned should be in proportion to the degree of sweetness. Subjects were periodically instructed not to swallow the solution and were also instructed to rinse their mouth twice with water before each tasting. The sucrose concentrations were presented on 24 trials with 4 presentations of each of the 6 stimulants. On each block of 6 trials, the tastants were presented in a different sequence. There was an interstimulus interval of not less than 30 sec.

For hedonic testing, the subject's task was to rate the solutions according to how much he/she liked or disliked the taste. The subjects were instructed to point to a spot on an "unmarked ruler" which reflected like/dislike markings. A 0 to 100 point scale was visible on the experimenter's side of the ruler, thus allowing a ready conversion of the subject's estimation into the appropriate metric. Specifically, if one taste was liked (disliked) twice as much as another, the subject was to point to the place on the ruler reflecting these preferences.

All 24 trials were given in a predetermined random order which was identical for all subjects.

Procedure — Olfactory Evaluation

The instructions for the olfactory test were similar to those of the taste test. That is, the subjects were instructed to rate the odors according to their strength and also according to their liking or disliking for the stimuli — after receiving one sniff of the odors — in a manner similar to that employed on the taste test. For this test also, an intermediate odor with a defined strength of 10 on a scale from 0 to 100 was given as an example.

Subjects were prevented from seeing the number and color of the solutions while testing, and the total amount of test time was approximately 40 min. A counter-balanced procedure was used in which some subjects were tested with the sucrose solutions followed by the essence of orange while for other subjects this order was reversed. No apparent differences resulted from these sequences.

RESULTS

Although subjects provided magnitude estimates as well as hedonic (preference) data to each presentation of the olfactory and gustatory stimuli, for purposes of this report we have decided to limit our focus to the preference data. Before dismissing the magnitude estimates completely, we can report that generally the diabetic groups (insulin, oral, and diet treated patients) tended to provide larger estimates than the controls to the taste and smell stimuli. However, the slopes of individual subject regression lines (betas) were similar among these groups. When R^2 values were examined, these tended to be low for many subjects, and thus we are not confident that the beta measures reliably fit all of these data. Despite this disclaimer, it was found that the 65- to 74-year-old subjects (mean beta = 0.47) were less sensitive ($p < 0.05$) to increasing concentrations of the olfactory stimulus (essence of orange) than were the 55- to 64-year-old subjects (mean beta = 0.66). Similar results have been reported by Kleinschmidt.[27] In addition, there was an interaction between age and gender which approached significance, $F(1, 155) = 2.70$, $p < 0.10$, and indicated that the decline in these two age groupings was much larger for males (mean beta decline = 0.33) than for females (mean beta decline = 0.04).

With increasing concentrations of sucrose, females (mean beta = 1.33) were more sensitive, $F(1, 171) = 4.47$, $p < .05$, than males (mean beta = 1.07). Similar results have been reported by Doty.[28]

A series of correlational analyses indicated that beta scores for smell and taste were significantly correlated in the insulin, diet, and control groups ($r = 0.33$, 0.37, and 0.34, respectively) while in the oral-treated group this relationship was clearly not significant, $p > 0.30$.

Hedonics — Data Evaluation

Each stimulus concentration was presented once in each of the four trial blocks while the sequence of concentrations was randomized within each block. Thus we decided, as Murphy[29]

Table 1
MEAN OLFACTORY PREFERENCES[a] FOR AGE, TREATMENT, AND CONCENTRATION VALUES

Concentrations[b] (%)	Age	Treatment			
		Insulin	Oral	Diet	Control
0.4	Young	45.54	49.50	44.25	49.42
	Old	47.40	50.80	49.89	49.77
1.2	Young	49.14	44.94	42.83	52.18
	Old	44.98	51.95	52.14	52.96
3.7	Young	48.12	54.80	47.44	54.94
	Old	48.95	50.45	55.80	53.31
11.0	Young	49.44	48.38	45.66	58.26
	Old	48.25	55.84	54.65	54.69
33.0	Young	46.53	58.94	45.02	55.32
	Old	49.55	56.39	53.36	53.23
100.0	Young	50.16	65.23	46.27	55.85
	Old	49.68	51.54	51.67	60.62

[a] Arithmetic means of 1st and 4th test; scale 0 to 100 (see text).
[b] Essence of orange oil.

had done previously, to examine the initial preferences as well as a possible change in preferences from the first to the fourth presentation of each stimulus. It is reasonable to expect that initial preferences would generally be higher for all subjects and that these would decline with repeated presentations as a result of satiation. Differential rates of change in preference with aging, gender, and diabetic treatment would be interesting theoretically as well as practically. Thus, the gustatory and olfactory data were so treated. Indeed, Murphy[29] reported differential rates of change in preference to an olfactory stimulus (menthol). That is, young subjects (18 to 26 years) showed a significant decline in preference with stimulus repetition while aged subjects (> 65 years) failed to show satiation. Since diabetes often mimics accelerated aging, we have chosen to examine the hedonic data in the method described.

Preference Analyses — Olfactory

In response to the stimuli, subjects indicated their neutrality, likes, or dislikes by use of a qualitative scale. These judgments were readily convertible into numerical values on a 0 to 100 scale (as described previously). A score of 50 reflected indifference or neutrality with respect to the judgments. These data were examined by an analysis of variance with gender, age, and treatment conditions as between-subjects variables and stimulus concentrations as a within-subjects variable. Data for the first and fourth stimulus presentation were analyzed; thus replication served as an additional within-subjects variable.

Results indicated that increasing concentrations generally resulted in stronger preferences, $F(5, 865) = 5.26$, $p < 0.001$. More importantly, there was an interaction of Concentration $\times$ Age $\times$ Treatment, $F(15, 865) = 1.71$, $p < 0.05$. These data are presented in Table 1.

Overall, the results indicated that preferences were lower for the insulin-treated group as compared to oral, diet, and control groups. Moreover, as concentrations increased, preference tended to increase for young and old subjects with the exception of the insulin- and diet-treated groups. Finally, the more or less systematic differences between young and old subjects tended to be somewhat different in the oral-treated group. Surprisingly, the replication factor was not a significant main effect nor did it interact with the other variables in this analysis.

Table 2
MEAN GUSTATORY PREFERENCES[a] FOR
FEMALES, MALES, AND CONCENTRATION
VALUES

	Sucrose Concentrations (mol/ℓ)					
	0.031	0.062	0.125	0.25	0.5	1.0
Female	53.01	51.63	49.40	42.80	43.73	41.05
Male	46.78	48.11	50.18	50.53	48.96	48.30

[a] Arithmetic means of 1st and 4th test; scale 0 to 100 (see text).

In order to more closely examine the observed interaction, a series of ANOVAS were run in which pairs of groups were systematically compared. Generally, the results indicated two major findings, namely, the control group tended to show higher preferences than did the diabetic groups, and these preferences increased with increasing concentrations. Second, with replications of the stimuli, relatively stronger satiation effects were observed in the control than in the diet-treated group. This satiation effect occurred more strongly for young vs. old subjects in the oral-treated group, while in the control group satiation effects were seen with older subjects. By comparison, the insulin-treated group showed a very weak satiation effect for both young and old subjects.

In conclusion, these data suggest that preferences are somewhat blunted in diabetes and blunting is especially marked in the insulin and diet groups. The results tend to suggest that diabetics appear more similar to old (65 to 74 years) than to young (55 to 64 years) control subjects, although there are many exceptions to that generalization within these data.

Gustatory Preference

Similar to the olfactory preference analysis, an ANOVA was conducted on gustatory preferences. Concentration and replication (trials 1 and 4) were within-subjects variables while gender, treatment, and age were between-subjects variables. In addition to a main effect of concentration, $F(5, 860) = 2.46$, $p < 0.05$, there was also an interaction of gender $\times$ concentration, $F(5, 860) = 4.07$, $p < 0.01$. The results of this interaction are presented in Table 2 and indicate that females tend to prefer less sweet-tasting substances and show a systematic decline in preferences as concentration increases. By comparison, males tend to prefer a higher concentration of sweetness.

In addition, there was an age $\times$ concentration $\times$ replication interaction (see Table 3), $F(5, 860) = 2.74$, $p < 0.01$. Generally, these results suggest that the older subjects prefer the less sweet concentration, and this preference tends to be more marked on the fourth presentation than it is on the first. The interaction with replication suggests that it takes longer for satiation to build in old than in young subjects. With respect to the younger subjects, there is some evidence that their preferences tend to decline from the first presentation to the fourth presentation, and this effect is relatively stronger for the lower concentrations. That is, since the younger subjects tend to prefer the weaker concentrations on the initial presentation, their moderate preferences decline still further on the final presentation. The preferences of the younger subjects were on the negative side of neutral (mean ratings < 50). Generally, these results show that for older subjects, less sweet concentrations are preferred relative to the sweeter concentrations. Further, older subjects tend to rate the less sweet concentrations higher than the young subjects. Older subjects tend not to satiate on the less sweet concentrations as the young subjects do. With respect to the higher concentrations, the older subjects tended to show a satiation effect which was greater than

Table 3
MEAN GUSTATORY PREFERENCES FOR AGE,
REPLICATION, AND CONCENTRATION VALUES

Replication	Age	Sucrose Concentrations (mol/ℓ)					
		0.031	0.62	0.125	0.25	0.5	1.0
1							
	Young	49.35	48.80	51.09	45.80	44.81	45.39
	Old	51.96	51.82	49.49	48.92	48.34	46.01
4							
	Young	46.06	47.17	46.33	47.68	46.23	45.05
	Old	52.20	51.68	52.26	44.26	45.99	42.26

[a] Scale 0 to 100 (see text).

that for the younger subjects. This interaction is likely attributable to the relatively low ratings which the younger subjects provided on both the first and last set of presentations.

Neither a main effect or interaction with treatment was observed with taste stimuli. These results are similar to those of Dye et al.[24] who found only very weak effects of diabetes on either threshold or preference for sucrose. On the other hand, the results show some similarity to those of Murphy[29] with respect to satiation. That is, in the present study, young subjects tended to satiate faster than older subjects. Murphy reported similar effects with an olfactory stimulus (menthol), and by comparison employed much younger subjects (mean age 21) than those used in the present study.

In order to determine whether the degree of diabetic control, as measured by HbA_{1c} and fasting plasma glucose levels (FBS), was related to preference measures for taste and smell, the following calculations were performed. Since preference ratings for sucrose tended to be curvilinear with increasing concentrations, we obtained the average preference rating to the three lowest concentrations only. The preference rating on trials 1 and 4 were thus indexed to form a single taste preference measure for each subject. A similar analysis was performed with the olfactory data except that the average smell index for each subject was based on six concentrations since preferences tended to increase somewhat linearly with increasing concentrations. A series of zero-order correlations were performed and the results indicate that the taste and smell indices were highly correlated ($r = 0.56, p < 0.001$), thus lending some validity to this derived measure.

With respect to the olfactory data, it was found that in males, as HbA_{1c} and FBS levels increased, mean olfactory preference ratings decreased ($r = -0.37, p < 0.001$ and $r = 0.23, p < 0.05$, respectively). Also in males, as age increased, olfactory preference ratings increased ($r = 0.24, p < 0.01$).

With respect to the magnitude of taste preferences, there was a weak overall positive relationship between age and the strength of the taste index ($r = 0.18, p < 0.01$), suggesting that preference grew stronger with increasing age. Also, as FBS increased, taste preference decreased in males only ($r = -0.30, p < 0.01$).

Finally, we also examined obesity by the formula weight^3/height. In the diet-treated group, taste preference ratings decreased as the obesity index increased ($r = -0.26, p < 0.05$). For all other groups no such relationships were observed.

CONCLUSION

In conclusion, the results of the present experiment offer some preliminary information on the blunting effects of diabetes on olfaction. More importantly, they also suggest that associated with insulin and dietary control of diabetes are more marked effects than those seen with oral agents. This observation tends to eliminate severity of disease as an explanation since it is likely that diet-treated patients would have a milder form of the disease. Nevertheless, other disease-related variables will have to be examined in order to explain the observed relationship. These data offer some general support for the contribution of age-related effects to the decline in olfactory sensitivity.

With respect to gustatory responses, the present results appear to be consistent with those of Dye et al.[24] in showing that diabetes fails to alter sensitivity or preferences to sucrose concentrations. On the other hand, these findings suggest that females (55 and older) are more sensitive to changes in sucrose concentrations than are comparably aged males. This effect is seen with beta as well as with preference measures. Generally, the results suggest that younger subjects tend not to like sweet-tasting concentrations, and that their dislike appears to peak at lower concentrations. Furthermore, the relatively more favored weaker concentrations become less desirable with repetition of the stimuli. That is, a satiation-like effect is observed. On the other hand, older subjects tend to reach satiation a little less rapidly for weaker concentrations, while showing satiation effects to the stronger concentrations.

The preliminary observations relating the degree of metabolic control with olfactory and gustatory preferences are tantalizing. The correlations, albeit statistically significant, are nevertheless weak. Whether the relationship between metabolic control and these sensory modalities is causal and if so, due to a first or second degree effect remains to be determined.

There are a number of practical implications of this study which have shown that age, gender, and treatment modality appear to influence olfactory as well as gustatory functioning. Although the observed effects were statistically significant, the magnitude of the alterations in sensory functioning were generally slight. Nevertheless, the observed changes may be of practical significance because of the crucial role played by dietary habits and preferences in the treatment and control of Type II diabetes. Indeed, it has been recognized that food choices and dietary habits can be influenced by slight changes in sensory activity. Moreover, digestion itself can be impaired when deficits in sensory functioning dampen anticipatory responses to the ingestion of food. Finally, these results have implications for dietary/behavioral management in Type II diabetes. If preferences and satiation effects are different for the elderly Type II diabetic, as demonstrated here, teaching strategies will have to be altered accordingly.

ACKNOWLEDGMENT

Special thanks to Alan Wayler, Ph.D., for his thoughtful comments on an earlier version of this paper. Joanne Katz and Jay Ginsberg assisted with data collection and analyses and their contributions are gratefully acknowledged. This research was supported by grant RO1 AG02300 from the National Institute on Aging.

REFERENCES

1. **West, K. M.,** Prevention and therapy of diabetes mellitus, *Nutr. Rev.,* 33, 193, 1975.
2. **Grodsky, G. M. and Benoit, F. L.,** Effect of massive weight reduction on insulin secretion in obese subjects, in *Diabetes: Proc. 6th Congr. Int. Diabetes Fed., Stockholm,* Excerpta Medica Foundation, Amsterdam, 1967, 540.
3. **Gerritsen, G. C., Blanks, M. C., Miller, R. L., and Dulin, W. E.,** Effect of diet limitation on the development of diabetes in prediabetic Chinese hamsters, *Diabetologia,* 10, 559, 1974.
4. **Gerritsen, G. C.,** The role of nutrition to diabetes in relation to age, in *Nutrition, Longevity and Aging,* Rockstein, M. and Sussman, M. L., Eds., Academic Press, N.Y., 1976, 229.
5. **Perlmuter, L. C., Katz, J., Ginsberg, J., Singer, D. E., Harrington, C., and Nathan, D. M.,** Cognitive deficits in elderly Type II diabetics, *Diabetes,* 32 (Abstr.), 39A, 1983.
6. **Nathan, D. M., Singer, D. E., Perlmuter, L. C., Harrington, C., Ginsberg, J., Katz, J., and Hakami, M. K.,** Hemoglobin A_{1c} influences prevalence of complications in elderly Type II diabetics, *Diabetes,* 32(Abstr.), 99A, 1983.
7. **Hinchcliffe, R.,** Clinical quantitative gustometry, *Acta Oto-Laryngol.,* 49, 453, 1958.
8. **Richter, C. P. and Campbell, H. K.,** Sucrose taste thresholds of rats and humans, *Am. J. Physiol.,* 128, 291, 1940.
9. **Bartalena, G. and Bocci, C.,** L'acutezza e la fatica olfactoria in relazoine alla' etta anagrafica e biologica, *Clin. Otorinol.,* 16, 409, 1964.
10. **Venstrom, D. and Amoore, J. E.,** Olfactory threshold in relation to age, sex, or smoking, *J. Food Sci.,* 33, 264, 1968.
11. **Kent, S.,** Is diabetes a form of accelerated aging?, *Geriatrics,* 31, 140, 1976.
12. **Ellenberg, M.,** Diabetic neuropathy, in *Diabetes Mellitus,* Sussman, K. E. and Metz, R. J. S., Eds., American Diabetes Association, 1975, 201.
13. **Prockop, L. D.,** Diabetic neuropathy, in *Diabetes Mellitus: Diagnosis and Treatment,* Fajans, S. S. and Sussman, K. E., Eds., American Diabetes Association, 1971, 347.
14. **Schmeck, R. R.,** Learning styles of college students, in *Individual Differences in Cognition,* Dillon, R. and Schmeck, R. R., Eds., Academic Press, N.Y., 1983.
15. **Chalke, H. D. and Dewharst, J. R.,** Coal gas poisoning: loss of sense of smell as a possible contributing factor with old people, *Br. Med. J.,* 2, 1915, 1957.
16. **Springer, K. J. and Dietzmann, H. E.,** Correlation studies of diesel exhaust odor measured by instrumental methods to human odor panel ratings, paper presented at Stockholm Odor Conf., Sweden, 1970.
17. **Massler, M.,** Geriatric nutrition: the role of taste and smell in appetite, *J. Prosth. Dent.,* 43, 247, 1980.
18. **Schiffman, S., Orlandi, M., and Erickson, R. P.,** Changes in taste and smell with age: biological aspects, in *Sensory Systems and Communication in the Elderly,* Ordy, J. M. and Brizzee, K., Eds., Raven Press, N.Y., 1979, 247.
19. **Smith, C. G.,** Age incidence of atrophy of olfactory nerves in man. *J. Comp. Neurol.,* 77, 589, 1942.
20. **Liss, L. and Gomez, F.,** The nature of senile changes of the human olfactory bulb and tract, *Arch. Otolarygol.,* 67, 167, 1958.
21. **Graziadei, P. P. C. and Graziadei, G. A.,** Plasticity of connections in the olfactory sensory pathway: transplant, in *Olfaction and Taste VII,* Vander Starre, H., Ed., IRL Press, London, 1980, 155.
22. **Stevens, J. C., Plantinga, A., and Cain, W. S.,** Reduction of odor and nasal pungency associated with aging, *Neurobiol. Aging,* 3, 125, 1982.
23. **Bisht, D. B., Krishnamurthy, M., and Rangaswamy, R.,** Gustatory threshold for sugar in diabetics and their sibs utilizing sucrose, *J. Assoc. Physicians India,* 19, 431, 1971.
24. **Dye, C. J. and Koziatek, D. A.,** Age and diabetes effects on threshold and hedonic perception of sucrose solutions, *J. Gerontol.,* 36, 310, 1981.
25. **Abbasi, A. A.,** Diabetes: diagnostic and therapeutic significance of taste impairment, *Geriatrics,* 36, 73, 1981.
26. **Porte, D., Jr., Robertson, R. P., Halter, J. B., Kulkosky, P. J., Makous, W. L., and Woods, S. C.,** Neuroendocrine recognition of glucose: the glucoreceptor hypothesis and the diabetic syndrome, in *Food Intake and Chemical Senses,* Katsuki, Y., Sato, M., Takagi, S. F., and Oomura, Y., Eds., University Park Press, Tokyo, 1977, 331.
27. **Kleinschmidt, E. G.,** Olfactometric results of the olfactory function in diabetics, *Z. Gesamte Inn. Med.,* 33, 901, 1978.
28. **Doty, R. L.,** Gender and reproductive state correlates of taste perception in humans, in *Sex and Behavior: Status and Prospectus,* McGill, T., Dewsbury, D. A., and Sachs, B., Eds., Plenum Press, N.Y., 1978, 337.
29. **Murphy, C.,** Age-related effects on the threshold, psychophysical function, and pleasantness of menthol, *J. Gerontol.,* 38, 217, 1983.

DIETARY PATTERNS AND PRACTICES WHICH AFFECT THE INCIDENCE OF CANCER IN THE ELDERLY

Bernard S. Linn and Margaret S. Linn

INTRODUCTION

This chapter includes information about cancer in old age, nutrition and cancer in general, and dietary influences on cancer in old age. Since other chapters in this book deal with nutrition in old age specifically, this subject will not be covered. The purpose is to characterize cancer as it appears in the elderly, to focus on some of the findings regarding the relationship of diet to cancer in animals and man, and to speculate on some of the questions that are raised in application of these data to the elderly. Data concerning effects of alcohol and tobacco as carcinogens are also excluded as beyond the scope of this chapter.

It is well-known that rates of cancer increase with aging. Statistics on cancer incidence and mortality in the elderly, however, are less than exact because of inaccurate and incomplete reporting practices in many countries. Accumulating evidence also suggests that many types of cancer may be caused in part by nutritional factors. Assessment of nutritional status in the elderly also has its limitations. Even for the healthy old, there is a question of whether evaluation against the same standards as those used for younger people is appropriate. For the impaired elderly, assessments must take into account effects of disease. More reliable and valid means of assessing nutrition and dietary practices are needed in order to determine whether or not impairment of health is the result of inadequate or inappropriate diet.

CANCER IN OLD AGE

Cancer is known to increase with age. The chances of developing cancer during the next 5 years rises from 1 in 700 at age 25 to 1 in 14 at age 65.[1] During the past few decades in the U.S., nearly half of all new cancers occurred in persons over age 65.[2] Certain cancers (stomach, colon and rectum, prostate in men, and breast in women) account for over half of the cancers in patients over 60 years of age.[3] GI cancers comprise between 25 and 44% of all cancers occurring after 75 and 85 years of age.[3] Cancer of the prostate is very common in elderly men and has been detected at the time of autopsy in 40 to 57% of those over 70 years of age.[4] Breast cancer in women occurs one half of the time after they reach age 65.[5] Other malignancies are also common in old age. Cancer of the cervix is becoming more frequent than reported in previous years in elderly women.[6] Epidermoid cancer of the vulva is especially a disease of elderly women.[7] Risk of recurrence from thyroid cancer increases with age and anaplastic carcinoma of the thyroid occurs predominantly in older patients.[8] Chronic lymphocytic leukemia is a monoclonal accumulation of malignant B-lymphocytes that occurs mostly in older people.[9] Lung cancer is continuing to increase in the general population; and since most are related to carcinogens in tobacco smoke, it is likely that the duration of regular exposure — rather than age — is associated with the incidence.[10]

Recently, more accurate statistics on mortality in older people in the U.S. have become available. Most published death rates at the older ages had previously been recorded for ages 75 and above combined. Problems in obtaining death rates at advanced ages included: misstatements of age, incomplete or inaccurate registration of deaths, and lack of comparability between population figures taken from the census and mortality statistics for the death registration areas.

With the advent of medicare, more accurate data on cancer have also been obtained. For most cancers, the incidence is known to rise with age, but there appears to be a decline following 85 to 90 years of age.[2] Among white males, death rates from all types of cancer continue to increase past age 65, although at a diminishing rate, up to about age 90. In white females, the rates rise up to about age 85. Death rates for all types of cancer after age 90 decline in both sexes. Data from England and Wales show the same decreases after age 85. In Sweden, they begin dropping after age 70 in males.[11]

Some of the decrease in rates could result from errors in reporting causes of death. Often the elderly are recorded as dying from intercurrent diseases such as heart conditions or pneumonias rather than from cancer, especially when cancer symptoms are silent or metastasis not apparent. Thus, this preference for other diseases over cancer as causes of death influences the accuracy of cancer mortality statistics. Incidence of cancer in old age could be a better measure than death rates of the frequency of cancer. However, in the early stages, cancer of the prostate, thyroid, colon, and kidney are frequently not detected on examination of older patients. On the other hand, the decrease in cancer rates in very old age is plausible. Those who live into extreme old age constitute a unique group who are biologically elite.[12,13] They may form a separate population from those who die earlier from cancer or other leading killer diseases.[14]

Care must be taken in comparing U.S. cancer rates in older persons with those in other countries. Butler[15] pointed out that an apparent decline in cancer incidence in Ibadan, Nigeria, after age 45 seemed actually to be an artifact introduced by older persons' preference in these countries for traditional medicine over hospital care, therefore by-passing that country's system for reporting disease.

Some research has focused on the relationship between aging and cancer, but many questions remain. It is unclear whether aging leads to cancer, whether aging and cancer share a common process, or whether the development of cancer simply requires a long time to become evident. If one takes the view that neoplasia is not part of the aging process but rather a disease that strikes more often in middle and later years, this raises the question of the facilitating effect which biological age seems to have on carcinogenesis. Several plausible theories in regard to the potential for such facilitating effects of aging have been offered, including accumulation of somatic mutations, increased sensitivity to oncogenic viruses, increased sensitivity to external carcinogens, decreased immunological resistance against neoplastic cells, increased tendency to hormone imbalances, and intrinsic age-related cell alterations which could influence cellular susceptibility to neoplastic transformation.[16]

NUTRITION AND CANCER

Animal studies have provided valuable information about the role of nutrition in cancer development. Caloric restriction, underfeeding, and deletion or addition of certain nutrients to the diet have been studied for long-term effects on the spontaneous onset of cancer. The development of models in which cancer is induced through the use of known carcinogens is making it possible to examine whether the onset of cancer is enhanced or delayed by manipulations of nutrients. Lastly, whether growth of established tumors can be significantly changed by dietary factors has been studied.

Most studies of cancer in man have been epidemiological in nature. Rates of cancer incidence and mortality have been associated with food intake of certain countries and groups of people. A growing number of studies have examined dietary practices elicited from personal interviews in regard to cancer incidence.

Data on nutrition and cancer will be presented as it relates to animal studies and the development of cancer, animal studies and the growth of established cancers, and epidemiological information on diet and cancer in humans.

ANIMAL STUDIES OF THE DEVELOPMENT OF SPONTANEOUS AND EXPERIMENTALLY INDUCED CANCERS

As Table 1 shows, a number of nutrients have been tested in regard to their effect on development of either spontaneous or induced cancers in animals. One of the most consistent findings has been the positive relationship between underfeeding and undernutrition and that of lower cancer rates.[17] It is interesting that underfeeding in rats has also been associated with longevity in studies of experimental gerontology. In mice, a decreased incidence of cancer was found with low caloric intake.[17] Even when rats were fed an identical diet, the rates of tumors were greater in heavier as compared to lighter weight animals.[17] This raises the question of whether it is the reduced caloric intake or decreased body weight that provides the protection. Animals underfed immediately after weaning for 7 weeks were found to have reduced long-term growth and incidence of neoplasms even though animals followed an *ad libitum* diet during the rest of their lives.[18] Heavier body weight of the rat during early life was a consistent predictor of tumor development.

High protein diets and obesity have also been associated with specific types of tumors.[19] Protein deficiency may protect the rat from the effects of chemicals but later appears to increase the rates of cancer of the kidney.[20] High protein diets may increase carcinogenicity of other agents. In all of these studies, however, heavier weight of the animal was almost always also associated with increased cancer incidence whether protein diets were high or low. Deficiencies in lipotropes have been associated with increased liver cancer in rats subjected to carcinogenic doses of aflatoxin B_1.[21] Deficiencies in lipotropes with high fat diets also increased accelerated induction of liver cancer[22] with other agents as well.

Rats have been more susceptible to colon tumors when fed high fat diets. Results vary in regard to type of fat. Several studies have shown no difference in effect of fats depending on type or degree of saturation, particularly in rates of colon and liver tumors.[22,23] However, polyunsaturated fats[24] and vegetable polyunsaturated fats[25] seem to enhance cancer development, particularly mammary cancers. One theory suggests that an excessive fat intake modifies metabolism of cholesterol, bile salts, and neutral steroids in the intestine and metabolism and secretion of steroid hormones in circulation. Bile salts could be degraded by bacteria in the intestine to form carcinogenic substances. This process could then be affected by fiber in the diet or other conditions. Altered metabolism of steroid hormones could at the same time affect the uterus or breast, causing cancers at these sites. High wheat bran diets reduced induced colon but not duodenal cancers.[26] High levels of cellulose, however, did not decrease tumors of the colon but did decrease the incidence of small intestinal cancers.[27]

Vitamin A has been of interest because of its role in epithelial cell differentiation.[28-35] Results of studies of experimentally induced cancers in Vitamin A deficient rats have varied by the type of experimental carcinogen used.[22,29] However, a rather impressive number of studies have shown that the use of Vitamin A as a supplement inhibited development of various types of cancers, such as respiratory,[30] bladder,[31] stomach and cervix,[32] tracheo-bronchial,[33] and skin and breast[34] in hamsters and rats.

Studies of addition of ascorbic acid to diets showed inhibition of induction of tumors in association with nitrite.[35,36] Use of ascorbic acid along with nitrosatable drugs and nitrate-containing foods may reduce the hazards associated with these substances, since ascorbic acid reacts rapidly with nitrite and effectively competes with amines for nitrate.

Other deficiencies, such as iodine,[37] magnesium,[38,39] choline[40] and zinc[41] have been related to the development of specific types of cancer. The addition of tryptophan[42] and thamine[43] to the diets of animals showed increased development of bladder cancer but no effect of thiamine on induced intestinal cancers. Early animal studies[44] suggested that artificial sweeteners in large doses were associated with bladder cancer.

Table 1
**THE ROLE OF DIET IN ANIMAL STUDIES OF DEVELOPMENT OF
SPONTANEOUS AND INDUCED CANCER**

Variables	Cancer development			
	Increased	**No effect**	**Decreased**	**Ref.**
Caloric intake				
Underfeeding			X	17
Early weight			X	18
Proteins				
High *ad libitum*				
Adenoma anterior pituitary	X			19
Deficient				
Protects against dismethylnitrosamide but kidney cancer	X			20
Lipotropes				
Deficiency				
Induced liver cancer	X			21
With high fat diet-induced hepatoma	X			22
Fats				
20% vs. 5% fat diets				
Induced colon	X			22
Degree of saturation		X		22
All types of fat		X		23
High polyunsaturates				
Transplanted adenocarcinoma	X			24
High vegetable polyunsaturates				
Induced tumors	X			25
High fat				
Breast	X			25
Fiber				
High wheat bran				
Induced colon			X	26
Induced duodenal		X		26
High refined cellulose				
Induced colon		X		27
Induced small intestine			X	27
Vitamins and minerals				
Vitamin A deficiency				
Salivary gland	X			28
Induced colon (aflatoxin)	X			22
Induced colon (DMH)		X		22
Induced colon (*N*-Methyl-*N*-Nitro-*N*-Nitrosoquandine)			X	29
Induced respiratory	X			30
Induced bladder	X			31
Induced stomach/cervix	X			32
Induced tracheobronchial	X			33
Induced skin/breast	X			34
Ascorbic acid added				
With nitrate containing food			X	35
Induced tumors by nitrite with urea, amines, or *N*-nitroso			X	36
Iodine deficiency				
Thyroid/pituitary	X			37
Magnesium deficiency				
Thymoma	X			38
Leukemia	X			39

Table 1 (continued)
THE ROLE OF DIET IN ANIMAL STUDIES OF DEVELOPMENT OF
SPONTANEOUS AND INDUCED CANCER

Variables	Cancer development			Ref.
	Increased	No effect	Decreased	
Choline deficiency				
Hepatoma	X			40
Zinc deficiency				
Induced esophageal	X			41
Tryptophan added				
Induced bladder	X			42
Thiamin added				
Induced bladder	X			43
Induced intestinal		X		43
Artificial sweetener				
Bladder	X			44

Table 2
ANIMAL STUDIES OF DIETARY MANIPULATION ON GROWTH OF
ESTABLISHED TUMORS

Variables	Growth			Ref.
	Increased	No effect	Decreased	
Protein deprived				
Only with weight loss			X	17
Zinc deficient, no weight loss			X	45
High protein				
Hepatic metastasis and tumor size	X			46
Amino acids				
Some essential ones omitted + weight loss			X	47
Restriction of tryptophan, leucine, methionine + weight loss			X	48
Restriction of phenylalnine, valine, isoleucine without weight loss			X	48
Restriction of lycine, no weight change		X		48
Restriction of phenylalanine-tyrosine, isoleucine or leucine with questionable weight loss	X			49
Vitamins				
Folic acid deficiency			X	50
Pyridoxine deficiency			X	51
Riboflavin deficiency			X	52

ANIMAL STUDIES ON GROWTH OF EXPERIMENTALLY INDUCED TUMORS

Table 2 shows that restriction of certain nutrients have been successful in retarding tumor growth. Again, underfeeding and protein-spare diets have been found to decrease tumor growth.[17] However, in zinc-deficient rats, transplanted tumors were reduced even when weight was not diminished.[45] High protein diets in rats were associated with increased hepatic metastasis and larger tumor size than in control animals fed lower protein diets.[46] Restriction of essential amino acids also has been associated with slower tumor growth in most instances[47,48] but not all.[49] Generally, deprivation had to be accompanied by weight loss in order to achieve reduction of tumor growth. Deficiencies in folic acid,[50] pyridoxine,[51] and riboflavin[52] have inhibited growth of some tumors.

STUDIES IN HUMANS

Epidemiological data have provided many clues to the association of dietary practices and certain types of cancer and cancer incidence. These studies have some limitations that should be pointed out. Often they are based on international food-disappearance data which is then correlated with cancer mortality rates. This assumes that all foods produced, imported, not used for livestock, or not exported are eaten by the population. No adjustments are made for homegrown foods, waste, or differences in consumption known to exist among various age groups. As pointed out earlier, mortality data, particularly for the elderly, may not be accurate or complete depending on the data information systems of the countries involved. Some have suggested that better data might be obtained from individual interviews concerning dietary practices correlated with natural mortality statistics. Problems encountered here relate again to accuracy of national statistics as well as to whether a representative sample of the population has been drawn. Furthermore, 24-hr or weekly recall of food intake, as is often collected, does not take into account whether the intake is representative of the person's usual or prior habits. Neither does it provide details about very early life food patterns that may be important.

Despite these limitations, there is an accumulating and convincing amount of evidence that diet and cancer in man are associated. Table 3 shows that a number of nutrients and substances are correlated with cancer development in humans. Diet has been said to correlate significantly with half of all cancers in women and at least a third in men.

Studies of migrants have been revealing in that rates of certain cancers change dramatically within a few generations from that of those of the original country to those of the new country. For example, in Japan, the incidence of colon and breast cancer are low and stomach cancer high. The opposite is true in the U.S. Japanese migrants to this country change within a few generations to reflect the U.S. statistics.

Within a given population, groups with different dietary patterns differ in rates of cancer, and those with similar patterns have similar rates. For example, Mormons and Seventh Day Adventists have different dietary patterns and different rates of cancer from the U.S. population as a whole.

Ethnic groups provide data on different dietary patterns. Jewish people in the U.S. have higher rates of cancer of the stomach, colon, pancreas, and kidney than other Americans. Changes in rates of cancer over time within the same country also provide clues that link these changes to changes in diet. The decrease in stomach cancer in the U.S. over the last few decades may be related to specific dietary components. Also, the use of more milk and milk products in Japan during the last 10 years appears to have effected a decrease in gastric cancer there.

High fat diets have been studied in regard to cancers of the colon, breast, stomach, ovaries, prostate, and testes.[53-57] Although most studies show colon cancer to be associated with high fat diets,[54,56] a few[55,57] have not. The high incidence in the U.S. compared with Japan corresponds with fat intake between these two countries. Further, the increased rates seen in Japanese migrants to the U.S. support the change to the Western diet. Still further, Japan iself has changed to more Western dietary practices and the rates of colon cancer are increasing there. Recently, interest has focused on cholesterol levels as predictors of colon cancer, and some[58,59] have found high levels associated with colon cancer; others[60] found high levels to be predictive only for men, and others[57] found no association between cholesterol levels and colon cancer.

Women with affluent lifestyles are at higher risk for breast cancer. Again, intake of fats has been implicated in the postmenopausal differences in rates of breast cancer in American

Table 3
HUMAN STUDIES OF DIETARY PRACTICES AND NUTRITION AND RATES OF CANCER

Variables	Cancer development			Ref.
	Increased	No effect	Decreased	
High fat				
Gastric cancer (fish only)	X			53
Colon	X			54
Colon		X		55
Colon	X			56
Colon		X		57
Breast	X			53, 56
Ovary, prostate, testis	X			56
Cholesterol high				
Colon	X			58
Colon	X			59
Colon (men only)	X			60
Colon		X		57
High animal protein				
Breast	X			53
Colon	X			56
Colon (+ fat)	X			61
Colon (beef)	X			62, 63
Colon (beef + bean)	X			64
High fiber				
Colon (cereal but low in meat)			X	56
Colon (crude fiber)		X		55, 65
Colon (bran with loss of acid, steroid, bile salts)			X	66
Colon (among Utah residents)		X		67
High starch				
Gastric	X			68
Iron deficiency				
Hypopharynx (females)	X			69
Iodine deficiency				
Thyroid	X			70
Nitrites				
Gastric (preserved meat/fish)	X			71
Gastric (in fertilizer)	X			72
Gastric (low in stomach of women with pernicious anemia)			X	73
Exposure to aflatoxin				
Liver	X			74
Specific foods				
Cycad nuts	X			75
Bracken fern	X			76
Broccoli, cabbage, brussel sprouts, cauliflower — gastric cancer			X	77
Dried fish — nasopharyngeal, gastric	X			78
Other substances				
Selenium blood levels high			X	79
Zinc blood levels high	X			80
Artificial sweeteners				
Bladder (males)	X			81
Bladder		X		82
Bladder		X		83
Beer (colon cancer)	X			84
Coffee (bladder cancer)	X			85
Coffee		X		86
Vitamin A (colon cancer)		X		87

and Japanese women.[53,56] Dietary fat promotes human prolactin production. Vegetarians show reduced plasma prolactin production. However, pregnant and lactating women who have higher prolactin levels do not have increased rates of breast cancer.

In countries where meat consumption is low and among Seventh Day Adventists and Mormons, colon cancer is also lower than the rest of the U.S. population. It is unclear whether it is the reduced meat intake or dietary fat that is related to low incidence. Several studies point to animal protein as a factor in colon cancer.[56,61] Evidence that beef may be a factor in colon cancer comes from the high rates of this disease where beef consumption is high.[62,63,64] Vegetarians have diets 25% less in fat and 50% less in fiber than nonvegetarians and show low colon cancer rates.

The observation that colon cancer is very rare in primitive and developing countries whose diets had high fiber content led to the theory that fiber was protective in that it increased stool bulk and decreased intestinal transit time.[56,66] However, some have not found that fiber prevented colon cancer.[55,65,67] Bile acids in the colon are largely dependent on dietary intake. Populations with a high incidence of colon cancer generally excrete high concentrations of bile acids as well as have a high fat diet. Patients with colon cancer excrete more bile acids than noncancer patients. It appears that diet acts as a modifying factor by increasing bile acid and affecting stool bulk, which relates to fiber intake. Dietary fat intake in Finland is similar to high colon cancer rate countries, yet they have low rates of colon cancer. Finland has a high intake of fiber, but the fat in the Finnish diet comes from dairy products rather than meat.

Diets high in starch may be predictive of gastric cancer.[68] Iron-deficient diets[69] are associated with cancer of the hypopharynx in women and iodine-deficient diets[70] with thyroid cancers. Nitrates used in preserved meat and fish[71] and in fertilizer in the soil[72] have been related to higher rates of gastric cancer. Low nitrate levels in the stomachs of women with pernicious anemia may also account for their lower rates of gastric cancer than other persons.[73]

High levels of aflatoxin, produced by *Aspergillus flavus*, are contaminants found in dried peanuts and cereal grain stored under warm, moist conditions. In a number of countries, aflatoxin levels are associated with hepatic cancer.[74] Protein deficiency increases toxic effects on the liver.[74]

The cycad nut is carcinogenic[75] and eaten in some countries. Bracken fern, frequently eaten in Japan and a salad delicacy in the U.S., appears to be carcinogenic. Rats have developed cancer of the intestine, kidney, and bladder when fed milk from cows that eat bracken fern.[76] Japanese who eat the fern have a fivefold increase of esophageal cancer.

Where colon cancer and breast cancer appear to be associated with high income, beef- and fat-eating populations, gastric cancer occurs in those with high starch and low fat diets.[68] No single food has been found that is common to countries where gastric cancer is high; however, pickled vegetables, dried salt fish, and nitrite preservatives are suspect. Green vegetables, such as broccoli, cabbage, brussel sprouts, and cauliflower seem to have a protective effect against gastric cancer, possibly because of a group of indols involved.[77] Nasopharyngeal cancers have also been associated with diets high in dried salt fish.[78]

Selenium supplements appear to inhibit tumors in animals. In humans, there is an inverse correlation between blood levels of selenium and cancer mortality, particularly cancers of the colon, rectum, breast, and ovaries.[79] Blood zinc levels, however, have been positively correlated with gastric cancer development.[80]

The use of artificial sweeteners and increased cancer of the bladder in men was reported;[81] however, later studies[82,83] found no association between the two. Beer has been implicated in cancer of the colon,[84] and coffee was found to correlate with cancer of the bladder in one study[85] but not in another.[86] A recent comparison of a large group of colon cancer patients with noncancer patients revealed no differences in prior use of Vitamin A in their diets.[87]

DIETARY INFLUENCES OF CANCER IN OLD AGE

Since cancer is a disease of old age, any dietary practices that are associated with the onset of cancer might also be considered those associated with old age and cancer. At the same time, there is the question of how early dietary practices affect development of cancer in later life. Aging is a process that begins with conception and ends with death. Preventive nutrition for aging should ideally begin before birth and continue through childhood and on into old age. There is evidence, at least from animal studies, that nutrition in very early life may predispose to cancer as well.

Older people vary more from each other than do persons in any other age group. It is not clear at the present time whether the healthy elderly have any special nutritional needs that are different from younger adults, assuming that caloric intake is proportional to energy expenditure. When nutritional needs differ, it is usually the result of medical, psychological, social, and economic factors that interact with aging.

The relationship of age and incidence of cancer has several implications for nutrition. It has been suggested[88] that a given cancer may have several factors contributing to its cause. It may be the end of many events occurring over a person's lifetime in which the nutritional environment in early life may help to explain the role of dietary factors in cancer. Therefore, the incidence of certain common cancers may be partly determined by nutritional habits in childhood. Lastly, it could take years before an increased incidence of a particular cancer calls attention to the dangers of dietary factors. The fact that a large number of cancers may be related to dietary practices also offers hope that there is the possibility of intervention that could lead to decreased rates of cancer in years to come.

Immune responses have been implicated in the process of aging. Immune competence declines with age and autoimmune responses increase. Protein-calorie malnutrition is the most common form of acquired immunodeficiency.[89] It is evident that decreased immune responses associated with nutritional deficiencies may have a deleterious effect upon resistance to infection. It has been suggested[90] that the role of nutrients in immunological processes may have broader significance to both aging and tumorogenesis. The growth and survival of a tumor may be influenced by the type and magnitude of the host's response to specific tumor antigens. This forms the ''surveillance'' reaction which prevents tumor initiation and proliferation via cell-mediated immune response. At the same time, the surveillance reaction may be inhibited by blocking antibodies resulting from the stimulus of tumor antigens. The complex relationship between nutrition and tumorigenesis is demonstrated by the possibility that specific nutrients may influence either or both of these immune processes. As Axelrod pointed out,[90] a decrease in tumor incidence in malnutrition could result from decreased formation of blocking antibody or from an increased activity of T cell function, with a resultant increased tumor surveillance. Experimental evidence exists to support either possibility.

Nutritional deficiency may produce abnormalities in the maturation of cells responsible for the immune response which may continue to affect the cells for a long time. Nutrition of the mother may affect the child. Prolonged or permanent impairment of immunocompetence may occur with possible long-term effects on susceptibility to infection, incidence of cancer, and even the aging process itself.

The interaction between longevity, immunity, and diet is complex. There appears to be a fine line between effects of frank malnutrition as opposed to marginal undernutrition. Malnutrition or severe nutritional stress, particularly early in life, is associated with increased incidence of disease[91] Lifespan is decreased when diets are severely low in protein and vitamins.[92] On the other hand, underfeeding or marginal undernutrition has produced enhanced cellular immunity, and life-long moderate dietary restrictions has increased longevity of laboratory animals.[93] Walford et al.[94] has suggested that dietary restriction causes a delay in maturation of the immune system and that it is this delay that increases longevity. Watson and Safranski[95] have pointed out that any agent that can slow the aging process can affect the incidence of the disease or for that matter any age-related functional change.

It is dangerous to try to extrapolate data from animal studies to man. It seems clear that obesity is a cancer risk in animals, and it is known to be a risk factor in cancer of the gastric fundus and cancer of the breast in older women. Data also indicate that about one fourth of the males and almost half of the females between age 60 to 79 in the U.S. are at least 20% over ideal weight.[96] Further, in a country like Greece where nutritional deficiencies are rare and where leanness is a characteristic of the rural population, the overall cancer mortality is among the lowest in the WHO ranking. It is important to clearly distinguish, however, between relatively low caloric intake in diets which include all major micronutrients and low caloric intakes in diets associated with specific vitamin deficiencies.

In taking an overview to animal and human studies of cancer and diets, it can be seen that conflicting findings occur frequently. Yet, some consistent patterns emerge. High fat diets, for example, are often associated with carcinogenesis. We are exposed to many substances in foods that occur naturally, mostly as contaminants, which can act as carcinogens or cofactors in inducing cancer. Dietary patterns exist which include *deficiencies* which predispose to one type of cancer and at the same time appear to protect against another. Dietary patterns are also found which include nutritional *excesses* which again predispose to one type of cancer and yet may protect against another.

The only reasonable approach would seem to be one of a moderate dietary pattern. This would include moderate restriction of calories and animal fat and a higher ratio of polyunsaturated to saturated fats. Diets should provide the necessary vitamin and mineral requirements and calories sufficient to meet energy expenditures without adding pounds to body weight. This prescription would apply to persons of all ages, not just the elderly, and might apply not only to prevention of cancer but to atherosclerosis as well. Existing evidence points to intrauterine and postnatal periods of life as those during which nutrition has the greatest impact on aging and possibly development of cancer. Prevention of chronic disease by nutritional planning from adolescence through late maturity offers some hope in terms of health in old age. Adequate nutrition in old age is important to the well-being of the older person. Whether intervention at that stage of the life cycle, when earlier dietary practices have been less than ideal, is able to alter the adverse effects of aging or forestall the development of cancer has not as yet been demonstrated.

Although diet appears to be associated with cancer development, this does not mean that other factors do not interact synergistically with diet to compound its effect. Research should be focused on defining desirable dietary intake and nutrients necessary to maintain health as well as dietary practices that produce chronic effects that lead to disease and cancer in particular.

REFERENCES

1. **Peto, R., Roe, F. J., Lee, P. N., Levy, L., and Clack, J.,** Cancer and ageing in mice and men, *Br. J. Cancer,* 32, 411, 1975.
2. **Cutler, S. J. and Young, J. L., Jr.,** Third National Cancer Survey: Incidence Data, Natl. Cancer Institute Monograph No. (NIH) 75-787, U.S. Government Printing Office, Washington, D.C., 1975.
3. **Cutler, S. J. and Eisenberg, H.,** Cancer in the aged, *Ann. N.Y. Acad. Sci.,* 114, 771, 1964.
4. **Waisman, J. and Mott, L. J. M.,** Pathology of neoplasms of the prostate gland, in *Genitourinary Cancer,* Skinner, D. G. and deKernion, J. B., Eds., W. B. Saunders, Philadelphia, 1978, 310.
5. **Schottenfeld, D. and Robbins, C. F.,** Breast cancer in elderly women, *Geriatrics,* 26, 121, 1971.

6. **Siegler, E. E.,** Cervical carcinoma in the aged, *Am. J. Obstet. Gynecol.,* 103, 1093, 1969.

7. **Breen, J. L.,** Gynecologic oncology in the aged, *Clin. Obstet. Gynecol.,* 10, 498, 1967.

8. **Cady, B., Sedgwick, C. E., Meissner, W. A., Wool, M. S., Salzman, F. A., and Werber, J.,** Risk factor analysis in differentiated thyroid cancer, *Cancer,* 43, 810, 1979.

9. **Boggs, D. R, Sofferman, S. A., Wintrobe, M. M., and Cartwright, G. E,** Factors influencing the duration of survival of patients with chronic lymphocytic leukemia, *Am. J. Med.,* 40, 243, 1966.

10. **Doll, R.,** The age distribution of cancer: implications for models of carcinogenesis, *J. R. Stat. Soc.,* 134, 133, 1971.

11. **Lew, E. A.,** Cancer in old age, *Cancer,* 28, 2, 1978.

12. **Linn, M. W., Linn, B. S., and Gurel, L.,** Physical resistance in the aged, *Geriatrics,* 22, 134, 1967.

13. **Linn, B. S., Linn, M. W., and Gurel, L.,** Physical resistance and longevity, *Gerontol. Clin.,* 11, 362, 1969.

14. **Linn, M. W., Linn, B. S., and Gurel, L.,** Patterns of illness in persons who lived to extreme old age, *Geriatrics,* 27, 67, 1972.

15. **Butler, R. N. and Gastel, B.,** Aging and cancer management. II. Research perspectives, *Cancer,* 29, 333, 1979.

16. **Ponten, J.,** Abnormal cell growth (neoplasia) and aging, in *Handbook of the Biology of Aging,* Finch, E. E. and Hayflick, L., Eds., Van Nostrand Reinhold, N.Y., 1977, chap. 22.

17. **Tannenbaum, A.,** The genesis and growth of tumors. Effects of a high fat diet, *Cancer Res.,* 2, 468, 1942.

18. **Ross, M. H. and Bras, G.,** Tumor incidence patterns and nutriton in the rat, *J. Nutr.,* 87, 245, 1965.

19. **Ross, M. H., Bras, G., and Ragbeer, M. S.,** Influence of protein and caloric intake upon spontaneous tumor incidence of the anterior pituitary gland of the rat, *J. Nutr.,* 100, 177, 1970.

20. **McLean, A. E. and Magee, P. N.,** Increased renal carcinogenesis by dimethylnitrosamine in protein deficient rats, *Br. J. Exp. Pathol.,* 51, 587, 1970.

21. **Rogers, A. E., Kula, N. S., and Newberne, P. M.,** Absense of an effect of partial hepatectomy on aflatoxin B1 carcinogenesis, *Nature (London),* 229, 62, 1971.

22. **Rogers, A. E. and Newberne, P. M.,** Dietary effects on chemical carcinogenesis in animal models for colon and liver tumors, *Cancer Res.,* 35, 3427, 1975.

23. **Carroll, K. K.,** Experimental evidence of dietary factors and hormone dependent cancer, *Cancer Res.,* 35, 3374, 1975.

24. **Hopkins, G. J. and West, C. E.,** Effect of dietary polyunsaturated fat on the growth of transplantable adenocarcinoma in C3HAvyfB mice, *J. Natl. Cancer Inst.,* 58, 753, 1977.

25. **King, M. M., Bailey, D. M., Gibson, D. D, Pitha, J. V., and McCay, P. B.,** Effect of antioxidant in diets containing different types and amounts of fat on mammary tumor incidence induced by a single dose of 7, 12-dimethyl-benzane-thracene, *Fed. Proc.,* 36, 1148, 1977.

26. **Barbolt, T. A. and Abraham, R.,** The effect of bran on dimethylyydrazine-induced colon carcinogenesis in the rat, *Proc. Soc. Exp. Biol. Med.,* 157, 656, 1978.

27. **Ward, J. M., Yamamoto, R. S, and Weisburger, J. H.,** Cellulose dietary bulk and azoxymethane-induced intestinal cancer, *J. Natl. Cancer Inst.,* 51, 713, 1973.

28. **Rowe, N. H., Grammer, F. C., and Watson, F. R.,** A study of environmental incidence upon salivary gland neoplasia in rats, *Cancer,* 26, 436, 1970.

29. **Reddy, B. S., Weisburger, J. H., and Wynder, E. L.,** Animal models for the study of dietary factors and cancer of the large bowel, *Cancer Res.,* 35, 3426, 1975.

30. **Nettesheim, P. and Williams, M. L.,** The influence of vitamin A on the susceptibility of the rat lung to 3-methylcholanthrene, *Int. J. Cancer,* 17, 351, 1976.

31. **Cohen, S. M., Wittenberg, J. F., and Bryan, G. T.,** Effect of avitaminosis A and hypervitaminosis A on urinary bladder carcinogencity of *N*-(4-[5-Nitro-a Furyl]-2-Thiazolyl) Formamide, *Cancer Res.,* 36, 2334, 1976.

32. **Chu, E. W. and Malmgren, R. A.,** An inhibitory effect of vitamin A on the induction of tumor of forestomach and cervix in the Syrian hamster by carcinogenic polycyclic hydrocarbons, *Cancer Res.,* 25, 884, 1965.

33. **Saffiotti, U., Montesano, R., Sellakumar, A. R., and Borg, S. A.,** Experimental cancer of the lung: inhibition by vitamin A of the induction of tracheobronchial squamous metaplasia and squamous cell tumors, *Cancer,* 20, 857, 1967.

34. **Shamberger, R. J.,** Inhibitory effect of vitamin A on carcinogenesis, *J. Natl. Cancer Inst.,* 47, 667, 1971.

35. **Newberne, P. M.,** Nitrate promotes lymphoma incidence in rats, *Science,* 204, 1079, 1979.

36. **Mirvish, S. S., Cardesa, A., Wallcave, L., and Shubik, P.,** Induction of mouse lung adenomas by amines or ureas plus nitrite and by *N*-nitroso compounds: effects of ascorbate, gallic acid, thiocyanate, and caffeine, *J. Natl. Cancer Inst.,* 55, 633, 1975.

37. **Axelrod, A. A. and Leblond, C. P.,** Induction of thyroid tumors in rats by low iodine diet, *Cancer,* 8, 339, 1955.

38. **Bois, P., Sandborn, E. B., and Messier, P. E.,** A study of thymic lymphosarcoma development in magnesium deficient rats, *Cancer Res., 29, 763, 1969.*

39. **Battifora, H. A., McCreary, P. A., Hahneman, B. M., Laing, G. H., and Hass, G. M.,** Chronic magnesium deficiency in the rat. Studies of chronic myelogenous leukemia, *Arch. Pathol., 86, 610, 1968.*

40. **Copeland, D. H. and Salmon, W. D.,** Occurrence of neoplasms in liver, lungs, and other tissues of rats as a result of prolonged choline deficiency, *Am. J. Pathol., 22, 1059, 1946.*

41. **Fong, L. Y., Sivak, A., and Newberne, P. M.,** Zinc deficiency and methylbenzylnitrosamine-induced esophageal cancer in rats, *J. Natl. Cancer Inst., 61, 145, 1978.*

42. **Dunning, W. F., Curtis, M. R., and Mann, M. E.,** Effects of added dietary trypophane on occurrence of Z-acetylaminoflourine-induced liver and bladder cancer in rats, *Cancer Res., 10, 454, 1950.*

43. **Pamukcu, A. M., Yalciner, S., Price, J. M., and Bryan, G. T.,** Effects of the coadministration of thiamin on the incidence of uinary bladder carcinomas in rats fed bracken fern, *Cancer Res., 30, 2671, 1970.*

44. **Price, J. M. and Biava, C. G.,** Bladder tumors in rats fed cyclohexylamine or high doses of a mixture of cyclamate and saccharin, *Science, 167, 1131, 1970.*

45. **DeWys, W., Pories, W. J., Richter, M. C., and Strain, W. H.,** Inhibition of walker 256 carcinosarcoma growth of dietary zinc deficiency, *Proc. Soc. Exp. Biol., 135, 17, 1970.*

46. **Fisher, B. and Fisher, C. R,** Experimental studies of factor influencing hepatic metastases: effect of nutrition, *Cancer, 14, 547, 1961.*

47. **Sugimura, T., Birnbaum, S. M., Winitz, M., and Greenstein, J. P.,** Quantitative nutritional studies with water soluble chemically defined diets: forced feeding of diets each lacking one essential amino acid, *Arch. Biochem. Biophys., 81, 439, 1959.*

48. **Theuer, R. C.,** Effect of essential amino acid restriction on the growth of female C57BL mice and their implanted BW10232 adenocarcinomas, *J. Nutr., 101, 223, 1971.*

49. **Worthington, B. S., Surotuck, J. A., and Ahmed, S. I.,** Effects of essential amino acid deficiencies on syngeneic tumor immunity and carcinogenesis in mice, *J. Nutr., 108, 1402, 1978.*

50. **Herbert, V., Colman, N., and Jacob, E.,** Folic acid and Vitamin B_{12}, in *Modern Nutrition in Health and Disease,* Goodhart, R. S. and Shils, M. E., Eds., Lea & Febiger, Philadelphia, 1980, 229.

51. **Rosen, R., Mihich, E., and Nichol, C. A.,** Selective metabolic and chemotherapeutic effects of vitamin B_6 antimetabolites, *Vitam. Horm., 22, 609, 1964.*

52. **Morris, H. P. and Robertson, W. V.,** Growth rate and number of spontaneous mammary carcinoma and riboflavin concentration of liver, muscle, and tumor of C3H mice as influenced by dietary riboflavin, *J. Natl. Cancer Inst., 3, 479, 1943.*

53. **Kolonel, L. N., Hankin, J. H., Lee, J., Chu, S. Y., Nomura, A. M., and Hinds, M. W.,** Nutrient intakes in relation to cancer incidence in Hawaii, *Br. J. Cancer, 44, 332, 1981.*

54. **Wynder, E. L.,** Nutrition and cancer, *Cancer Res., 35, 3388, 1975.*

55. **Bingham, S., Williams, D. R., Cole, T. J., and James, W. P.,** Dietary fibre and regional large bowel cancer mortality in Britain, *Br. J. Cancer, 40, 546, 1979.*

56. **Armstrong, B. and Doll, R.,** Environmental factors and cancer incidence and mortality in different countries, with special reference to dietary practices, *Int. J. Cancer, 15, 617, 1975.*

57. **Kolonel, L. N., Hankin, J. H., Nomura, A. M., and Chu, S. Y.,** Dietary fat intake and cancer incidence among five ethnc groups in Hawaii, *Cancer Res., 41, 3727, 1981.*

58. **Cruse, P., Lewin, M., and Clark, C. G.,** Dietary cholesterol is cocarcinogenic for human colon cancer, *Lancet, 7, 752, 1979.*

59. **Liu, K., Moss, D., Persky, V., Stamler, J., Garside, D., and Soltero, I.,** Dietary cholesterol, fat, and fibre and colon-cancer mortality, *Lancet, 13, 782, 1979.*

60. **Williams, R. R., Sorlie, P. D., Feinleib, M., McNamara, P. M., Kannel, W. B., and Dawber, T. R.,** Cancer incidence by levels of cholesterol, *JAMA, 245, 247, 1981.*

61. **Gregor, O., Toman, R., and Prusova, F.,** Gastrointestinal cancer and nutrition, *Gut, 10, 1031, 1969.*

62. **Berg, J. W. and Howell, M. A.,** The geographic pathology of bowel cancer, *Cancer, 34S, 807, 1974.*

63. **Enstrom, J. E.,** Colorectal cancer and consumption of beef and fat, *Br. J. Cancer, 32, 432, 1975.*

64. **Haenszel, W., Berg, J. W. Segi, M., Kurihara, M., and Locke, F. B.,** Large bowel cancer in Hawaiian-Japanese, *J. Natl. Cancer Inst., 51, 1765, 1973.*

65. **Drasar, B. S. and Irving, D.,** Environmental factors and cancer of the colon and breast, *Br. J. Cancer, 27, 167, 1973.*

66. **Kern, F., Jr., Birkner, H. J., and Ostrower, V. S.,** Binding of bile acids by dietary fiber, *Am. J. Clin. Nutr., 31, 5175, 1978.*

67. **Lyon, J. L. and Sorenson, A. W.,** Colon cancer in a low-risk population, *Am. J. Clin. Nutr., 31, S227, 1978.*

68. **Graham, S.,** Future inquiries into the epidemiology of gastric cancer, *Cancer Res., 35, 3464, 1975.*

69. **Larrson, L., Sandstrom, G., and Westling, P.,** Relationship of Plummer-Vinson's disease to cancer of the upper alimentary tract in Sweden, *Cancer Res., 35, 3303, 1975.*

70. **Cowdry, E. V.,** Malignant neoplasms of thyroid gland, in *Etiology and Prevention of Cancer in Man*, Cowdry, E. V., Ed., Appleton, N.Y., 1968, 277.
71. **Issenberg, P.,** Nitrite, nitrosamines, and cancer, *Fed. Proc.*, 35, 1322, 1976.
72. Nitrate and human cancer (editorial), *Lancet*, 6, 281, 1977.
73. **Ruddell, W. S., Bone, E. S., Hill, M. J., and Walters, C. L.,** Pathogenesis of gastric cancer in pernicious anaemia, *Lancet*, 11, 521, 1978.
74. **Linsell, C. A. and Peers, F. G.,** Field studies on liver cell cancer, in *Origin of Human Cancer*, Hiatt, H. H., Watson, J. D., and Winsten, J. A., Eds, Cold Spring Harbor Lab., N.Y., 1977, 549.
75. **Miller, J. A. and Miller, E. C.,** Carcinogens occurring naturally in foods, *Fed. Proc.*, 35, 1316, 1976.
76. **Pamukcu, A. M., Erturk, E., Yalciner, S., Milli, U., and Bryan, G. T.,** Carcinogeneic and mutagenic activities of milk from cows fed bracken fern, *Cancer Res.*, 38, 1556, 1978.
77. **Wattenberg, L. W., Loub, W. D., Lam, L. K., and Spier, J. L.,** Dietary constituents altering responses to chemical carcinogens, *Fed. Proc.*, 35, 1327, 1976.
78. **Lanier, A., Bender, T., Talbot, M., Wilmeth, S., Tschopp, C., Henle, W., Ritter, D., and Terasaki, P.,** Nasopharyngeal carcinoma in Alaskan Eskimos, Indians, and Aleuts, *Cancer*, 46, 2100, 1980.
79. **Schrauzer, G. N. and Ishmael, D.,** Effects of selenium and of arsenic on the genesis of mammary tumors in inbred C3H mice, *Ann. Clin. Lab. Sci.*, 4, 441, 1974.
80. **Stocks, P. and Davies, R. I.,** Zinc and copper content of soils associated with the incidence of cancer of the stomach and other organs, *Br. J. Cancer*, 18, 14, 1964.
81. **Howe, G. R., Burch, J. D., Miller, A. B., Morrison, B., Gordon, P., Weldon, L., Chambers, L. W., Fodor, G., and Winsor, G. M.,** Artificial sweeteners and human bladder cancer, *Lancet*, 17, 578, 1977.
82. **Morrison, A. S. and Buring, J. E.,** Artificial sweeteners and cancer of the lower urinary tract, *N. Engl. J. Med.*, 302, 537, 1980.
83. **Hoover, R. N. and Strasser, P. H.,** Artificial sweeteners and human bladder cancer, *Lancet*, 19, 837, 1980.
84. **McMichael, A. J., Potter, J. D., and Hetzel, B. S.,** Time trends in colorectal cancer mortality in relation to food and alcohol consumption, *Int. J. Epidemiol.*, 8, 295, 1979.
85. **Howe, G. R., Burch, J. D., Miller, A. B., Cook, G. M., Esteve, J., Morrison, B., Gordon, P., Chambers, L. W., Fodor, G., and Winsor, G. M.,** Tobacco use, occupation, coffee, various nutrients, and bladder cancer, *J. Natl. Cancer Inst.*, 64, 701, 1980.
86. **Morrison, A. S.,** Geographic and time trends of coffee imports and bladder cancer, *Europ. J. Cancer*, 14, 51, 1977.
87. **Smith, P. G. and Jick, H.,** Cancers among users of preparations containing vitamin A, *Cancer*, 42, 808, 1978.
88. **Marks, P. A.,** Nutrition and the cancer problem, *Curr. Concepts Nutr.*, 6, 7, 1977.
89. **Watson, R. R. and McMurray, D. N.,** Effects of malnutrition on secretory and cellular immunity, *CRC Crit. Rev. Food Sci. Nutr.*, 12, 113, 1979.
90. **Axelrod, A. E.,** Nutrition in relation to immunity, in *Modern Nutrition in Health and Disease*, Goodhart, R. S. and Shils, M. E., Eds., Lea & Febiger, Philadelphia, 1973, chap. 18.
91. **Scrimshaw, N. S., Taylor, C. E., and Gordon, J. E.,** Interactions of nutrition and infection, *WHO Monogr. Ser.*, 57, 3, 1968.
92. **Watson, R. R., Rister, M., and Baehner, R. L.,** Superoxide dismutase activity in polymorphonuclear leukocytes and alveolar macrophages of protein malnourished rats and guinea pigs, *J. Nutr.*, 106, 1801, 1976.
93. **Cooper, W. C., Good, P. A., and Mariani, T.,** Effect of protein insufficiency on immune responsiveness, *Am. J. Clin. Nutr.*, 27, 647, 1974.
94. **Walford, R. L., Liu, R. K., Gerbase-DeLima, M., Mathies, M., and Smith, G. S.,** Long-term dietary restriction and immune function in mice: response of sheep red blood cells and to mitogenic agents, *Mech. Ageing Dev.*, 2, 447, 1972.
95. **Watson, R. R. and Safranski, D. V.,** Dietary restrictions and immune responses in the aged, in *Handbook of Immunology in Aging*, Kay, M. and Makinodan, T., Eds., CRC Press, Boca Raton, 1981, 125.
96. Metropolitan Life Insurance Co., N.Y., Frequency of overweight and underweight, *Stat. Bull.* 41, 4, 1960.

NUTRITIONAL SUPPORT THERAPY IN GERIATRIC CANCER PATIENTS

Nathaniel Ching, Carlo Grossi, Helen Zurawinsky,
Christopher Mills, and Thomas Nealon, Jr.

INTRODUCTION

Cancer depresses the nutritional status of many patients in the geriatric group over 60 years of age. To aid in identifying the special needs of these patients, a study was carried out on nutritional defects and their correction in a group of 60 patients undergoing cancer treatment. A nutritional support program has been developed to help cancer patients tolerate the rigors of surgical therapy. A significant percentage of our cancer patients are in the geriatric age group and the design and maintenance of nutritional support plays an important role in their care.

MATERIALS AND METHODS

A group of 60 patients undergoing cancer therapy in the hospital were studied. They ranged in age from 60 to 86 years and were undergoing various modalities of cancer therapy, i.e., surgery, radiation, or chemotherapy. They all showed evidence of malnutrition and hypoalbuminemia.

The hospital's computer facility provided our nutritional chemotherapy unit with a daily printout of all serum albumin levels below 3.5 g/100 mℓ. This enabled us to identify the patients in need of nutritional support and to follow their progress. The charts for all patients appearing on the computer list were examined to identify those with a final diagnosis of cancer, and whether or not they were treated by our unit.

The study included 60 patients (aged 60 to 86) who received intensive nutritional therapy to help support primary surgical, chemotherapy, or radiation therapy for cancer. The patients in this age group represented 65% of the total 85 patients undergoing the same protocol of nutritional support therapy as an adjuvant to cancer treatment.

Among the geriatric patients, 40 of 60 were in the 60 to 69 age group, 17 of 60 in the 70 to 79 age group, and 3 of 60 in the 80+ group. The primary cancers were chiefly located as follows: (1) GI tract; 29 of 60, (2) head and neck region; 11 of 60, and (3) lung; 10 of 60. There were only a few cancers of the breast (3 of 60), lymphoma (4 of 60), and GU tract (3 of 60). Most of the patients (75%) had metastatic cancer and were undergoing palliative chemotherapy and/or radiotherapy.

The nutritional status of these patients was maintained with (1) the hospital's regular diet, (2) peripheral amino acids (with or without sources of nonprotein energy), (3) total parenteral nutrition (TPN), or (4) commercially available, high carbohydrate, high protein, low residue, enteral diets (Vivonex-HN or Precision-HN).[1,2] Combinations of the regular diet and supplements with the oral enteral diets were necessary in some instances to maintain adequate nutrition. In general, a daily intake of at least 2000 to 3000 kcal was the ultimate goal. Body weight, the levels of serum albumin, serum fatty acids and plasma amino acids, and the 24-hr urinary excretion of creatinine and urea nitrogen were sequentially determined to gauge the efficacy of the nutritional therapies.

Total parenteral nutrition (TPN) was provided under the hospital's standard TPN protocol, utilizing 500 mℓ travasol 8.5% amino acid solution mixed with 500 mℓ of 50% G/W; electrolytes were added as indicated.[3] This hypertonic solution was infused via standard subclavian central venous catheter. The enteral low residue diets[4,5] employed were Vivonex-HN (Eaton Laboratories, crystalline amino acids were the source of nitrogen) or Precision-

HN (The Doyle Co., egg albumin was the source of nitrogen). These were instilled via continuous silastic feeding tubes (Keofeed, HEDECO Co., Palo Alto, Calif.) when they provided the entire caloric intake. When these diets were employed as supplements the patients drank them *ad libitum*; 300 cal would be offered between meals and at bedtime. Crystalline amino acids (4.25%) and 5% glucose were infused at 2 to 3 ℓ volume daily via peripheral vein. A nonprotein energy was provided in the form of 1 unit of intralipid (Cutter Laboratory) via peripheral infusion, the oral intake of 100 kcal of the elemental diet Vivonex-HN and/or the *ad libitum* intake of the standard diet.

RESULTS

Serum albumin — Approximately 10% of the patients admitted to the hospital have a low concentration of serum albumin, and 25% of these hypoalbuminemic levels are associated with the diagnosis of cancer.[6] Hypoalbuminemia was chiefly associated with (1) GI tract tumor (32.2%), (2) lung tumors (18.1%), (3) hematopoietic cancers (14.1%), and (4) pharyngeal-laryngeal tumors (11.8%).

Clinical relationships — Serum albumin levels lower than 3.5 g/100mℓ were noted in 50% of the patients being treated for cancer and receiving nutritional support. A similar percentage was noted in the entire group of 85 patients, irrespective of age. The depression in serum albumin was not related to the age of the patient, but to the bulk and extent of spread of the tumor.

Total parenteral nutrition therapy was of value in the preoperative management of surgical candidates who had various degrees of obstruction to oral intake. TPN provided adequate nutrition to replenish depleted patients and maintain their nutritional status through the catabolic effects of surgical procedures. Despite some degree of GI obstruction, the low-residue elemental diet (Vivonex-HN) proved equal to TPN in preserving serum albumin levels in patients undergoing surgical therapy. There was no significant difference between the serum albumin levels of surgical patients treated with TPN and those treated with Vivonex-HN (elemental diet) (Figure 1). The beneficial effects of the preoperative nutritional support could be maintained only if the tumor and the resultant obstruction could be removed and corrected by the surgical procedure.

Not all diets had the same nutritional effect when employed to support metastatic cancer patients undergoing palliative chemotherapy and/or radiotherapy. Serum albumin concentration was best preserved when adequate levels of a regular diet could be ingested, or a regular diet supplemented with the artificial enteral diets (Vivonex-HN or Precision-HN). Significant decreases in serum albumin levels occurred during the periods of chemotherapy and/or radiotherapy for the patients nutritionally supported only with peripheral amino acids ($\overline{3.5}$ g/100 mℓ ± 0.4 to $\overline{3.0}$ g/100 mℓ ± 0.4; $p < 0.05$), total parenteral nutrition ($\overline{3.7}$ g/100 mℓ ± 0.3 to $\overline{2.8}$ g/100 mℓ ± 0.4; $p < 0.02$), or the enteral diets ($\overline{3.3}$ g/100 mℓ ± 0.4 to $\overline{2.8}$ g/100 mℓ ± 0.7; $p < 0.02$). There was no significant change in serum albumin levels when these patients were maintained on the hospital's regular diet ($\overline{3.8}$ g/100 mℓ ± 0.4 to $\overline{3.6}$ g/100 mℓ ± 0.3), peripheral amino acids supplemented with Vivonex-HN or a regular diet ($\overline{3.4}$ g/100 mℓ ± 0.4 to $\overline{3.3}$ g/100mℓ ± 05.), or the regular diet plus elemental diet supplement ($\overline{3.5}$ g/100 mℓ ± 0.7 to $\overline{3.4}$ g/100 mℓ ± 0.6).

The serum albumin levels in the first 30 patients were analyzed and related to the mortality rates. Table 1 shows the changes. The serum albumin levels of the survivors averaged 3.4 g/100 mℓ ± 0.5 on admission and 3.5 g/100 mℓ ± 0.5 on discharge. There was no statistical difference between the admission and discharge values. This was in contrast to the data on patients who did not survive the therapies; their serum albumin levels averaged 3.5 g/100 mℓ ± 0.3 on admission and decreased to an average of 2.9 g/100 mℓ ± 0.6 upon discharge. This decrease was significant at the $p < 0.01$ level.

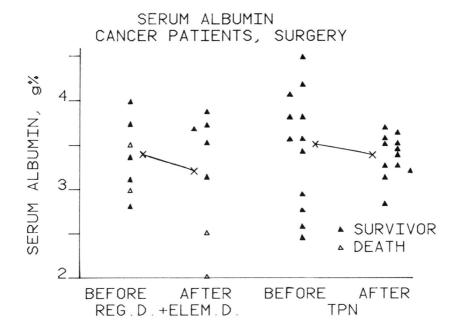

FIGURE 1. The changes in serum albumin levels of cancer patients undergoing surgical treatment. The nutrition was maintained with (1) the standard diets supplemented with enteral elemental diets (Reg. D. + Elemental D.) or Total Parenteral Nutrition (TPN). There was a minimal but not significant decrease in serum albumin utilizing either dietary regimen.

Table 1
**SERUM ALBUMIN LEVELS AND
SURVIVAL**

	No. of patients	Serum albumin levels (g/100 mℓ)	
		Admission	Discharge
Survivors	20	3.4 ± 0.5	3.5 ± 0.5
Nonsurvivors	10	3.5 ± 0.3	2.9 ± 0.6

Only four deaths could be associated with nutritional therapy. One was that of a woman in whom hyperchloremic acidosis developed as a result of excessive oral ingestion of an elemental diet; more careful monitoring could have prevented such an occurrence. The other 3 patients had hyperbilirubinemia and were unable to tolerate the infusion of an amino acid solution; they were virtually in a terminal state before the start of therapy.

DISCUSSION

The persistence of malnutrition which develops in the course of the care of the cancer patient is well demonstrated by the high proportion (25%) of such patients admitted with depressed serum albumin levels. Without adequate nutritional support, any surgical therapy, chemotherapy, or radiotherapy results in further lowering of the serum albumin concentration. Analysis of the data on serum albumin of cancer patients with values lower than 3.5 g/100 mℓ revealed that only 20% were attributable to the terminal state of the patients; many factors were responsive to nutritional support as an adjunct to conventional cancer therapy. The course of the malnutrition seen in many cancer patients has been demonstrated by several observers.[7-11]

Cancer patients who are older than 60 show the same nutritional deficits as other cancer patients and require the same nutritional support. They can tolerate nutritional support but require more careful monitoring and special encouragement. Fortunately, serum albumin levels can be best preserved in those patients who can tolerate adequate amounts of a regular diet or a regular diet supplemented with a special low-residue liquid diet. Both of these techniques are applicable in the outpatient situation or a nursing-home environment and do not require frequent hospitalization for specialized care. When spontaneous oral intake decreases, special silastic feeding tubes can be used with minimal discomfort to the patient and no compromise of respiratory function. Chemotherapy or radiotherapy outpatients can be adequately supported nutritionally by using combinations of these enteral diets. The more dangerous and expensive total parenteral nutrition technique can be reserved for selected and more critical phases in the care of these patients.

SUMMARY

Nutritional deficits were studied in 60 cancer patients over the age of 60 who were undergoing various modalities by a variety of special diets. These deficits were corrected by the use of appropriate enteral formulas. This group of older patients tolerated nutritional support therapy, especially when it was carefully monitored. This method lends itself to successful use in the outpatient or nursing home situation in patients with advanced cancer.

ACKNOWLEDGMENT

This study was supported by Grant No. CA 19632, awarded by the National Cancer Institute, Department of Health, Education and Welfare of the John F. Lindsley and the John V. Mara Fund.

REFERENCES

1. **Nealon, T. F., Jr., Grossi, C. E., and Steir, M.,** Use of elemental diets to correct catabolic states prior to surgery, *Ann. Surg.,* 180, 9, 1974.
2. **Mitty, W. F., Jr., Nealon, T. F., Jr., and Grossi, C. E.,** Use of elemental diets in surgical cases. *Am. J. Gastroenterol.,* 65, 297, 1976.
3. **Caldwell, M. D., O'Neil, J. A., Meng, H. C., and Stahlman, M. H.,** Evaluation of a new amino acid source for use in parenteral nutrition, *Ann. Surg.,* 185, 153, 1977.
4. **Copeland, E. M., III, Daly, J. M., Ota, D. M., and Dudrick, S. J.,** Nutrition, cancer and intravenous hyperalimentation, *Cancer,* 43, 2108, 1979.
5. **Stephens, R. V. and Randall, H. T.,** Use of concentrated, balanced liquid elemental diet for nutritional management of catabolic states, *Ann. Surg.,* 170, 642, 1969.
6. **Ching, N., Grossi, C. E., Zurawinsky, H., Jham, G., Angers, J., Mills, C., and Nealon, T. F., Jr.,** Nutritional deficiencies and nutritional support therapy in geriatric cancer patients, *J. Am. Ger. Soc.,* 27, 491, 1979.
7. **Schwartz, G. F., Green, H. L., Benton, M. L. et al.,** Combined parenteral hyperalimentation and chemotherapy in the treatment of disseminated solid tumors, *Am. J. Surg.,* 121, 169, 1971.
8. **Dematteis, R. and Herman, R. E.,** Supplementary parenteral nutrition in patients with malignant disease, *Cleveland Clin. Q.,* 40, 139, 1973.
9. **Copeland, E. M., III, MacFayden, B. V., and Dudrick, S. J.,** Intravenous hyperalimentation as an adjunct to cancer chemotherapy, *Am. J. Surg.,* 129, 167, 1975.
10. **Copeland, E. M., Couchon, E. A., MacFadyen, B. V., Jr. et al.,** Intravenous hyperalimentation as an adjunct to radiation therapy, *Cancer,* 39, 609, 1977.
11. **Brennan, M. F.,** Total parenteral nutrition in the cancer patient. *N. Engl. J. Med.,* 305, 375, 1981.

OBESITY IN THE AGED

Bertil Steen

INTRODUCTION

Despite certain controversies regarding the relative epidemiological and clinical significance of obesity in relation to morbidity and mortality at different ages, the fact remains that obesity is a very common type of malnutrition in developed countries in the elderly. Most authors agree that the definition of obesity is "too much body fat in relation to body weight".The discrepancies concern the definition of "too much".

The problem and management of obesity in man is, in most respects, very much the same whether the obese individuals are middle aged or old. In this respect the relevance of a special review on obesity at higher age groups can be questioned. Furthermore, the group of elderly — defined as 65 years of age or older — is a very heterogenous one, concerning such factors as prevalence of disease, socioeconomic conditions, and age *per se* with an age span of more than 40 years. Therefore, the grouping together of these heterogenous groups of people may also be questioned.

Recent longitudinal studies[1,2] have shown that many functional parameters remain rather constant through middle life until age 70 to 75, where a more pronounced decrease is seen. To this might be added that there is obviously an increasing variability in functional capacity among individuals during life up to at least age 75. From these points of view the classification of elderly people into "young elderly" (age 65 to 75) and "old elderly" (over 75) seems, therefore, to bear a physiological meaning.

The literature on obesity is abundant. Excellent up-to-date reviews on selected topics can be found (e.g., the proceedings from the three international congresses on obesity).[3,4,5]

The aim of the present chapter is to review some aspects on obesity of special gerontological and geriatric interest.

BODY FAT

The measurement or estimation of body fat can be performed in several ways.[6-8] These include cadaver analyses[9] and methods using absorption of fat-soluble gases such as krypton.[10] Isotope dilution methods to calculate total body water[11] and whole body potassium counting[12] to estimate body cell mass[6] can be used to estimate body fat.[6,13] Furthermore, measurements of subcutaneous fat thickness with roentgenography[14] and calipers[15] are frequently used. It is advantageous to use combined measurements in this respect.

The correlations between simple anthropometric parameters and body fat are usually rather high in the elderly. In a Swedish population study of 70-year-olds,[16] it was shown that such correlations to body fat estimated with a combined approach of total body water measurement using tritiated water and whole body potassium counting, were higher in females than in males regarding body weight, waist girth, upper arm girth, thigh skinfold, and body cell mass. Regarding subscapular and suprailiac skinfolds, the correlations were of the same order of magnitude in both sexes. The correlation coefficients between body fat and body weight, waist girth, and subscapular and triceps skinfold, respectively, ranged from $r = 0.47$ (triceps skinfold in males) to $r = 0.83$ (body weight in females) in that study. The correlation coefficient between body fat and the difference between body weight and "ideal body weight"[17] was $r = 0.71$ in males and $r = 0.91$ in females. A multiple regression analysis with body fat as the dependent variable revealed body weight as the most important independent variable in both sexes.

Energy intake and expenditure vary markedly between different populations, which means that the average amount of body fat — in absolute as well as in relative terms — varies accordingly. In the above-mentioned Swedish study of body composition,[16] the average percentage of body fat relative to body weight at age 70 was 16.1 and 26.6 for males and females, respectively. As judged from comparisons of body weight/height measurements and skinfold measurements between Sweden on the one hand and the U.K. and the U.S. on the other, Swedish people seem to have a lower amount of body fat than people in the two other countries mentioned.[18,19,20]

A longitudinal population study of the same subjects from age to 70 to age 79 in Göteborg, Sweden,[21] showed that the absolute amount of body fat did not change in any noteworthy way between age 70 and 75, but decreased in an order of magnitude of 2 kg between age 75 and 79 in females. A loss of body weight occurred in both sexes between age 70 and 79; this loss corresponded more to a loss of body cell mass in males than in females, and to a decrease of body fat in females.

It is well known that very old people are thin. Many clinicians have interpreted this fact as caused by a selective mortality of obese people. However, at least to some degree, it seems that the explanation is a decreasing degree of fatness in the same individuals. This decrease seems to start somewhat earlier (age 75) in females than in males. This fact has an obvious bearing when discussing indications for the therapy of obesity in the elderly. Some age-dependent changes have been described concerning the distribution of body fat. Thus, Skerlj et al.[22] found that subcutaneous fat was deposited more on the trunk than on the extremities in old compared to young women. Furthermore, Durnin and Womersley[18] described that there was an increase of deep adipose tissue relative to subcutaneous fat with age.

DEFINITIONS OF OVERWEIGHT AND OBESITY

The risks of obesity are still under debate. While it is a general agreement that obesity is associated with a number of major conditions such as coronary heart disease,[23] hypertension,[24] cerebrovascular disease,[25] glucose intolerance,[26] and osteoarthritis of the weight-bearing joints (especially the knees),[27] the relation of obesity to mortality is still somewhat unclear. Again, there is a need for definition of obesity and overweight.

Obesity in women can be defined as a body fat content greater than 30% of body weight, and in males greater than 25%.[28] However, the percentage of body fat to body weight seems to increase to about age 70 to 75 as judged from cross-sectional studies,[16] the clinical significance of this change being unclear.

According to Keys et al.,[29] a common clinically and epidemiologically useful way to express the degree of overweight and obesity is the use of weight/height indices of different kinds. They claim that — on the basis of more than 7000 "healthy" males in 12 cohorts in 5 countries — the ponderal index is the poorest of the relative weight indices studied and the body mass index (weight to height squared) is slightly better than the simple ratio of weight to height. Bray[30] defines overweight as a body mass index (Quetelet's index) between 24 or 25 and 30, and obesity as a body mass index above 30.

RISKS OF OVERWEIGHT AND OBESITY

The well-known Build and Blood Pressure Study from 1959[31] describes an almost linear relation between overweight and excess mortality in both sexes starting from 10% overweight. That study has, however, been subjected to certain criticism.[32] Available data also indicates that the impact of obesity on longevity is most pronounced when the onset of obesity occurs in early life.[33]

Recently, Andres[32] reviewed a number of population studies regarding the relation between obesity in the aged and mortality, including the Framingham study,[34] the Chicago Peoples Gas Company Study,[35] and the Baltimore Longitudinal Study of Aging.[36] He found that the hypothesis that obesity is a graded variable in its impact on mortality, and that even minor degrees of obesity are detrimental, could not be supported by data from the studies reviewed. However, results from crude relation analyses between overweight and mortality have to be interpreted with great caution. They can certainly not tell us about the impact of obesity on morbidity; also, the impact on mortality may be obscure because of possible confounding factors, which can also be different in different populations. A major confounding factor in this respect is tobacco smoking, which is known to be related on the one hand positively to mortality, and on the other negatively to obesity. Preliminary data from the gerontological and geriatric population studies in Göteborg, Sweden, suggests this confounding to be pronounced in the relation analyses between body weight and mortality[37]

ETIOLOGICAL FACTORS

As pointed out by several leading authors in the field such as Björntorp[38] and Bray,[30] the understanding of obesity requires knowledge of the many different etiological and patho-physiological aspects of this condition. Bray suggested an anatomical classification of the syndromes of obesity in generalized or diffuse accumulation of fat and localized fat accumulation. This has no special geriatric implication. He also suggested an etiological classification into hypothalamic, endocrine, nutritional, physical inactivity induced, predominantly genetic- and drug-induced obesity, where some have an obvious geriatric importance. These are the high fat nutritional, the physical inactivity, and the drug-induced obesity types of Bray.

The nutrient density of the food is especially important in the elderly, who on the average constitute a low-energy-consumer group. The major explanation for the reduction of energy expenditure with age is probably the decreasing number of cells in the organs, loss of metabolizing tissue, and reduced physical activity,[39] although some intracellular aged-induced changes on the enzyme level might also be responsible. In the Baltimore longitudinal study of males,[36] the average total energy expenditure amounted to 2811 kcal/11.8 M and 1924 kcal/8.2 MJ at age 30 and 80, respectively.

Ordinary food in most industrialized countries has a low nutrient density because of a high proportion of fat and refined sugar. Therefore, intake of the same type of food in old age as earlier in life might obviously give either too much energy, too little essential nutrients, or both.

The energy "recommendations" of the latest Recommended Dietary Allowances[40] for the age group 51 years and over (2400 kcal/10.1 M and 1800 kcal/7.6 MJ for males and females, respectively) is obviously too difficult to apply to such a heterogenous group of individuals as the elderly. Body weight, among many other things, varies widely between age groups above 50 and between different populations. Thus, as an example, average body weight at age 70, 75, and 79 in a Swedish population[21] were 73, 70, and 69 kg for males and 67, 65, and 63 kg for females, which, for females means much more than the reference woman aged 51 or over of the Recommended Dietary Allowances.[40]

In this context it is important to point out that malnutrition means undernutrition, overnutrition, or both at the same time, as undernutrition of essential nutrients can be combined with overnutrition of fat, refined sugar, and energy. Physical inactivity is not an uncommon cause of overweight and obesity as shown in many studies such as the seven country study of Keys et al.[41] This is a general phenomenon and not related to age, however. Obese subjects are significantly less active than nonobese people.[42] Physical inactivity in the elderly, however, may — more than in young age groups — be the result of not only sedentary occupation but also due to physical handicap.

Social factors are of importance to obesity of high age for several reasons. Especially in females, the socioeconomic state is inversely related to the prevalence of obesity.[30,43] Furthermore, although many elderly suffering from loneliness have a bad appetite, there is also a tendency in the opposite direction with an overrepresentation of increasing appetite in lonely elderly as judged from a Swedish study.[44] This has also been found by others.[45]

OTHER COMPLICATIONS OF OBESITY FOR HEALTH CARE AND NURSING OF THE ELDERLY

Some special problems at advanced obesity are diagnostic difficulties when examining neck, thorax, and abdomen. Jugular venous pressure has been reported to be difficult to assess because of cervical adiposity.[46] The apex beat is not easily felt in obese patients and heart sounds and murmurs are muffled.[47] The amplitude of precordial T-waves in the electrocardiogram has been reported to be reduced to half the normal.[48] These diagnostic difficulties are common in all age groups, but adds to already existing diagnostic difficulties in elderly patients. Symptoms are often vague and uncharacterisic in geriatric medicine, and many diseases do not reveal themselves by presenting symptoms at all (silent disease).

The practical handling of obese elderly patients in, for instance, long-term care medicine is often a difficult task. Changing position from bed to chair and even walking may be difficult, and must be looked upon as potential hazards for falls, the result of which can be a fractured hip due to co-existing osteoporosis.

Intertrigo is commonly found on various parts of the body and is often difficult to treat effectively. Toileting is difficult especially together with co-existing urinary incontinence — a very common condition in geriatric medicine. Pressure sores, hypostatic congestion of the lungs, and leg vein thrombosis are more likely to develop in obese patients if they have to stay in bed for some time.[47]

TREATMENT

Treatment of obesity in elderly patients does not require other methods than those which are used in other age groups. However, many existing methods are not applicable to elderly people, and only a few, therefore, remain to be discussed in this context.

As was mentioned earlier, the indications for treatment might be different in the higher age groups, the tendency being generally to have more rigorous indications before intervening. This is especially true in the highest age groups — above age 75 to 80 — where there seems to be an ongoing decrease of body fat and body weight apparently related to normal aging.[21]

Surgical methods to treat obesity seem to be indicated only in exceptional cases. Nonsurgical methods include jaw fixation, intragastric balloons, diets, exercise, behavior modification, drugs, and hypnosis.[49] The National Board of Health and Welfare withdrew anorectic drugs from use in Sweden in 1980. The reasons given were that the efficiency of the drugs was doubtful, that there were indications that overprescription occurred, and that the drugs were abused.[49] Therefore, of the methods mentioned, those of choice in geriatric medicine are obviously to try changes of diet, physical activity, and sometimes behavior. An attractive principle in especially elderly obese patients is the use of large amounts of dietary fiber as an adjunct in the treatment.[50] A rather slow, progressive weight loss is preferable, especially in the elderly.[51] A daily reduction of 500 kcal can be mentioned as an optimal example. Acute intermittent fasts should be avoided in these age groups. Most fad diets are unbalanced and, therefore, to be avoided in the elderly, because of the need for high quality food in these low energy consumers.

Physical activity is of special importance in the treatment of overweight in the elderly, because of many other beneficial effects. Exercise should, however, be light to moderate.

Every "exercise candidate" should have a thorough physical examination before any exercise program.[52] This is obviously especially true in the higher age groups.

REFERENCES

1. **Rinder, L., Roupe, S., Steen, B., and Svanborg, A.,** 70-year-old people in Gothenburg. A population study in an industrialized Swedish city. I. General design of the study, *Acta Med. Scand.,* 198, 397, 1975.
2. **Svanborg, A., Landahl, S., and Mellström, D.,** Basic issues of health care, in *New Perspectives on Old Age: A Message to Decision Makers,* Thomae, H. and Maddox, G. L., Eds., Springer-Verlag, N.Y., 1982, 31.
3. **Howard, A., Ed.,** *Recent Advances in Obesity Research.* I, Newman, London, 1975.
4. **Bray, G. A., Ed.,** *Recent Advances in Obesity Research.* II, Newman, London, 1978.
5. **Björntorp, P., Cairella, M., and Howard, A., Eds.,** *Recent Advances in Obesity Research.* III, John Libbey, London, 1981.
6. **Moore, F. D., Olesen, K. M., McMurray, J. D., Parker, H. V., Ball, M. R., and Boyden, C. M.,** The body cell mass and its supporting environment, W. B. Saunders, Philadelphia, 1963.
7. **Steinkamp, R. C., Cohen, N. L., Siri, W. E., Sargent, T. W., and Walsh, H. E.,** Measures of body fat and related factors in normal adults. I. Introduction and methodology, *J. Chron. Dis.,* 18, 1279, 1965.
8. **Steen, B.,** Nutrition in 70-year-olds. Dietary habits and body composition. A report from the population study "70-year-old people in Gothenburg, Sweden", *Näringsforskning,* 21, 201, 1977.
9. **Forbes, R. M., Cooper, A. R., and Mitchell, H. H.,** The composition of the adult human body as determined by chemical analysis, *J. Biol. Chem.,* 203, 359, 1953.
10. **Hytten, F. E., Taylor, K., and Taggart, N.,** Measurement of total body fat in men by absorption of 85 Kr, *Clin. Sci.,* 31, 111, 1966.
11. **Lindholm, B.,** Changes in body composition during long-term treatment with cortisone and anabolic steroids in asthmatic subjects, *Acta Allerg.,* 22, 261, 1967.
12. **Sköldborn, H., Arvidsson, B., and Andersson, M.,** A new whole body monitoring laboratory, *Acta Radiol.,* Suppl. 313, 213, 1972.
13. **Berg, K. and Isaksson, B.,** Body composition and nutrition of school children with cerebral palsy, *Acta Paediat. Scand. Suppl.,* 204, 41, 1970.
14. **Garn, S. M.,** Roentgenogrammetric determinations of body composition, *Human Biol.,* 29, 337, 1957.
15. **Pascale, L. R., Grossman, M. J., Sloane, H. S., and Frankel, T.,** Correlations between thickness of skin folds and body density in 88 soldiers, *Human Biol.,* 27, 165, 1955.
16. **Steen, B., Bruce, Å., Isaksson, B., Lewin, T., and Svanborg, A.,** Body composition in 70-year-old males and females in Gothenburg, Sweden. A population study, *Acta Med. Scand., Suppl.,* 611, 87, 1977.
17. **Lindberg, W., Natvig, H., Rygh, A., and Svendsen, K.,** Høyde - og vektundersøkelser hos voksne menn og kvinner. Forslag til nye norske høyde - vektnormer, *Tidsskr. Nor. Laegeforen.,* 76, 361, 1956.
18. **Durnin, J. V. G. A. and Womersley, J.,** Body fat assessed from total body density and its estimation from skinfold thickness: measurements on 481 men and women aged from 16 to 72 years, *Br. J. Nutr.,* 32, 77, 1974.
19. **Montoye, H. J., Epstein, F. H., and Kjelsberg, M. O.,** The measurement of body fatness. A study in a total community, *Am. J. Clin. Nutr.,* 16, 417, 1965.
20. **Seltzer, C. C., Stoudt, H. W., Jr., Bell, B., and Mayer, J.,** Reliability of relative body weight as a criterion of obesity, *Am. J. Epidemiol.,* 92, 339, 1970.
21. **Steen, B., Isaksson, B., and Svanborg, A.,** Body composition at 70, 75, and 79 years of age. A longitudinal population study, *Proc. 12th Int. Congr. Nutr.,* San Diego, 1981, 102.
22. **Skerlj, B., Brozek, J., Hunt, E. E.,** Subcutaneous fat and age changes in body built and body form in women, *Am. J. Phys. Anthropol.,* 11, 577, 1953.
23. **Gordon, T. and Kannel, W. B.,** The effects of overweight on cardiovascular disease, *Geriatrics,* 28, 80, 1973.
24. **Kannel, W. B., Brand, N., Skinner, J. J., Jr., Dawber, T. R., and McNamara, P. M.,** The relation of adiposity to blood pressure and development of hypertension, *Ann. Intern. Med.,* 67, 48, 1967.
25. **Heyden, S., Hames, C. G., Bartel, A., Cassel, J. C., Tyroler, H. A., and Cornoni, S. C.,** Weight and weight history in relation to cerebrovascular and ischaemic heart disease, *Arch. Intern. Med.,* 128, 956, 1971.
26. **Gordon, E. S.,** Obesity: gluttony or genes?, *Postgrad. Med.,* 45, 95, 1969.
27. **Leach, R. E., Baumgard, S., and Bloom, J.,** Obesity: its relationship to osteoarthritis of the knee, *Clin. Ortop.,* 93, 271, 1973.

28. **Bray, G. A.,** *The Obese Patient,* W. B. Saunders, Philadelphia, 1976.

29. **Keys, A., Fidanza, F., Karvonen, M. J., Kimura, N., and Taylor, H. L.,** Indices of relative weight and obesity, *J. Chron. Dis.,* 25, 329, 1972.

30. **Bray, G. A.,** Definition, measurement, and classification of the syndromes of obesity, *Int. J. Obesity,* 2, 99, 1978.

31. Society of Actuaries, Build and Blood Pressure Study, The Society of Actuaries, Chicago, 1960, 1:1.

32. **Andres, R.,** Influence of obesity on longevity in the aged, in *Aging, Cancer and Cell Membranes,* Borek, C., Fenoglio, C. M., and King, D. W., Eds., Thieme, Stuttgart, 1980, 238.

33. **Marks, H. H.,** Influence of obesity on morbidity and mortality, *Bull. N.Y. Acad. Med.,* 36, 296, 1960.

34. **Kannel, W. B. and Gordon, T.,** Obesity and cardiovascular disease. The Framingham Study, in *Obesity Symposium,* Burland, W. L., Saunel, P. D., and Yudkin, J., Eds., Churchill Livingstone, Edinburgh, 1974, 24.

35. **Dyer, A. R., Stamler, J., Berkson, D. M., and Lindberg, H. A.,** Relationship of relative weight and body mass index to 14-year-mortality in Chicago Peoples Gas Company Study, *J. Chron. Dis.,* 28, 109, 1975.

36. **McGandy, R. B., Barrows, C. H., Jr., Spanias, A., Meredith, A., Stone, J. L., and Norris, A. H.,** Nutrient intakes and energy expenditure in men of different ages, *J. Gerontol.,* 21, 581, 1966.

37. **Steen, B., Mellström, D., and Sundh, W.,** Cigarette smoking as a confounding factor in relation analyses between body weight and mortality. A longitudinal population study, in preparation.

38. **Björntorp, P.,** Effects of age, sex, and clinical conditions on adipose tissue cellularity in man, *Metabolism,* 23, 1091, 1974.

39. **Shock, N. W.,** Energy metabolism, caloric intake and physical activity of the aging, in *Nutrition in Old Age,* Ed., Carlsson, L. A., 10th Symp. Swedish Nutr. Found., Almqvist & Wiksell, Uppsala, 1972, 12.

40. Food and Nutrition Board, *Recommended Dietary Allowances,* 9th edition, National Academy of Sciences, Washington, D.C., 1980.

41. **Keys, A. Ed.,** Coronary heart diease in seven countries, *Circulation,* 41, 1, 1970.

42. **Chirico, A. M. and Stunkard, A. J.,** Physical activity and human obesity, *N. Engl. J. Med.,* 263, 935, 1960.

43. **Burnight, R. G. and Marden, P. G.,** Social correlates of weight in an aging population, *Milbank Mem. Fund Q.,* 45, 75, 1967.

44. **Mellström D. and Steen, B.,** Some examples of relations between social factors and dietary habits in 70-year-old people, in *Nordisk Gerontolgi,* Beverfelt, E., Julsrud, A. C., Kjørstad, H., and Nygård, A. M., Eds., Hammerstad Boktrykkeri, Oslo, 1981, 176.

45. **Price, J. H.,** Nutrition for the elderly, *J. Nurs. Care,* 12, 14, 1979.

46. **Alexander, J. R.,** Chronic heart disease due to obesity, *J. Chron. Dis.,* 18, 895, 1965.

47. **Haleem, M. A.,** The problem of obesity in the elderly, *Br. J. Clin. Pract.,* 32, 45, 1978.

48. **Jaffe, H. L., Corday, E., and Master, A. M.,** Evaluation of the precordial leads of the electrocardiogram in obesity, *Am. Heart J.,* 39, 911, 1948.

49. **Rössner, S.,** Examples of non-surgical methods to treat overweight, *Näringsforskning,* 26, 125, 1982.

50. **Van Itallie, T. B.,** Dieary fibers and obesity, *Am. J. Clin. Nutr.,* 31, 543, 1978.

51. **Albanese, A. A.,** *Nutrition for the Elderly,* Alan R. Liss, N.Y., 1980.

52. **Stuart, R. B.,** Obesity, *Geriatric Med.,* 5, 84, 1976.

GOVERNMENT NUTRITION PROGRAMS FOR THE AGED

H. Smiciklas-Wright and G. J. Fosmire

INTRODUCTION

Government nutrition programs for the elderly have emerged relatively recently. The impetus for their development and the goals of the programs can best be understood when discussed in the context of sociodemographic and health status of the elderly and in the context of overall government programs for the elderly.

The 25 million people aged 65 and over in the U.S. today make up 11% of the total population. The percentage of the population over 65 years of age grew from 3% after the Civil War to 8.2% after World War II to the current 11%.[1] Projections through 2030 estimate that 18% of the total population will be age 65 and over.[2]

The steady rise in the numbers and proportion of the elderly has brought increased awareness of their health and social needs. The elderly are by no means a homogeneous population with common needs. There is considerable heterogeneity in their health and well-being. Most elderly are ambulatory and have the physical capacity for personal care. Approximately 5% of older Americans are institutionalized. Shanas[3] reported that 2% were bed-ridden at home, 6% were house-bound and 6% were ambulatory, but with some incapacity.

For most people health deteriorates with age. Chronic diseases, dental problems, poor vision, and hearing impairments increase with age. They do not impair the ability of all elderly to function independently but they are incapacitating for some and they do increase sharply around age 75.[4] Low income and minority elderly are the most likely to be in poor health.

Individuals differ in their ability to cope and adapt to the realities of aging. Emotional disturbances are not inevitable concomitants of aging but poor health, sensory deficits, losses of work, income, and close interpersonal relationships may exhaust coping mechanisms leading to severe depression and social isolation.[5]

The elderly differ in economic as well as physical and mental well-being. Not all elderly persons are poor. In 1979, about 1 out of 7 elderly persons had incomes of $10,000 and one half of these had incomes of $15,000 or more.[6] Most elderly live in homes they own, pay lower taxes than do younger persons with comparable incomes, and are eligible for various "senior citizen discounts." Yet the elderly are also one of the poorest segments of the population. In 1978, persons 65 and older had a median income of $5630.[6] In 1979, 15% of the elderly lived below the poverty level.[7] People over 65 often suffer an income loss that thrusts many into poverty for the first time in their lives.[8] The problem of poverty is particularly severe for elderly women and minorities, with black women 65 years and over experiencing a 42% rate of poverty in 1979.[7]

The growing population of elderly has brought not only increased awareness of their needs but also consideration of appropriate policies and programs to meet those needs. Beattie[9] reported that social policies and programs to meet the specialized needs of the aged did not really emerge until the 1940s and that much of the lag was due to historical and cultural ideologies which influenced the health and social services. He noted that when public health, social work, and medical care were developing professionally early in this century, the major emphasis was on the early stages of the lifespan. High infant and maternal mortality rates, infectious and communicable diseases, and psychosocial theories which emphasized the child and parent relationship directed both medical and social care. It is only recently that social planning has been directed to the needs and conditions of older persons and that those needs have been given national recognition. Conferees to the first National Conference on Aging, which was initiated by President Truman in 1950, urged both government and voluntary agencies to accept greater responsibility for the welfare of older people.[10] The 1981 White House Conference on Aging was the fourth such national conference. Delegates to the Nutrition Section of the 1971 Conference called for concerted action of the federal, state, and local governments to make more efficient use of existing programs related to the needs of the elderly and to develop new ones in providing better services to the elderly.[11]

The balance of this chapter is divided into three sections. The first will provide an overview of government assistance to the elderly. The second section will examine government nutrition programs affecting the elderly. The final section will address implications and future needs.

GOVERNMENT ASSISTANCE TO THE ELDERLY

Beattie[9] notes that the majority of older persons throughout the world are "unreached and unserved by social service programs." The emergence of social services for the aging is a 20th century phenomenon, a phenomenon which is most evident in countries such as the U.S., where there is a firmly established notion that the government should intervene in the social condition of the elderly.[12]

The federal government of the U.S. has assumed increasing responsibility for the welfare of the elderly. Its first important role was the introduction of the Social Security System in the 1930s. Over the years, social security coverage has broadened. Additional benefits have been provided through the enactment of a large number of federal programs which benefit the elderly. Some 48 major federal programs are directed specifically at the elderly.[8] As many as 200 additional programs, principally in the areas of health care and retirement benefits, affect the elderly although not designed specifically for them.

The broad range of government programs have been classified into two major groups: programs for the elderly or age-entitlement programs and programs that affect the elderly or needs-entitlement programs.[1] Age-entitlement programs are universal programs for the elderly in that benefits depend upon age alone. Social Security retirement benefits and Medicare payments are the best known age-entitlement programs for the elderly.[13] The Nutrition Program for Older Americans to be discussed later in this chapter is another age-entitlement program. Needs-entitlement programs such as Food Stamps and Medicaid are contingent on financial need.

Of the many programs for or affecting the elderly, eight have been described as major strategies employed by the Federal Government to help the elderly.[1] Of the eight — Social Security Old Age Insurance and Medicare which provide assistance to the majority of the elderly and Supplemental Security Income and Medicaid which assist low-income elderly — have been described as the cornerstones of federal assistance to the elderly. Others — taxation practices, housing subsidies, food stamps, and Older American Act Services — are described as significant to many elderly but of incidental importance to most. It should be noted that the Nutrition Program for Older Americans was authorized under the Older Americans Act.

Approximately one quarter of the federal budget in fiscal year 1980 was allocated to the elderly. Most of that money was allocated to retirement benefits and health care (71.4 and 22.7¢, respectively, of each federal dollar spent on the elderly).[14] The remaining 5.9¢ are distributed among income maintenance (welfare) programs, employment and social services, and nutrition programs. The amount allocated to nutrition programs (0.5¢) represents a fairly reasonable allocation given the wide variety of other services and benefits which must be provided out of the 5.9¢. This is not to argue that it can provide for the nutrition needs of the elderly but that it is in competition with a host of other social service needs.

The number of programs and the size of the budget suggest that the elderly are well provided for by the federal government. There has been an undeniable growth of federal assistance to older Americans. Federal assistance programs have, however, come under increasing criticism. A major criticism is that many elderly do not benefit because of a lack of coordination and fragmentation of the available services. Demonstration "channeling" projects are underway to attempt to deal with the lack of coordination. The projects are based on providing a single place for a person to go to enter the health and social system rather than trips to numerous departments scattered all over a community.[13] A second criticism is that the system, particularly in its health care benefits, is more crisis than prevention oriented and that there is more support for the highly technical help administered to the acutely ill rather than for home health care support services.[8] The benefits of the nutrition program are exempt from this criticism because they do promote independent living.

Some critics of government assistance to the elderly charge that the elderly are receiving more than their appropriate share of public benefits, yet the growing budgetary outlay is not necessarily well-directed to those in most need.[1] Such a criticism has most implications for age-entitlement programs. This point will be discussed later when we consider the Nutrition Program for older Americans. Unquestionably, we are entering a period of considerable public debate on government assistance to the elderly. Political activism has gained considerable public support for older Americans. However, the growing federal budgetary allocation for the support will be scrutinized closely with implications for all programs, including nutrition programs.

GOVERNMENT NUTRITION PROGRAMS

The 1950s and 1960s was a time of growing awareness of the elderly as a nutritionally vulnerable sector of the American population. The Senate Select Committee on Nutrition and Human Needs reviewed data from nutrition surveys and found that elderly, especially low-income elderly, were nutritionally vulnerable.[15] The Panel on Aging of the 1969 White House Conference on Food, Nutrition, and Health, a 1970 President's Task Force on Aging,[17] and the Nutrition section of the 1971 White House Conference on Aging[11] were important forums calling for expanded nutritional programs for the elderly. Recommendations from the conferences recognized that the nutritional problems of the elderly are a complex phenomenon comprised of many factors — insufficient income, social isolation, and inadequate knowledge of nutrition, to name only a few. Thus, the recommendations dealt with income, revisions of the Food Stamp Program, provisions of meals in social settings, home-delivered meals, as well as with research and nutrition education. The President's Task Force on Aging included the following remarks in its report:

In examining the incidence of malnutrition among the elderly the Task Force concluded that insufficient income was only one of several causes. The lonely older person who can afford an adequate diet but does not eat properly, the older person who finds going to the store too great a burden, the older person who is nutritionally ignorant, the chronically ill older person unable to prepare a hot meal — all are part of the problem. The Task Force believes that programs can be designed which not only provide adequate nutrition to older persons, but, equally important, combat their loneliness, channel them into the community, educate them about proper nutrition, and afford some of them an opportunity for paid community service.[17]

Two food assistance programs, the Food Stamp Program and the Direct Food Distribution Program were already operating during the conferences of the late 1960s and early 1970s. The Conferences were, however, a significant impetus to Congressional Hearings from which evolved legislation authorizing the Nutrition Program for Older Americans. In the remainder of this section we will review the two food assistance programs available before 1972 with particular emphasis on the Food Stamp Program and will review the Nutrition Program for Older Americans authorized in 1972.

We will limit our discussion to the programs listed above although we recognize that these are not the only government activities which impact on the nutritional well-being of the elderly. Amendments in 1972 to the Social Security Act (P.L. 92-603) required that the department of Health, Education, and Welfare develop uniform standards for facilities participating under Medicare and Medicaid. The standards specify requirements for professional staffing, nutritional adequacy of menus, and preparation and service of food. Smith[19] and Roe[20] have provided excellent reviews of the standards which will not be further discussed in this paper.

Food Stamp Program

The Food Stamp and Direct Food Distribution Programs were available in the 1960s to low-income elderly. Both programs were under the administration of the U.S. Department of Agriculture (USDA) working cooperatively with state and local governments. The Direct Food Distribution Program provided free food to people determined eligible by local social service or welfare agencies.[21] Complaints about the program included problems that recipients had in acquiring the food, in transporting it to their homes, as well as criticisms about food quality and lack of variety. Some communities organized volunteer groups to overcome difficulties by delivering foods to the home-bound elderly.[21] The program has been largely replaced by the Food Stamp Program, although recently the USDA has been distributing surplus butter and cheese to the needy. The USDA is also pilot-testing a food distribution program for low-income elderly. The intent of the program would be the delivery of a food package approved by the Food and Nutrition Service of the Department of Agriculture to the homes of participants. The target population are thus low-income, home-bound people, 60 years and older.

The Food Stamp Act was enacted in 1964 with two objectives: to raise nutrition levels in low-income households and to strengthen the agricultural economy. Food Stamp Program benefits are in the form of ''stamps'' which may be used to buy food, plants, or seeds in authorized retail stores. Food stamps cannot be used to buy alcoholic beverages, pet food, tobacco, or other nonfood items. Food stamp eligibility is determined on the basis of income and resources. Allotments are determined on the basic premise that a family should be able to buy a nutritionally adequate diet for no more than 30% of its net income.

Weimer[22] reported recently that households headed by the elderly participated proportionately less than other households in the Food Stamp Program. Major obstacles to participation were the cash payments needed before 1977 to purchase food stamps and the welfare connotation of the program.

Several major revisions made to the Food Stamp Act have helped to increase the number of elderly participants. An important revision was the elimination of the purchase price. Prior to that revision, participants made a cash payment and received stamps equal to that payment plus an additional bonus. When the cash payment was eliminated participants received the bonus only. Weimer[22] reported that participation by households headed by an elderly person increased by approximately 32% from February 1978 to April 1979. During the same time period participation by nonelderly households increased about 14%.

Other revisions to the 1964 Act have also led to an increase in the number of elderly participants. Most households are eligible to be considered for the program if their resources or assets are below the $1500 resource limit and income is below 150% of the poverty level. Households with at least two members, if one is age 60 or older, have a resource limit of $3000. Additional provisions which have made the program more attractive to the elderly are mail certification or a home visit instead of an office interview, medical deductions that can be applied to determining the net monthly income and thus the food stamp allotment, and the use of stamps to pay for home-delivered meals, and group meals for the elderly.

There have been proposed changes to the Food Stamp Program which could have a negative impact on elderly participants. Elderly participants at present are allowed an unlimited excess shelter deduction in addition to the standard monthly deduction of $85. That is, all households are allowed a standard deduction of $85 in determining their net monthly income. They are also allowed a shelter deduction based on shelter expenses (rent, mortages, utilities, and telephone) that exceed 50% of income after other deductions have been subtracted. Households which do not contain a person 60 years of age or older cannot have shelter and/or dependent care deductions which exceed $115. A similar ceiling has not been imposed on households with an elderly member. One proposed change was for a combined monthly deduction of $150. It is estimated that with the changes, most elderly households would have lost about $11 per month and some up to $18 a month in benefits.[23]

Currently, about 10% of food stamp participants are age 60 or older. There is some work to show whether food stamps have enhanced the nutritional status of the elderly. Weimer[22] referred to a recent USDA Study which found that the program had a positive effect on the intake of selected nutrients.

Nutrition Program for Older Americans

The Nutrition Program for Older Americans (NPOA) was established by Congress in 1972. The foundation for the 1972 program was a 1968 congressional authorization for $2 million annually for a 3-year nutrition research and demonstration project under Title IV of the Older Americans Act of 1965 to deliver meals in group or congregate settings. Under Title IV, grants were made for research and development projects on ways to develop and evaluate alternative approaches to meal delivery and to the delivery of health, transportation, and outreach services, as well as the provision of consumer and nutrition education. Reviews of the research and demonstration projects which were carried out in 1971 found them to be feasible approaches for delivering food to the elderly, and approaches which appeared to attract minority and low-income elderly. Posner[24] and Binstock and Levin[12] point out that there were no quantitative evaluations of the projects, but that their popularity among participants, the public, and politicians was overwhelming. Estes and Freeman[25] go further to say that relevant research did not provide sufficient data on what programs "worked" and should be implemented nationally and, in fact, research on the nutrition projects illustrated some negative findings. Nevertheless, advocacy for a nutrition program for the elderly was well established, resulting ultimately in the 1972 Title VII Amendment of the Older Americans Act, which established the NPOA.

The Older Americans Act of 1965 is itself considered to be a major piece of legislation providing programs for older Americans. The Act was established by Congress in response to a lack of community social services designed specifically for older persons. The Older Americans Act established the Administration of Aging (AOA) within the Department of Health, Education, and Welfare to administer programs in consultation with other departments of the federal government. The Act authorizes grants for social and nutrition services, multipurpose senior activities facilities, training, research and demonstration projects, and public service employment projects. Federal dollars appropriated under the Act have grown from $65 million in 1966 to $914 million in fiscal year 1982.[10]

The NPOA, established under Title VII,[26] and now provided under Title III[27] of the Older Americans Act is major legislation providing nutrition programs for older Americans. The statute setting out the findings of the Title VII project and the purpose of the Title VII legislation reads as follows:

Many elderly persons do not eat adequately because (1) they cannot afford to do so; (2) they lack the skills to select and prepare nourishing and well-balanced meals; (3) they have limited mobility which may impair their capacity to shop and cook for themselves; and (4) they have feelings of rejection and loneliness which obliterate the incentive necessary to prepare and eat a meal alone. These and other physiological, psychological, social and economic changes that occur with aging result in a pattern of living which causes malnutrition and further physical and mental deterioration.

...there is an acute need for national policy which provides older Americans, particularly those with low incomes, with low cost, nutritionally sound meals served in strategically located centers such as schools, churches, community centers, senior citizen centers, and other public or private nonprofit institutions where they can obtain other social and rehabilitative services. Besides promotion of health among the older segment of our population through improved nutrition, such a program would reduce the isolation of old age, offering older Americans an opportunity to live their remaining years in dignity.[26]

Both congregate and home-delivered nutrition services are available to those aged 60 and over. They are also available to handicapped or disabled persons who live in housing facilities which are primarily occupied by elderly persons and where congregate meals are served. It is interesting to note that the eligibility age for the program is 60. That is also the age at which food stamp benefits such as the higher resource limit of $3000 become available to the elderly. Many legislative programs for the elderly are defined by age 65. Kovar[4] points out that there are no biological reasons for defining elderly in terms of a specified calendar age. The younger the age of eligibility, the more potential there is for a program to fulfill the purpose of preventive health care. The age definition may change, however, in response to economic pressures. Tentative recommendations have already been made to raise the age by which elderly is defined for food stamp benefits.

The NPOA is primarily, then, an age-entitlement program. There is no test of eligibility based on income. There is no cost for meals. Individuals may make a voluntary contribution toward the cost of the meals. Income collected is to be used to increase the number of meals served by the project, to facilitate access to meals, and to provide supportive services directly related to nutrition services.

No one is required to make a donation and contributions are to be anonymous. Problems do exist in maintaining confidentiality during the collection of monies. A recent General Accounting Office report found that some sites provided envelopes to help protect confidentiality but others use receptacles placed where it was easy to observe what, if anything, was paid for the meal.[28] In some cases participants perceive payments as charges and not as voluntary donations. This is particularly true at sites where anonymity of donations is not protected. The Administration on Aging recently encouraged a nationwide initiative to increase voluntary contributions. AOA officials reported a good response to the initiative. It is uncertain, however, whether the more overt emphasis on volunteer donations will affect program participation.[29]

Congregate meals are served in settings such as senior centers, public housing facilities, schools, and churches. Sites should be established in as close a proximity to a majority of participants as possible, preferably where there is a concentration of low income, minority individuals, and preferably within walking distance or close to public transportation. Projects must develop plans for providing transportation when public transportation is not available.

Federal guidelines require that nutrition projects serve at least one hot or other appropriate meal per day, 5 days a week, and may serve additional meals.[10] Each meal must provide one third of the daily recommended dietary allowances established by the Food and Nutrition Board of the National Academy of Sciences-National Research Council.[30]

There are many home-bound and isolated elderly who cannot attend congregate meal sites. Home meal delivery services designated as "Meals on Wheels" have been established in many communities under the auspices of voluntary, nonprofit organizations and human service agencies. Funding for the programs has come from a variety of community sources such as United Fund Contributions, church or civic group support, and payment by program participants when possible. Meals on Wheels programs, which are primarily community-funded, are to be differentiated from the home-delivered meals program which is associated with NPOA. The former are not required to adhere to dietary guidelines, whereas NPOA home-delivered meals must meet the requirements for the one third daily recommended dietary allowances. Home-delivered meal projects are required to serve at least one hot, cold, frozen, dried, canned, or supplemental food meal per day, 5 or more days per week, and may serve additional meals.[10]

The cost of transportation for meal delivery and the needs of those who cannot participate in regular NPOA programs have led to a number of experimental and innovative ways of meeting NPOA goals.[20] Some projects use volunteers or shopping aides to purchase food and deliver it to home-bound clients. The food may supplement home-delivered meals which the client is receiving. Other projects have evaluated a weekly delivery of five frozen meals in place of daily deliveries[31] or provided freeze-dried foods which can be stored at room temperature and reconstituted with hot water or formula-type foods consisting of milk powder and added nutrients in place of traditional congregate and home-delivered meals.

The NPOA has grown rapidly, available at more than 10,000 sites with allocations that exceed $300 million. Posner[24] reported that since its implementation in 1973, NPOA has become the major operating program within the Administration on Aging, accounting for half of the Administration's total expenditures. One of the important features of the program has been its political popularity. Reports from each congressional district on the number and location of programs and number of meals served are quick and tangible results that politicians like,[12] and have generated considerable support for the program. Posner summarized the growth of NPOA as follows:

The growth in the number of meal sites and aged served has provided the short-run evidence of the logistical feasibility of the national congregate meal delivery system to the aged. What has failed to develop during this same period is a monitoring and evaluation scheme which can assess on an on-going basis the extent to which program goals were being achieved across Title VII programs and the relative merits of alternative program delivery schemes.[24]

Schneider observed that evaluating the effectiveness of the NPOA is "a complex task due to ambiguous goal, goals that are difficult to measure, such as better health and living with dignity."[32] Evaluators have addressed the ambiguities by considering such factors as the target groups served, program satisfaction, impact on socialization, life satisfaction, financial savings, and nutritional well-being, and by analyses of outreach, referral services and nutrition education.

Program Participation

As already stated, NPOA is an age-entitlement program and as such cannot deny services to those who qualify by nature of age. At the same time, the program's mandate is to serve minority and low-income elderly. In the early years of the program the Administration on Aging emphasized serving as many meals as possible to those 60 years and older without concentrating on specific target populations. This was hastened by Administration directives which required the projects to begin operations the first day of their budget years and to be fully operational within 90 days of the beginning of the budget year. In his analysis of the early NPOA experience, Watkin[33] described these as moves made deliberately to assure prompt utilization of appropriated funds, to have data on meals served daily quickly available, and to gain national visibility for the program.

While some modifications were subsequently made to the initial directives, a consequence was lower participation by potentially needy participants, including the very poor, the very old, those who were isolated, disabled, or members of minority groups. Projects were able to reach quotas quickly and placed limited efforts on outreach activities. Recent reports suggest that the low-income, eldest of the elderly, and minorities are being reached. At the same time, those reached are most likely to be socially active, involved with the community, enjoying out-of-home and structured activities.[34]

Outreach is challenging, given the many barriers to its effectivness.[32] Potential participants may be fearful of leaving their homes even when informed about a program; outreach workers may be fearful of moving into areas where the most isolated people live. Racial prejudice, location of a site in a church, stigma of participating in a government program, and isolation from community networks are additional barriers to participation. Programmatic barriers include quality of meals as well as inadequate transportation and limited resources to pursue active outreach. Schneider[32] recommended careful attendance to geographic location, choice of premises, and training of outreach workers as suggestions to overcome the barriers, but cautioned that "comprehensive solutions appear to be very remote."

Recently, Posner[24] proposed a triage and referral system in which priority be given to those in greatest need in order that the program meet its long-term health and nutrition goals. Such a system would depend on establishing eligibility criteria based on socioeconomic and demographic characteristics as well as measures of physical and psychological health and nutritional status. Other government programs provide services on a triage and referral system. The Supplemental Food Program for Women, Infants, and Children (WIC) has such a triage system based on economic, nutritional, and medical needs categorized into a priority system, and involving interviews and assessment of needs with potential clients. Subsequent to assessment, clients would be matched with appropriated services. Such a system would be a departure from the NPOA as it now operates. Easy access to the program has probably been attractive to many elderly. Elderly are often reluctant to participate in programs which require a public declaration of poverty, are perceived as welfare programs, and present confusion over eligibility.[35] Nevertheless, Posner[24] argues that funding limitations and the limited number of elderly to whom services can be provided warrant a review of current patterns of program participation. Should the nutrition program, however, move from an age-entitlement to some type of needs-entitlement program, considerable care will need to be given to the "marketing" of the program and to outreach efforts.

Program Impact

Significant impacts, or at least the best documented impacts, appear to be sociological and economic benefits: socialization, morale, and financial savings. Some data are to be found for diet-related, health, and nutritional status measures.

Kohrs et al.[36,37] evaluated the impact of the NPOA on the nutritional status of 547 rural aged in Central Missouri. They were evaluated according to whether they had discontinued participation, attended less than twice a week, or attended 2 to 5 times a week. Dietary intakes of Vitamins A, C, and Riboflavin and biochemical assessments of Vitamins A and C were positively associated with program participation. Although the NPOA meal is planned to provide one third of the recommended dietary needs for the day, program meals in the Missouri study provided considerably more than that level. Nutritive values calculated from menus showed that meals provided more than 70% of the RDA for protein, Vitamin A, and Vitamin C and at least 40% of the RDA for other nutrients examined. The investigators concluded that participation in the program is related to improvement in nutritional status.

Posner[24] cautions that a confounding factor which may limit conclusions to be drawn from the comparisons of Missouri participants and nonparticipants is that measurements were taken after but not prior to intervention. Such a design does not allow for adequate documentation of group comparability before intervention. Nevertheless, she concludes that the research makes a valuable contribution in assessing the contribution of the program meal to daily nutrient intakes.

Meal Quality

Little work has been done on the quality of meals served in project sites, even though meal quality is a factor affecting program participation.[24] Kincaid[38] set up taste panels and microbiological evaluations of both catered and on-site prepared meals. She found that overall on-site prepared meals scored higher in organoleptic tests and microbiological safety. There was evidence that participants hesitated to complain about food quality for fear of losing the program. Kincaid concluded that the generally positive acceptance of the program coupled with its costliness necessitate close scrutiny of food quality and safety.

Nutrition Education

Assessments of both the early Title IV demonstration projects and later Title VII projects stress the variability of nutrition education efforts.

Successful educational programs have actively involved the elderly and minimized traditional pedagogical techniques.[39] Often, however, nutrition efforts have been sporadic and haphazard, reflecting a lack of commitment to and support for nutrition education. One barrier to effective education is the lack of professionals to conduct programs. The use of the elderly as peer educators is an approach that may be useful in providing nutrition education if careful attention is paid to recruitment and to the availability of a support system providing sound nutrition information to peer educators.[40]

In summary, the Nutrition Program for Older Americans is a major nutrition intervention for the elderly. The legislation which established the program recognized that the many changes which may occur with aging can contribute to nutritional problems. Reduction of the social isolation of old age and the promotion of better health through improved nutrition are among the stated goals of the programs to be achieved through providing meals, opportunities for socialization, and opportunities to gain information and master new tasks.

The program has considerable national visibility and popularity with legislators. One reason for the political support is that it lends itself well to short-term measures of accomplishment such as the number and location of sites and number of meals served. More difficult to measure are long-term, less tangible objectives. It is fair to ask whether elderly participants of the program fare any better than nonparticipants. Participants tend to speak favorably of the program, recommending it both for its meals and social function. General improvement on outlook on life, diminished loneliness, and financial savings are among the specific benefits reported by participants.

There have been several small scale evaluations and a nationwide long-term evaluation in progress to determine nutrition benefits to be derived from the program. Weimer[22] reported that first-wave findings from the long-term evaluation indicate better dietary intake for participants but the differences in dietary intake between participants and nonparticipants is more evident on days when meals are eaten at the site.

The nutrition program has grown rapidly since it became funded in 1973. The program has been protected to the extent that separate funding has been maintained for congregate nutrition services and for home-delivered nutrition services, respectively. It is uncertain whether this directed funding will continue. There are efforts to consolidate funding into block grants whereby single amounts of money would be allocated to cover congregate and home-delivered meals as well as supportive services, senior centers, and possibly employment programs for the elderly. A consolidated or block grant fund would give states more flexibility in allocating funds but could weaken nutrition programs.

IMPLICATIONS AND FUTURE NEEDS

Focus on the needs of the elderly is not likely to diminish given the demographic trends. We live in an aging society, a society in which government support both in number of programs and budgeting outlay to older persons has grown markedly in the past decade. The growth in provision of services and implication for the future is addressed in Campion's review of White House Conferences:

The 1961 conference focused largely on health care and was a major force leading toward the creation of Medicare. The 1971 conference led to the Older Americans Act as a means to expand, improve, and coordinate home-care services. The 1981 WHCOA, in all likelihood, will be remembered as the conference that zeroed in on the issue of the elderly's competing as a group for limited economic resources in an era of fiscal restraint. This meeting also demonstrated that the elderly have become politically aware and politically adept and will not be easily manipulated.[41]

His reference to the 1981 Conference addresses two issues which are likely to be significant in considering future public-sector services for the elderly: a politically aware constituency and fiscal restraint.

Earlier in the paper we discussed two general types of public social service programs for the elderly, that is, age-entitlement and needs-entitlement programs. Nelson[42] discusses the two-tiered policy orientation that guides such programs. The one he calls a constituent or interest-group policy, the second a welfare-based policy. The former, exemplified by Older Americans Act programs, puts a priority on low income, minority elderly but not as a central thrust. Indeed, programs are likely to be directed at middle and lower-middle class elderly, whereas welfare-based, means-tested policies are directed primarily at poor elderly. Nelson goes on to make the point that most interest in social services is on the interest-group policies since "these represent a more middle class constituency and, therefore, have more political clout, less stigma associated with them, and more community visibility."

As program costs for elderly have grown, questions have been raised about the justification for age-based policy programs for the elderly. Kutza[1] suggests that the justification rests on four rationales, broadly categorized as economic, political, psychosociological, and ethical. The economic rationale, for example, stresses that the elderly are disadvantaged in the job market and are subject to reduced income with retirement and high medical expenses. Age-based programs do not measure economic need individual by individual but rather attribute a certain need to the group as a whole.

There are arguments for and against the use of chronological age for determining public benefits. Those who favor universal age-entitlement programs argue that they are administratively simple (fewer eligibility disputes), less stigmatizing than means-tested programs, and more likely to receive sustained political support. Arguments against such programs are based on the fact that chronological age is not necessarily a useful indicator of changes, that fiscal restraints will mean that public monies will need to be more effectively targeted, and finally, that means-tested programs do not have to be punitive and demeaning.[1] Kutza is quite right when she writes that there will be impassioned public debate on social policy and programs for aged. The future will be marked by competition between elderly and nonelderly populations and competition within the elderly population for scarce resources.[42] Social service providers will be challenged to clarify the purpose of nutrition programs, to develop a needs criteria, and to increase program evaluation.

REFERENCES

1. **Kutza, E. A.,** *The Benefits of Old Age: Social Welfare Policy for the Elderly,* The University of Chicago Press, Chicago, 1981, chap. 3, 7.
2. U.S. Bureau of the Census; Projections of the Population of the United States 1977—2050, Current Population Reports, Ser. P-25, No. 704, U.S. Government Printing Office, Washington, D.C.
3. **Shanas, E.,** Health status of older people: cross-national implications, *Am. J. Public Health,* 64, 261, 1974.
4. **Kovar, M. G.,** Health of the elderly and use of health services, *Public Health Rep.,* 92, 9, 1977.

5. **Furukawa, C.,** Coping and decision making, in *Community Health Services for the Aged,* Furukawa, C. and Shomaker, D., Eds., Aspen Systems Corp., Rockville, 1982, chap. 10.

6. **Pitts, J. M.,** Economic well-being of the elderly: recommendations from the White House Conference on Aging, *Fam. Econ. Rev.,* 4, 23, 1982.

7. U.S. Department of Commerce, Bureau of the Census, Characteristics of the population below the poverty level: Ser. P-60, No. 130, U.S. Government Printing Office, Washington, D.C., 1979.

8. **Barberis, M.,** America's elderly: policy implications, *Popul. Bull., Policy Suppl.,* January 1981.

9. **Beattie, W. M.,** Aging and the social services, in *Handbook of Aging and the Social Sciences,* Binstock, R. H. and Shanas. E., Eds., Van Nostrand Reinhold, N.Y., 1976, chap. 24.

10. U.S. Congress, House, Older Americans Act: A Staff Summary, Committee Publication No. 97-352, 97th Congress, U.S. Government Printing Office. Washington, D.C., 1982.

11. Towards a National Policy on Aging, *Proc. 1971 White House Conf. Aging,* 2, 1971.

12. **Binstock, R. H. and Levin, M. A.,** The political dilemmas of intervention policies, in *Handbook of Aging and the Social Sciences,* Binstock, R. H. and Shanas, E., Eds., Van Nostrand Reinhold, N.Y., 1976, chap. 20.

13. **Soldo, B. J.,** America's elderly in the 1980's, *Popul. Bull.,* 35, 1980.

14. U.S. House of Representatives, Select Committee on Aging, Subcommittee on Human Services, Future Directions for Aging Policy: A Human Service Model, Committee Publication No. 96-226, U.S. Government Printing Office. Washington, D.C., 1980.

15. Hearings Before the Select Committee on Nutrition and Human Needs, U.S. Senate, Part 14, Nutrition and the Aged, U.S. Government Printing Office, Washington, D.C., 1969.

16. White House Conference on Food, Nutrition and Health, U.S. Government Printing Office, Washington, D.C., 1969.

17. Toward a Brighter Future for the Elderly, Report of the President's Task force on Aging, U.S. Govenment Printing Office, Washington, D.C., 1970.

18. **Shannon, B. and Smiciklas-Wright, H.,** Nutrition education in relation to the needs of the elderly, *J. Nutr. Educ.,* 11, 85, 1979.

19. **Smith, C. E.,** Influence of standards in the nutritional care of the elderly, *J. Am. Diet. Assoc.,* 73, 115, 1978.

20. **Roe, D. A.,** *Geriatric Nutrition,* Prentice-Hall, Englewood Cliffs, 1983, chap. 10.

21. **Luhrs, C. E.,** Feeding the elderly, *Am. J. Clin. Nutr.,* 26, 1150, 1973.

22. **Weimer, J. P.,** The nutritional status of the elderly, *National Food Review,* Summer 1982, p. 7.

23. Elderly hit by cuts in FSP and Older Americans Act, *Foodlines,* 1, 13, 1983.

24. **Posner, B. M.,** *Nutrition and the Elderly,* Lexington Books, D. C. Heath, Lexington, 1979, chap. 6—8.

25. **Estes, C. L. and Freeman, H. E.,** Strategies of design and research interventions, in *Handbook of Aging and the Social Sciences,* Binstock, R. H. and Shanas, E., Eds., Van Nostrand Reinhold, N.Y., 1976, chap. 21.

26. PL 92-258, 92d Congress, S1163, March 22, 1972.

27. U.S. Congress. Conference Committees, Older Americans Act Amendments of 1981. Conference Report to accompany S1086. Ser. Rep. 97-293, 9th Congress, U.S. Government Printing Office, Washington, D.C., 1981.

28. General Accounting Office, Actions Needed to Improve the Nutrition Program of the Elderly, HRD-78-58, February 23, 1978.

29. AoA cites progress in contribution initiative, *CNI Weekly Report,* XII, 2, Nov. 18, 1982.

30. Food and Nutrition Board: Recommended Dietary Allowances, ed. 9, National Research Council, National Academy of Sciences, Washington, D.C., 1980.

31. **Osteraas, G. and Posner, B.,** Alternative approaches to nutrition services for the elderly, Fifteenth Annual Meeting, Soc. Nutr. Educ. 7 (Abstr.), 36, 1982.

32. **Schneider, R. L.,** Barriers to effective outreach in Title VII nutrition programs, *Gerontologist,* 19, 163, 1979.

33. **Watkin, D. M.,** The Nutrition Program for Older Americans. A successful application of current knowledge in nutrition and gerontology, *Wld. Rev. Nutr. Diet.,* 26, 27, 1977.

34. **Hassen, A. M., Meima, N. J., Buckspan, L. M., Henderson, B. E., Helbig, T. L., and Zarit, S. H.,** Correlates of Senior Center participation, *Gerontologist,* 18, 193, 1978.

35. **Moen, E.,** The reluctance of the elderly to accept help, *Soc. Probl.,* 25, 293, 1978.

36. **Kohrs, M. B., O'Hanlon, P., and Eklund, D.,** Title VII — Nutrition Program for the Elderly, *J. Am. Diet. Assoc.,* 72, 487, 1978.

37. **Kohrs, M. B., Nordstrom, J., Plowman, E. L., O'Hanlon, P., Moore, C., Davis, C., Abrahams, D., and Eklund, D.,** Association of participation in a nutritional program for the elderly with nutritional status, *Am. J. Clin. Nutr.,* 33, 2643, 1980.

38. **Kincaid, J. W.,** Acceptance of on-site prepared versus catered meals in Title VII Programs, *J. Nutr. Elderly,* 1, 27, 1981.

39. **Bechill, W. D. and Wolgamut, I.,** Nutrition for the Elderly: The Program Highlights of Research and Development Nutrition Projects, Social and Rehabilitation Service, Administration on Aging, U.S. Department of Health, Education, and Welfare, DHEW Publ. No. (SRS) 73-20236, Washington, D.C., 1973.

40. **Shannon, B. M., Smiciklas-Wright, H., Davis, B. W., and Lewis, C.,** A peer educator approach to nutrition for the elderly, *Gerontologist,* 23, 123, 1983.

41. **Campion, E. W.,** Observations from the 1981 White House Conference on Aging, *N. Engl. J. Med.,* 306, 373, 1982.

42. **Nelson, G.,** A role for Title XX in the aging network, *Gerontologist,* 22, 18, 1982.

Index

INDEX

A

I

N

O

Q

R

W